FUTURE TRENDS IN MICROELECTRONICS

THE WILEY BICENTENNIAL–KNOWLEDGE FOR GENERATIONS

Each generation has its unique needs and aspirations. When Charles Wiley first opened his small printing shop in lower Manhattan in 1807, it was a generation of boundless potential searching for an identity. And we were there, helping to define a new American literary tradition. Over half a century later, in the midst of the Second Industrial Revolution, it was a generation focused on building the future. Once again, we were there, supplying the critical scientific, technical, and engineering knowledge that helped frame the world. Throughout the 20th Century, and into the new millennium, nations began to reach out beyond their own borders and a new international community was born. Wiley was there, expanding its operations around the world to enable a global exchange of ideas, opinions, and know-how.

For 200 years, Wiley has been an integral part of each generation's journey, enabling the flow of information and understanding necessary to meet their needs and fulfill their aspirations. Today, bold new technologies are changing the way we live and learn. Wiley will be there, providing you the must-have knowledge you need to imagine new worlds, new possibilities, and new opportunities.

Generations come and go, but you can always count on Wiley to provide you the knowledge you need, when and where you need it!

William J. Pesce
President and Chief Executive Officer

Peter Booth Wiley
Chairman of the Board

FUTURE TRENDS IN MICROELECTRONICS

Up the Nano Creek

Edited by

SERGE LURYI
JIMMY XU
ALEX ZASLAVSKY

A Wiley-Interscience Publication
JOHN WILEY & SONS, INC.

Published by John Wiley & Sons, Inc., Hoboken, New Jersey.
Published simultaneously in Canada.

Library of Congress Cataloging-in-Publication Data is available.

ISBN 978-0-470-08146-4

Printed in the United States of America.

10 9 8 7 6 5 4 3 2 1

CONTENTS

Preface

S. Luryi
Dept. of Electrical and Computer Engineering
SUNY–Stony Brook, Stony Brook, NY 11794, U.S.A

J. M. Xu
Division of Engineering, Brown University, Providence, RI 02912, U.S.A.

A. Zaslavsky
Division of Engineering, Brown University, Providence, RI 02912, U.S.A.

This book is a brainchild of the fifth workshop in the *Future Trends in Microelectronics* series (FTM-5). The first of the FTM conferences, "*Reflections on the Road to Nanotechnology*", had gathered in 1995 on Ile de Bendor, a beautiful little French Mediterranean island.[1] The second FTM, "*Off the Beaten Path*", took place in 1998 on a larger island in the same area, Ile des Embiez.[2] Instead of going to a still larger island, the third FTM, "*The Nano Millennium*" went back to its origins on Ile de Bendor in 2001.[3] To compensate, the next FTM, "*The Nano, the Giga, the Ultra, and the Bio*" took place on the biggest French Mediterranean island of them all, Corsica.[4] Normally, the FTM workshops gather every three years; however, the FTM-4 was held one year ahead of the usual schedule, in the summer of 2003, as a one-time exception. Keeping in line with its inexorable motion eastward, the latest FTM workshop, "*Up the Nano Creek*", had convened on Crete, Greece, in June of 2006.

The FTM workshops are relatively small gatherings (less than 100 people) by invitation only. If you, the reader, wish to be invited, please consider following a few simple steps outlined on the conference website. The FTM website at *www.ece.sunysb.edu/~serge/FTM.html* contains links to all past and planned workshops in the series, their programs, publications, sponsors, and participants. Our attendees have been an illustrious lot. Suffice it to say that among FTM participants we find five Nobel laureates (Zhores Alferov, Herbert Kroemer, Horst Stormer, Klaus von Klitzing, and Harold Kroto) and countless others poised for a similar distinction. To be sure, high distinction is not a prerequisite for being invited to FTM, but the ability and desire to bring fresh ideas is. All participants of FTM-5 can be considered authors of this book, which in this sense is a collective treatise.

The main purpose of FTM workshops is to provide a forum for a free-spirited exchange of views among the leading professionals in industry, academia, and government. It is a common view among the leading professionals in micro-electronics, that its current explosive development will likely lead to profound paradigm shifts in the near future. Identifying the plausible scenarios for the future evolution of microelectronics presents a tremendous opportunity for constructive action today.

For better or worse our civilization is destined to be based on electronics. Ever since the invention of the transistor and especially after the advent of integrated circuits, semiconductor devices have kept expanding their role in our lives. Electronic circuits entertain us and keep track of our money, they fight our wars and decipher the secret codes of life, and one day, perhaps, they will relieve us from the burden of thinking and making responsible decisions. Inasmuch as that day has not yet arrived, we have to fend for ourselves. The key to success is to have a clear vision of where we are heading.

Some degree of stability is of importance in these turbulent times and should be welcome. Thus, although the very term "*microelectronics*" has been generally re-christened "*nanoelectronics*", we have stuck to the original title of FTM workshop series.

The present volume contains a number of original papers, some of which were presented at FTM-5 in oral sessions, other as posters. From the point of view of the program committee, there is no difference between these types of contributions in weight or importance. There was, however, a difference in style and focus – and that was intentionally imposed by the organizers. All speakers were asked to focus on the presenter's views and projections of future directions, assessments or critiques of important new ideas/approaches, and *not* on their own achievements. This latter point is perhaps the most innovative and distinguishing feature of FTM workshops. Indeed, we are asking scientists not to speak of their own work! This has proven to be successful, however, in eliciting powerful and frank exchange. The presenters were asked to be provocative and/or inspiring. Latest advances made and results obtained by the participants could be presented in the form of posters and group discussions.

Each day of the workshop was concluded by an evening panel or poster session that attempted to further the debates on selected controversial issues connected to the theme of the day. Each such session was chaired by a moderator who invited two or three attendees of his or her choice to lead with a position statement, with all other attendees serving as panelists. The debate was forcefully moderated and irrelevant digressions cut off without mercy. Moderators were also assigned the hopeless task of forging a consensus on critical issues.

All FTM workshops adhered to these principles in the past and, hopefully, will do so in the future. To accommodate these principles, the FTM takes a format that is less rigid than usual workshops to allow and encourage uninhibited exchanges and sometimes confrontations of different views. A central theme is designed together with the speakers for each day. By the tradition of FTM, the first day belongs to "*Captains of Industry*" who set the themes for the day, culminating with the "*Captains' Roundtable*" panel in the evening. At FTM-5 this panel was called "*Prospecting Up the Nano Creek*".

Another traditional feature of FTM workshops is a highly informal vote by the participants on the relative importance of various fashionable current topics in modern electronics research. This tradition owes its origin to Horst Stormer, who composed the original set of questions and maintained the results over four conferences. These votes are perhaps too bold and irreverent for general publication, but they are carefully maintained and made available to every new generation of FTM participants. Unfortunately, Horst missed the Crete gathering, but the tradition was maintained in his absence. Perhaps, one day we shall convince him to edit these votes and open them to general public. Another traditional vote concerned the best poster. The 2006 winning poster was "Formation of three-dimensional SiGe quantum dot crystals" by Detlev Grützmacher.

From all the deliberations and discussion at FTM-5 the following trends could be discerned, with the caveat that our crystal ball is as muddy as ever.

Firstly, although silicon is undoubtedly still full of steam, the word "post-CMOS" has become a commonplace. Somehow, the general perception by FTM attendees of CMOS Technology Roadmap predictions went from "*will be fulfilled ahead of schedule*" to "*should not be taken too seriously*", in a very short time.

A clearly discernible trend is the quest for novel and exotic materials. It looks like we are back to fundamentals. The big-brother silicon is clearly pressed (at least in terms of the conference publicity) by its much nimbler sibling carbon, who is capable of self-organizing into nanotubes and buckyballs, not to speak of graphene sheets. Ah well, it is all in column IV. One of the inventors of C_{60}, the Nobelist Harold Kroto, gave an inspiring lecture on Crete, titled "Architecture in nanospace".

Not every contribution presented at FTM-5 has made it into this book (not for the lack of persistence by the editors). Besides the paper version of Kroto's talk, we sorely miss the exciting contributions by Federico Capasso and Cees Dekker. Abstracts of these and all other presentations can be found on the workshop program webpage, *http://www.ee.sunysb.edu/~serge/ARW-5/program.html*

Besides the technical sessions and voting on the future trends, the FTM-5 program had room for events of general culture. Harold Kroto gave a passionate lecture, defending the Age of Reason, which in his view is under serious attack. Dr. Anna Zdanovich delivered a delightful lecture on the "Mystical symbolic language of byzantine icons". These were very well-received digressions and we intend to continue such practices at FTM as another trend of the future.

The FTM meetings are known for the professional critiques – or even demolitions – of fashionable trends, that some may characterize as hype. The previous workshops had witnessed powerful assaults on quantum computing, molecular electronics, and spintronics. The majority of FTM participants did not consider quantum computing a realistic future technology, but gave it credit as an interesting playground for physicists with some hope of settling old debates about the wavefunction collapse and other fundamental issues. It seems that by now most of the hype associated with quantum computing has dissipated and perhaps we can take some credit for the more balanced outlook that has emerged since. The assault continued at FTM-5 where the very concept of fault-tolerant computing was put in question, on theoretical grounds (Dyakonov). Of course, this is not an issue to be resolved by a vote and both points of view on quantum computing are presented in this book.

We have grouped all contribution into four parts, titled very generally *Physics, Biology, Electronics,* and *Photonics.* The breakdown could not be uniquely defined, because some papers fit all four categories! The discussion of quantum computing, spintronics, molecular electronics and quantum wires went into Part I (*Physics*). The list of controversial papers in Part I includes discussions of "perfect lensing" in negative refraction materials (Efros) and the foundations of laser theory (Spivak and Luryi).

Part II (*Biology*) is first in FTM workshop treatises. Of course, we had talks related to biology and medicine at the previous meetings, but they never comprised a critical mass. One of the papers is entirely biological and contributed by a biologist (Wimmer). This, however, should not scare the physics/ electronics/photonics reader, because Eckard Wimmer made it entirely understandable to a layman. Indeed, let us take a random quote from his paper: "UUAAAACAGCUCUGGGGUUGUACCCACCCCAGAGGCCCACGUGG ..." As will become clear from the paper itself, the fact it is so understandable constitutes a grave danger to this world! This is not the only biological danger described in Part II, as we are at least equally scared by micro-array brain implants (Nurmikko).

Parts III (*Electronics*) and IV (*Photonics*) are less dangerous but no less important. In *Electronics* we learn, perhaps with some dismay, how nano-manufacturing is making money, nano-dollar by nano-dollar. We find, perhaps with some sense of *deja vu*, that lithography is again at the cross-roads and that solutions may come from carbon nanotubes. The *Photonics* part will declare terahertz a form of light, even though it was only recently an ultrafast electronic oscillation. This is what happens when one merges nano with nano ...

To produce a coherent collective treatise out of all of this, the interaction between FTM participants had begun well before their gathering at the workshop. All the proposed presentations were posted on the web in advance and could be subject to change up to the last minute to take into account peer criticism and suggestions. After the workshop is over, these materials (not all of which have made it into this book) remain on the web indefinitely, and the reader can peruse them starting at the *www.ece.sunysb.edu/~serge/FTM.html* home page.

Acknowledgments

The 2006 FTM workshop on Crete and therefore this book were possible owing to support from:

- U.S. Department of Defense: AFOSR, ARO, ONR, AFRL;
- U.S. DoD European offices: EOARD, ONRIFO;
- NRC of Canada (Institute for Microstructural Sciences);

- NASA Langley Research Center;
- Industry: Applied Materials Inc., Philips Electronics Nederland, SAIC (Science Applications International Corporation);
- Academia: SUNY–Stony Brook;
- IEEE (EDS and LEOS, technical co-sponsor institutions).

On behalf of all Workshop attendees sincere gratitude is expressed to the above organizations for their generous support and especially to the following individuals whose initiative was indispensable:

- Chagaan Baatar
- William Clark
- James DeCorpo
- Marie D'Iorio
- Gail Habicht
- Michael Milligan
- Mark Pinto
- Daniel Purdy
- Howard Schlossberg
- Donald Silversmith
- Upendra Singh
- Trey Smith
- Henk van Houten
- Colin Wood

Finally, the organizers wish to thank all of the contributors to this volume and all the attendees for making the workshop a rousing success.

References

1. S. Luryi, J. M. Xu, and A. Zaslavsky, eds., *Future Trends in Microelectronics: Reflections on the Road to Nanotechnology*, NATO ASI Series E Vol. 323, Dordrecht: Kluwer Academic, 1996.
2. S. Luryi, J. M. Xu, and A. Zaslavsky, eds., *Future Trends in Microelectronics: The Road Ahead*, New York: Wiley Interscience, 1999.
3. S. Luryi, J. M. Xu, and A. Zaslavsky, eds., *Future Trends in Microelectronics: The Nano Millennium*, New York: Wiley Interscience/IEEE Press, 2002.
4. S. Luryi, J. M. Xu, and A. Zaslavsky, eds., *Future Trends in Microelectronics: The Nano, The Giga, and The Ultra*, New York: Wiley Interscience/IEEE Press, 2004.

Part I

Physics: The Foundations

1 Physics: The Foundations

The first part of this volume covers a range of fundamental physics-related issues, from quantum computing to quantum wires. While fault-tolerant quantum computing may be mathematically allowed, is it physically possible? Or, is the quantum error-correction scheme itself flawed by depending on error-free correction means? If a quantum computing system of many qubits is unattainable physically, would a system of a few qubits still be uniquely useful?

Spin control may be difficult to ascertain but is essential for spintronics. While spin control by magnetic fields has been popular, would electrical control via the spin-orbit coupling be inherently more efficient in semiconductors? Conceptually, both charge and spin can carry information, but one is conserved and the other is not, would that necessitate a fundamentally different approach for processing information by spintronics?

Phenomena at larger scales, a thousand times larger than a nanometer, have looked much clearer by comparison. So much so that the notion of perfect lens has cropped up in popular literature. A perfect lens could certainly remove one lithography limit of microelectronics among other things. But can a perfect lens exist, according to Maxwell? Furhter, imperfect lenses are not the only source of limitations the microelectronics industry is facing. Imperfect light sources comprise another. The laser linewidth is a measure of this imperfection and has been well modeled. But is it actually understood?

Interesting questions for sure. Profound insights too. All questions of basic physics, all questioning the basis of some particular or particularly popular directions in microelectronics research. All following the long-standing tradition of the FTM, and presented by some of the luminaries of our profession.

Contributors

1.1 M. I. Dyakonov
1.2 P. Hawrylak
1.3 E. I. Rashba
1.4 I. Zutic and J. Fabian
1.5 N. B. Zhitenev
1.6 A. L. Efros
1.7 B. Spivak and S. Luryi
1.8 E. Levy, A. Tsukernik, M. Karpovski, A. Palevski, B. Dwir, E. Pelucchi, A. Rudra, E. Kapon, and Y. Oreg

Future Trends in Microelectronics. Edited by Serge Luryi, Jimmy Xu, and Alex Zaslavsky

Is Fault-Tolerant Quantum Computation Really Possible?

M. I. Dyakonov
Laboratoire de Physique Théorique et Astroparticules
Université Montpellier II, France

1. Introduction

The answer that the quantum computing community currently gives to this question is a cheerful "yes". The so-called "threshold" theorem says that, once the error rate per qubit per gate is below a certain value, estimated as 10^{-4}–10^{-6}, indefinitely long quantum computation becomes feasible, even if all of the 10^{3}–10^{6} qubits involved are subject to relaxation processes, and all the manipulations with qubits are not exact. By active intervention, errors caused by decoherence can be detected and corrected during the computation. Though today we may be several orders of magnitude above the required threshold, quantum engineers may achieve it tomorrow (or in a thousand years). Anyway large-scale quantum computation is possible *in principle,* and we should work hard to achieve this goal.

The enormous literature devoted to this subject (Google gives 29300 hits for "fault-tolerant quantum computation") is purely mathematical. It is mostly produced by computer scientists with a limited understanding of physics and a somewhat restricted perception of quantum mechanics as nothing more than unitary transformations in the Hilbert space plus "entanglement". On the other hand, the heavy machinery of the theoretical quantum computation with its specific terminology, lemmas, *etc.*, is not readily accessible to most physicists (including myself). The vast majority of researchers, who start their articles with the standard mantra that the topic is pertinent to quantum computation, do not really understand, nor care, what stands behind the threshold theorem and how quantum error correction is supposed to work. They simply accept these things as a proven justification of their activity.

Meanwhile, even the most ardent proponents of quantum computing recognize today that it is impossible to build a useful machine without implementing efficient error correction. Thus the question in the title is equivalent to asking whether quantum computing is possible altogether.

In a previous publication,[1] I too accepted the threshold theorem but argued that the required enormous precision will not be achieved in any foreseeable future. The purpose of this article, intended for physicists, is to outline the ideas of quantum error correction and take a look at the technical instructions for fault-tolerant quantum computation, first put forward by Shor and elaborated by other mathematicians. It seems that the mathematics behind the threshold theorem is

Future Trends in Microelectronics. Edited by Serge Luryi, Jimmy Xu, and Alex Zaslavsky

somewhat detached from the physical reality, and that some flawless elements are inevitably present in the construction. This raises serious doubts about the possibility of large-scale quantum computation, even as a matter of principle.

2. Brief outline of ideas

The idea of quantum computing is to store information in the values of 2^N amplitudes describing the wavefunction of N two-level systems, called *qubits*, and to process this information by applying unitary transformations (*quantum gates*), that change these amplitudes in a very precise and controlled manner, see the clear and interesting review by Steane.[2] The value of N needed to have a useful machine is estimated as 10^3–10^6. Even for $N = 1000$, the number of continuous variables (the complex amplitudes of this grand wavefunction) that we are supposed to control, is $2^{1000} \sim 10^{300}$. For comparison, the total number of protons in the Universe is only about 10^{80} (give or take a couple of orders of magnitude).

The interest in quantum computing surged after Shor[3] invented his famous algorithm for factoring very large numbers and showed that an ideal quantum computer can solve this problem much faster than a classical computer that would require exponentially great time and resources.

Generally, there are many interesting and useful things one could accomplish with *ideal* machinery, not necessarily quantum. For example, one could become younger by *exactly* reversing all the velocities of atoms in one's body (and in the immediate vicinity), or write down the full text of all the books in the world in the *exact* position of a single particle, or store information in the 10^{23} vibrational amplitudes of a cubic centimeter of a solid. Unfortunately, unwanted noise, fluctuations, and inaccuracies of our manipulations impose severe limits to such ambitions. Thus, while the ideas of quantum computing are fascinating and stimulating, the possibility of actually building a quantum computer, even in some distant future, was met from the start with a healthy scepticism.[4-6]

Unlike the digital computer employing basically the on/off switch, which is stable against small-amplitude noise, the quantum computer is an analog machine where small errors in the values of the continuous amplitudes are bound to accumulate exponentially. So, it seems that a quantum computer of a complexity sufficient to be of any practical interest will never work.

In response to this challenge, Shor[7] and Steane[8] proposed the idea of quantum error correction – an ingenious method designed primarily to overcome the so-called "no cloning" theorem: an unknown quantum state cannot be copied. At first glance, this theorem prevents us from checking for errors in the quantum amplitudes, so that one can correct them. The idea of quantum error correction is to spread the information contained in a *logical* qubit among several *physical* qubits and to apply a special operator, which detects errors in physical qubits (the *error syndrome*) and writes down the result by changing the state of some auxiliary (*ancilla*) qubits. By measuring the ancilla qubits only, we can see the error in the original quantum state and then correct it (see Section 6).

However, this method assumed that the ancilla qubits, the measurements, and the unitary transformations to be applied, remain ideal. It is said that this type of error correction is not fault-tolerant, whatever this may mean. (If the ancilla qubits are flawless, why not use them in the first place?) The ultimate solution, the *fault-tolerant quantum computation*, was advanced by Shor[9] and further developed by other mathematicians, see Refs. 10-14 and references therein. Now, nothing is ideal: all the qubits are subject to noise, measurements may contain errors, and our quantum gates are not perfect. Nevertheless, the threshold theorem says that arbitrarily long quantum computations are possible, so long as the errors are not correlated in space and time and the noise level remains below a certain threshold. In particular, with error correction a single qubit may be stored in memory, *i.e.* it can be maintained arbitrarily close to its initial state for an indefinitely long time.

This striking statement implies among other things that, once the spin resonance is narrow enough, it can be made *arbitrarily* narrow by active intervention with imperfect instruments. This contradicts all of our experience in physics. Imagine a pointer that can rotate in a plane around a fixed axis. Fluctuating external fields cause random rotations of the pointer, so that after a certain relaxation time the initial position gets completely forgotten. How can the use of other identical pointers (also subject to random rotations) and some external fields (that cannot be controlled perfectly) make it possible to maintain indefinitely a given pointer close to its initial position? The answer we get from experts in the field, is that it may work because of quantum mechanics: "*We fight entanglement with entanglement*"[10] or, in the words of the Quantum Error Correction Sonnet by Gottesman:[11]

> *With group and eigenstate, we've learned to fix*
> *Your quantum errors with our quantum tricks.*

This does look suspicious, because in the physics that we know, quantum-mechanical effects are more susceptible to noise than classical ones. Before going into the details of the proposed fault-tolerant computation, the following section will present a slight divertissement relevant to our subject.

3. Capturing a lion in the desert

The scientific folklore knows an anecdote about specialists in various fields proposing their respective methods of capturing a lion in a desert. (For example, the Philosopher says that a captured lion should be defined as one who is separated from us by steel bars. So, let us go into the cage and the lion will be captured). Here, we are concerned with the Mathematician's method:[15]

> The desert D being a separable topological space, it contains a countable subset S that is everywhere dense therein. (For example, the set of points with rational coordinates is eligible as S.) Therefore, letting $x \in D$ be the point at which the lion is located, we can find a sequence $\{x_n\} \subset S$ with

$\lim_{n\to\infty} x_n = x$. This done, we approach the point x along the sequence $\{x_n\}$ and capture the lion.

This method, illustrated in Fig. 1, assumes that the only relevant property of the lion is to be located at a given point in 2D space. Note also that neither *time* nor what can happen to the lion and the hunter during the process is a point of concern. And finally, it is not specified how the sequence $\{x_n\}$ should be chosen, nor what the limit as $n \to \infty$ could mean in practice. These points are left to be elaborated by the practical workers in the field.

Certainly, mathematics is a wonderful thing, both in itself, and as a powerful tool in science and engineering. However, we must be very careful and reluctant in accepting theorems, and especially technical instructions, provided by mathematicians in domains outside pure mathematics.[16] Whenever there is a complicated issue, whether in many-particle physics, climatology, or economics, one can be almost certain that no theorem will be applicable or relevant, because the explicit or implicit underlying assumptions will never hold in reality. The Hunter must first explain to the Mathematician, what a lion looks like.

Figure 1. The Mathematician's method of capturing the lion in the desert.

4. Spin relaxation or decoherence

While the relaxation of two-level systems was thoroughly studied during a large part of the 20th century, and is quite well understood, in the quantum computing literature there is a strong tendency to mystify the relaxation process and make it look as an obscure quantum phenomenon:[9,10] "*The qubit (spin) gets entangled with the environment ...*" or "*The environment is constantly trying to look at the state of a qubit, a process called decoherence*", *etc.*

In a way, this sophisticated description may be true, however it is normally quite sufficient to understand spin relaxation as a result of the action of a time-dependent Hamiltonian $H(t) = \boldsymbol{A}(t)\boldsymbol{\sigma}$, where $\boldsymbol{A}(t)$ is a random vector function, and $\boldsymbol{\sigma}$ are Pauli matrices. In simple words, the spin continuously performs precession around magnetic fields that fluctuate in time. In most cases these magnetic fields are not real, but rather effectively induced by some interactions. A randomly fluctuating field is characterized by its correlation time τ_C and by the average angle of spin precession α during the time τ_C. For the most frequent case when $\alpha << 1$, the spin vector experiences a slow angular diffusion. The RMS angle after a time $t >> \tau_C$ is $\varepsilon \sim \alpha(t/\tau_C)^{1/2}$. Hence the relaxation time τ is given by $\tau \sim \tau_C/\alpha^2$. If one chooses a time step t_0, such that $\tau_C << t_0 << \tau_C/\alpha^2$, it can be safely assumed that $\varepsilon << 1$, and that rotations during successive time steps are not correlated.

These random rotations persist *continuously* for *all* the qubits inside the quantum computer. It is important to understand that the wavefunction describing an arbitrary state of N qubits will deteriorate much faster than any individual qubit. The reason is that this wavefunction

$$\Psi = A_0|000...00\rangle + A_1|000...01\rangle + A_2|000...10\rangle + ... + A_{2^N-1}|111...11\rangle$$

describes complicated correlations between the N qubits, and correlations always decay more rapidly. For simplicity, suppose that all qubits are subject to random and uncorrelated rotations around the z axis only. Then during one step the state of the jth qubit will change accordingly to the rule: $|0\rangle \to |0\rangle$, $|1\rangle \to \exp(i\varphi_j)|1\rangle$, where φ_j is the random rotation angle for this qubit. Then the amplitudes A will acquire phases $\Sigma\varphi_j$, where the sum goes over all qubits that are in the state $|1\rangle$ in a given term of Ψ. The typical RMS value of this phase is $\sim \varepsilon N^{1/2}$. Thus the time it takes for unwanted phase differences between the amplitudes A to become large is $\sim N$ times shorter than the relaxation time for an individual qubit.

For this reason, it seems that if we choose the time step so that $\varepsilon^2 = 10^{-6}$ (the most cautious of existing estimates for the noise threshold), but the quantum state contains 10^6 qubits, then the computer is likely to crash during a single time step. I am unaware of anybody discussing this problem.

5. Quantum computation with decoherence-free subspaces

This is a flourishing and respectable branch of quantum computing mathematics (Google gives 24800 hits for "decoherence-free subspaces"). The idea is that there may exist some special symmetry of the relaxation processes, due to which certain many-qubit states do not relax at all. (The simplest model is to consider a relaxation process, in which all the qubits are rotated collectively). It is then discussed how information can be hidden in this decoherence-free subspace, and what would be the best way to proceed with quantum computation. The conditions for the existence of such subspaces are given[17] by the following:

Theorem 4. If no special assumptions are made on the coefficient matrix $a_{\alpha\beta}$ [Eq. (8)] and on the initial conditions ρ_{ij} [Eq. (13)] then a necessary and sufficient condition for a subspace $\tilde{\mathcal{H}} = Span[\{|\tilde{k}>\}]_{k=1}^{N}$ to be decoherence-free is that all the basis states $|\tilde{k}\rangle$ are degenerate eigenstates of all Lindblad operators $\{\mathbf{F}_\alpha\}$,

$$\mathbf{F}_\alpha|\tilde{k}> = c_\alpha|\tilde{k}> \quad \forall \alpha, k.$$

This gives the reader an idea of what the quantum computing literature looks like.

Although it is not difficult to construct artificial models with special symmetries, my guess is that in any *real* situation the Lindblad operators do not have common eigenstates at all. Obviously, the simplest way to fight noise is to suppose that at least *something is ideal* (noiseless). Unfortunately, this is not what happens in the real world.

6. Quantum error correction by encoding

Below is a simplified example[12,18] of quantum error correction using *encoding*. The simplification results from the assumption that the only errors allowed are rotations around the x axis, described by the matrix $E = \cos(\theta/2)I - i\sin(\theta/2)\sigma_X$, where I is the unit matrix and σ_X is the Pauli matrix. For small rotation angles $\theta = 2\varepsilon \ll 1$, this gives $E = I - i\varepsilon\sigma_X$. In the case when these are the only possible errors for individual qubits, it is sufficient to encode the logical $|0\rangle$ and $|1\rangle$ by three physical qubits: $|0\rangle \to |000\rangle$, $|1\rangle \to |111\rangle$. The encoding procedure requires the following steps:

1) The general state of a qubit, $a|0\rangle + b|1\rangle$, is encoded as $\psi = a|000\rangle + b|111\rangle$. Suppose that the three physical qubits experience small and uncorrelated rotations E_1, E_2, and E_3. Let us see how the initial state can be recovered. For example, suppose there is an error in the second qubit only. The wavefunction becomes:

$$E_{2\psi} = [a|000\rangle + b|111\rangle] - i\varepsilon_2[\,a|010\rangle + b|101\rangle].$$

2) We now mechanically add 3 auxiliary ancilla qubits, obtaining the state:

$$E_{2\psi} = [\,a|000\rangle + b|111\rangle]\,|000\rangle - i\varepsilon_2[\,a|010\rangle + b|101\rangle]\,|000\rangle.$$

3) Next we introduce the syndrome *extraction operator S*, defined as follows:

$$S|000\rangle|000\rangle = |000\rangle|000\rangle, \quad S|111\rangle|000\rangle = |111\rangle|000\rangle,$$

$$S|100\rangle|000\rangle = |100\rangle|100\rangle, \quad S|011\rangle|000\rangle = |011\rangle|100\rangle,$$
$$S|010\rangle|000\rangle = |010\rangle|101\rangle, \quad S|101\rangle|000\rangle = |101\rangle|101\rangle,$$
$$S|001\rangle|000\rangle = |001\rangle|001\rangle, \quad S|110\rangle|000\rangle = |110\rangle|001\rangle.$$

The first 3 qubits, containing the data, are left intact. If one of them is flipped, then the ancilla bits are changed accordingly. The operator S writes down the error into the ancilla, allowing us to identify the error location. Now:

$$SE_{2\psi} = [a|000\rangle + b|111\rangle]\,|000\rangle - i\varepsilon_2[\,a|010\rangle + b|101\rangle]\,|101\rangle.$$

4) Finally, we measure the three ancilla qubits. If we get (000), then we do nothing, since this result shows that the state automatically has been reduced to the initial state $\psi = a|000\rangle + b|111\rangle$. If (with a small probability equal to ε_2^2) we obtain the result (101), then we know that there is an error in the second qubit, and that the state of the remaining (data) qubits is $a|010\rangle + b|101\rangle$. This error is easily corrected by applying the operator σ_X to the second qubit. Thus, in both cases we recover the original state $\psi = a|000\rangle + b|111\rangle \rightarrow a|0\rangle + b|1\rangle$!

The method works equally well if all three data qubits have errors, provided that the second-order terms proportional to $\varepsilon_1\varepsilon_2$, $\varepsilon_1\varepsilon_3$, and $\varepsilon_2\varepsilon_3$ in the $E_1E_2E_3\psi$ wavefunction can be neglected. This is justified by the small probability for the admixture of the corresponding states, proportional to ε^4 (compared to the ε^2 probability of one-qubit errors). However, this method does not work if errors in different qubits are correlated, *i.e.* if there is an admixture of states with two errors with an amplitude ε, rather than ε^2. This is why the requirement that errors are uncorrelated is crucial.

There is, indeed, a remarkable "quantum trick" here: the measurement of the ancilla qubits automatically reduces the wavefunction representing a large superposition of states to only one of its terms! Because of this, knowing how to correct a bit-flip ($|0\rangle \rightarrow |1\rangle$, $|1\rangle \rightarrow |0\rangle$), referred to as "fast error", we can also correct "slow errors", $E = I - i\varepsilon\sigma_X$, for *arbitrary* (but small) values of ε. This property is called "digitization of noise".

To correct general one-qubit errors a more sophisticated encoding[7,8] by a greater numbers of qubits is needed. For example, the logical $|0\rangle$ might be encoded as: $|0\rangle \rightarrow (1/\sqrt{8})\ \{|0000000\rangle + |0001111\rangle + |0110011\rangle + |0111100\rangle + |1010101\rangle + |1011010\rangle + |1100110\rangle + |1101001\rangle\}$. However the principle of error correction is the same. It is believed that it may be advantageous to use *concatenated* encoding, in which each encoding qubit should be further encoded in the same manner, and so on ... It is supposed that the future quantum engineer[19] might wish to encode the logical $|0\rangle$ and $|1\rangle$ by complicated superpositions of the states of $7^3 = 343$ physical qubits!

What if we apply the same method of error correction not just to one qubit, but to some N-qubit state? Making exactly the same assumptions, we encode each of the N logical qubits by three physical ones, add an appropriate number of ancilla qubits, and proceed in the same way as above. Then, in accordance with the remark at the end of Section 4, the condition for the method to work will be $N\varepsilon^2 \ll 1$, not simply $\varepsilon^2 \ll 1$. Indeed, after applying the syndrome extraction operator, the number of one-qubit error terms with probabilities ε^2 is N, while the number of two-qubit error terms with probabilities ε^4 is $N(N-1)/2$. In order for the method to work, the total probability of obtaining *any* two-error state during measurements should be small, which is true when $N\varepsilon^2 \ll 1$.

7. The imperfect two-qubit gate

For quantum computation one needs to apply one-qubit gates, but also two-qubit and three-qubit gates. Application of two-qubit gates is necessary from the outset to perform the first step, encoding. The problem is not only that individual qubits are subject to relaxation, as described in Section 4, but also that the quantum gates are also not perfect, because neither the Hamiltonian that should be switched on at the desired moment, nor the duration of its action can be controlled exactly. While an error in a one-qubit gate can be simply added to the random rotations that exist anyway, errors in two-qubit gates requires more care.

We should first decide what an imperfect two-qubit gate is. It seems that the generally accepted model is the following:[9] "*For the error model in our quantum gates, we assume that with some probability p, the gate produces unreliable output, and with probability 1–p, the gate works perfectly.*"

A more detailed description[2,13] also specifies the exact meaning of an unreliable output: "*The failure of a two-qubit gate is modeled as a process where, with probability* $1-\gamma_2$ *no change takes place before the gate, and with equal probabilities* $\gamma_2/15$ *one of the 15 possible single- or two-qubit failures take place.*"

In other words, the faulty gate is supposed to act as an *ideal* one (with high probability) or to act as an ideal one preceded by *uncorrelated* errors in the two qubits involved (with low probability). In reality, there will always be some more or less narrow probability distribution around their desired values of the 16 real parameters defining the unitary transformation. Never, under any circumstances, will an ideal gate exist. The crucial difference is that any real gate will introduce *correlated* errors of the two qubits. Such correlated errors are not correctable within the error-correcting scheme described in Section 6.

Here is a more sophisticated model:[20] "*The noise model we will consider can be formulated in terms of a time-dependent Hamiltonian H that governs the joint evolution of the system and the bath. We may express H as* $H = H_S + H_B + H_{SB}$, *where* H_S *is the time-dependent Hamiltonian of the system that realizes the ideal quantum circuit,* H_B *is the arbitrary Hamiltonian of the bath, and* H_{SB} *couples the system to the bath.*"

This is a quite general approach, especially if the arbitrary (?) Hamiltonian of the "bath" also describes the electronic equipment and the quantum engineer himself. However, in reality there is no such thing as a "time-dependent Hamiltonian of the system that realizes the ideal quantum circuit", just as there is no such thing as square root of 2 with all the infinite number of its digits. True, such abstractions are routinely used in mathematics and theoretical physics. However the whole issue at hand is to understand whether the noisy nature of the *real* Hamiltonian does, or does not, make it possible to realize anything sufficiently close to the "ideal quantum circuit". Thus, it could well happen that the supposed success of fault-tolerant quantum computation schemes is entirely due to the uncontrolled use of innocent-looking abstractions and models (see Section 3). A very careful analysis is needed to understand the true consequences of any simplifications of this kind.

Another implicit assumption, which may be not quite innocent, is that the gates are infinitely fast. In fact, new errors may appear *during* error correction.[21]

8. The prescription for fault-tolerant quantum computation

The error correction scheme, briefly described in Section 6, assumes that encoding, syndrome extraction, and recovery are all ideal operations, that ancilla qubits are error-free, and that measurements are exact. Fault-tolerant methods are based on the same idea, but are supposed to work even if all these unrealistic assumptions (making, in fact, error correction unnecessary) are lifted. The full instructions[9-13] are extraordinary complicated, details that may be important are often omitted, and the statements are not always quite clear. The basic ideas are as follows.

1) There exists a universal set of three gates, sufficient for quantum computation. "*The proof of this involves showing that these gates can be combined to produce a set of gates dense in the set of 3-qubit gates*"[9] (see Section 3). In other words, any gate can be approximated to any desired accuracy by application of a large enough number of the three special gates belonging to the universal set.

2) These three gates can be used in a fault-tolerant manner, which means in such a way that only uncorrelated, and thus correctable, errors are produced. Fault-tolerance is achieved by encoding the logical qubits, using specially prepared states of ancilla qubits, and some rules designed to avoid error propagation. Thus application of a single 3-qubit gate fault-tolerantly amounts to a mini-quantum computation with thousands of elementary operations.

3) The encoding itself cannot be done fault-tolerantly:[10] "*Therefore, we should carry out a measurement that checks that the encoding has been done correctly. Of course, the verification itself may be erroneous, so we must repeat the verification a few times before we have sufficient confidence that the encoding was correct.*" A similar prescription concerns ancilla states:[21] "*However, the process of creating the ancilla blocks may introduce correlated errors, and if those errors enter the data, it will be a serious problem. Therefore, we must also verify the ancilla blocks to eliminate such correlated errors. Precisely how we do this is not important for the discussion below, but it will certainly involve a*

number of additional ancilla qubits." It is recommended[10] that one construct auxiliary "cat states": $(1/\sqrt{2})(|00\ldots00\rangle + |11\ldots11\rangle)$, where the size of the cat depends on the number of qubits used for encoding. Again, since there may be (one might better say: there *always* will be) errors in the cat state, it must be *verified* before being used.

4) Some operations should be repeated:[10] "*If we mistakenly accept the measured syndrome and act accordingly, we will cause further damage instead of correcting the error. Therefore, it is important to be highly confident that the syndrome measurement is correct before we perform the recovery. To achieve sufficient confidence, we must repeat the syndrome measurement several times.*"

5) All these precautions still do not guarantee that the computer will not crash. However, what matters is the *probability* of crash. Once this probability is small enough, which is supposed to happen at a low enough noise rate, we can repeat the whole quantum calculation many times to get reliable results. By estimating the crash probability, one obtains an estimate for the threshold noise level.

A detailed description of the fault-tolerant computation rules can be found in Ref. 10. I do not find this description clear and/or convincing enough. Taking into account the continuous nature of random qubit rotation and gate inaccuracies, even with all the verifications and repetitions, there seems no way to avoid small admixtures of unwanted states.

In fact, the pure spin-up state can never exist in reality (one reason is that we never know the exact direction of the z axis). Similarly, in the classical world, one can never have a pointer pointing exactly in the z direction. Generally, no desired state can ever be achieved *exactly*. Rather, whatever we do, we will *always* have an admixture of unwanted states, more or less rich. One can never have an exact $(|0\rangle + |1\rangle)/\sqrt{2}$ state, let alone more complicated "cat" states like $(|0000000\rangle + |1111111\rangle)/\sqrt{2}$. Such abstractions must be used with extreme caution when discussing the role of errors and inaccuracies.

When the small undetected and unknown admixture of unwanted states together with the "useful" state is fed into the subsequent stages of the quantum network, it is most likely that the error will grow exponentially in time. Accordingly, the crash time will depend only logarithmically on the initial error value. This is what happens when one tries to reverse the evolution of a gas of hard balls. At a given moment one reverses the direction of all the velocities, but oops, the gas will never return to its initial state (even in computer simulation, let alone reality). The reason is that however small the initial (and subsequent) computer errors are, they will increase exponentially (the Lyapunov exponent). It is a great illusion to think that things are different in quantum mechanics.

Related to this, there is another persistent misunderstanding of quantum mechanics that plagues the quantum error correction literature. Using the classical language, it is said that the qubit "decoheres" with probability $p = \sin^2\theta$, instead of saying that the qubit is in the state $\psi = \cos\theta|0\rangle + \sin\theta|1\rangle$. This makes only a semantic difference if we are going to immediately *measure* the qubit, since the probability of finding it in a state $|1\rangle$ is indeed p. However, this language becomes completely wrong if we consider some further evolution of our qubit described by some unitary matrix R. The common thinking (applied, for example, to estimating the noise threshold) is that we will have the state $R|0\rangle$ with probability $1-p$, and the state $R|1\rangle$ with probability p. In reality, we will have the state $R\psi$, which is not the same thing. The former line of reasoning gives the probability of measuring $|0\rangle$ in the final state as $(1-p)|\langle 0|R|0\rangle|^2 + p|\langle 0|R|1\rangle|^2$, while the latter (and correct) one will give $|\langle 0|R|\psi\rangle|^2$, and these results are very different. As an exercise, the reader can take for R a rotation of our qubit around the x axis by some angle and compare the results. A quantum-mechanical surprise lies in store. In quantum mechanics, one cannot calculate probabilities by considering what happens to ideal states. Instead, one must look at the evolution of *real* states, which always differ from ideal ones by some admixture of unwanted states.

Another important point is that the finite time required to do anything at all is usually not taken into account (see Section 3). According to the procedure described in Section 6, measuring the syndrome and obtaining (000) indicates the correct state that requires no further action. In fact, while we were making our measurements, the data qubits have experienced their random rotations. And this will happen no matter how many times we repeat the syndrome measurement. So why bother with error correction?

Alicki[21] has made a mathematical analysis of the consequences of finite gate duration. I am not in a position to check his math, but I like his result: the fidelity exponentially decreases in time. He writes: "*... unfortunately, the success of existing error correction procedures is due to the discrete in time modeling of quantum evolution. In physical terms, discrete models correspond to unphysical infinitely fast gates.*"

9. Designing perpetual motion machines of the second kind

This is certainly *not* equivalent to achieving fault-tolerant quantum computation, during which we will put some energy into the system by applying external fields and performing measurements.

However there is a certain similarity between the two problems in the sense that what we are trying to do is to maintain a reversible evolution of a large system with many degrees of freedom in the presence of noise and using noisy devices.[23] People who have had the opportunity of examining proposals for perpetual motion machines, know their basic principle: insert at least *one* ideal (*i.e.* not sensitive to thermal fluctuations) element somewhere deep within a complicated construction. Finding and identifying such an ideal element may be a daunting task.

Naively, one starts by proposing a valve that preferentially lets through only fast molecules. Next, one understands that the valve itself is "noisy", so that it will not work as expected. However, if one adopts a noise model in which the valve is faulty with a probability p but works perfectly with a probability $(1-p)$, or makes a sophisticated construction involving many valves connected by wheels and springs, and if just one of these elements is considered as *ideal* (or even working perfectly with some probability), one can immediately arrive at the conclusion that a perpetual motion machine feeding on thermal energy is possible.

This lesson should make us extremely vigilant to the explicit or implicit presence of ideal elements within the theoretical error-correcting schemes.

10. Challenge

After ten years of doing mathematics devoted to fault-tolerant quantum computation, maybe the time is ripe for a simple numerical test. Let us focus on the simplest, almost "trivial" task of storing just one qubit and let us verify the statement that its initial state can be maintained indefinitely in the presence of low-amplitude noise. Take, for example, an initial state $(|0\rangle + |1\rangle)\sqrt{2}$ – a spin pointing in the x direction.

Assume that all qubits experience continuous random uncorrelated rotations, the RMS rotation angle being small during a single time step. As a further simplification, one could restrict the random rotations to the xy plane only. The two- and three-qubit gates have a narrow probability distribution for all of the parameters, defining the corresponding unitary transformation, around their desired values. Errors in successive gates are not correlated. We have a refrigerator containing an unlimited number of ancilla qubits in the pure state $|0\rangle$. Once the ancillas are out of the refrigerator, they become subject to the same random rotations. Measurements involve errors: when the state $a|0\rangle + b|1\rangle$ is measured with the result 0, the quantum state is not reduced to exactly $|0\rangle$, but rather to $|0\rangle + c|1\rangle$ with some unknown, but hopefully small c. Allocate a certain time for measurements and duration of gates and take into account that all qubits continue to be rotated randomly during this time. This simple model cannot be relaxed further without entering an imaginary world where something is ideal.

Presumably, to maintain our single qubit close to its initial state, a certain sequence of operations (with possible branching depending on the result of intermediate measurements) should be applied periodically. *Provide a complete list of these elementary operations*, so that anybody can use a PC to check whether qubit storage really works. The future quantum engineer will certainly need such a list! *If* it works, this demonstration would be a convincing, though partial, proof that the idea of fault-tolerant quantum computation is sound.

Steane[13] undertook a thorough numerical simulation of error propagation in a quantum computer with results confirming the threshold theorem. Since an exact simulation of quantum computing on a classical computer is impossible, he used a partly phenomenological model, based on the (questionable) assumption that "*it is sufficient to keep track of the propagation of errors, rather than the evolution of the complete computer state*". Since the real errors always consist in admixtures of unwanted states, considering the quantum evolution of the complete computer state seems to be the only way to respect quantum mechanics (see Section 8).

The task of maintaining a single qubit in memory is much simpler and, once the list of required operations is provided, it hopefully can be simulated in a straightforward manner. I predict that the result will be negative.

11. Conclusions

It is premature to accept the threshold theorem as a proven result. The state of a quantum computer is described by the monstrous wavefunction with its 10^{300} complex amplitudes, all of which are continuously changing variables. If left alone, this wavefunction will completely deteriorate during $1/N$ of the relaxation time of an individual qubit, where $N \sim 10^3$–10^6 is the number of qubits within the computer. It is absolutely incredible, that by applying external fields, which cannot be calibrated perfectly, doing imperfect measurements, and using converging sequences of "fault-tolerant", but imperfect, gates from the universal set, one can continuously repair this wavefunction, protecting it from the random drift of its 10^{300} amplitudes, and moreover make these amplitudes change in a precise and regular manner needed for quantum computations. And all of this on a time scale greatly exceeding the typical relaxation time of a single qubit.

The existing prescriptions for fault-tolerant computation are rather vague, and the exact underlying assumptions are not always clear. There are several subtle issues, some of which are discussed in this chapter, that should be examined more closely. It seems likely that the (theoretical) success of fault-tolerant computation is due not so much to "quantum tricks", but rather to the backdoor introduction of ideal (flawless) elements in an extremely complicated construction. Previously, this view was expressed by Kak.[24]

It would be useful to check whether the fault-tolerant methods really work by numerically simulating quantum evolution during the proposed recovery procedures for a single qubit using a realistic noise model that does not contain any ideal elements.

References

1. M. I. Dyakonov, "Quantum computing: A view from the enemy camp," in: S. Luryi, J. Xu, and A. Zaslavsky, eds., *Future Trends in Microelectronics. The Nano Millennium*, New York: Wiley, 2002, pp. 307–318; arxiv.org/cond-mat/0110326.
2. A. Steane, "Quantum computing," *Reports Prog. Phys.* **61**, 117 (1998); arxiv.org/quant-ph/9708022.
3. P. W. Shor, "Polynomial-time algorithms for prime factorization and discrete logarithms on a quantum computer," in: S. Goldwasser, ed., *Proc. 35th Annual Symp. Found. Computer Sci.*, Los Alamos: IEEE Computer Society Press, 1994, p.124; arxiv.org/quant-ph/9508027.
4. W. G. Unruh, "Maintaining coherence in quantum computers," *Phys. Rev. A* **51**, 992 (1995).

5. R. Landauer, "The physical nature of information," *Phys. Lett. A* **217**, 188 (1996).
6. S. Haroche and J.-M. Raimond, "Quantum computing: Dream or nightmare?" *Phys. Today* **49**, August issue (1996), p. 51.
7. P. W. Shor, "Scheme for reducing decoherence in quantum computer memory," *Phys. Rev. A* **52**, 2493 (1995).
8. A. M. Steane, "Error correcting codes in quantum theory," *Phys. Rev. Lett.* **77**, 793 (1996).
9. P. Shor, "Fault-tolerant quantum computation", in: *Proc. 37th Symp. Found. Computer Sci.*, Los Alamos: IEEE Computer Society Press, 1996, pp. 56–65; arxiv.org/quant-ph/9605011.
10. J. Preskill, "Fault-tolerant quantum computation," in: H.-K. Lo, S. Papesku, and T. Spiller, eds., *Introduction to Quantum Computation and Information*, Singapore: World Scientific, 1998, pp. 213–269; arxiv.org/quant-ph/9712048.
11. D. Gottesman, "An introduction to quantum error correction," in: S. J. Lomonaco, Jr., ed., *Quantum Computation: A Grand Mathematical Challenge for the Twenty-First Century and the Millennium*, Providence, RI: American Mathematical Society, 2002, pp. 221–235; arxiv.org/quant-ph/ 0004072.
12. A. M. Steane, "Quantum computing and error correction," in C. Vorderman, A. Gonis, and P. E. A. Turchi, eds., *Decoherence and Its Implications in Quantum Computation and Information Transfer*, Amsterdam: IOS Press, 2001, pp. 284–298; arxiv.org/quant-ph/0304016.
13. A. M. Steane, "Overhead and noise threshold of fault-tolerant quantum error correction," *Phys. Rev. A* **68**, 042322 (2003).
14. D. Kribs, R. Laflamme, and D. Poulin, "A unified and generalized approach to quantum error correction," *Phys. Rev. Lett.* **94**, 180501 (2005).
15. I thank K. M. Dyakonov for providing the following text and E. M. Diakonova for drawing Figure 1.
16. Somebody has proved a theorem concerning the existence of bound states for the one-dimensional Schrödinger equation. The surprise comes for *repulsive* potentials $U(x) = -Ax^n$: it is proved that for $n = 4, 6 ...$ bound states always exist. The reader might be interested to find out a) why this theorem is true; b) what is its physical meaning; and c) why it is irrelevant to the real world.
17. D. A. Lidar and K. B. Whaley, "Decoherence-free subspaces and subsystems," in F. Benatti and R. Floreanini, eds., *Irreversible Quantum Dynamics*, Berlin: Springer Lecture Notes in Physics Vol. 622, 2003, pp. 83–120; see also arxiv.org/quant-ph/0301032.
18. E. G. Rieffel and W. Polak, "An introduction to quantum computing for non-physicists," arxiv.org/quant-ph/9809016.
19. The *future quantum engineer* is a mythical personage, who will not only develop and implement the hardware for the future quantum computer, but also design the sequence of the necessary operations based on the ideas presented in Refs. 9–14 (see Ref. 1 for the job description).
20. D. Aharonov, A. Kitaev, and J. Preskill, "Fault-tolerant computation with long-range correlated noise," *Phys. Rev. Lett.* **96**, 050504 (2006).

21. R. Alicki, "Quantum error correction fails for Hamiltonian models," arxiv.org/quant-ph/0411008.
22. D. Gottesmann, "Fault-tolerant quantum computation with local gates," *J. Modern Opt.* **47**, 333 (2000); arxiv.org/quant-ph/9903099.
23. Because of this similarity, it is quite probable that the design and the theory of such machines will become the next hot topic, especially if a good name (beginning with the magic word "quantum") for this activity can be invented.
24. S. Kak, "General qubit errors cannot be corrected," *Information Sci.* **152**, 195 (2003); arxiv.org/quant-ph/0206144.

Quantum Computation – Future of Microelectronics?

P. Hawrylak
Institute for Microstructural Sciences
National Research Council of Canada, Ottawa, K1A OR6 Ontario, Canada

1. Introduction

Current silicon technology has steadily improved our ability to compute by increasing the number of bits and gates. Yet with the existing technology many problems are likely to remain unsolved: properties of quantum materials, multiscale problems inherent to nanoscience, drug design and discovery, hard mathematical problems such as factorization of prime numbers essential for security, to name a few. Quantum instead of classical computation has been suggested as a possible solution.[1,2] We will attempt to present our perspective of how quantum computing might fit into microelectronics.

Quantum computation attempts to take advantage of the same property that makes some of the problems so difficult to solve – quantum-mechanical behavior of many-particle systems. In quantum mechanics the state of the system, *e.g.* composed of a number of electrons and nuclei, is described by a superposition of electronic configurations. Imagine such electronic configuration, $|1,1,0,0,1,0,0,1,0,1,0,0\rangle$, where $N_e = 5$ electrons are distributed on $N_S = 12$ possible states. Here the states could be atomic orbitals of the quantum material, but we can think of them equally well as "quantum registers", where "0" means empty and "1" means occupied. The problem with simulating a quantum material is the number of possible configurations in which this material can be found. For $N_S = 12$ and $N_e = 5$, the number of configurations is $12!/[5!(12–5)!] = 792$, a manageable number. However, doubling the number of both states and electrons to $N_S = 24$ and $N_e = 10$ gives rise to over a million of configurations. Increasing the number of states tenfold to 240, and the number of electrons to 100, leads to 10^{70} possible configurations, *i.e.* a number comparable to the number of atoms on Earth. Hence it is clear that, on one hand, it is impossible to know all the quantum configurations of even a very small quantum system, and on the other hand, if we learn how to harness these quantum states new avenues for computation become possible. Here the abundance of configurations is treated as a resource and the computation is facilitated by interference and entanglement. Of course, these ideas are not new: they have long been studied in condensed matter physics as "correlated electron physics". Here the entanglement and quantum properties of materials are due to "correlations" among electrons. Correlations found so far, such as superfluidity and superconductivity, exist only at very low temperatures. At higher temperatures, the correlations cease to be important and the classical behavior

Future Trends in Microelectronics. Edited by Serge Luryi, Jimmy Xu, and Alex Zaslavsky

dominates. This brings us to a notion of decoherence. Decoherence implies interaction of the quantum system with environment and destruction of its quantum state. Hence, quantum computation, just like it classical counterpart, is prone to errors. Detecting errors requires measurement, and measurement destroys, or collapses, the quantum state. So, at a first glance, the error correction appears to be impossible in quantum computation. Answering this challenge, *i.e.* showing how to correct errors in a quantum computer, was a very important achievement.[3] Error correction can be crudely described as correcting the phase of a qubit using auxiliary qubits, and correcting the amplitude by using classical error correcting means. While possible in principle, error correction leads to an enormous increase in complexity. Hence it seems we should initially attempt quantum information processing with devices that require only a small number of coherent qubits.

2. Applications of few-qubit systems

There are several potential applications of few-qubit systems, from quantum economics, quantum cryptography, quantum metrology, to quantum nuclear magnetic resonance (NMR). Quantum economics is perhaps less known than the rest.[4] It relies on quantum games and quantum strategies of conflict resolution.[5] Quantum bidding is attractive because it involves only a few players and the pay-off for the winner is high. An example of an experimentally realized two-qubit quantum game is the "prisoner dilemma", where the prisoners chances of escape are enhanced by using a "quantum game" to reach cooperation.[6]

The quantum key distribution systems employed in quantum cryptography are another example of a few-qubit system. This field is very advanced, and a number of commercial quantum cryptography systems are already on the market. Here the main challenge is a technical one, *i.e.* generating reliable sources of single photons and pairs of entangled photons[7] with wavelength compatible with current telecommunication wavelength of 1.5 μm.[8]

3. Few-qubit systems based on electron spin

There are many proposals and several rudimentary implementations of a quantum computer. They include: (a) nuclear spin combined with commercial NMR techniques,[9] (b) superconductive qubits,[10] (c) atom and ion traps,[11] (e) polarization states of a photon and linear optics,[12] (f) topological qubits in fractional quantum Hall effect (FQHE),[13] (g) solid state nuclear spins,[14] and (h) quantum dot and electron-spin based qubits.[15,16] Here the references are not exhaustive but rather indicative of research activities in each of the subfields. Out of these different proposals, the quantum dot and electron-spin based implementation starts with the field effect transistor (FET) structures and appears the most compatible with current microelectronic technology. In this chapter we will explore whether micro-electronics may evolve into quantum information technology, and investigate the current state-of-the-art. While the examples below will be drawn primarily from our own work, there is a significant and parallel effort worldwide.

With $|0\rangle$ and $|1\rangle$ states of the computational electron spin qubit basis equivalent to the two states of the electron spin S, the quantum computer Hamiltonian can be simply written as:

$$H = \sum_{i} \boldsymbol{S}_{\mathrm{i}} \cdot \boldsymbol{B}_{\mathrm{i}} + \sum_{ij} \boldsymbol{S}_{\mathrm{i}} \cdot J_{\mathrm{ij}} \cdot \boldsymbol{S}_{\mathrm{j}} \,. \qquad (1)$$

Here $\boldsymbol{S}_{\mathrm{i}}$ is the spin of electron localized on the ith quantum dot, $\boldsymbol{B}_{\mathrm{i}}$ is the local magnetic field acting on the ith spin and allowing single-qubit operations, and J_{ij} is the exchange coupling of two different spins. To build such a device one needs to develop technology that can: (a) localize electrons, (b) control precisely their numbers, (c) control their spin, and (d) control coupling between electrons. We like to think of this emerging enabling technology as nano-spintronics.[17] We now proceed to describe the current status of nano-spintronics with quantum dots.

- *Single-electron spin qubit*

A prerequisite for nano-spintronics is an ability to localize a single electron in a specific location, and "functionalize" this location. Localizing a single electron has been accomplished by using donors in semiconductors. However, the control of the position of individual donors and hence individual electrons is yet to be realized. Initial progress has been made with vertical quantum dots. Ashoori and co-workers build vertical quantum dots which were charged with few electrons.[18] Their properties were measured and understood using single electron capacitance spectroscopy.[18,19] In the next step, Tarucha and co-workers demonstrated a vertical quantum dot device with electron numbers down to one.[20] This device was connected both to source and drain, and the electronic states of this device were measured using the Coulomb blockade spectroscopy. The scalability of single electron devices is, however, best accomplished by using lateral metallic gates in a planar technology familiar in the microelectronics. Prior to 1999, lateral gating was not entirely successful in controlling electron numbers, but in 1999 Hawrylak and co-workers showed how to use spin flips to determine the number of electrons that are already in the lateral quantum dot.[21] This theory and experiment suggested that the application of moderate magnetic fields leads to spin polarization of edge states in both source and drain. Hence, a lateral quantum dot can be connected to spin-polarized contacts and current through such a device can be blocked not only by the charge (as in Coulomb blockade) but also by the spin – hence the concept of spin blockade spectroscopy (SBS). In the following year, Sachrajda and co-workers emptied a lateral quantum dot,[22] shown in Fig. 1, and filled it with a controlled number of electrons. The properties of such a device were probed by SBS and it was demonstrated that one can manipulate the spin state of a quantum dot. When a dot was connected to spin-polarized leads, the current was switched on and off by changing the spin state of the quantum dot device at a single-electron level, demonstrating the single spin transistor.[23] The lateral quantum dot device shown in Fig.1 allows to localize a single electron spin and represents a single qubit.

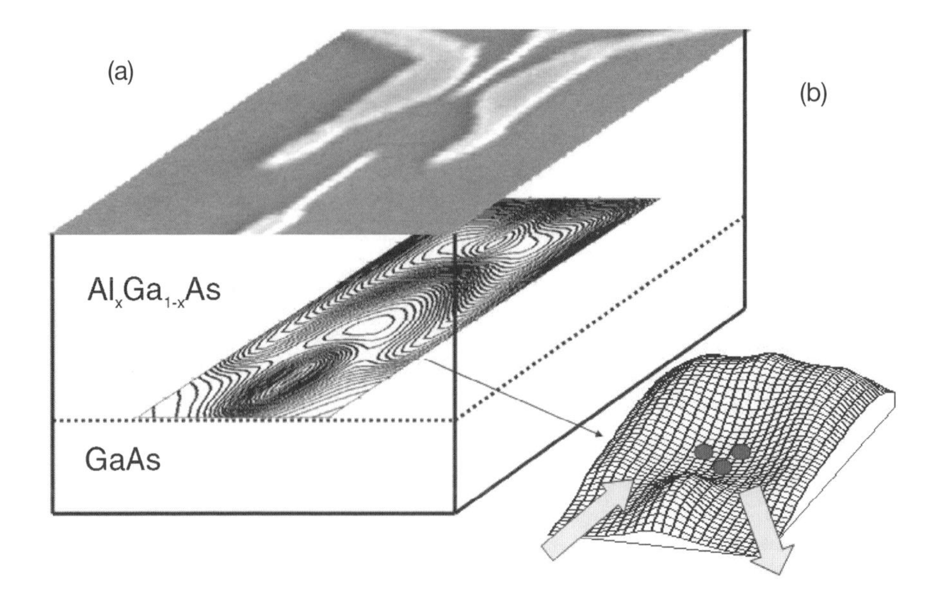

Figure 1. Schematic structure (a) and the energy profile (b) of the single-electron lateral quantum dot with controlled electron numbers.

- *Two-electron spin qubits*

In Fig. 2(a) we show a double-dot device,[24-27] studied by Pioro-Ladriere *et al.*,[24,25] which makes it possible to localize two electrons in two specific locations, as indicated by arrows. A quantum point contact (QPC) has been build in the vicinity of the double dot device. The current I_{QPC} through the QPC is a sensitive function of total charge and its distribution inside the double quantum dot: the closer the electrons in a double dot are to the QPC, the smaller the current. Figure 2(b) shows the stability diagram of the double dot device,[25] *i.e.* the derivative of QPC current as a function of two plunger gate voltages, after Ref. 25. Changing the potential applied to the gates changes the electron numbers and their location in the device. In Fig. 2(b), lines correspond to changes in total electron numbers, and different slope of these lines allows us to deduce in which quantum dot the electrons can be found. In this way, the quantum-dot occupation numbers N_1 and N_2 can be assigned to dot 1 and dot 2. We see that one can have no electrons in the device (0,0), one electron in the left dot (1,0) or one electron in the right dot (0,1), and one electron in each dot (1,1).

The (1,1) configuration is schematically represented by arrows in Fig. 2(a). One of the objectives is the ability to hybridize the two dots, *i.e.* to control the exchange coupling constant J_{ij} in Eq. (1). There are two measures of hybridization: the quantum-mechanical tunneling t of a single electron between the two dots, and the exchange coupling J_{ij} for two electrons.

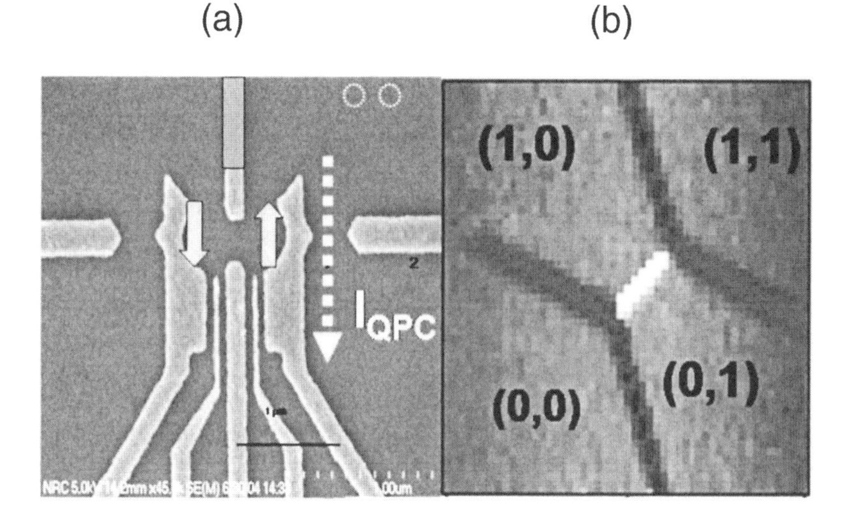

Figure 2. Lateral quantum dot implementation of the two-qubit system. (a) SEM picture of metallic gates of a double dot device integrated with a quantum point contact (QPC) charge readout device. (b) Charging diagram of a double dot device – derivative of the I_{QPC} as a function of plunger gate voltages for a fixed voltage applied to the barrier gate; (N_1, N_2) indicate electron numbers in dot 1 and dot 2.

The single-particle tunneling contributes to the exchange coupling via the super-exchange $\sim 2t^2/U$, where U is the on-site Coulomb energy. The quantum-mechanical tunneling is indirectly responsible for the width of the transition region between the (1,0) and (0,1) configurations: the stronger the tunneling, the wider and smoother the transition. Typical values of t measured in lateral dots approach 100 μeV. A number of groups have measured and tuned the exchange coupling J by applying bias to one of the dots. Moreover, Petta *et al.* demonstrated coherent Rabi oscillations between the singlet and one of the triplet two-electron states.[28] The two states served as the two states of a "coded qubit". However, the operation of a double dot as a controlled-NOT (CNOT) gate remains to be demonstrated.

- *Three-electron spin qubits*

The double quantum dot device is the first step toward a scalable quantum processor. It can, however, serve at best as a CNOT gate. The realization of a class of nontrivial algorithms, such as quantum teleportation,[29] requires at least three spins.[30] It has been also shown that three spins are needed to realize a coded qubit,[31,32] which combines long coherence times associated with electron spin with ease of operation associated with voltages. Inset to Fig. 3(a) shows both a schematic representation of a triple-dot device with one spin per dot, as well as the SEM picture of a triple dot device investigated by Gaudreau *et al.*[33] This triple-dot device combines a double-dot metallic gate layout with an impurity located in the left dot. The presence of three potential minima being filled with electrons localized in three spatially distinct regions is demonstrated by three different slopes of addition lines shown in Fig. 3(a). The three families of lines correspond to electron addition to a particular dot. An analysis similar to our treatment of the double-dot device allows us to assign particle numbers (N_1,N_2,N_3) for each of the dots. One can empty the device, *i.e.* drive it into the (0,0,0) state, and then fill it up with three electrons, one electron per dot, indicated by the (1,1,1) region in Fig. 3(a). Figure 3(b) shows the manipulation of the three-electron complex using gate voltages V_A and V_B. The goal here is to bring the three dots into resonance and manipulate their exchange coupling. This is difficult because it is hard to visualize, understand, and control a complex network with just two external gate voltages.

While we are focusing here on quantum information, the fabrication of artificial networks, where electrons are localized in specific locations, and tunnel from dot to dot, is equivalent to the experimental realization of the Hubbard model. The physics of the Hubbard model is a cornerstone of the physics of "quantum materials",[34–36] from high-T_C superconductors to unusual magnetic properties of oxides. We feel that the artificial quantum materials on a chip, which one is beginning to explore, will be very useful in answering many of the outstanding questions in the physics of quantum materials.

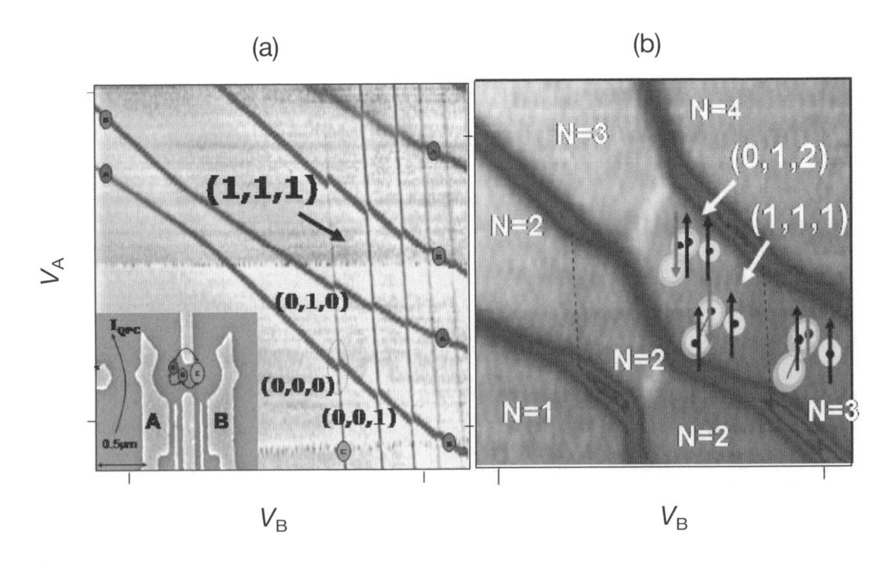

Figure 3. Triple-dot implementation of the three-qubit system: (a) stability diagram showing two dots on resonance, with inset showing an SEM picture of device, including gates A and B together with the I_{QPC}; (b) stability diagram and charge reorganization induced by gate voltages of a three-electron complex.[33]

4. Challenges in electron-spin qubits and conclusions

There has been a significant progress in the realization of electron-spin based qubits, as summarized above. A single-, double-, and a three-qubit systems have been realized in lateral quantum dots, starting from a FET structure. Single qubit operation has been demonstrated very recently by Koppens *et al.*[37] This progress bodes well for future few-qubit systems compatible with microelectronics.

However, there are still many challenges left. In particular, the CNOT gate still needs to be demonstrated. The CNOT gate requires both the ability to control the exchange coupling, as well as the single qubit operation. The CNOT gate, when integrated into the triple dot design, would allow the realization of simple algorithms. In addition, the problem of decoherence, and in particular the coupling of electron and nuclear spins, needs addressing. Perhaps a solution might lie in carbon-based quantum dot systems. Recent incorporation of graphene into FET structures opens up this possibility.[38]

Finally, the progress in single photon sources and sources of entangled photon pair production looks promising for quantum cryptography.

Acknowledgments

This work was supported in part by the Canadian Institute for Advanced Research. I thank A. Sachrajda, M. Korkusinski, and R. Williams for collaboration.

References

1. R. P. Feynman, "Quantum mechanical computers," *Found. Physics* **16,** 507 (1986).
2. C. H. Bennett and D. P. DiVincenzo, "Quantum information and computation," *Nature* **404**, 247 (2000).
3. A. M. Steane, "Introduction to quantum error correction," *Phil. Trans. Royal Soc. A* **356**, 1739 (1998).
4. K.-Y. Chen, T. Hogg, and R. Beausoleil, "A quantum treatment of public goods economics," *Quantum Information Processing* **1**, 449 (2002).
5. J. Eisert, M. Wilkens, and M. Lewenstein, "Quantum games and quantum strategies," *Phys. Rev. Lett.* **83**, 3077 (1999).
6. J. Du, H. Li, X. Xu, M. Shi, J. Wu, X. Zhou, and R. Han, "Experimental realization of quantum games on a quantum computer," *Phys. Rev. Lett.* **88,** 137902 (2002).
7. R. M. Stevenson, R. J. Young, P. Atkinson, K. Cooper, D. A. Ritchie, and A. J. Shields, "A semiconductor source of triggered entangled photon pairs," *Nature* **439**, 179 (2006).

8. D. Chithrani, M. Korkusinski, S.-J. Cheng, *et al.*, "Electronic structure of the p-shell in single, site-selected InAs/InP quantum dots," *Physica E* **26,** 322 (2005).
9. L. M. Vandersypen, M. Steffen, G. Breyta, C. S. Yannoni, M. H. Sherwood, and I. L. Chuang, "Experimental realization of Shor's quantum factoring algorithm using nuclear magnetic resonance," *Nature* **414**, 883 (2001).
10. Yu. A. Pashkin, T. Yamamoto, O. Astafiev, Y. Nakamura, D. V. Averin, and S. Tsai, "Quantum oscillations in two coupled charge qubits", *Nature* **421**, 823 (2003).
11. D. Kielpinski, C. Monroe, and D. J. Wineland, "Architecture for a large-scale ion-trap quantum computer," *Nature* **417**, 709 (2002).
12. E. Knill, R. Laflamme, and G. J. Milburn, "A scheme for efficient quantum computation with linear optics," *Nature* **409**, 46 (2001).
13. A. Yu. Kitaev, "Fault-tolerant quantum computation by anyons," *Ann. Phys.* **303**, 2 (2003).
14. B. E. Kane, "A silicon-based nuclear spin quantum computer," *Nature* **393**, 133 (1998).
15. J. A. Brum and P. Hawrylak, "Coupled quantum dots as quantum exclusive-OR gate," *Superlatt. Microstruct.* **22**, 431 (1997).
16. D. Loss and D. P. DiVincenzo, "Quantum computation with quantum dots," *Phys. Rev. A* **57**, 120 (1998).
17. A. Sachrajda, P. Hawrylak, and M. Ciorga, "Nano-spintronics with lateral quantum dots," in: J. Bird, ed., *Transport in Quantum Dots*, Dordrecht: Kluwer, 2003.
18. R. C. Ashoori, "Electrons in artificial atoms," *Nature* **379**, 413 (1996).
19. P. Hawrylak, "Single electron capacitance spectroscopy of artificial atoms: Theory and experiment," *Phys. Rev. Lett.* **71**, 3347 (1993).
20. S. Tarucha, D. G. Austing, T. Honda, R. J. van der Hage, and L. P. Kouwenhoven, "Shell filling and spin effects in a few electron quantum dot," *Phys. Rev. Lett.* **77**, 3613 (1996).
21. P. Hawrylak, C. Gould, A. Sachrajda, Y. Feng, and Z. Wasilewski, "Collapse of Zeeman gap in quantum dots due to electronic correlations," *Phys. Rev. B* **59**, 2801 (1999).
22. M. Ciorga, A. Sachrajda, P. Hawrylak, *et al.*, "Addition spectrum of a lateral dot from Coulomb and spin-blockade spectroscopy," *Phys. Rev. B* **61**, 16315 (2000).
23. M. Ciorga, A. Wensauer, M. Pioro-Ladriere, *et al.*, "Collapse of the spin-singlet phase in quantum dots," *Phys. Rev. Lett.* **88**, 256804 (2002).
24. M. Pioro-Ladriere, M. Ciorga, J. Lapointe, *et al.*, "Spin-blockade spectroscopy of a two-level artificial molecule," *Phys. Rev. Lett.* **91**, 026803 (2003).
25. M. Pioro-Ladreiere, R. Abolfath, P. Zawadzki, *et al.*, "Charge sensing of an artificial H_2^+ molecule in lateral quantum dots," *Phys. Rev. B* **72**, 125307 (2005).

26. J. R. Petta, A. C. Johnson, C. M. Marcus, M. P. Hanson, and A. C. Gossard, "Manipulation of a single charge in a double quantum dot," *Phys. Rev. Lett.* **93**, 186802 (2004).
27. A. K. Hüttel, S. Ludwig, H. Lorenz, K. Eberl, and J. P. Kotthaus, "Direct control of the tunnel splitting in a one-electron double quantum dot," *Phys. Rev. B* **72,** 081310 (2005).
28. J. R. Petta, A. C. Johnson, J. M. Taylor, *et al.*, "Coherent manipulation of coupled electron spins in semiconductor quantum dots," *Science* **309**, 2180 (2005).
29. C. H. Bennett, G. Brassard, C. Crepeau, R. Jozsa, A. Peres, and W. K. Wootters, "Teleporting an unknown quantum state via dual classical and Einstein-Podolsky-Rosen channels," *Phys. Rev. Lett.* **70**, 1895 (1993).
30. M. A. Nielsen, E. Knill, and R. Laflamme, "Complete quantum teleportation using nuclear magnetic resonance," *Nature* **396**, 52 (1998).
31. P. Hawrylak and M. Korkusinski, "Voltage-controlled coded qubit based on electron spin," *Solid State Commun.* **136**, 508 (2005).
32. D. P. DiVincenzo, D. Bacon, J. Kempe, G. Burkard, and K. B. Whaley, "Universal quantum computation with the exchange interaction," *Nature* **408**, 339 (2000).
33. L. Gaudreau, S. Studenikin, A. Sachrajda, *et al.*, "Stability diagram of a few-electron artifical triatom," *Phys. Rev. Lett.* **97**, 036807 (2006).
34. M. Coey, "Charge ordering in oxides," *Nature* **430**, 155 (2004).
35. I. S. Elfimov, S. Yunoki, and G. Sawatzky, "Possible path to a new class of ferromagnetic and half-metallic ferromagnetic materials," *Phys. Rev. Lett.* **89**, 216403 (2002).
36. F. C. Zhang and T. M. Rice, "Effective Hamiltonian for the superconducting Cu oxides," *Phys. Rev. B* **37**, 3759 (1988).
37. F. H. L. Koppens, C. Buizert, K. J. Tielrooij, *et al.*, "Driven coherent oscillations of a single electron spin in a quantum dot," *Nature* **442**, 766 (2006).
38. K. S. Novoselov, A. K. Geim, S. V. Morozov, *et al.*, "Electric field effect in atomically thin carbon films," *Science* **306**, 666 (2004).

Semiconductor Spintronics: Progress and Challenges

Emmanuel I. Rashba
Dept. of Physics, Harvard University, Cambridge, MA 02138, U.S.A.

1. Introduction

Spin is the only internal degree of freedom of the electron, and utilizing it in the new generations of semiconductor devices is the main goal of semiconductor spintronics. Contemporary semiconductor electronics is based on electron charge only. It is expected that spin-based phenomena will provide electronic devices with new functionality, and achieving quantum computing with electron spins is among the most ambitious goals of spintronics.[1,2] During the last five years, there has been impressive progress in this field, both in experiment and in developing theoretical concepts. Since the goals are highly challenging and require overcoming a number of difficult problems, research is developing along several avenues. For example, the prospects of spin-based computing with quantum dots require an increase in the spin coherence time of gate-controlled double quantum dots by several orders of magnitude, and a great advance in this direction has been achieved recently by applying spin echo techniques.[3] In what follows, we concentrate on fundamentals and on recent developments related to a different branch of spintronics: the use of spin-orbit coupling to achieve direct electrical control of electron spins in semiconductor nanostructures. Compared with magnetic control, electrical control holds promise of much higher efficiency, as well as access to electron spins on the nanoscale.

Strong enhancement of spin-orbit coupling in crystals as compared to vacuum originates from the large gradients of crystal field $\nabla V(\boldsymbol{r})$ and high electron velocities $\boldsymbol{v}$ near nuclei. The enhanced spin-orbit coupling affects the wave-functions and energy spectra of Bloch states. In vacuum, the dimensionless parameter of spin-orbit coupling is about $E(k)/m_0c^2 \sim 10^{-6}$, with the electron energy $E(k) \sim 1$ eV and the Dirac gap $m_0c^2 \sim 1$ MeV. In semiconductors, the similar parameter is $\Delta_{SO}/E_G \sim 1$, where $\Delta_{SO} \sim 1$ eV is the spin-orbit splitting of valence bands and the $E_G \sim 1$ eV is the bandgap. This enhancement makes semiconductors promising for electrical manipulation of electron spins.[4]

2. Basic concepts of semiconductor spintronics

Apparently, the first practical application of electrically-driven spin transitions belongs to laser physics and dates back as far as 1971.[5] More recent research was

Future Trends in Microelectronics. Edited by Serge Luryi, Jimmy Xu, and Alex Zaslavsky

initiated by the 1990 paper of Datta and Das, who advanced the idea of a spin-transport device based on spin interference in media with spin-orbit coupling.[6] Afterwards, this device became known as a spin transistor. Despite the fact that the attempts to fabricate such a device have not been successful to date and its feasibility has been questioned,[7] the basic principles underlying it strongly influenced subsequent research.

A toy-model spin-orbit Hamiltonian describing electrons in asymmetric two-dimensional (2D) systems (Rashba term[8,9]) is

$$H_\alpha = \alpha(\boldsymbol{\sigma}\times\boldsymbol{k})\cdot z_0, \tag{1}$$

where α is the spin-orbit coupling constant, $\boldsymbol{\sigma}$ is the Pauli matrix vector, $\boldsymbol{k}$ is electron wavevector, and z_0 is a unit vector perpendicular to the confinement plane. When rewritten as

$$H_\alpha = g\mu_B(\boldsymbol{B}_\alpha\times\boldsymbol{\sigma})/2, \quad \boldsymbol{B}_\alpha(\boldsymbol{k}) = (2\alpha/g\mu_B)(\boldsymbol{k}\times z_0), \tag{2}$$

where μ_B is the Bohr magneton and $\boldsymbol{B}_\alpha(\boldsymbol{k})$ is an effective momentum-dependent spin-orbit field, the Hamiltonian H_α describes spin precession in the field $\boldsymbol{B}_\alpha(\boldsymbol{k})$. The same phenomenon can be also understood in terms of two eigenstates of the Hamiltonian H_α with the same propagation direction $\boldsymbol{k}_0$ and energy E, but with different momenta $k_\pm$ depending on the spin-orbit coupling constant α (spin birefringence). Therefore, if spin-polarized electrons are injected at $x = 0$ along the direction $\boldsymbol{k}_0$ in a spin state that is not an eigenstate for the field $\boldsymbol{B}_\alpha(\boldsymbol{k})$, the resistance of the device is controlled by the α-dependent phase of the electron wave function near the spin-polarized drain at $x = L$.

The proposed Datta and Das device is based on the following principles:

- spin injection from a ferromagnetic source and spin detection by a ferromagnetic drain;
- electrical control of spin-orbit coupling α by a Schottky gate;
- spin precession in the spin-orbit field $\boldsymbol{B}_\alpha$; and
- spin interference.

Lately, there has been impressive progress in developing ferromagnetic injectors, including better understanding of the role of contacts between spin injectors and semiconductor microstructures. At the same time, much research effort has been focused on generating and injecting spin populations electrically, by means of spin-orbit coupling, as avoiding ferromagnetic elements would allow the elimination of stray magnetic fields. Electrical control of spin-orbit coupling[10,11] and spin precession[12] in the field $\boldsymbol{B}_\alpha$ have been reported long ago, while spin interference has been observed only recently, as discussed in Section 4 below.

All-semiconductor electrically controlled spintronics needs a better understanding spin transport in media with spin-orbit coupling that is rather nontrivial. In what follows, we review some of the recent progress in this field.

3. The origin of spin coupling to external electric field

Electrically induced quantum transitions are usually described in terms of oscillator strengths that are subject to the oscillator sum rule (Thomas-Kuhn-Reiche theorem). This sum rule follows from the standard commutation relation

$$i[k, x] = 1 \; . \tag{3}$$

When this commutation relation is written as a sum over the intermediate states, it becomes a sum of the terms

$$f_{n\leftarrow\ell} = i\{\langle\ell|k|n\rangle\langle n|x|\ell\rangle - \langle\ell|x|n\rangle\langle n|k|\ell\rangle\}, \tag{4}$$

which are oscillator strengths of $n \leftarrow \ell$ transitions. In the absence of spin-orbit coupling, calculating the commutator of the coordinate x and the Hamiltonian, we find

$$\langle n|x|\ell\rangle\,(E_\ell - E_n) = i\hbar^2 \langle n|k|\ell\rangle / m_0, \tag{5}$$

and after substitution into Eq. (4) we arrive at

$$f_{n\leftarrow\ell} = (2\hbar^2/m_0)\,|\langle\ell|k|n\rangle|^2 / (E_n - E_\ell), \tag{6}$$

where m_0 is the electron mass in vacuum. For local states, the oscillator sum rule

$$\Sigma_n\, f_{n\leftarrow\ell} = 1 \tag{7}$$

includes nondiagonal $n \neq \ell$ terms only, as the diagonal $n = \ell$ terms vanish because matrix elements of the coordinate x in Eq. (5) are finite.

However, because Bloch states are extended, diagonal matrix elements of x diverge. Hence, diagonal matrix elements of k may survive. Observing that the oscillator strengths $f_{n\leftarrow\ell}$ of Eq. (6) coincide, with accuracy up to the factor m_0, with the summands in the standard expression of $\boldsymbol{k}\cdot\boldsymbol{p}$ theory for the inverse effective mass m_ℓ in the ℓth Bloch band, one arrives at the equation

$$m_0/m_\ell + \Sigma_{n\neq\ell}\, f_{n\leftarrow\ell} = 1. \tag{8}$$

Therefore, m_0/m_ℓ is the oscillator strength $f_{\ell\leftarrow\ell}$ for the transition from the state ℓ "into itself".[13] This is precisely the oscillator strength that manifests itself in the Drude and cyclotron absorption.

The problem we need to solve is what happens to this oscillator strength in a noncentrosymmetric system when spin-orbit coupling enters the picture, $\alpha \neq 0$, and a spin-degenerate band splits into two subbands. This situation is illustrated in Fig. 1. For each state, the total oscillator strength m_0/m_ℓ is divided between the transition "into itself" and the transition between branches. For the transitions from the bottom of the band, the inter-branch transition energy equals $2E_{SO}$, with $E_{SO} = m_\ell\alpha^2/\hbar^2$, and the oscillator strength is divided equally between both transitions. For a given wavevector $\boldsymbol{k}$, electron spins have opposite directions on two branches of the energy spectrum and hence inter-branch transitions are spin-flip transitions. These transitions have high oscillator strengths, comparable to the oscillator strength of the cyclotron resonance. With increasing electron energy, intensities of inter-branch transitions decrease, but only as k_α/k_F, where $k_\alpha = m_\ell\alpha/\hbar^2$

is the spin precession momentum in the field $B_\alpha(k)$, and k_F is Fermi momentum. Therefore, their intensities remain high for reasonable α values. Outside the spectral region of interbranch transitions, their Kramers-Kronig transform describes spin coupling to electric fields;[14] spectral dependence of the corresponding response can be only found from detailed transport equations. In a strong magnetic field B, inter-branch transitions transform into the electric dipole spin resonance (EDSR), whose intensity is usually much higher than the intensity of electron paramagnetic resonance (EPR).[4,8]

The important role that intrabranch transitions play in spin transport will be discussed in Section 6 below.

4. Experimental achievements: Spin populations and spin interference

Because of the spin coupling to electric field, propagation of electric current across a sample is accompanied by spin accumulation in the bulk of three-dimensional (3D) and 2D systems.[15,16] For thin 3D layers and 2D systems, this was recently observed by Kato *et al.*,[17] Silov *et al.*,[18] and Ganichev *et al.*[19] Another related phenomenon is spin Hall effect[20-22] that manifests itself in spin accumulation near the edges of a sample, for review see Refs. 23 and 24. Spin Hall effect observed in *n*-GaAs[25] was attributed to an extrinsic mechanism, whereas the effect observed in *p*-GaAs[26] was taken to be intrinsic – that is, the former effect originates from impurity scattering, whereas the latter is due to spin-orbit coupling in the bulk.

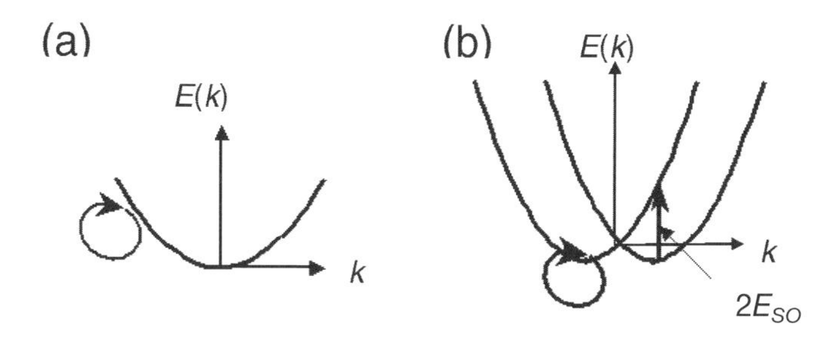

Figure 1. Spin-degenerate spectrum in the absence of spin orbit coupling (a) and spin-split spectrum of 2D electrons with spin-orbit coupling of Eq. (1) (b). In (a), the oscillator strength for transitions "into itself" (a circle with an arrow) equals m_0/m_ℓ. In (b), this oscillator strength is divided between transitions "into itself" and transitions between two branches (vertical arrow). For the transitions from the bottom of the band (shown in the figure), the oscillator strength m_0/m_ℓ is equally divided between both types of transitions.

Spin interference phenomena, besides their promise for applications, are important from the fundamental point of view because they are related to quantum phases that essentially depend on the shape of the electron paths. In particular, these phases differ for rings, where electron motion is close to adiabatic, and polygons where motion near vertices is strongly nonadiabatic. Spin interference on a large array of InGaAs square loops was reported by Koga *et al.*,[27] and on a single HgTe/HgCdTe ring by Koenig *et al.*[28]

5. Enhancing spin responses to electric fields

Long delay in the experimental observation of electrically driven spin populations in 2D systems after their prediction was caused by the small magnitude of these effects. The weakness of the effect also hinders its utilization in semiconductor devices. In this Section, we discuss some options for enhancing spin response.

- Spin response to an inhomogeneous dc electric field $\boldsymbol{F}(\boldsymbol{r}) = \boldsymbol{F}_0\exp(i\boldsymbol{q}\cdot\boldsymbol{r})$ diverges when $q \rightarrow 2k_\alpha$.[29] This behavior can be easily understood if one takes into account that adding wavevector $\boldsymbol{q}$ results in a mutual displacement of two Fermi surfaces with the same energy, and for $q = 2k_\alpha$ these surfaces touch. This results in "*spin breakdown*" because spin can be flipped at zero cost in energy. Therefore, one should choose inhomogeneous fields with large $q \approx 2k_\alpha$ spectral components. It also suggests that optimal sizes of elements with $\alpha \neq 0$ (rings or diamonds) designed for injecting spins into $\alpha = 0$ wires should be about the spin precession length $\ell_\alpha = 1/k_\alpha$.

- Frequencies of time-dependent fields should be nearly resonant, *i.e.* for $B = 0$ the frequency should be about $2\alpha k_F$, whereas for EDSR conditions it should be close to the EDSR frequency $g\mu_B B$. It is also important that strong scattering, $\hbar/\tau >> \alpha k_F$ where τ is the momentum relaxation time, results in the narrowing of the EDSR line to the inverse spin relaxation time $1/\tau_S$.[30] The explicit shape of the line, with τ_S equal to the Dyakonov-Perel' relaxation time,[31] has been recently derived by Duckheim and Loss.[32]

- In quantum wells, the in-plane polarization of the electric field $\boldsymbol{F}$ is more efficient than the out-of-plane polarization by a factor $(\omega_0^2/\omega_C\omega_S)^2$, where ω_0, ω_C, and ω_S are, respectively, the confinement frequency and the frequencies of the cyclotron and spin resonances.[33]

- Spin response is stronger in *p*-type materials.[34]

- Surface states on the (111) face of Bi[35] and on Bi/Ag(111) monolayer alloys[36] show giant spin-orbit splittings with E_{SO} up to 0.4 eV. Thin layers of such materials can provide ultra-short spin precession lengths ℓ_α.

New options arise if two different spin-orbit coupling mechanisms are combined. The symmetry group C_{2V} of (001) quantum wells in $A_{III}B_V$ materials, in addition to an invariant of Eq. (1), has a different invariant (Dresselhaus term) that is linear in k:

$$H_\beta = \beta(\sigma_x k_x - \sigma_y k_y). \quad (9)$$

Pikus noticed that the 3D Dresselhaus k^3-spin splitting[37] reduces to Eq. (9) in the limit of narrow quantum wells.[38] Combining H_α and H_β provides new options for spintronic devices,[39,40] especially in the vicinity of the magic points $\alpha = \pm\beta$. At these points, stable spin superstructures with a k_α-dependent period have been predicted recently.[41]

6. Conceptual theoretical problems

Generation of spin populations by a driving electric field is possible only due to spin nonconservation. As a result, theory of spin transport essentially differs from theory of charge transport. This difference is already obvious from Maxwell's equations that include four electric variables $\boldsymbol{F}$, $\boldsymbol{D}$, charge density ρ and current $\boldsymbol{J}$, but only two magnetic variables, $\boldsymbol{B}$ and $\boldsymbol{H}$ (or magnetization $\boldsymbol{M} = (\boldsymbol{B} - \boldsymbol{H})/4\pi$). Therefore, absence of magnetic monopoles results not only in the absence of a magnetic analog of ρ, but also in the absence of magnetization current. Introduction of such a current is justified only under some special conditions, particularly, in the framework of the Mott two-fluid theory of electron transport in ferromagnets *without* spin-orbit coupling.[42] Spin-orbit coupling results in spin nonconservation. As a result, time derivative of spin magnetization, $\partial\boldsymbol{S}/\partial t$, cannot be represented as a divergence of any vector. Therefore, there is no unambiguous definition of spin current, and the form of the extra term depends on the spin current definition; this term is known as torque.[43] Usually, ith component of the spin current j_i^ℓ is defined as

$$j_i^\ell = < v_i\sigma_\ell + \sigma_\ell v_i >/2, \quad (10)$$

where an anticommutator is taken because in the media with spin orbit coupling the velocity $\boldsymbol{v}$ depends on Pauli matrices σ_ℓ, and $< \ldots >$ stands for averaging over the electron distribution; however, different definitions for j_i^ℓ have also been proposed.[44] The concept of spin currents has been used in literature for many years, but it attracted more attention after Murakami *et al.*[45] and Sinova *et al.*[46] reported some unexpected properties of these currents for 3D holes and 2D electrons, respectively. Since then, these quantities have become a popular playground for comparing spin responses of particles described by various spin-orbit Hamiltonians to dc and ac electric fields.

In particular, dc spin-Hall conductivity defined as $\sigma_{SH} = j_x^z/F_y$, when calculated using the Kubo formula for a perfect system with spin-orbit Hamiltonian H_α, comes out as $\sigma_{SH} = e/4\pi\hbar$ for arbitrary chemical potential $\mu > 0$. This "universal conductivity" raised hopes that there might be possible to find some simple results for spin accumulation near the sample edges; spin accumulations are the only quantities currently accessible for experimental detection. However, calculations of σ_{SH} that properly account for electron scattering have shown that σ_{SH} vanishes, for a review see Refs. 23 and 24. The simplest formal argument, explaining this spin current cancellation, was provided by Dimitrova,[47] who noticed that j_x^z is proportional to the mean value of the derivate $d\sigma_y/dt$ that should vanish in a stationary state. A different argument demonstrating that this cancellation follows from the form of the free Hamiltonian only, irrespective of the potentials of non-magnetic scatterers, is based on vanishing the spin current j_x^z in a perfect sample placed in an external magnetic field perpendicular to the confinement plane.[48] Physically, the vanishing of σ_{SH} comes from the fact that there exists an intrabranch contribution to σ_{SH}, similar to the intrabranch oscillator strength of Section 3, that cancels the universal contribution $e/4\pi\hbar$. From this standpoint, impurities and the magnetic field play a similar role: by violating momentum conservation, they give rise to the intrabranch contribution.

It is currently well understood that the above cancellation is an exceptional property of the terms H_α and H_β in conjunction with a quadratic nonrelativistic Hamiltonian $H_0 = \hbar^2k^2/2m$, and it underscores the fact that while spin response to electric fields *per se* originates from the spin-orbit coupling term in the free electron Hamiltonian (Section 3), the specific form of the spin response can be found only by rigorous solution of the proper transport equations.

Boltzmann equations for systems with a spin-split energy spectrum were derived in a number of papers.[49-51] In principle, they can describe transport problems for arbitrary values of the parameter $\alpha k_F\tau/\hbar$, but they are usually solved in the diffusive limit $\alpha k_F\tau/\hbar \ll 1$.[43,52,53] In this limit, the problem of boundary conditions becomes nontrivial because of spin nonconservation. Indeed, spin is not conserved even on a perfect boundary between $\alpha \neq 0$ and $\alpha = 0$ regions because the currents defined by Eq. (10) persist in thermodynamic equilibrium in the $\alpha \neq 0$ region but vanish in the adjacent $\alpha = 0$ region.[54] Numerical work shows that these "equilibrium currents", which are not related to any real spin transport, are especially strong near boundaries.[55] Therefore, boundary conditions for diffusive equations cannot be derived from spin conservation, but only from a consistent solution of the transport equations near boundaries on a scale small compared with the spin diffusion length L_{SD}. For the Dyakonov-Perel' spin relaxation mechanism[31] L_{SD} is roughly equal to ℓ_α. This problem is still awaiting a solution.[56] For an H_α semiconductor, it is expected that a dc current flowing along a perfect hard-wall boundary would produce only tiny spin accumulation near the edge.[52,57]

A different problem concerns the relative influence of extrinsic and intrinsic mechanisms and their interplay. Extrinsic mechanisms are related to impurity scattering and are traditionally discussed in terms of skew scattering and side jump contributions. Intrinsic mechanisms are usually attributed to the spin-orbit coupling terms in the Hamiltonian $H(\boldsymbol{k})$. A similar problem has existed in the theory of anomalous Hall effect (AHE) for over 50 yeas,[58] but still remains somewhat controversial. The early period has been summarized in the paper by Nozieres and Lewiner,[59] where a set of competing (and partly canceling) terms had been derived and compared for a centrosymmetric semiconductor. They attributed AHE to extrinsic mechanisms. Remarkably, mean free time τ drops out from the side jump that therefore depends only on the parameters of a perfect crystal; this conclusion agrees with the previous result by Luttinger.[60] More recently, AHE has been related to a Berry phase in $\boldsymbol{k}$-space that is essentially intrinsic,[61-63] and it seems probable now that Berry curvature is an elegant mathematical language for describing the mechanism that in simplified models was understood as a side jump contribution. Within the framework of the Boltzmann equation, side jump appears as the next order correction in the small parameter $\hbar/E_F\tau$ to the skew scattering term in the Hall conductivity σ_{xy} (E_F is the Fermi energy). In the meantime, some experimental data suggest that this correction term dominates in the dirty regime.[60,64] A topological interpretation of the side jump contribution to σ_{xy} seems to be the most natural explanation of its remarkable ubiquity. However, the fact that the side jump contribution has, in the framework of Ref. 59, the same magnitude but opposite sign in the clean and dirty limits, indicates that the problem still persists.

The problems that make theory of AHE so tricky are also inherent in the theory of the spin Hall effect. Moreover, while the definition of the anomalous Hall current is straightforward, the ambiguity of the spin current concept makes calculating the spin Hall effect much trickier. It has been shown that the data of Ref. 25 can be reasonably described by the extrinsic mechanism,[65,66] while the data of Ref. 26 seem to point to the role of intrinsic mechanisms.[67] Remarkably, the side jump term of Ref. 65 coincides with Berry curvature ($\nabla_k \times \boldsymbol{r}_{SO}$), where $\boldsymbol{r}_{SO}$ is the spin-orbit contribution to the coordinate operator in the crystal-momentum representation. From this standpoint, side jump can be understood as an intrinsic effect that originates from the operator $\boldsymbol{r}$ rather than the Hamiltonian $H(\boldsymbol{k})$. Meanwhile, there is no doubt that in noncentrosymmetric crystals $H(\boldsymbol{k})$ contributes to spin transport, and this contribution cannot be expressed in terms of Berry curvature. Indeed, $\boldsymbol{r}_{SO} = (u_k|i\nabla_k|u_k)$ is exactly the same for the Hamiltonians H_α with α = constant and $\alpha = \alpha(k^2)$, while spin currents do vanish in the first case and not in the second[68] (u_k are eigen-spinors). The same holds for Hamiltonians with parabolic and nonparabolic H_0 parts.[69] Also, $\nabla_k u_k$ is not defined at $\boldsymbol{k} = 0$ for the Hamiltonians of H_α type.

It has been shown recently,[70] that the joint effect of intrinsic and extrinsic terms in $H(\boldsymbol{k})$ on spin currents is singular. In H_α semiconductors, spin current j_x^z defined according Eq. (10) vanishes for arbitrary $\alpha \neq 0$, *i.e.* spin precession in the field $\boldsymbol{B}_\alpha$ nullifies the extrinsic spin current after the integration over the whole sample. This can be understood as the result of averaging the spins, polarized by skew scattering, over the electron trajectories, and seems to underscore the fact that spin accumulation near boundaries cannot be derived from spin currents of Eq. (10). The same conclusion comes from the observation that spin relaxation on the boundary leads to the spin Hall effect even when bulk spin currents vanish.[71,72]

Analysis of the existing data on spin currents and spin Hall effect suggest that, at a qualitative level, they can be better related if instead of $q = 0$ components of spin currents corresponding to averaging over the entire infinite homogeneous space, one considers their Fourier components at $q \approx k_\alpha$. Such an approach corresponds to the idea that when it comes to spin accumulation S at the edge, only the adjacent layer of width ℓ_α matters. Fourier components $j_i^z(k_\alpha)$ do not vanish for the H_α Hamiltonian and have the same magnitude of about $eE/\hbar$ as for the Hamiltonians that are nonlinear in k (*e.g.* the k^3 heavy hole spin-orbit Hamiltonian),[73] as long as the definition of k_α is generalized by expressing it in terms of the spin-orbit splitting δ_{SO} at the Fermi level, $k_\alpha \rightarrow k_{SO} = m\delta_{SO}/2\hbar^2 k_F$. Moreover, spin Hall currents j_{SH} defined in this way can be related to spin accumulations as follows:

$$S/\hbar \sim k_{SO}\tau j_{SH}(k_{SO}), \quad j_{SH}(k_{SO}) \sim eE/4\pi\hbar \,. \tag{11}$$

After such redefinition of spin currents, they acquire some universality in establishing the basic scales and connection to spin accumulations near the edges.[72] Equation (11) shows that $j_{SH}(k_{SO})$ coincides in order of magnitude with the spin current calculated by Sinova *et al.*,[46] but has a somewhat different physical meaning. Numerical constants in Eq. (11) essentially depend on the specific form of the spin-orbit coupling and on the boundary conditions and can be only found from detailed transport equations. There is no doubt that physical quantities like S are continuous functions of all parameters, including α. Also, near the sample edge, spin magnetization $S(x)$ is an oscillating function of x (with a period about k_{SO}^{-1}) that usually changes sign, x being separation from the edge. Hence, it is difficult to expect any universal relation even between the signs of the bulk spin current and the spin accumulation at the edge.

7. Conclusions

Spin-orbit coupling is currently considered a key to creating and manipulating spin populations electrically, on a nanometer scale. Recent years have witnessed impressive progress in this field, both in experiment and in theory, as described in this chapter. The very possibility of creating nonequilibrium spins by electric fields is based on spin nonconservation. This fact, in turn, results in an essential difference between the spin-transport theory in media with spin-orbit coupling and the traditional theory of charge transport.

Acknowledgments

This work was supported by the Harvard Center for Nanoscale Systems. Inspiring discussions with H.-A. Engel, B. I. Halperin, A. H. MacDonald, C. M. Marcus, D. Loss, and Q. Niu are gratefully acknowledged.

References

1. D. D. Awschalom, D. Loss, and N. Samarth, *Semiconductor Spintronics and Quantum Computation*, Berlin: Springer, 2002.
2. I. Zutic, J. Fabian, and S. Das Sarma, "Spintronics: Fundamentals and applications," *Rev. Mod. Phys.* **76**, 323 (2004)
3. J. R. Petta, A. C. Johnson, J. M. Taylor, *et al.*, "Coherent manipulation of coupled electron spins in semiconductor quantum dots," *Science* **309,** 2180 (2005).
4. E. I. Rashba and V. I. Sheka, "Electric-dipole spin resonances", in: G. Landwehr and E. I. Rashba, eds., *Landau Level Spectroscopy*, Amsterdam: North-Holland, 1991, pp. 131–206.
5. S. R. J. Brueck and A. Mooradian, "Efficient, single-mode, cw, spin-flip Raman laser," *Appl. Phys. Lett.* **18**, 229 (1971).
6. S. Datta, and B. Das, "Electronic analog of the electro-optic modulator," *Appl. Phys. Lett.* **56**, 665 (1990).
7. M. Cahay and S. Bandyopadhyay, "Phase-coherent quantum mechanical spin transport in a weakly disordered quasi-one-dimensional channel," *Phys. Rev. B* **69**, 045303 (2004).
8. E. I. Rashba, "Properties of semiconductors with an extremum loop," *Sov. Phys. Solid State* **2**, 1109 (1960).
9. Yu. A. Bychkov and E. I. Rashba, "Properties of a 2D electron-gas with lifted spectrum degeneracy," *JETP Lett.* **39**, 78 (1984).
10. J. Nitta, T. Akazaki, H. Takayanagi, and T. Enoki, "Gate control of spin-orbit interaction in an inverted $In_{0.53}Ga_{0.47}As/In_{0.52}Al_{0.48}As$ heterostructure," *Phys. Rev. Lett.* **78**, 1335 (1997).
11. G. Engels, J. Lange, Th. Schaepers, and H. Lueth, "Experimental and theoretical approach to spin splitting in modulation-doped $In_xGa_{1-x}As/InP$ quantum wells for $B \to 0$," *Phys. Rev.* B **55**, R1958 (1997).
12. V. K. Kalevich and V. L. Korenev, "Effect of electric field on optical orientation of 2D electrons," *JETP Lett.* **52**, 230 (1990).
13. A. Sommerfeld and H. Bethe, "Electronentheorie der Metalle," in: *Handbuch der Physik*, 2nd ed., Berlin: Springer, 1933, Vol. 24, Part 2.
14. E. I. Rashba, "Spin currents, spin populations, and dielectric function of noncentrosymmetric semiconductors," *Phys. Rev. B* **70**, 161201(R) (2004).
15. E. L. Ivchenko and G. E. Pikus, "New photogalvanic effect in gyrotropic crystals," *JETP Lett.* **27**, 604 (1978).
16. V. M. Edelstein, "Spin polarization of conduction electrons induced by electric current in two-dimensional asymmetric electron system," *Solid State Commun.* **73**, 233 (1990).
17. Y. K. Kato, R. C. Myers, A. C. Gossard, and D. D. Awschalom, "Current-induced spin polarization in strained semiconductors," *Phys. Rev. Lett.* **93**, 176601 (2004).
18. A. Yu. Silov, P. A. Blajnov, J. H. Wolter, R. Hey, K. H. Ploog, and N. S. Averkiev, "Current-induced spin polarization at a single heterojunction," *Appl. Phys. Lett.* **85**, 5929 (2004).

19. S. D. Ganichev, S. N. Danilov, P. Schneider, *et al.*, "Electric current-induced spin orientation in quantum wells structures," *J. Magn. Magn. Mater.* **300**, 127 (2006).
20. M. I. Dyakonov and V. I. Perel', "Current-induced spin polarization of electrons in semiconductors," *Phys. Lett. A* **35**, 459 (1971).
21. E. N. Bulgakov, K. N. Pichugin, A. F. Sadreev, P. Streda, and P. Seba, "Hall-like effect induced by spin-orbit interaction," *Phys. Rev. Lett.* **83**, 376 (1999).
22. J. E. Hirsch, "Spin Hall effect," *Phys. Rev. Lett.* **83**, 1834 (1999).
23. J. Schliemann, "Spin Hall effect," *Intern. J. Modern. Phys.* **20**, 1015 (2006).
24. H.-A. Engel, E. I. Rashba, and B. I. Halperin, "Theory of spin Hall effects," *cond-mat*/0603306.
25. Y. K. Kato, R. C. Myers, A. C. Gossard, and D. D. Awschalom, "Observation of the spin Hall effect in semiconductors," *Science* **306**, 1910 (2004).
26. J. Wunderlich, B. Kaestner, J. Sinova, and T. Jungwirth, "Experimental observation of the spin-Hall effect in a two-dimensional spin-orbit coupled semiconductor system," *Phys. Rev. Lett.* **94**, 047204 (2005).
27. T. Koga, Y. Sekine, and J. Nitta, "Experimental realization of a ballistic spin interferometer based on the Rashba effect using a nanolithographically defined square loop array," *Phys. Rev. B* **74**, 041302 (2006).
28. M. A. Tschetschetkin, E. Hankiewicz, *et al.*, "Direct observation of the Aharonov-Casher phase," *Phys. Rev. Lett.* **96**, 076804 (2006).
29. E. I. Rashba, "Spin dynamics and spin transport," *J. Supercond.* **18**, 137 (2005).
30. V. I. Mel'nikov and E. I. Rashba, "Influence of impurities on combined resonance in semiconductors," *Sov. Phys. JETP* **34**, 1353 (1972).
31. M. I. Dyakonov and V. I. Perel', "Spin relaxation of conduction electron in noncentrosymmetric semiconductors," *Sov. Phys. Solid State* **13**, 3023 (1972).
32. M. Duckheim and D. Loss, "Electric-dipole-induced spin resonance in disordered semiconductors," *Nature Phys.* **2**, 195 (2006).
33. E. I. Rashba and A. L. Efros, "Efficient electron spin manipulation in a quantum well by an in-plane electric field," *Appl. Phys. Lett.* **83**, 5295 (2003).
34. D. M. Gvozdic and U. Ekenberg, "Superefficient electric-field-induced spin-orbit splitting in strained *p*-type quantum wells," *Europhys. Lett.* **73**, 927 (2006).
35. Yu. M. Koroteev, G. Bihlmayer, J. E. Gayone, *et al.*, "Strong spin-orbit splitting on Bi surfaces," *Phys. Rev. Lett.* **93**, 046403 (2004).
36. C. R. Ast, D. Pacile, M. Falub, *et al.*, "Giant spin-splitting in the Bi/Ag(111) surface alloy," *cond-mat*/0509509.
37. G. Dresselhaus, "Spin-orbit coupling effects in zincblende structures," *Phys. Rev.* **100**, 580 (1955).
38. G. E. Pikus, private communication (1984), see also Refs. 74–76.
39. J. Schliemann, J. C. Egues, and D. Loss, "Nonballistic spin-field-effect transistor," *Phys. Rev. Lett.* **90**, 146801 (2003).
40. D. Z. Y. Ting and X. Cartoixa, "Bulk inversion asymmetry enhancement of polarization efficiency in nonmagnetic resonant-tunneling spin filters," *Phys. Rev. B* **68**, 235320 (2003).

41. B. A. Bernevig, J. Orenstein, and S.-C. Zhang, "An exact SU(2) symmetry and persistent spin helix in a spin-orbit coupled system," *cond-mat*/0606196.
42. N. F. Mott, "The electrical conductivity of transition metals," *Proc. Roy. Soc. London A* **153**, 699 (1936).
43. A. A. Burkov, A. S. Nunez, and A. H. MacDonald, "Theory of spin-charge-coupled transport in a two-dimensional electron gas with Rashba spin-orbit interactions," *Phys. Rev. B* **70**, 155 (2004).
44. J. Shi, P. Zhang, and Q. Niu, "Proper definition of spin current in spin-orbit coupled systems," *Phys. Rev. Lett.* **96**, 076604 (2006).
45. S. Murakami, N. Nagaosa, and S.-C. Zhang, "Dissipationless quantum spin current at room temperature," *Science* **301**, 1348 (2003).
46. J. Sinova, D. Culcer, Q. Niu, N. A. Sinitsyn, T. Jungwirth, and A. H. MacDonald, "Universal intrinsic spin Hall effect," *Phys. Rev. Lett.* **92**, 126603 (2004).
47. O. V. Dimitrova, "Spin-Hall conductivity in a two-dimensional Rashba electron gas," *Phys. Rev. B* **71**, 245327 (2005).
48. E. I. Rashba, "Sum rules for spin-Hall conductivity cancellation," *Phys. Rev. B* **70**, 201309 (2004).
49. A. G. Aronov, Y. B. Lyanda-Geller, and G. E. Pikus, "Spin polarization of electrons due to the electric current," *Sov. Phys. JETP* **73**, 537 (1991).
50. A. V. Shytov, E. G. Mishchenko, H.-A. Engel, and B. I. Halperin, "Small-angle impurity scattering and the spin Hall conductivity in two-dimensional semiconductor systems," *Phys. Rev. B* **73**, 075316 (2006).
51. A. Khaetskii, "Intrinsic spin current for an arbitrary Hamiltonian and scattering potential," *Phys. Rev. B* **73,** 115323 (2006).
52. E. G. Mishchenko, A. V. Shytov, and B. I. Halperin, "Spin current and polarization in impure two-dimensional electron systems with spin-orbit coupling," *Phys. Rev. Lett.* **93**, 226602 (2004).
53. C. S. Tang, A. G. Mal'shukov, and K. A. Chao, "Generation of spin current and polarization under dynamic gate control of spin-orbit interaction in low-dimensional semiconductor systems," *Phys. Rev. B* **71**, 195314 (2005).
54. E. I. Rashba, "Spin currents in thermodynamic equilibrium," *Phys. Rev. B* **68**, 241315 (2003).
55. B. K. Nikolic, L. P. Zarbo, and S. Souma, "Imaging mesoscopic spin Hall flow: Spatial distribution of local spin currents and spin densities in and out of multiterminal spin-orbit coupled semiconductor nanostructures," *Phys. Rev. B* **73**, 075303 (2006).
56. V. M. Galitski, A. A. Burkov, and S. Das Sarma, "Boundary conditions for spin diffusion," *Phys. Rev. B* **74**, 115331 (2006).
57. O. Bleibaum, "Boundary conditions for spin-diffusion equations with Rashba spin-orbit interaction," *Phys. Rev. B* **74**, 113309 (2006).
58. R. Karplus and J. M. Luttinger, "Hall effect in ferromagnetics," *Phys. Rev.* **95**, 1154 (1954).
59. P. Nozieres and C. Lewiner, "A simple theory of the anomalous Hall effect in semiconductors," *J. Phys. (Paris)* **34**, 901 (1973).
60. J. M. Luttinger, "Theory of the Hall effect in ferromagnetic substances," *Phys. Rev.* **112**, 739 (1958).

61. M. Onoda and N. Nagaosa, "Topological nature of anomalous Hall effect in ferromagnets," *J. Phys. Soc. Japan* **71**, 19 (2002).
62. T. Jungwirth, Q. Niu, and A. H. MacDonald, "Anomalous Hall effect in ferromagnetic semiconductors," *Phys. Rev. Lett.* **88**, 207208 (2002).
63. F. D. M. Haldane, "Berry curvature on the Fermi surface: Anomalous Hall effect as a topological Fermi-liquid property," *Phys. Rev. Lett.* **93**, 206602 (2004).
64. N. Nagaosa, "Anomalous Hall effect – a new perspective," *J. Phys. Soc. Japan* **75**, 042001 (2006).
65. H.-A. Engel, B. I. Halperin, and E. I. Rashba, "Theory of spin Hall conductivity in *n*-GaAs," *Phys. Rev. Lett.* **95**, 166605 (2005).
66. W.-K. Tse and S. Das Sarma, "Spin Hall effect in doped semiconductor structures," *Phys. Rev. Lett.* **96**, 056601 (2006).
67. K. Nomura, J. Wunderlich, J. Sinova, B. Kaestner, A. H. MacDonald, and T. Jungwirth, "Edge-spin accumulation in semiconductor two-dimensional hole gases," *Phys. Rev. B* **72**, 245330 (2005).
68. S. Murakami, "Absence of vertex correction for the spin Hall effect in *p*-type semiconductors," *Phys. Rev. B* **69**, 241202 (2004).
69. P. L. Krotkov and S. Das Sarma, "Intrinsic spin Hall conductivity in a generalized Rashba model," *Phys. Rev. B* **73**, 195307 (2006).
70. W.-K. Tse and S. Das Sarma, "Intrinsic spin Hall effect in the presence of extrinsic spin-orbit scattering," cond-mat/0602607.
71. I. Adagideli and G. E. W. Bauer, "Intrinsic spin Hall edges," *Phys. Rev. Lett.* **95**, 256602 (2005).
72. E. I. Rashba, "Spin-orbit coupling and spin transport," *Physica E* **34**, 31 (2006).
73. R. Winkler, *Spin-Orbit Coupling Effects in Two-Dimensional Electron and Hole Systems,* Berlin: Springer, 2003.
74. G. Lommer, F. Malcher, and U. Roessler, "Reduced *g* factor of subband Landau levels in AlGaAs/GaAs heterostructures," *Phys. Rev. B* **32**, 6965 (1985).
75. Yu. A. Bychkov and E. I. Rashba, "Effect of *k*-linear terms on electronic properties of 2D systems," in: J. D. Chadi and W. A. Harrison, eds., *Proc. 17th Intern. Conf. Phys. Semicond.*, New York: Springer, 1985, p. 321.
76. M. I. Dyakonov and V. Y. Kachorovskii, "Spin relaxation of two-dimensional electrons in noncentrosymmetric semiconductors," *Sov. Phys. Semicond.* **20**, 110 (1986).

Towards Semiconductor Spin Logic

Igor Zutic
Dept. of Physics, SUNY–Buffalo, Buffalo, NY 14260, U.S.A.

Jaroslav Fabian
Institute for Theoretical Physics
University of Regensburg, 93040 Regensburg, Germany

1. Introduction

The anticipated limits of scaling for silicon CMOS transistors have spurred researchers to explore more exotic schemes for computation. A promising approach comes from the field of spin electronics or spintronics,[1,2] which seeks to manipulate and use the spin of carriers and not just their charge. An impressive success of spintronic applications has been achieved in metal-based structures that utilize magnetoresistive effects for substantial improvements in the performance of computer hard drives and magnetic random access memories (MRAMs).[3,4] Correspondingly, the theoretical understanding of spin transport is usually limited to the metallic regime in linear response, which, while providing a good description for data storage and magnetic memory devices, is not sufficient for signal processing and digital logic. In contrast, much less is known about possible applications of semiconductor spintronics and spin transport in related structures that could utilize strong intrinsic nonlinearities in current-voltage characteristics to implement spin-based logic. Spin transport differs from charge transport in that spin orientation, even in the absence of magnetic impurities, is a nonconserved quantity due to spin-orbit and hyperfine coupling.[1]

Many basic questions, such as how would spintronic applications be faster or require less power than their charge-logic counterparts, are only beginning to be investigated. What are the advantages of pure spin currents,[5] could they indeed be dissipationless and still contribute to spin transport in actual devices? There are also practical challenges in the choice of suitable spintronic materials, detection of nonequilibrium spin populations, and understanding the design of spin-based devices. However, it is encouraging to note experimental findings that the spin degrees of freedom lead to very fast switching[6] or to reducing pump power for lasing operation,[7] as well as theoretical schemes for logic gates with smaller numbers of elements than needed for their charge-based counterparts.[8]

Future Trends in Microelectronics. Edited by Serge Luryi, Jimmy Xu, and Alex Zaslavsky

2. From spin-valves to spin field effect transistors

A simple implementation of the prototypical two-terminal spintronic device is a spin-valve, a key element for read-heads of computer hard drives and MRAMs, which consists of a non-magnetic material sandwiched between two ferromagnetic electrodes. The flow of carriers through a spin-valve is determined by the direction of their spin (up or down) relative to the magnetic polarization of the device's electrodes. Typically, when the magnetizations of the two electrodes are parallel (P), current flow is permitted, and when they are antiparallel (AP) it is restricted.

Since magnetization in ferromagnets persists even when the power is switched off, these applications have significant advantage of being nonvolatile. The corresponding difference in resistance (between AP and P orientation) is also known as the magnetoresistance (MR). A spin-valve can be implemented as a magnetic tunnel junction (MTJ), in which the non-magnetic region is a tunnel barrier. Then the normalized tunneling magnetoresistance (TMR) figure of merit is $(R_{AP} - R_P)/R_P$, which can be simply expressed as[1,3]

$$\mathrm{TMR} = 2P_L P_R/(1 - P_L P_R), \tag{1}$$

where $P_{L,R}$ are polarizations of the spin-resolved density of density in the left and the right ferromagnets, respectively. Similar to the signal-to-noise ratio, it is desirable to have a large TMR, which is equivalent to finding highly spin-polarized materials. While most of the previous efforts were centered on finding (often exotic) highly spin-polarized ferromagnets, a different strategy, highlighting the importance of interfaces, was recently proved more effective. A dramatic increase of TMR was demonstrated with conventional ferromagnetic electrodes (CoFe showing only a modest bulk spin polarization) by simply replacing Al_2O_3 tunnel barrier by MgO.[9,10]

What about spin-based logic? While valuable for magnetic storage and memory, two-terminal spin valves are of limited use for digital logic. An early semiconductor-based spin-logic device[11] is the Datta-Das spin field effect transistor (FET), depicted in Fig. 1. Despite extensive experimental efforts, there remain important challenges to its realization and it is not clear that it would ever a have practical use. However, the spin FET has an important conceptual value. It illustrates a generic scheme for a spin logic device together with challenges to realize its basic elements, such as spin injection and detection, as well as spin transport and manipulation. One of these elements – efficient room temperature spin injection, – was recently demonstrated[13] using the same approach as for enhancing the TMR. The CoFe/MgO tunnel injector enabled spin injection into GaAs with the carrier spin polarization exceeding 70%. Consequently, there is a shift of focus in semiconductor spintronics. Previous questions about the feasibility of an efficient spin injection are gradually being replaced by the questions on how would spin injection lead to improved devices.

The spin FET can be viewed as gate-controlled (via spin-orbit coupling) three-terminal generalization of a spin-valve. However, in contrast to extensive efforts to

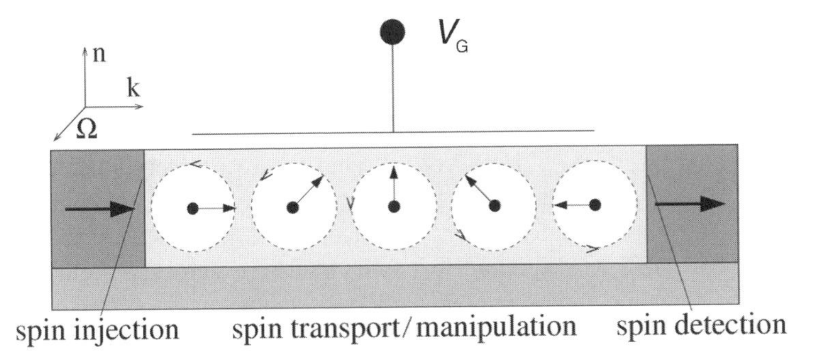

Figure 1. Datta-Das spin field effect transistor.[1,11] The source and drain are ferromagnetic, while the channel is formed at a heterojunction interface. The gate modifies the Bychkov-Rashba field[12] **Ω**, which is perpendicular to both the growth direction ***n*** and electron momentum ***k***. The electron either enters the drain or bounces off depending on whether the precessing spin direction is parallel or antiparallel (as depicted) to that of the drain, giving ON and OFF states, respectively.

realize a semiconductor-based spin FET, a similar functionality has been recently realized in a very different implementation. A carbon nanotube (CNT) was used as the nonmagnetic material sandwiched between the ferromagnetic source and drain with tunneling contacts. While a CNT has negligible spin-orbit coupling, both the magnitude and the sign of TMR in such a CNT spin-valve can be controlled by a gate voltage that changes the on- or off-resonance condition.[15,16]

3. Materials challenges

Two distinct designs for spin-logic devices that would operate at room temperature are currently pursued: (i) hybrid structures that combine metallic ferromagnets and semiconductors; and (ii) all-semiconductor structures. In the first approach, the wide range of metallic ferromagnetic materials with high Curie temperature T_C provides a possible advantage. Nominally highly spin-polarized materials (for example, CrO_2 and Fe_3O_4) could provide both effective spin injection in semiconductors and large magnetoresistive effects, important for nonvolatile applications. However, similar to the discussion of enhancing TMR in MTJs, bulk properties of ferromagnetic metals can be strongly modified at interfaces with semiconductors (effective spin polarization can be strongly suppressed or enhanced). Interfacial defects arising from lattice mismatch between the metal and semiconductor could suppress the efficiency with which spin-polarized carriers can be injected into a semiconductor.

Optical detection of injected spin in direct band gap semiconductors is usually realized with spin light-emitting diodes (LEDs).[1] The radiative recombination of spin-polarized carriers will lead to circularly polarized light which in turn allows for quantifying the degree of injected carrier spin polarization. In Fe/GaAs junctions, a combination of spin LED studies and direct atomic-scale information about the interfacial structure showed a 44% increase of spin injection efficiency for an annealed interface.[16] However, even in the extensively studied Fe/GaAs junctions,[17,18] with good lattice matching, important uncertainties remain (instead of an abrupt interface, an interlayer of Fe_3GaAs could be formed). Furthermore, a comprehensive theoretical picture of spin transport across Schottky barriers between ferromagnetic metals and conventional semiconductors is still missing.

In the second approach it would be desirable to have ferromagnetic semiconductors with sufficiently high T_C. An exchange coupling between the carrier and magnetic impurity spins can lead to carrier-mediated ferromagnetism. Selective doping and quantum confinement can provide a substantial increase in T_C,[19] as compared to the uniformly doped bulk ferromagnetic semiconductors – thus, in (Ga,Mn)As $T_C \sim 170$ K in bulk material, which can be increased to $T_C \sim 250$ K by δ-doping. With selective doping and in heterojunctions of ferromagnetic semiconductors one could potentially utilize all-semiconductor device designs with external control of T_C. A possible realization of such tunable T_C is shown in Fig. 2.

For example, when the number of carriers is changed, either by shining light[20] or by applying a gate bias in a field effect transistor geometry,[21] the material can be switched between the paramagnetic and ferromagnetic states. These experiments suggest the prospect of nonvolatile multifunctional devices with tunable optical, electrical, and magnetic properties. Furthermore, the demonstration of optically or electrically controlled ferromagnetism provides a method for distinguishing carrier-induced semiconductor ferromagnetism from ferromagnetism that originates from metallic magnetic inclusions (such as MnAs nanoclusters[22] in GaAs). While there is a good support for room-temperature ferromagnetism in

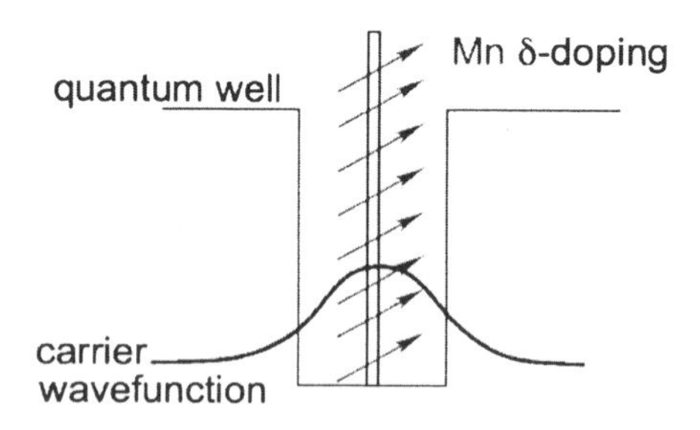

Figure 2. External control of T_C in ferromagnetic semiconductors. By applying bias in Mn δ–doped GaAs quantum well it is possible to shift the carrier wavefunction and change its overlap with the spins of magnetic impurities. Such a change alters the exchange coupling and correspondingly the T_C.

(Zn,Cr)Te,[23] it is still unclear how many of the other reported semiconductors[23,24] (some with claimed T_C as high as 900 K) have genuine carrier-mediated ferromagnetism above the room temperature. The role of confinement to enhance the T_C in quantum wells could be even more favorable in quantum wires and quantum dots.

Most of the progress in fabrication of hybrid structures or magnetic doping of semiconductors is based on direct bandgap materials, such as GaAs or InAs.[2] Additionally, they allow the use of standard optical techniques for spin injection and detection.[25,26] Circularly polarized light can be used to polarize carriers in semiconductors with a direct band gap. Moreover, both the direction and the magnitude of optically generated charge currents and pure spin currents can be controlled optically.[1] In the reverse process, the presence of polarized carriers in a direct-gap semiconductor can be detected by measuring the circular polarization of the recombination light, as discussed for spin LEDs.

What about silicon, which has desirable spin-dependent properties, such as long spin relaxation and decoherence times? Even after several decades of effort, a robust spin injection and detection scheme in silicon remains elusive.[1] In ferromagnetic metal/Si hybrid structures, the formation of silicides and poor interface quality are compounded by the lack of a theoretical description of nonequilibrium spin transport that accounts for the complexity of such structures. All-semiconductor Si-compatible schemes face additional challenges, since a reliable ferromagnetic Si-based semiconductor remains to be found, while the indirect band gap and weak spin-orbit coupling preclude the direct use of optical spin injection and detection. A possible solution would be to carefully fabricate semiconductor heterojunctions combining silicon with the closely lattice-matched

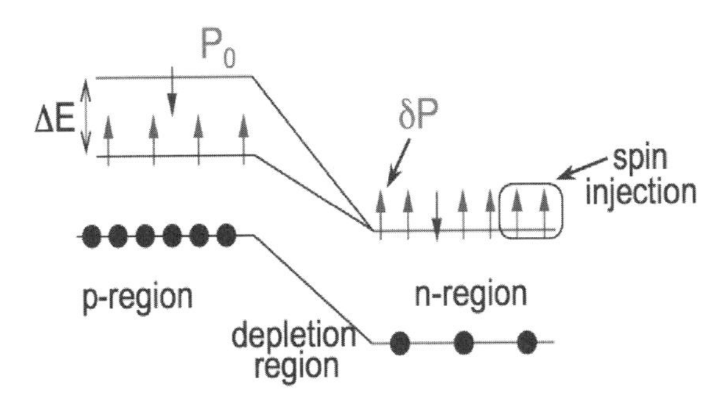

Figure 3. A simplified band diagram of a magnetic heterojunction for detection of injected spin in silicon (*n*-region). Spin-polarized electrons are denoted by arrows and unpolarized holes by circles. Spin splitting (due to ferromagnetism or an applied magnetic field) in the *p*-region implies equilibrium spin polarization P_0. The injected spins lead to the nonequilibrium spin polarization δP at the edge of the depletion region. Spin splitting acts as a spin filter and the change of relative sign and/or magnitude of the injected and equilibrium spin polarizations directly modifies current-voltage characteristics, leading to a spin-voltaic effect.

ferromagnetic or direct band gap semiconductor.[27,28] An interplay between the equilibrium spin polarization in the magnetic region and injected nonequilibrium spin polarization in silicon, P_0 and δP, could lead to the spin-voltaic effect[29], a spin analog of the photovoltaic effect that does not rely on a direct bandgap. A simple scheme is depicted in Fig. 3. The symmetry properties of the charge current can be exploited to detect spin injection in Si using different contributions of the charge current or open-circuit voltage under magnetization reversal. The experimental feasibility is further supported by the recent experiments in III-V heterojunctions.[30]

4. Spin transistors

Future spin logic devices should be capable of amplifying signals and not just limited to magnetoresistive effects as in spin-valves. Important challenges for spin transistors would be to provide improved speed and power consumption compared to conventional electronics, as well as to extend their functionality by utilizing the control of spin and magnetism. The feasibility of such goals will not only depend on overcoming the materials challenges, but also on the understanding of how spin transport and spin-charge coupling could lead to novel device concepts and designs. For example, there are proposals of nonvolatile transistors that, while appearing to operate normally, would be turned off most of the time and require less power.[23]

Several existing spin transistor designs are also known as hot-electron spin transistor, spin-valve transistor, and magnetic tunneling transistor.[4,31] While they include a semiconductor region (sandwiched between Schottky or tunnel barriers), they are intended for magnetic storage or sensing rather than spin logic. The term transistor characterizes their three-terminal structure, rather than to ability to provide amplification or current gain. Efforts that are more directly related to spin logic were often stimulated by the proposal for spin FET, discussed in Section 2. Substantially different realizations of semiconductor-based spin transistors have been considered during the last several years.

Spin control of current gain can be illustrated on the example of magnetic bipolar junction transistor[32,33] (MBT) – a generalization of conventional bipolar junction transistor. In the MBTs there is at least one magnetic region – emitter (E), base (B), or collector (C) – with spin splitting, *i.e.* with equilibrium spin polarization. In the forward active mode (forward biased EB and reversed bias BC junction), an external change of that spin splitting via magnetic field or tunable ferromagnetism could act like an effective change of related energy gap and change the charge current and current gain. Similar to magnetic heterojunctions, spin injection could also give rise to the spin-voltaic effect. A change of nonequilibrium spin polarization, with the contribution of the form $P_0\delta P$, could even lead to the reversal of direction of collector current and the sign of current gain.[34] An prototype of an MBT, limited to low-temperature operation, was fabricated using GaAs/(Ga,Mn)As junctions.[35]

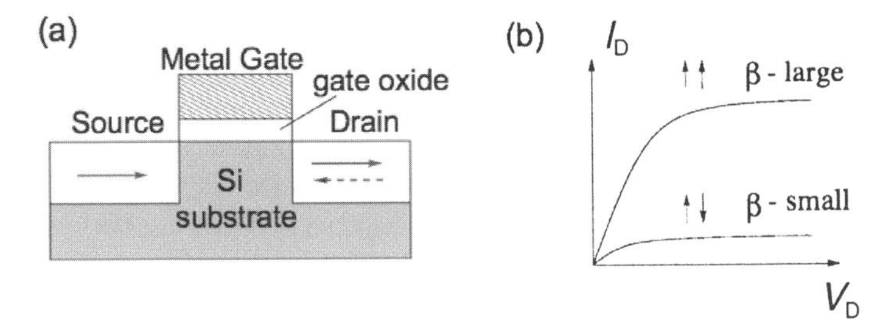

Figure 4. (a) A schematic diagram of a spin MOSFET. By changing the orientation of magnetization in the drain (parallel or antiparallel, with the respect to the one in the source) there is a relative difference of the current gain, depicted schematically in (b). Completely spin-polarized source and drain provide the best performance.

Another proposed three-terminal device that provides a magnetic control of gain is the spin MOSFET.[8] The source and drain, analogous to the spin FET, consist of ferromagnetic metals, as shown in Fig. 4. Similar to an MBT, the relative spin polarization in the two regions (in the MBT one was nonequilibrium) changes the gain β. A spin MOSFET combines the spin-valve effect with the possibility of amplifying signals, which may be useful for reconfigurable logic gates and high density nonvolatile memories. Circuit analysis of logic gates including a spin MOSFET suggested smaller number of elements needed than for implementation using CMOS.[8] It is also proposed that spin MOSFET would have very good scaling performance in the sub-100 nm regime, low-power delay product, and low off-current. However, the feasibility of simple Si integration has yet to be demonstrated. Optimal performance requires highly spin-polarized ferromagnetic contacts with silicon, which appears very challenging, although there has been some encouraging progress recently.[36,37] As discussed in Section 4, robust spin injection in silicon has yet to be demonstrated and there is no guarantee that materials with predicted high spin polarization in the bulk would be capable of maintaining it at an interface with silicon. However, progress towards implementing silicon-based spintronics continues. An early device demonstration of silicon spin diffusion transistor reported both the current gain and TMR of the order of 10 % at room temperature.[38]

5. Conclusions

While efforts to develop spin-based logic are still at an early stage, important elements have already been demonstrated. Large magnetoresistive effects, efficient spin injection and detection, and ferromagnetism tunable by light and gate voltage have been realized in several direct-gap semiconductors.[1] To utilize the nonvolatility in scaled-down devices, it will also be important to provide efficient magnetization switching. Passing spin-polarized current through a ferromagnet can reverse magnetization direction[1,2] (also known as current-induced magnetization reversal or spin torque). This may prove to be a method of choice requiring no external magnetic field. The success of metal-based applications for magnetic storage and sensing, combined with the progress in the active manipulation of spin degrees of freedom in semiconductors, offers an intriguing possibility to explore the potential for spin logic devices.

Acknowledgments

This work was supported by the U.S. ONR and NSF-ECCS CAREER, and SFB 689.

References

1. I. Zutic, J. Fabian, and S. Das Sarma, "Spintronics: Fundamentals and applications," *Rev. Mod. Phys.* **76**, 323 (2004).
2. S. Maekawa, ed., *Concepts in Spin Electronics*, Oxford: Oxford University Press, 2006.
3. S. Maekawa and T. Shinjo, eds., *Spin Dependent Transport in Magnetic Nanostructures,* New York: Taylor & Francis, 2002.
4. S. S. P. Parkin, X. Jiang, C. Kaiser, A. Panchula, K. Roche, and M. Samant, "Magnetically engineered spintronic sensors and memory," *Proc. IEEE* **91**, 661 (2003).
5. E. I. Rashba, "Spin currents in thermodynamic equilibrium: The challenge of discerning transport currents," *Phys. Rev. B* **68**, 241315 (2003).
6. Y. Nishikawa, A. Tackeuchi, S. Nakamura, S. Muto, and N. Yokoyama, "All-optical picosecond switching of a quantum well etalon using spin-polarization relaxation," *Appl. Phys. Lett.* **66**, 839 (1995).
7. J. Rudolph, D. Hagele, H. M. Gibbs, G. Khitrova, and M. Oestreich, "Laser threshold reduction in a spintronic device," *Appl. Phys. Lett.* **82**, 4516 (2003).
8. M. Tanaka, "Spintronics: Recent progress and tomorrow's challenge," *J. Crystal Growth* **278**, 25 (2005).
9. S. S. P. Parkin, C. Kaiser, A. Panchula, P. Rice, M. Samant, and S.-H. Yang, "Giant room temperature tunneling magnetoresistance in magnetic tunnel junctions with MgO(100) tunnel barriers," *Nature Mater.* **3,** 862 (2004).
10. S. Yuasa, T. Nagahama, A. Fukushira, Y. Suzuki, and K. Ando, "Giant room-temperature magnetoresistance in single-crystal Fe/MgO/Fe magnetic tunnel junctions," *Nature Mater.* **3**, 868 (2004).
11. S. Datta and B. Das, "Electronic analog of the electro-optic modulator," *Appl. Phys. Lett.* **56**, 665 (1990).
12. Yu. A. Bychkov and E. I. Rashba, "Properties of a 2D electron-gas with lifted spectral degeneracy," *JETP Lett.* **39**, 78 (1984).

13. G. Salis, R. Wang, X. Jiang, R. M. Shelby, S. S. P. Parkin, S. R. Bank, and J. S. Harris, "Temperature independence of the spin-injection efficiency of a MgO-based tunnel spin injector," *Appl. Phys. Lett.* **87**, 262503 (2005).
14. S. Sahoo, T. Kontos, J. Furer, C. Hoffman, M. Graber, A. Cottet, and C. Schonenberger, "Electric control of spin transport," *Nature Phys.* **1**, 99 (2005).
15. I. Zutic and M. Fuhrer, "A path to spin logic," *Nature Phys.* **1**, 85 (2005).
16. T. J. Zega, A. T. Hanbicki, S. C. Erwin, *et al.*, "Determination of interface atomic structure and its impact on spin transport using Z-contrast microscopy and density-functional theory," *Phys. Rev. Lett.* **96**, 196101 (2006).
17. A. T. Hanbicki, O. M. J. Van 't Erve, R. Magno, *et al.*, "Analysis of the transport process providing spin injection through an Fe/AlGaAs Schottky barrier," *Appl. Phys. Lett.* **82**, 4092 (2003).
18. S. C. Erwin, S.-H. Lee, and M. Scheffler, "First-principles study of nucleation, growth and interface structure of Fe/GaAs," *Phys. Rev. B* **65**, 205422 (2002).
19. A. M. Nazmul, T. Amemiya, Y. Shuto, S. Sugahara, and M. Tanaka, "High temperature ferromagnetism in GaAs-based heterostructures with Mn δ–doping," *Phys. Rev. Lett.* **95**, 017201 (2005).
20. S. Koshihara, A. Oiwa, M. Hirasawa, *et al.*, "Ferromagnetic order induced by photogenerated carriers in magnetic III-V semiconductor heterostructures of (In,Mn)As/GaSb," *Phys. Rev. Lett.* **78**, 4617 (1997).
21. H. Ohno, D. Chiba, F. Matsukura, *et al.*, "Electric-field control of ferromagnetism," *Nature* **408**, 944 (2000).
22. J. De Boeck, R. Oesterholt, A. Van Esch, *et al.*, "Nanometer-scale magnetic MnAs particles in GaAs grown by molecular beam epitaxy," *Appl. Phys. Lett.* **68**, 2744 (1996).
23. K. Ando, "Seeking room-temperature ferromagnetic semiconductors," *Science* **312**, 1883 (2006).
24. S. J. Pearton, C. R. Abernathy, M. E. Overberg, *et al.*, "Wide bandgap ferromagnetic semiconductors and oxides," *J. Appl. Phys.* **93**, 1 (2003).
25. F. Meier and B. P. Zakharchenya, eds., *Optical Orientation,* New York: North-Holand, 1984.
26. C. Weisbuch, *Contribution à l'Etude du Pompage Optique dans les Semiconducteurs III-V*, Ph. D. Thesis, Paris, 1977, unpublished.
27. I. Zutic, J. Fabian, and S. C. Erwin, "Spin injection and detection in silicon," *Phys. Rev. Lett.* **97**, 026603 (2006).
28. S. C. Erwin and I. Zutic, "Tailoring ferromagnetic chalcopyrites," *Nature Mater.* **3,** 410 (2004).
29. I. Zutic, J. Fabian, and S. Das Sarma, "Spin-polarized transport in inhomogeneous magnetic semiconductors: Theory of magnetic/nonmagnetic p-n junctions," *Phys Rev. Lett.* **88**, 066603 (2006).
30. T. Kondo, J. Hayafuji, and H. Munekata, "Investigation of spin voltaic effect in p-n heterojunction," *Jpn. J. Appl. Phys.* **45**, L663 (2006).

31. R. Jansen, "The spin-valve transistor: A review and outlook," *J. Phys. D. Appl. Phys.* **36**, R289 (2003).
32. J. Fabian, I. Zutic, and S. Das Sarma, "Theory of magnetic bipolar transistor," cond-mat/0211639.
33. J. Fabian and I. Zutic, "Spin-polarized current amplification and spin injection in magnetic bipolar transistors," *Phys. Rev. B* **69**, 115314 (2004).
34. J. Fabian and Zutic, "The Ebers-Moll model for magnetic bipolar transistors," *Appl. Phys. Lett.* **86**, 133506 (2005).
35. M. Field, Rockwell Scientific, Thousand Oaks, private communication.
36. B.-C. Min, K. Motohashi, J. C. Lodder, and R. Jansen, "Tunable spin-tunnel contacts to silicon using low-work function ferromagnets," *Nature Mater.* **5**, 817 (2006).
37. I. Zutic, "Gadolinium makes good spin contacts," *Nature Mater.* **5**, 771 (2006).
38. C. L. Dennis, C. V. Tiusan, J. F. Gregg, G. J. Ensell, and S. M. Thompson, "Silicon spin diffusion transistor: Materials, physics and device characteristics," *IEE Proc. Circ. Dev. Syst.* **152**, 340 (2005).

Molecular Meso- and Nanodevices: Are the Molecules Conducting?

N. B. Zhitenev
Bell Laboratories
Lucent Technologies, 600 Mountain Ave., Murray Hill, NJ 07974, U.S.A.

1. Introduction

As Si-based microelectronics approaches fundamental limits to continued scaling, organic-inorganic hybrid electronics has become one of several options that are proposed to not only extend performance but increase functionality.[1-3] Molecular electronics is projected to introduce the vast capabilities of synthetic chemistry into the realm of nanoelectronic devices. The rich phenomena of self-organization, self-assembly and area-specific growth that occur at the molecular scale are proposed[4-6] to replace some of the most prohibitive and expensive processes in fabrication, such as nanopatterning[7] and film growth. Yet, despite noticeable progress, it is premature to claim that a fundamental understanding of the conductance mechanisms in nanoscale molecular organic or inorganic devices has been achieved; certainly no feasible manufacturing scheme is available or even on the horizon.

The general approach for molecular device fabrication is to perform the most critical patterning of nanometer features without molecules, to assemble the molecules, and then to complete the structure with a hopefully noninvasive processing step. Material and chemistry unpredictability, as well as diffusion and defect generation in the different fabrication processes, inevitably result in significant variations in the final devices.

The experiments can be classified in a number of ways, one of which is geometrical configuration of the contacts. In one scheme, the molecular layer itself defines the separation between the metal contacts. The molecules are usually assembled on metal or semiconductor as self-assembled monolayers (SAMs) or Langmuir-Blodgett mono- or multi-layers. The second contact is then made by evaporation through a small mask[6,8,9] or by electrochemical metal deposition in nanopores,[10] or by etching structures in top metal layer,[6] or by contact printing.[7,11]

A second group of experiments that have demonstrated some of the most spectacular physics is realized by trapping molecules between laterally separated contacts.[12,13] The reliable fabrication of contacts with reproducible gaps of a few nanometers is too demanding for even the most advanced lithographic techniques. Small gaps have been created by: breaking connected metal contacts using electromigration,[12-14] electrochemical etching and/or deposition,[15,16] evaporating through nanoscale shadow masks,[17] and advanced e-beam lithography.[18] The

Future Trends in Microelectronics. Edited by Serge Luryi, Jimmy Xu, and Alex Zaslavsky

microscopic configuration of the gaps between electrodes strongly varies depending on the fabrication technique, the granularity and ductility of the metal contacts, and the interaction between the metals, substrates and molecules.

There is noticeable quantitative and qualitative discrepancy between the modeling of the transport through a single molecule and the majority of the experimental results. For typical metal-molecule contacts, the metal Fermi level falls in the HOMO-LUMO gap (referring to the highest occupied and lowest unoccupied molecular orbitals). In many experiments, the attempts to determine the energy scale for the appropriate transport mechanism leads to unexpectedly small values, typically of the order of 30–300 meV[14,19,20] that are smaller than expected from calculations.[21-23] At the same time, conductance values measured in experiments are typically much lower than calculated ones.[21]

Research directed toward a comprehensive characterization of defects and traps[24,25] and their microscopic understanding[26] is just emerging. It is known that the coherence of the molecular packing is preserved at best only within single grains (perfectly crystalline terraces) of the underlying metal. Grain boundaries can induce defects in the molecular order.[27] The molecular packing can itself affect the morphology of the metal surface.[28] Metal electrodes can undesirably react with the molecules[29] and additional defects within the SAM can be created[30] during contact fabrication. Currently, every experiment is unique in that the atomic placement of relevant nanoscale device constituents cannot be reproduced.

The use of a gate electrode to capacitively couple to a device has been essential for MOSFET technology and is being explored as a method to change the properties of molecules. In the study of nanostructures, the use of a gate has likewise been indispensable since it allows the energy levels of the structure to be changed independently of the bias applied across a device, thus allowing experiments that would otherwise not have been possible.[12,13,31] However, in molecular electronics the placement of the gate is particularly challenging if a short molecule is being electrically probed. Two current-carrying electrodes (source and drain) need to be quite close to the molecule to probe its conductivity; the resulting electrostatic screening of the gate reduces the gate-molecule coupling.[9,32]

2. Nano- and mesojunctions with evaporated top contacts

In our research, we have focused on the issue of non-invasive fabrication of nano- and mesoscale molecular devices, followed by structural and electrical characterization. Metal-SAM-metal junctions have been made using three original techniques. The smallest junctions, with sizes down to single molecules, were made on sharp ends of quartz tips.[33] We have further studied conducting properties of these junctions by varying the size, the topography, the grain structure of the metal electrode used for molecular assembly,[34] and the material of the contacts[35]. The dominant conductance mechanism in junctions smaller than the size of metal grain is hopping, based on studies of the temperature dependence of charge transport and nonlinear current-voltage curves. In larger molecular junctions containing SAMs assembled on multiple metal grains, the conductance does not vanish at low temperatures. A residual tunneling conductance exists in addition to a thermally activated conductance channel. Electrodes ranging from noble metals that weakly react with the organics, to more reactive metals, to conductive metal oxides have been explored. Somewhat surprisingly, the characteristic energy obtained from T-dependent studies is usually determined by the composition of the metal contacts, rather than the molecules themselves. The hopping energies are small compared to the expected energies of the molecular levels for all combinations of metal contacts. This suggests the presence of low-lying energy states within the monolayers (not usually considered in the theoretical models).

A second fabrication scheme[36] that we employ uses small shadow masks defined within a stack of $SiO_2/SiN_x/SiO_2$ layers grown on doped Si substrates to obtain features below the lithographic limit. The fabrication of metal-SAM-metal junctions using such masks is illustrated in Fig. 1. We have compared two types of organic SAM-forming molecules, representing opposite ends of expected

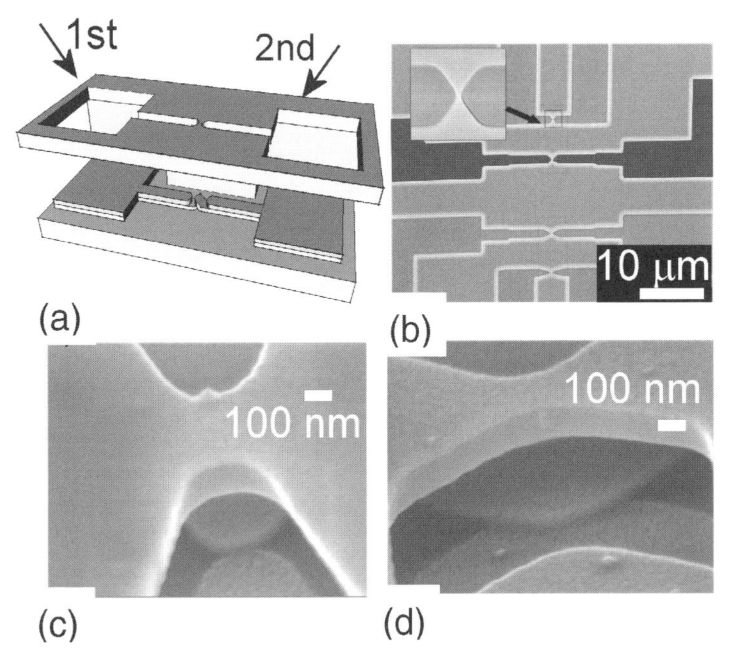

Figure 1. Fabrication of templates and molecular junctions. (a) The stencil mask is defined within an insulating stack of SiO_2 (~150 nm)/Si_3N_4 (~400 nm) layers grown on Si substrate. The mask is defined by photolithography and etching of the top SiO_2 layer, with Si_3N_4 layer then selectively removed underneath the SiO_2. Bottom electrodes are defined on the substrate by angled evaporation (1st arrow), followed by molecular layer deposition from solution, and a second angled evaporation (2nd arrow) to complete the junction. (b) SEM images of the template (inset shows close-up of one of the four devices). (c) and (d) Examples of small and large junctions, with SEM micrographs taken at a tilt angle of 60°.

electronic functionality. Terthiophenedithiol (T3), synthesized using previously described methods,[37] is a conjugated molecule with thiol groups responsible for chemical attachment to metal electrodes and is often considered as a candidate for conducting molecular wires. Decanedithiol (C10) is fully saturated with a molecular length of 1.5 nm, similar to that of T3.

We have studied electronic transport in the junctions systematically varying the growth conditions at the metal-molecule interface. Some results of these experiments are shown in Fig. 2. In the initial experiments, we assembled SAMs on an as-deposited Au electrode. All junctions were shorted. The top metal easily diffuses through the SAMs, presumably at monolayer packing defects propagated into the film by the grain structure of bottom metal electrode. In the next set, the bottom Au contact is annealed to make the grains smoother and the grain size larger before the molecular assembly. The distribution of device resistances measured at room temperature is shown in Fig. 2 (top row). Although the variation

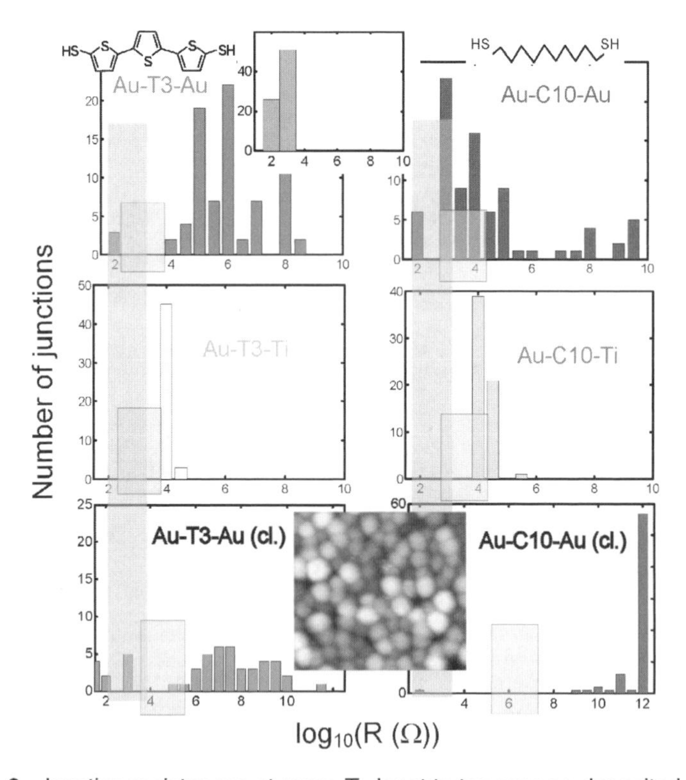

Figure 2. Junction resistances at room *T*. Inset to top row: as-deposited Au bottom electrode. In all other cases Au is annealed before SAM deposition. Resistance ranges of shorted junctions are shown with long gray bar. Lighter-colored bars show the ranges of resistance based literature data. Bottom row: Au clusters are created at the top metal-molecule interface. Inset to bottom row: STM image of clusters on top of a T3 SAM (50x50 nm).

is very broad, the yield of non-shorted devices is above ~90–95%; this is much higher than in previous experiments by others (0.5%–5%) with similar electrode arrangements.[38,39] Surprisingly, the apparent median resistance for T3 devices is higher than for C10 devices, implying that that the conductance of real junctions is not directly determined by the electronic structures of the molecules.

To study bond formation at the top interface, we used overlayers of metals with different chemical reactivities: Au, Ag and Ti. Silver is more reactive than Au, and titanium is known to strongly react not only with thiol groups at the top interface but also with carbon atoms[29] of molecular backbones. The electrical properties of Ag junctions are generally similar to those of Au junctions. The conductance of the junctions with Ti overlayers is consistently higher, with overall histograms being very similar for both T3 and C10 SAMs (Fig. 2, middle row).

In an attempt to control the microscopic topography at the top SAM-metal interface and the number of connected molecules, we intentionally created clusters at the top interface. First, 0.3–0.5 nm of Au was evaporated on top of the SAM, resulting in clusters with an average diameter ~6 nm. The junction fabrication was completed by evaporating an additional 8 nm of Au to form a continuous film. The electrical properties of the C10 junctions change dramatically in comparison with the uninterrupted Au evaporation, as shown in Fig. 2 (bottom row).

One of the important results was the range of values we obtained for the overall conductance through the SAM. Commonly accepted tunneling conductances per conjugated molecule[20,22] of comparable length lie in the 10^{-6}–10^{-8} Ω^{-1} range, whereas the conductance per alkane molecule[40,41] is 10^{-8}–10^{-9} Ω^{-1}. Ranges of expected tunnel resistances for a median ~100x100 nm junction containing ~$5\text{x}10^4$ molecules are shown in Fig. 2. In the top four panels, we assumed that every molecule is well bonded on both sides, while in bottom two panels a conservative assumption of having just a single bond per metal cluster of the top contact (~300 well-bonded molecules) was made.

Residual impurities, clusters, other low-energy defects and interface topography leading to different conductance mechanisms and broad distribution of junction resistances enormously complicate a reliable determination of tunneling conductance through the molecular levels. While a significant disagreement with the literature data can be easily demonstrated based on the full data set, to estimate the molecular conductance we select the more resistive junctions with *I–V* curves displaying negligible contribution from the low-energy transport channels. The estimates give $R_{T3} > 10^{11}$ Ω and $R_{C10} > 10^{14}$–10^{15} Ω for T3 and C10 respectively.

3. Mesojunctions with imprinted top contacts

The third fabrication technique we have used is based on the transfer of patterned metal directly to the molecules terminated with appropriate bonding groups.[11] A pattern is formed on a flexible rubber stamp. The stamp is then covered by the metal such that the raised and recessed areas are disconnected and is brought in contact with the monolayer. If bonding between the metal and the monolayer is stronger than between the stamp and the metal, withdrawing the stamp transfers the metal pattern to the monolayer.[42] Compared to evaporation techniques, the transfer fabrication method can lead to very different inter-diffusion and microscopic topography at metal-molecular interfaces.

Conductive AFM in force feedback mode was used to contact ~100 nm Au pads transferred on top of octanedithiol (C8) and decanedithiol (C10) SAMs. Some results of electrical characterizations are illustrated in Fig. 3. As expected, C10 devices are significantly less conductive than C8 devices. The *I–V* curves display significant dependence on force applied to the AFM tip – see inset to Fig. 3. Different force regimes can be identified. At the lowest force (1–3 nN) the current is extremely small. We suggest that only small fraction (~1–3%) of the molecules is simultaneously bonded to bottom and top electrodes. Larger force can increase the percentage of "wired" molecules. The steep rise of the current in the 3–5 nN force range is likely associated with the device deformation of this type. Further force increase can lead to the tilting of molecules and distortion of interfacial bonds that should decrease the conductance according to some calculations.[32] Importantly, even at the highest applied force the measurement pressure is 1–2 order of magnitude lower than in the previous experiments by others,[43,44] resulting in a smaller distortion of molecular properties.

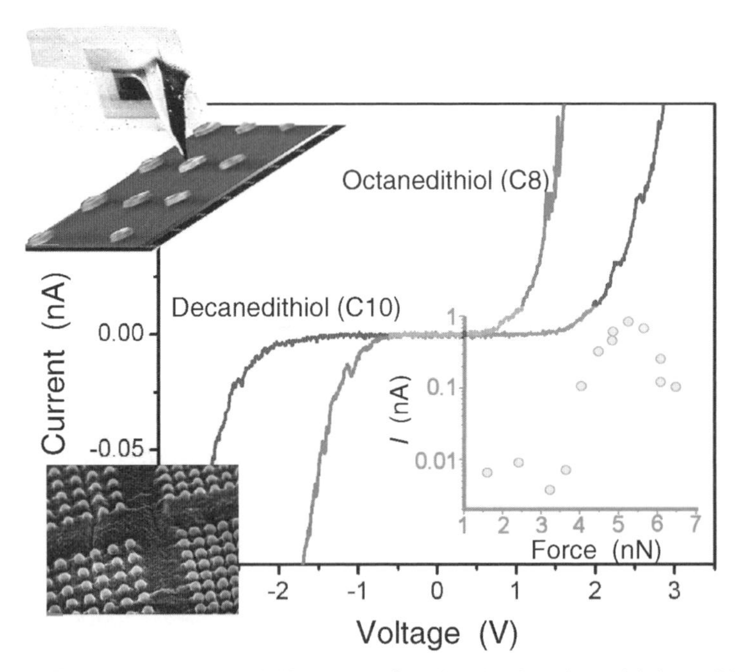

Figure 3. Room temperature *I–V* curves of molecular junctions fabricated by the nanotransfer technique, with the tilted SEM image of the polymer (PDMS) stamp carrying metal dots separated by 120 nm shown in the bottom left inset. The metal pads on top of the self-assembled molecular layer are contacted by conductive AFM (top left inset). Bottom right inset: current measured at 2.3 V in C8 junction as a function of contact force.

The highest conducting state of the devices as a function of pressure is used to estimate the conductance of the molecules. The current at small tip-substrate voltage is significantly below the noise floor (1 pA), yielding a conductance below 10^{-11}–10^{-12} Ω^{-1}. We note that this conductance is much smaller than 10^{-9} Ω^{-1}, the typical literature value[41] of conductance of *single* alkane molecule.

4. Conclusions

Simple approximation of tunneling through a rectangular barrier is often useful to relate the values of tunneling resistances with the energy structure. In such a model, $R = R_0\exp(\beta L)$ where R_0 is of the order of the quantum of resistance, L is the thickness of the barrier, and $\beta = 2(2m^*U)^{1/2}/\hbar$ is the tunneling decay parameter, where U is the barrier height and m^* is the effective mass. Based our the estimates of resistances, we find $\beta_{C10} \sim 1.5$ Å^{-1} and $\beta_{T3} \sim 1.1$ Å^{-1}. Assuming $U_{C10} \sim 5$ eV and $U_{T3} = 1.4$ eV (~half the bandgap), we obtain $m^*_{C10} \sim 0.4m_0$ and $m^*_{T3} \sim 0.8m_0$ where m_0 is the free electron mass. We note that in other well-studied tunnel barriers, such as AlO_x and SiO_2, the effective mass usually comes out close to $0.5m_0$. Calculations of molecular tunneling, when reduced to the simple single barrier approximation, typically predict much smaller m^* values such as $m^* \sim 0.2m_0$ for alkanes and $m^* \sim (0.06$–$0.25)m_0$ for various conjugated molecules.

The results show that several key assumptions fundamental to nanoscale molecular electronics and the current approaches for molecular integration have to be reexamined. Our most important experimental conclusions can be summarized as follows: i) although our devices are substantially less defective than in other experiments, the conductance of nano- and mesoscale devices is controlled by defects; ii) the molecules (both conjugated and saturated) are much less conductive (4–5 orders of magnitude for 1.5 nm molecules) than is commonly believed. The types of molecules/interfaces targeted to date for electronics applications appear to be much less conductive than anticipated and they are not robust against the creation of defects. Our results shed light on a variety of material transformations and self-organization processes occurring during integration of organic and inorganic components in nanoscale devices.

References

1. M. R. Stan, P. D. Franzon, S. C. Goldstein, J. C. Lach, and M. M. Ziegler, "Molecular electronics: From devices and interconnect to circuits and architecture," *Proc. IEEE* **91**, 1940 (2003).
2. D. B. Strukov and K. K. Likharev, "Prospects for terabit-scale nanoelectronic memories," *Nanotechnol.* **16**, 137 (2005).
3. J. E. Brewer, V. V. Zhirnov, and J. A. Hutchby, "Memory technology for the post CMOS era," *IEEE Circ. Dev.* **21**, 13 (2005).
4. J. Chen, M. A. Reed, A. M. Rawlett, and J. M. Tour, "Large on-off ratios and negative differential resistance in a molecular electronic device," *Science* **286**, 1550 (1999).
5. C. P. Collier, G. Mattersteig, E. W. Wong, *et al.*, "A [2]catenane-based solid state electronically reconfigurable switch," *Science* **289**, 1172 (2000).
6. Y. Chen, G.-Y. Jung, D. A. A. Ohlberg, *et al.*, "Nanoscale molecular-switch crossbar circuits," *Nanotechnol.* **14**, 462 (2003).
7. N. A. Melosh, A. Boukai, F. Diana *et al.*, "Ultrahigh-density nanowire lattices and circuits," *Science* **300**, 112 (2003).
8. N. B. Zhitenev, A. Erbe, H. Meng, and Z. Bao, "Gated molecular devices using self-assembled monolayers," *Nanotechnol.* **14**, 254 (2003).
9. C. R. Kagan, A. Afzali, R. Martel, *et al.*, "Evaluations and considerations for self-assembled monolayer field-effect transistors," *Nano Lett.* **3**, 119 (2003).
10. L. T. Cai, H. Skulason, J. G. Kushmerick, *et al.*, "Nanowire-based molecular monolayer junctions: Synthesis, assembly, and electrical characterization," *J. Phys. Chem. B* **108**, 2827 (2004).
11. Y.-L. Loo, D. V. Lang, J. A. Rogers, and J. W. P. Hsu, "Electrical contacts to molecular layers by nanotransfer printing," *Nano Lett.* **3**, 913 (2003).
12. J. Park, A. N. Parsupathy, J. I. Goldsmith, *et al.*, "Coulomb blockade and the Kondo effect in single-atom transistors," *Nature* **417**, 722 (2002).
13. W. Liang, M. P. Shores, M. Bockrath, J. R. Long, and H. Park, "Kondo resonance in a single-molecule transistor," *Nature* **417**, 725 (2002).
14. Y. Selzer, M. A. Cabassi, T. S. Mayer, and D. L. Allara, "Temperature effects on conduction through a molecular junction," *Nanotechnol.* **15**, S483 (2004).
15. Y.-V. Kervennic, D. Vanmaekelbergh, L. P. Kouwenhoven, and H. S. J. van der Zant, "Planar nanocontacts with atomically controlled separation," *Appl. Phys. Lett.* **83**, 3782 (2003).
16. A. F. Morpurgo, C. M. Marcus, and D. B. Robinson, "Controlled fabrication of metallic electrodes with atomic separation," *Appl. Phys. Lett.* **74**, 2084 (1999).
17. S. Kubatkin, A. Danilov, M. Hjort, *et al.*, "Single-electron transistor of a single organic molecule with access to several redox states," *Nature* **425**, 698 (2003).
18. M. A. Guillorn, D. W. Carr, R. C. Tiberio, E. Greenbaum, and M. L. Simpson, "Fabrication of dissimilar metal electrodes with nanometer interelectrode distance for molecular electronic device characterization," *J. Vac. Sci. Technol. B* **18**, 1177 (2000).

19. C. Zhou, M. R. Deshpande, M. A. Reed, L. Jones, and J. M. Tour, "Nanoscale metal/self-assembled monolayer/metal heterostructures," *Appl. Phys. Lett.* **71**, 611 (1997).
20. C. Kergueris, J.-P. Bourgoin, S. Palacin, *et al.*, "Electron transport through a metal-molecule-metal junction," *Phys. Rev. B* **59**, 12505 (1999).
21. M. D. Ventra, S. T. Pantelides, and N. D. Lang, "First-principles calculation of transport properties of a molecular device," *Phys. Rev. Lett.* **84**, 979 (2000).
22. J. Heurich, J., J. C. Cuevas, W. Wenzel, and G. Schon, "Electrical transport through single-molecule junctions: From molecular orbitals to conduction channels," *Phys. Rev. Lett.* **88**, 256803 (2002).
23. S. Hong, R. Reifenberger, W. Tian, *et al.*, "Molecular conductance spectroscopy of conjugated, phenyl-based molecules on Au(111): The effect of end groups on molecular conduction," *Superlatt. Microstruct.* **28**, 289 (2000).
24. Y. S. Yang, S. H. Kim, J.-I. Lee, *et al.*, "Deep-level defect characteristics in pentacene organic thin films," *Appl. Phys. Lett.* **80**, 1595 (2002).
25. D. V. Lang, X. Chi, T. Siegrist, A. M. Sergent, and A. P. Ramirez, "Bias-dependent generation and quenching of defects in pentacene," *Phys. Rev. Lett.* **93**, 076601 (2004).
26. J. E. Northrup and M. L. Chabinyc, "Gap states in organic semiconductors: Hydrogen- and oxygen-induced states in pentacene," *Phys. Rev. B* **68**, 041202 (2003).
27. Z. J. Donhauser, B. A. Mantooth, K. F. Kelly, *et al.*, "Conductance switching in single molecules through conformational changes," *Science* **292**, 2303 (2001).
28. C. Schonenberger, J. A. M. Sondaghuethorst, J. Jorritsma, and L. G. J. Fokkink, "What are the holes in self-assembled monolayers of alkanethiols on gold," *Langmuir* **10**, 611 (1994).
29. A. V. Walker, T. B. Tighe, J. Stapleton, *et al.*, "Interaction of vapor-deposited Ti and Au with molecular wires," *Appl. Phys. Lett.* **84**, 4008 (2004).
30. G. Philipp, G., Müeller-Schwanneke, M. Burghard, S. Roth, and K. von Klitzing, "Gold cluster formation at the interface of a gold/Langmuir-Blodgett film/gold microsandwich resulting in Coulomb charging phenomena," *J. Appl. Phys.* **85**, 3374 (1999).
31. H. Park, J. Park, A. K. L. Lim, E. H. Anderson, A. P. Alivisatos, and P. L. McEuen, "Nano-mechanical oscillations in a single-C60 transistor," *Nature* **407**, 57 (2000).
32. A. M. Bratkovsky, and P. E. Kornilovitch, "Effects of gating and contact geometry on current through conjugated molecules covalently bonded to electrodes," *Phys. Rev. B* **67**, 115307 (2003).
33. N. B. Zhitenev, H. Meng, and Z. Bao, "Conductance of small molecular junctions," *Phys. Rev. Lett.* **88**, 226801 (2002).
34. N. B. Zhitenev, A. Erbe, and Z. Bao, "Single- and multigrain nanojunctions with a self-assembled monolayer of conjugated molecules," *Phys. Rev. Lett.* **92**, 186805 (2004).

35. N. B. Zhitenev, A. Erbe, Z. Bao, W. Jiang, and E. Garfunkel, "Molecular nano junctions formed with different metallic electrodes," *Nanotechnol.* **16**, 495 (2005).
36. N. B. Zhitenev, W. Jiang, A. Erbe, *et al.*, "Control of topography, stress and diffusion at molecule-metal interfaces," *Nanotechnol.* **17**, 1272 (2006).
37. B. de Boer, H. Meng, D. F. Perepichka, *et al.*, "Synthesis and characterization of conjugated mono- and dithiol oligomers and characterization of their self-assembled monolayers," *Langmuir* **19**, 4272 (2003).
38. M. D. Austin, and S. Y. Chou, "Fabrication of a molecular self-assembled monolayer diode using nanoimprint lithography," *Nano Lett.* **3**, 1687 (2003).
39. W. Y. Wang, T. Lee, and M. A. Reed, "Electron tunnelling in self-assembled monolayers," *Rep. Progr. Phys.* **68**, 523 (2005).
40. J. K. Tomfohr and O. F. Sankey, "Simple estimates of the electron transport properties of molecules," *Physica Status Solidi B* **233**, 59 (2002).
41. A. Salomon, D. Cahen, S. Lindsay, J. Tomfohr, V. B. Engelkes, and C. D. Frisbie, "Comparison of electronic transport measurements on organic molecules," *Adv. Mater.* **15**, 1881 (2003).
42. W. Jiang, N. B. Zhitenev, Z. Bao, *et al.*, "Structure and bonding issues at the interface between gold and self-assembled conjugated dithiol monolayers," *Langmuir* **21**, 8751 (2005).
43. J. M. Beebe, V. B. Engelkes, L. L. Miller, and C. D. Frisbie, "Contact resistance in metal-molecule-metal junctions based on aliphatic SAMs: Effects of surface linker and metal work function," *J. Amer. Chem. Soc.* **124**, 11268 (2002).
44. D. J. Wold, R. Haag, M. A. Rampi, and C. D. Frisbie, "Distance dependence of electron tunneling through self-assembled monolayers measured by conducting probe atomic force microscopy: Unsaturated versus saturated molecular junctions," *J. Phys. Chem. B* **106**, 2813 (2002).

The Problem of a Perfect Lens Made From a Slab With Negative Refraction

A. L. Efros
Dept. of Physics, University of Utah, Salt Lake City, UT 84112, U.S.A.

1. Introduction

Recently there has been growing interest in the creation of lenses with unusually sharp foci. These lenses come in two varieties.

- *Veselago lens*

The original proposal by Moscow physicist Victor Veselago dates back to 1967.[1] Veselago studied a medium where for some hypothetical reason both electric permittivity ε and magnetic permeability μ are negative at some frequency ω_0. Since the $\varepsilon\mu$ product is positive, the light velocity remains real and the wave equation is unchanged. However, the vectors **k**, **E**, **H** of a plane wave now form a left-handed rather than a right-handed set, so that the Poynting vector **S** points opposite to the wavevector **k**. This anomaly is not forbidden by any general principle and such a medium is often called a left-handed medium (LHM). Veselago predicted negative refraction (minus sign in Snell's law) at the interface of the LHM and a regular medium (RM) and some other interesting manifestations of the LHM.

It is important to understand that the negative refraction is a property of the interface that follows from the regular boundary conditions. One should be cautious in ascribing a negative value to the bulk refractive index n of the LHM[2,3] even though this definition would restore a positive sign to Snell's law. The point is that neither Maxwell's equations nor the boundary conditions contain n. The algorithms required for the solution do not include the operation of taking the square root of $\varepsilon\mu$. Therefore, we do not believe the sign of n has any physical meaning. One can define n as a negative branch of the square root for either LHM or RM, but then some standard equations, like $\omega = ck/n$ and the expression for group velocity, should also be changed and this may lead to errors (see Ref. 4 for details). In this chapter, n is taken to be a positive number.

Veselago proposed a lens, based on negative refraction. As illustrated in Fig. 1, the lens consists of a long slab of a material with negative ε and μ embedded in an RM with positive ε and μ that have the same absolute values. Since impedances are matched, there are no reflected waves.

The ideas of Veselago were completely forgotten for many years, but they got a new life after J. Pendry published in 2000 a paper with a startling title: "Negative refraction makes a perfect lens".[5]

Future Trends in Microelectronics. Edited by Serge Luryi, Jimmy Xu, and Alex Zaslavsky

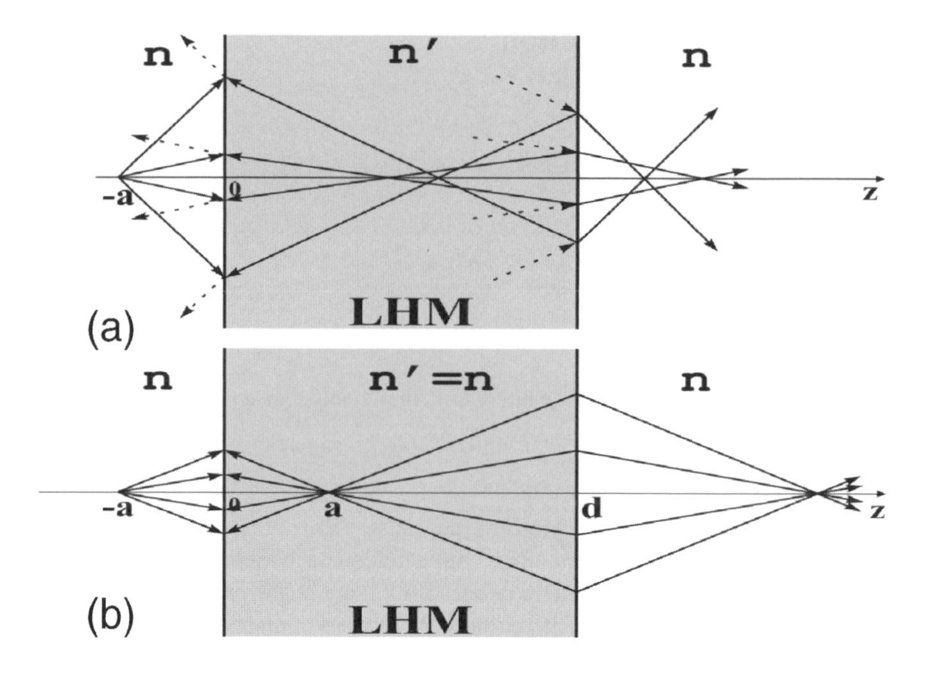

Figure 1. Schematic of the Veselago lens. The source is at a point $z = -a$. The width of the LHM slab is d. The arrows of rays show direction of **k**. (a) If refractive indices and impedances of the LHM and the RM are not matched, there is no focusing. (b) If they are matched, the lens has two foci, one at $z = a$, and another at $z = 2d-a$, where it is assumed that $d > a$.

- *Electrostatic or quasistatic lens*

Research on this lens goes back to the paper by Nicorovici *et al.*[6] of 1994 that predates the plane lens proposed by Pendry.[5] Consider a metallic slab and an electromagnetic field with such a low frequency that the wavelength is much larger than the width of the slab. Then one can neglect retardation. In this approximation magnetic field is small and the electrostatic potential $\phi(r)$ obeys the Laplace equation:

$$\nabla[\varepsilon(r)\nabla\phi(r)] = 0\ . \tag{1}$$

On the other hand, if the frequency obeys condition $\omega\tau >> 1$, where τ is the relaxation time of electrons, the dielectric constant of metallic slab may be written in a form

$$\varepsilon(\omega) = 1 - (\omega_P/\omega)^2\ , \tag{2}$$

where ω_p is the plasma frequency. To get $\varepsilon = -1$, we need $\omega = \omega_p/\sqrt{2}$.

All these conditions may be fulfilled only if the slab is thin enough (20–40 nm). So the new physics of this approach is closely connected to the recent development of nanotechnology and can rarely be found in classical textbooks. The force lines of electric field reproduce Fig. 1 if the point charge is at $z = -a$. So, the electrostatic slab with $\varepsilon = -1$ works in the same way as the Veselago lens. This was discovered by Nicorovici *et al.*[6] for the cylindrical lens and by Pendry[5] for the plane lens. Note, however, that the wavelength criterion for the width of the focus, which should be used for the Veselago lens, is irrelevant for the electrostatic lens because in the electrostatic approximation the wavelength is infinite.

Pendry's result for both lenses is that the fields near the foci are exactly the same as near the source (leading to perfect imaging). According to Pendry a small absorption inside the slab may only slightly smear the foci.

The ideas behind his paper are very straightforward. In the case of the Veselago lens they are as follows. If the slab of the Veselago lens is infinite, the lens has an infinite aperture. A source produces both propagating and evanescent waves (EWs). The latter waves exist usually in the near field region only. In the regular lens, these waves would never reach the focus, and that is why the image is not ideal. However, Pendry has shown that EWs are amplified by the LHM. Moreover, he claimed that this amplification leads to a complete restoration of the EWs in a focal point. Therefore, the image should perfectly repeat the source.

In the case of the electrostatic lens all waves are evanescent. Pendry claims that they also decay in the regular material and get amplified by the metallic slab up to the ideal restoration of the image in the foci.

Pendry demonstrated the amplification of a single EW, but while the argument is enticing, it is nonetheless incorrect. The perfect lens with a real focus is forbidden by the laws of electrodynamics. This can be easily seen from the following argument. The field near the point source at $z = -a$ decays as r^{-1}, where r is the distance from the source. But the field near the focus would not obey Maxwell's equations if it behaved the same way, because there is no source near the focus. The perfect lens scenario in the electrostatic regime is even more problematic. Not only is the r^{-1} singularity forbidden without the source, but also any potential with a maximum or minimum cannot be a solution of the Laplace equation.

The flaw in Pendry's arguments was found immediately at least by three groups:[7-9] the solution in coordinate space, summed over all EWs, diverges exponentially in a three-dimensional domain near the rear face of the slab. Thus, this solution does not obey Maxwell's equations. Note that Nicorovici *et al.*[6] understood this problem in 1994.

There were numerous attempts to fix up Pendry's solution by using small imaginary parts of ε and μ[10,11] or by considering time-dependent instead of stationary solutions.[12,13] However, the divergences of the fields are too strong to be removed and still maintain the physical picture. Recently Milton *et al.* have shown[11,14] that if $a < d/2$ (see Fig. 1) the absorption of energy in the slab $\mathrm{Im}\{\varepsilon|E|^2\}$

diverges as Im{ε} vanishes. This result is valid for both types of lenses. It reflects a strong divergence of the electric field as Im{ε} vanishes. Note that the increase of the fields in the case of non-zero Im{ε} occurs near both faces of the slab due to reflection of the wave.

Both the power that is going through the slab and the power that is dissipated inside are provided by the source. However, real sources have a finite power. Milton *et al.*[14] argued that for $a < d/2$ all the energy provided by a source is dissipated in the slab. Then, in the limit of small Im{ε} the slab turns into the opposite of a perfect lens: it cloaks the source, makes it invisible behind the slab.

In the case $d/2 < a < d$, the divergences of the fields are weaker so that absorption tends to zero as Im{ε} tends to zero. Nevertheless, the limit of the solution as Im{ε} → 0 does not obey Maxwell's equations, which is worrisome.

2. Existence of a perfect lens with virtual focus

Now I present a modification of the above geometry, such that Pendry's arguments on the amplification of EWs becomes perfectly correct, so that both types of the lenses under discussion are indeed perfect in some sense, and a solution of Maxwell's equations exists even without absorption.

Consider a lens with of thickness d and an object at a distance $d_1 > d$ from the lens, as shown in Fig. 2. In this case, the lens does not have any real focus but Fig. 2 shows that the lens has a virtual focus (VF) at $z = 2d$. If $d < d_1 < 2d$, this focus is within the slab, whereas if $2d < d_1$ (this case is shown in Fig. 2), the VF is to the left of the slab.

Significantly, since the fields do not have any singularity near the virtual focus there are no arguments forbidding perfect virtual focus. We show below that an observer behind the slab sees the image translated without any distortion a distance $2d$ toward the observer. Pendry's amplification of the EWs plays a crucial role in this derivation. We consider the case of the electrostatic lens, but the same arguments hold for the Veselago lens.

Following Pendry, we suppose that electrostatic potential at point S ($z = 0$) has the form

$$\phi(x,y,0) = (2\pi)^{-2} \iint dk_x dk_y\, V(k_x,k_y) \exp(ik_x x + ik_y y), \tag{3}$$

where the time exponent is omitted. For $z > 0$, the potential has a form

$$\phi(x,y,z) = (2\pi)^{-2} \iint dk_x dk_y\, V(k_x,k_y) \exp(ik_x x + ik_y y)\, F(z). \tag{4}$$

The function $F(z)$ should be found in the three different regions using the equation $(d^2/dz^2 - k_0^2)F(z) = 0$, with $k_0^2 = k_x^2 + k_y^2$, combined with the boundary conditions at the faces of the slab. Assuming that $\varepsilon = 1$ outside the slab and $\varepsilon = -1$ inside, one finds that $F(z)$ is continuous while dF/dz changes sign at the boundaries, as follows:

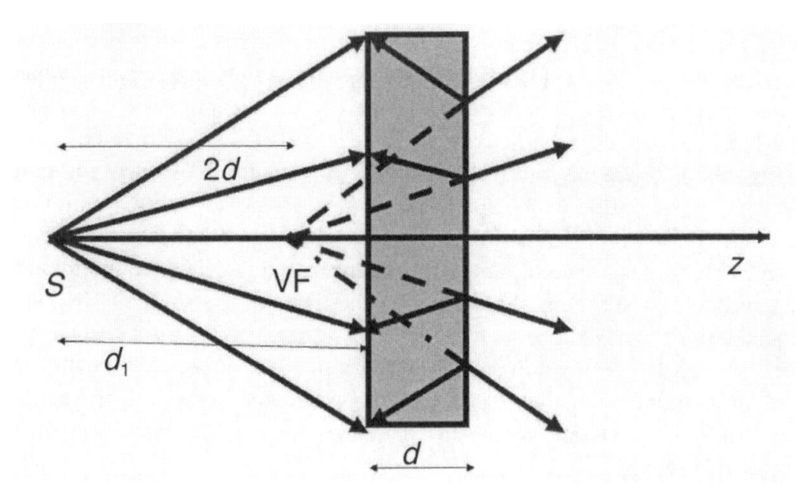

Figure 2. Veselago lens or electrostatic lens with the object at $z = 0$ at a distance $d_1 > d$ from the slab. One can see that the lens has a virtual focus (VF) at $z = 2d$, which may be inside or to the left of the slab. It is important that the field in the VF does not have a singularity. An observer behind the lens sees the image located not at point S ($z = 0$), where the object is, but at $z = 2d$. Thus the image looks shifted a distance $2d$ toward the observer. The lens is perfect since the image is only shifted but it is not distorted.

$$F(z) = \exp[-k_0 z] \text{ for } 0 < z < d_1 , \tag{5}$$

$$F(z) = \exp[k_0(z-2d)] \text{ for } d_1 < z < (d_1 + d) , \tag{6}$$

$$F(z) = \exp[-k_0 z] \text{ for } (d_1 + d) < z . \tag{7}$$

Here $k_0 = (k_x^2 + k_y^2)^{1/2} > 0$. One can see a single virtual focus. An observer at any point behind the lens, $z > (d_1 + d)$, sees potential $\phi(x,y,z)$ as if it were created by the same object, but shifted without any distortion from $z = 0$ to $z = 2d$. The absence of distortion is due to the amplification of the potential with increasing z inside the slab, as predicted by Pendry. Note that all the integrals over k_x and k_y are finite at $d_1 > d$, because $(z - 2d_1) < 0$ inside the slab and $(z - 2d) > 0$ behind the slab. Thus, in this case the solution of Maxwell's equations can be obtained without any regularizing procedure.

Interestingly, the VF may be located infinitesimally close to the rear face of the slab (at the point $z = d + d_1$) if the distance d_1 between the object and the lens is only infinitesimally larger than the thickness of the slab d. Let $d_1 = d + \eta$, where $\eta \to 0$. Now the virtual image is inside the slab at a distance η from the rear face. At negative η, the focus becomes real and the theory is ruined by divergent integrals, as happens with Pendry's theory.[5] But for positive η everything is fine. This means that the image without any distortions can be created infinitesimally close to the rear surface of the slab, but still inside the slab.

Finally, let us discuss how the spatial dispersion of the dielectric constant affects our results. In the case of the Veselago lens in a photonic crystal, where the working frequency is close to the Γ-point of the second Brillouin zone, the propagating waves are characterized by negative ε and μ, and they undergo negative refraction at the interface with an RM. But the EWs have different (**k**-dependent) ε and μ due to the spatial dispersion effect, so that the amplification may simply be absent.[15,16] Then there are no divergences without absorption for all real and virtual foci. There is no perfect focusing and the fields at the foci can be found simply by subtracting the EWs from the fields of the source. Note that the amplification of the EWs in photonic crystals need not arise from surface plasma modes that result from negative ε and μ, but also from surface modes that appear due to the termination of the crystal. The existence and properties of such modes depend on the way the surface is cut. These modes may provide superlensing in the near-field regime.[15]

In the particular case of an electrostatic lens made from a metallic slab, spatial dispersion cuts off all wavevectors above ω/v_F, where v_F is the velocity of electrons at the Fermi surface. Due to this limitation, the maximum resolution of this type of lens will be about 10 nm, as discussed in detail in Ref. 17.

3. Conclusions

This chapter examines whether a perfect lens and superlensing are possible in principle. We have demonstrated that in the case of a virtual focus, the idea of a perfect lens based on amplification of evanescent waves, as proposed by Pendry, is perfectly correct. This is not true for the real focus. We believe that some of the experimental results claiming superlensing can be explained in terms of the proposed theory for the case when the virtual focus is inside the lens but very close to the rear face.

Acknowledgements

I am grateful to Graeme Milton and Emmanuel Rashba for reading the manuscript and important comments. I appreciate multiple discussions with Serge Luryi and Michael Shur during my sabbatical stay at the Stony Brook University and during the FTM Workshop of 2006.

References

1. V. G. Veselago, "Properties of materials having simultaneously negative values of the dielectric and magnetic susceptibilities," *Sov. Phys. Solid State* **8**, 2854 (1967).
2. D. R. Smith and N. E. Kroll, "Negative refractive index in left-handed materials," *Phys. Rev. Lett.* **85,** 2933 (2000).
3. J. B. Pendry and D. Smith, "The quest for the superlens," *Sci. Amer.* **295**, 61 (2006).
4. A. L. Pokrovsky and A. L. Efros, "Sign of refractive index and group velocity in left-handed media," *Solid State Commun.* **124**, 283 (2002).
5. J. B. Pendry, "Negative refraction makes a perfect lens," *Phys. Rev. Lett.* **85**, 3966 (2000).
6. N. A. Nicorovici, R. C. McPhedran, and G. W. Milton, "Optical and dielectric properties of partially resonant composites," *Phys. Rev. B* **49**, 8479 (1994).
7. N. Garcia and M. Nieto-Vesperinas, "Left-handed materials do not make a perfect lens," *Phys. Rev. Lett.* **88**, 207403 (2002).
8. A. L. Pokrovsky and A. L. Efros, "Diffraction in left-handed materials and theory of Veselago lens," cond-mat/0202078; see also *Physica B* **338**, 333 (2003).
9. F. D. M. Haldane, "Electromagnetic surface modes at interfaces with negative refractive index make a 'not-quite-perfect' lens," cond-mat/0206420.
10. V. A. Poldolskiy and E. Narimanov, "Near-sighted superlens," *Optics Lett.* **30**, 75 (2005).
11. G. W. Milton, N.–A. P. Nicorovici, R. C. McPhedran, and V. A. Podolskiy, "A proof of superlensing in the quasistatic regime and limitations of superlenses in this regime due to anomalous localized resonance," *Proc. Royal Soc. London A* **461**, 3999 (2005).
12. G. Gomes-Santos, "Universal features of the time evolution of evanescent modes in a left-handed perfect lens," *Phys. Rev. Lett.* **90**, 077401 (2003).
13. A. D. Yaghjian and T. B. Hansen, "Plane wave solution to frequency domain and time-domain scattering from magnetodielectric slab," *Phys. Rev. E* **73** 046608 (2006).
14. G. W. Milton and N.–A. P. Nicorovici, "On the cloaking effect associated with anomalous localized resonance," *Proc. Royal Soc. London A* **462**, 3027 (2006).
15. C. Y. Li, J. M. Holt, and A. L. Efros, "Far-field imaging by Veselago lens made of photonic crystal," *J. Opt. Soc. Amer. B* **23**, 490 (2006).
16. C. Y. Li, J. M. Holt, and A. L. Efros, "Imaging by Veselago lens based upon two-dimensional photonic crystal with triangular lattice," *J. Opt. Soc. Amer. B* **23**, 936 (2006).
17. I. A. Larkin and M. I. Stockman, "Imperfect perfect lens," *Nano Lett.* **5**, 339 (2005).

Is There a Linewidth Theory for Semiconductor Lasers?

Boris Spivak
Dept. of Physics, University of Washington, Seattle, WA 98195, U.S.A.

Serge Luryi
Dept. of Electrical and Computer Engineering
SUNY–Stony Brook, Stony Brook, NY 11794, U.S.A.

1. Introduction

Laser linewidth theory was pioneered by Schawlow and Townes[1] and further developed in Refs. 2 and 3. In this chapter, we will discuss the status of the Schawlow-Townes-Lax-Henry (STLH) theory of laser linewidth in the instance of semiconductor injection lasers. At injection levels I below threshold $I < I_C$, one can introduce two spectra $g(\omega,I)$ and $\sigma(\omega,I)$, see Fig. 1(a), describing respectively the material gain and the loss at cavity mirrors of the electromagnetic field intensity. The gain $g(\omega,I)$ is generally an increasing function of I. At $I = I_C$, the two spectra touch each other, $g(\omega_0,I) = \sigma(\omega_0)$, and the generation begins. The STLH theory of laser linewidth is based on the assumption that in the mean-field approximation (*i.e.* without fluctuations), the laser generation remains singular in frequency for $I > I_C$, *i.e.* above threshold. In the framework of this approach, the laser line acquires a finite width Γ entirely due to fluctuations. In an ideal laser these fluctuations are due to the random discrete nature of spontaneous emission.

We shall refer to the property of the two spectral curves $g(\omega,I)$ and $\sigma(\omega)$ touching each other at a singular frequency for $I > I_C$ as *rigidity*, see Fig. 1(b). In principle, however, scenarios other than rigidity are also possible. For example, the curves may touch each other for $I > I_C$ in a finite interval of frequencies, as in Fig. 1(c), so that there is a finite linewidth even in the mean-field approximation. In this case, taking account of fluctuations would provide only a correction. This is not an unusual situation. For example, the conventional mean-field scenario for multimode laser generation, illustrated schematically in Fig. 1(d), involves oscillations at several discrete frequencies.

In this chapter we examine the validity of the assumption of rigidity. In Section 2 we briefly review the standard STLH linewidth theory. In Section 3 we derive a mean-field expression for the linewidth using Boltzmann's kinetic equation for electrons and photons. In this approach, the linewidth turns out to be an increasing function of injection, which violates the assumption of rigidity and is in contradiction with the STLH scenario. Curiously, however, it is not necessarily in contradiction with experiment, see the discussion in Section 4.

 Future Trends in Microelectronics. Edited by Serge Luryi, Jimmy Xu, and Alex Zaslavsky

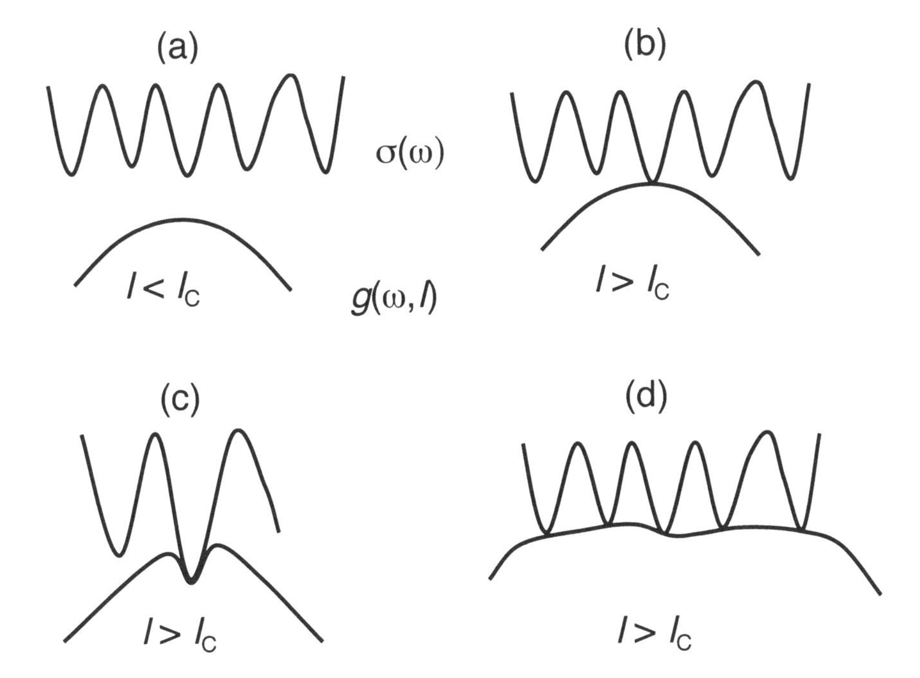

Figure 1. Relative configuration of the spectral curves corresponding to the material gain $g(\omega, I)$ and the loss $\sigma(\omega)$ of the electromagnetic field intensity at the cavity mirrors below (a) and above (c-d) the threshold I_C.

2. Standard model of a semiconductor laser

The simplest model of the laser is a pumped two-level electronic system, immersed in an electromagnetic wave resonator. It is described (see, for example, Ref. 4) by rate equations for the electron population difference equation $n = n_2 - n_1$ and the number and the number of photons N in the resonator:

$$dn/dt + (n(t) - n_0)/\tau = I - \gamma n N$$
$$dN/dt + \sigma N = \gamma n N \tag{1}$$

where the differential gain γ, defined by $g(\omega,I) = \gamma(\omega)n(I)$, is a coefficient independent of n, and τ is the characteristic time describing all non-stimulated recombination processes (in a high-quality material with negligible nonradiative recombination one has $\tau = \tau_{sp}$, where τ_{sp} is the characteristic time of spontaneous emission). The equilibrium population difference at $I = 0$ is denoted by n_0. Laser generation begins when the photon gain γn exceeds loss σ. In this case, the stationary solution of Eq. (1) is $\gamma n = \sigma$ and $N = (I - I_C)/\sigma$, where the threshold current is defined as $I_C \equiv (\sigma/\gamma\tau) - n_0/\tau$.

In this simplest model, the I dependence of gain $g(\omega,I)$ is parameterized by a single number n and the rigidity illustrated in Fig. 1(b) arises automatically. Above the threshold, the mean-field equations (1) describe a wide range of phenomena, including relaxation of an arbitrary initial state to the steady state at given I.

The standard STLH theory of laser linewidth is developed as follows. In the limit $N >> 1$, the electromagnetic field $\boldsymbol{E}(t) = \boldsymbol{E}_0\exp(i\omega_0 t)$ of a single resonator mode is considered classical, characterized by amplitude and phase. Here ω_0 is the mode frequency, and $\boldsymbol{E}_0$ is a complex vector that may be slowly varying in time. In the mean-field approximation, the phase φ of the field is definite, while its amplitude is proportional to $N^{1/2}$, that is $\boldsymbol{E}_0 \sim N^{1/2}\exp(i\varphi)$. Beyond the mean-field approximation, the quantities N, n, and φ fluctuate in time due to the randomness of recombination and relaxation processes. It is these fluctuations that determine the linewidth in the conventional STLH approach. In an idealized laser, the fluctuations arise from randomness of the spontaneous emission. All fluctuations of interest, including spontaneous emission, can be described classically in the sense that they are generated by δ-correlated Langevin forces (white noise). The reason for the classical description of fluctuations is that the time scale we are interested in (of the order of inverse linewidth) is long compared to all kinetic relaxation times.

In the $N >> 1$ limit, where the fluctuations in the number of photons are small, $\delta N << N$, the fluctuations of φ are decoupled from those of n and N. Fluctuations δN and δn give rise to the intensity noise, while only fluctuations of the phase $\delta\varphi$ contribute to the linewidth. These fluctuations correspond to a random walk of the complex variable $\boldsymbol{E}_0$ of a constant modulus (see Ref. 3). Each event of spontaneous emission adds to $\boldsymbol{E}_0$ a small vector $\delta\boldsymbol{E}_0 \sim (\hbar\omega)^{1/2}$. The angle between the two complex vectors $\boldsymbol{E}_0$ and $\delta\boldsymbol{E}_0$ is random, and both the amplitude and the phase of the sum $\boldsymbol{E}_0 + \delta\boldsymbol{E}_0$ are varying. The amplitude variation, $|\boldsymbol{E}_0 + \delta\boldsymbol{E}_0|^2 - |\boldsymbol{E}_0|^2$, corresponds to δN and, according to Eq. (1), it relaxes to its steady-state value, while $\delta\varphi \approx \delta\boldsymbol{E}_0/\boldsymbol{E}_0 \approx N^{-1/2}$. The diffusion coefficient describing the angular random walk, $D_\varphi = (\delta\varphi)^2/\tau_{SP}$, determines the laser linewidth $\Gamma \approx D_\varphi$, which thus turns out to be inversely proportional to the intensity of laser emission:

$$\Gamma_{STLH} = 1/(\tau_{sp}N) \, . \tag{2}$$

Thus, at large N, the linewidth is much smaller than any characteristic frequency of the system, such as the spectral width of the laser cavity $\sigma(\omega_0)$, the rate of electronic collisions $1/\tau_{ee}$ that determine the broadening of the quantum electronic levels in semiconductors, the spectral width of the gain $g(\omega)$, and the spontaneous emission rate $1/\tau_{sp}$. It is worth noting that the precise meaning of the spontaneous emission rate in this model is not clear. What is the spectral width for spontaneous emission events that appears in the derivation of Eq. (2)? For example, in some scenarios it may be reasonable to include only the emission into the linewidth Γ itself, in which case N would enter Eq. (2) with a different power.

We would like to stress that the STLH approach essentially relies on the assumption that the mean-field equations have a singular solution with no width at all. Discussion of this assumption requires a detailed analysis of the injection-level dependence $g(\omega,I)$, which in turn requires a consideration of energy and frequency dependences of the electron and photon distributions, n_ε and N_ω, respectively. In Section 3 we discuss such a description based on Boltzmann's kinetic equation. It turns out that singular solutions are ruled out in the kinetic description that yields a finite laser linewidth already in the absence of fluctuations.

3. Kinetic equation

The simplest kinetic equation describing the energy distribution of electrons n_ε and photons N_ω is of the form

$$dn_\varepsilon/dt = = -\gamma_\varepsilon n_\varepsilon N_\omega + I + S\{n_\varepsilon\} \tag{3a}$$

$$dN_\omega/dt = \gamma_\varepsilon n_\varepsilon N_\omega - \sigma_\omega N_\omega \tag{3b}$$

Here, the energy parameters ε and ω are related by $\hbar\omega = \varepsilon(k) + E_G$, where E_G is the bandgap and $\varepsilon(k) = \varepsilon_e(k) + \varepsilon_h(k)$ is the kinetic energy of carriers at wavevector $\boldsymbol{k}$ corresponding to the transition. In terms of the dimensionless n_ε, the total electron population difference n that enters Eq. (1) can be expressed as

$$n = \int n_\varepsilon\, \nu(\varepsilon)\, d\varepsilon, \tag{4}$$

where $\nu(\varepsilon)$ is the density of electronic states. Similarly, the total injection level I is given by the integral

$$I = \int I_\varepsilon\, \nu(\varepsilon)\, d\varepsilon, \tag{5}$$

where I_ε is the differential injection intensity.

The collision integral S comprises contributions from electron-electron and electron-phonon interactions, as well as non-stimulated recombination,

$$S\{n_\varepsilon\} = S_{ee} + S_{e\text{-}ph} + S_{rec}\,. \tag{6}$$

We consider the simplest situation when the electron-electron scattering rate $1/\tau_{ee}$ is fastest. This situation is also most relevant for semiconductor lasers operating at room temperature. The collision integral S_{ee} is nullified by the Fermi distribution function n_ε^F, parameterized by an arbitrary chemical potential μ_{eff} and temperature T_{eff}. These parameters are determined from the conservation laws for the number of particles and energy, which can be obtained from Eq. (3) by integrating over ε and ω. At room temperature, the energy relaxation rate is fast and one has $T_{eff} = T$.

The distribution function n_ε deviates from the Fermi shape in a narrow interval of energies of order the linewidth Γ, where $N_\omega \neq 0$ and $n_\varepsilon = n_\varepsilon^F + \delta n_\varepsilon$. The typical energy exchange involved in electron-electron scattering events is of the order of T and in the limit $\Gamma \ll T$ the relaxation time approximation for electron-electron scattering is exact,

$$S_{ee} = -\delta n_\varepsilon/\tau_{ee} \,. \tag{7}$$

The reason for this is that δn_ε in region Γ is formed by incoming and outgoing fluxes from a much larger region of order T_{eff} or μ_{eff}, whichever is larger. According to Eq. (3b), in a stationary state $dN_\omega/dt = 0$, the electron distribution function is pinned in region Γ and is independent of the injection level I or its energy distribution I_ε:

$$n_\varepsilon = \sigma_\omega/\gamma_\varepsilon. \tag{8}$$

On the other hand, the electronic distribution outside region Γ is not pinned because the escape rate from the outside region (where $N_\omega = 0$) into the active region Γ is finite and characterized by a time constant of order τ_{ee}. The total electron concentration outside region Γ hence grows with the injection I. This means that the width of Γ must itself increase with I.

To make this argument quantitative, we note that δn_ε vanishes at the edges of region Γ. Depending on the shape of the function $f(\varepsilon) = \sigma_\omega/\gamma_\varepsilon$ on the right-hand side, Eq. (7) may have many solutions which correspond to the existence of multiple lasing modes in the mean-field approximation. Let us focus on the single-mode case, when $f(\varepsilon)$ has a single minimum at $\varepsilon = \varepsilon_0$ and can be approximated by $f(\varepsilon) = f(\varepsilon_0) + a(\varepsilon - \varepsilon_0)^2$, where $f(\varepsilon_0) \approx 1$, see Fig. 2. The shape of $f(\varepsilon)$ can be characterized by a halfwidth $\Delta \approx 2a^{-1/2}$. In the case when σ_ω is a sharper function than γ_ω, the quantity Δ is the resonator linewidth. Within interval Γ we can write

$$\delta n_\varepsilon = a\Gamma^2/4 - a(\varepsilon - \varepsilon_0)^2, \tag{9}$$

where the coefficients are chosen such that $\delta n_\varepsilon = 0$ for $(\varepsilon - \varepsilon_0) = \pm\Gamma/2$.

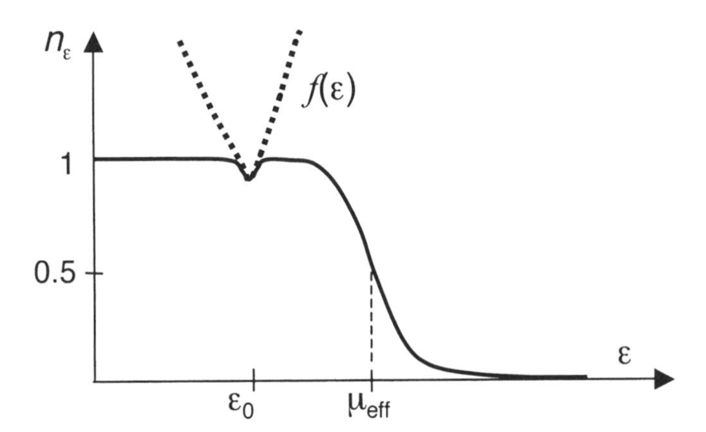

Figure 2. Schematic representation of the electron energy distribution n_ε and the function $f(\varepsilon) = \sigma_\omega/\gamma_\varepsilon$. These functions coincide in region Γ.

Integrating Eq. (3a) over all energies in the stationary case $dn_\varepsilon/dt = 0$, we find:

$$I - I_C = \int \sigma(\omega) N(\omega) d\omega \tag{10}$$

where the threshold injection I_C is given by

$$I_C = -\int S_{rec} \nu(\varepsilon) d\varepsilon \tag{11}$$

The terms $S_{e\text{-}ph}$ and S_{ee} drop out of Eq. (10) when integrated over all energies since they conserve the number of electrons. We note that the integrand in Eq. (9) is nonvanishing only in the small region Γ that is much narrower than either the effective temperature T_{eff} or the Fermi level μ_{eff}. Therefore, if we integrate Eq. (3a) over Γ, we find

$$I - I_C = -\int_\Gamma (\delta n_\varepsilon / \tau_{ee})\, \nu(\varepsilon) d\varepsilon \tag{12}$$

Substituting Eq. (8) into Eq. (11) we obtain an estimate of the laser linewidth:

$$\Gamma^3 = 6(I - I_C)\tau_{ee}/a\nu(\varepsilon_0) \tag{13a}$$

or, equivalently,

$$\Gamma = \Delta [3/2\nu(\varepsilon_0)\Delta]^{1/3}\, [(I - I_C)\tau_{ee}]^{1/3}. \tag{13b}$$

We see that the linewidth in the mean-field approximation increases with pumping. This is in drastic contradiction with the conventional STLH result of Eq. (2) that predicts a linewidth decreasing with I.

The fundamental reason for this discrepancy is the assumption by STLH of a singular frequency dependence of the field $\boldsymbol{E}(\omega) \sim \delta(\omega - \omega_0)$ in the absence of fluctuations. In contrast, the solutions of kinetic equations are smooth functions of ε and ω and do not exhibit any singularities. Consequently, an account of fluctuations would make only a small correction to our result.

4. Discussion

A caveat to our derivation is that validity of kinetic equations (3a) and 3(b) requires the uncertainty in electronic energies due to collisions to be smaller than the interval of electronic energies that we are interested in, that is $1/\tau_{ee} \ll \Gamma$. According to Eq. (13), this condition is satisfied at sufficiently high injection intensities. However, semiconductor lasers at room temperature are typically in the opposite regime $1/\tau_{ee} \gg \Gamma$. In this regime, we are concerned with the details of the electron distribution function resolved on a much finer scale than that on which the single electronic states themselves are well defined. We are not aware of any example in kinetic theory where a quantitative description of such a situation has been developed. Its qualitative physical aspects can to some extent be captured in a model that relaxes the strict energy conservation in single-electron transitions,

$$\delta(\varepsilon - \omega) \Rightarrow \Theta(\varepsilon - \omega) \sim \tau_{ee}/[\tau_{ee}^2((\varepsilon - \omega)^2 + 1]. \tag{14}$$

Although this model will give a somewhat different expression for the linewidth compared to Eq. (13), it is clear that Γ will remain an increasing function of I.

Available experiments lend conclusive support neither to our result nor to STLH theory. Experimentally, at low intensities above threshold one observes a decreasing linewidth, but at higher intensities the linewidth often saturates and then re-broadens, so that $\Gamma(I)$ exhibits a minimum (*e.g.* see Fig. 6.15 in Ref. 5, Fig. 9.11 in Ref. 6, or the more recent data by Su and co-workers[7]). One of the possible scenarios that would reconcile the two pictures is that at low injection, the mean-field linewidth given by the kinetic equation approach happens to be much smaller than the STLH linewidth given by Eq. (2), *i.e.* $\Gamma(I) < \Gamma_{\mathrm{STLH}}(I)$ at least near threshold. In this case, the initial decrease of the linewidth with I could be attributed to a STLH-like mechanism, whereas for larger I the increasing mean-field linewidth takes over and re-broadening is observed.

In the opposite limit, which we find more realistic, there is no range for which STLH can be expected to hold and we would have to conclude that the decreasing linewidth lacks a theoretical explanation. Development of a satisfactory linewidth theory would then require inclusion of additional phenomena that go beyond the kinetic description. We would like to mention here two such phenomena:

- If the spectral width of the laser oscillations is narrower than the energy width of single electron states, which is of order $1/\tau_{ee} \gg \Gamma$, then the energy conservation low should only be satisfied with the precision of $1/\tau_{ee}$. In this case, in Eq. (3a) the term nN, which is responsible for the electron-hole recombination rate, should be replaced by a term proportional to

$$n(t)|\boldsymbol{E}(t)|^2. \tag{15}$$

  When the electric field is monochromatic, these two expressions are identical, and we come back to Eq. (3a). However, for a finite frequency range, inclusion of the term (15) leads to interference between different frequency components of the field. With the electron concentration $n(t)$ beating in time, the problem becomes non-stationary and highly nonlinear. This applies both to the case of frequency-distributed field within a single mode and to the multimode case. In the latter case, a related problem also arises: the dependence of the number of lasing modes on the pumping intensity. This dependence is often nonmonotonic, increasing at small I and decreasing at large I. In a broad sense, it could be interpreted as a narrowing of the total spectral width of laser oscillations. An attempt has been made in Ref. 8 to explain this phenomenon by a mode competition arising from the term (15). As far as we know, however, this problem remains unsolved.

- Different harmonics of lasing radiation have different spatial dependencies. The electron recombination rate depends on the local amplitude of the electromagnetic field. Thus, different harmonics of the electromagnetic field can compete via the spatial dependence of the electron concentration $n(\mathbf{r})$ due to the spatial hole burning. As far as we know, the significance of this effect for the laser linewidth has not been elucidated.

5. Conclusion

We see that the standard theory of laser linewidth is unsatisfactory. The theory attributes the spectral width of laser oscillation to fluctuations brought about by random spontaneous emission events and is essentially based on the assumption that in the absence of fluctuations laser radiation is monochromatic. We have shown that this assumption is inconsistent and that already in the mean-field model the laser oscillations have a finite spectral linewidth that furthermore increases with pumping.

Our discussion was restricted to semiconductor lasers, but our conclusion is likely to be more general, applicable to other lasers as well, such as solid-state lasers and gas lasers. The question of why the laser linewidth can be much narrower than either the gain spectrum or the resonator linewidth is begging a theoretical explanation.

Acknowledgment

We are grateful to R. F. Kazarinov for useful discussions.

References

1. A. L. Schawlow and C. H. Townes, "Infrared and optical masers", *Phys. Rev.* **112**, 1940 (1958).
2. M. Lax, "Classical noise *vs*. noise in self-sustained oscillators", *Phys. Rev.* **160**, 290 (1967); R. D. Hempstead and M. Lax, "Classical noise VI. Noise in self-sustained oscillators near threshold", *Phys. Rev.* **161**, 350 (1967).
3. C. H. Henry, "Theory of linewidth of semiconductor lasers," *IEEE J. Quantum Electronics* **QE-18**, 259 (1982); "Theory of the phase noise and power spectrum of a single-mode injection laser," *IEEE J. Quantum Electronics* **QE-19**, 1391 (1983).
4. A. E. Siegman, *Lasers*, Sausalito, CA: University Science Books, 1986.
5. G. P. Agrawal and N. K. Dutta, *Semiconductor Lasers*, 2nd ed., New York: Van Nostrand, 1993.
6. G. Morthier and P. Vankwikelberge, *Handbook of Distributed Feedback Laser Diodes*, Boston: Artech House, 1997.
7. H. Su, L. Zhang, R. Wang, T. C. Newell, A. L. Gray, and L. F. Lester, "Linewidth study of InAs-InGaAs quantum dot distributed feedback lasers," *IEEE Photonics Technol. Lett.* **16**, 2206 (2004).
8. R. F. Kazarinov, C. H. Henry, and R. A. Logan, "Longitudinal mode self-stabilization in semiconductor lasers," *J. Appl. Phys.* **53**, 4631 (1982).

Fermi Liquid Behavior of GaAs Quantum Wires

E. Levy, A. Tsukernik, M. Karpovski, A. Palevski
School of Physics and Astronomy, Tel Aviv University, Tel Aviv 69978, Israel

B. Dwir, E. Pelucchi, A. Rudra, E. Kapon
Ecole Polytechnique Federale de Lausanne (EPFL)
CH-1015 Lausanne, Switzerland

Y. Oreg
Dept. of Condensed Matter Physics
The Weizmann Institute of Science, Rehovot 76100, Israel

1. Introduction

The electrical conductance through noninteracting clean quantum wires containing a number of one-dimensional subbands is quantized in the universal units $2e^2/h$.[1,2] As the number of the subbands is changed, the conductance varies in a step-like manner (with the plateaus at integer values of the universal unit), as was observed in narrow constrictions in 2D electron gas (2DEG) systems.[3,4] Typically, the size of the constrictions is comparable to the Fermi length λ_F, and is much shorter than the mean free path in the 2DEG. For such short and clean narrow wires, the electron-electron interactions described by the so-called Luttinger liquid (LL) model[5] do not affect the value of the conductance, namely it is temperature and length independent, as indeed was shown experimentally.[3,4] In the presence of disorder suppression of the conductance is expected at low temperatures in sufficiently long quantum wires (QWRs) containing at least a few electrons. A number of theoretical papers[6–9] predicted a negative correction to the conductance temperature dependence $G(T)$, which obeys a T^{g-1} power law, where $g < 1$ is an interaction parameter. Eventually, at $T = 0$, the above theories predict a vanishing conductance. However, these theories all use a perturbation treatment of the disorder and are thus limited to the cases of relatively small disorder. No theoretical treatment beyond perturbation theory has been suggested so far.

The validity of the implications arising from the LL theory has been recently demonstrated in a number of experiments.[10, 11] The most clear-cut proof of the theoretical predictions were shown in the *tunneling* experiments performed in T-shaped cleaved-edged overgrown GaAs quantum wires[10] and in carbon nanotubes.[11] Earlier *non-tunneling* experiments, in which suppression of conductance occurs in the linear response regime, did not unambiguously confirm the validity of the theory, and the value of the g parameter could not be deduced from the experimental data.[12–14] Several complications are encountered in such

Future Trends in Microelectronics. Edited by Serge Luryi, Jimmy Xu, and Alex Zaslavsky

experiments. For sufficiently disordered wires, where the correction to the conductance $G(T)$ is expected to be large, the value of the conductance at the plateau is not well defined due to the specific realization of the disordered potential in the wire, as was the case for the long wires of Tarucha *et al.*[12] Moreover, in the intermediate regime, namely, for the disorder level for which the plateau could be well defined but the corrections to $G(T)$ are already significant for a relatively narrow temperature range, the parameter g cannot be extracted by applying perturbation theory. The attempt to extract g from such samples by applying the theory in the limit of weak disorder results in reduced values of g, as we believe was the case in Ref. 14. If, however, the disorder is very weak so that the plateaus are well defined at all temperatures,[12,13] the variation of its value *vs.* temperature is so weak that the g parameter cannot be reliably determined. Therefore, if one wishes to compare $G(T)$ to the theory, a wire possessing just the right amount of disorder is needed in order to avoid the above difficulties. On one hand, the disorder should be weak enough to give well defined plateaus, while on the other hand, the overall weak variation of $G(T)$ should be observed over an extended temperature range so that the perturbation theory could be applied.

In this chapter, we present an experimental study of the conductance in single mode GaAs QWRs grown on a V-groove substrate. The variation of the conductance was measured over a wide temperature range. Our results are consistent with the theories[8,9] based on the LL model for weakly disordered wires, allowing us to deduce the value of $g = 0.66$, as expected for interacting electrons in GaAs and as was observed experimentally in tunneling experiments.[10] We show results coming from QWRs displaying varying amounts of disorder, thus enabling us to elucidate the importance of the degree of disorder in fitting to perturbation theory and to show its limits.

2. Sample preparation and the experimental setup

The QWRs studied in this chapter were produced by self-ordered growth of GaAs/AlGaAs heterostructures, using low pressure (20 mbar) metalorganic vapor phase epitaxy (MOVPE) on undoped (001) GaAs substrates with a few isolated V-grooves oriented in the $[01\bar{1}]$ direction, fabricated by electron-beam and UV lithography, followed by wet chemical etching.[15] The heterostructure consisted of a 230 nm GaAs buffer layer, followed by a 1 μm thick $Al_{0.27}Ga_{0.73}As$ layer. The carriers are supplied to the quantum well from 20 nm Si-doped ($\approx 1\times10^{18}$ cm^{-3}) $Al_{0.27}Ga_{0.73}As$ layers on both sides of the GaAs quantum well layer, spaced by 80 and 60 nm, respectively, in the growth order. The quoted thicknesses are "nominal", as calibrated on a planar (100) sample. Figure 1(a) shows a cross-sectional TEM image of the QWR region of a typical sample used in our experiments. We observe that the thickness of the quantum well varies along the direction perpendicular to the plane, thus strengthening the confinement. In this way, the QWR is formed at the bottom of the V-groove. For the transport measurements, multiple contact samples were fabricated using standard photolithography techniques with a mesa etched along the QWR and with Au/Ge/Ni ohmic contacts, as shown in Fig. 1(b). Additionally, narrow (0.5 μm) Ti/Au Schottky gates were formed using electron beam lithography in order to isolate the QWR and control the number of 1D subbands in it.

The conductance of the wire was measured using a four-terminal ac method and lock-in amplifier detection (schematically shown in Fig. 1(b)). If the 2DEG is not depleted in the sidewalls, the electronic transport in our system is carried by both the electrons in the 2D sidewalls and in the 1D quantum wire in parallel. By applying negative voltage to the Schottky gate deposited on top of the mesa, one can fully deplete the 2DEG in the sidewalls and thus isolate a section in which there are electrons only in the 1D wire.[16] Over a certain range of the negative gate voltage, a single populated 1D channel is realized. It would be natural to think that the length of the 1D region is determined by the width of the gate. However, as was demonstrated in our previous experiments,[17] the electrons remain in their one-dimensional state over a transition length Δ on both sides of the gate – see Fig. 1(b). This transition length arises from the poor coupling between the 1D states and the 2DEG that acts as an electron reservoir and is found to be as large as $\Delta = 2$ μm. From the above discussion it is reasonable to conclude that the effective length of the 1D wire (although not accurately defined) markedly exceeds the actual width of the gate (0.5 μm).

3. Experimental results

Figure 2 shows the variation of the conductance with gate voltage V_G, in the range where electrons populate only a single 1D subband, at temperatures between 100

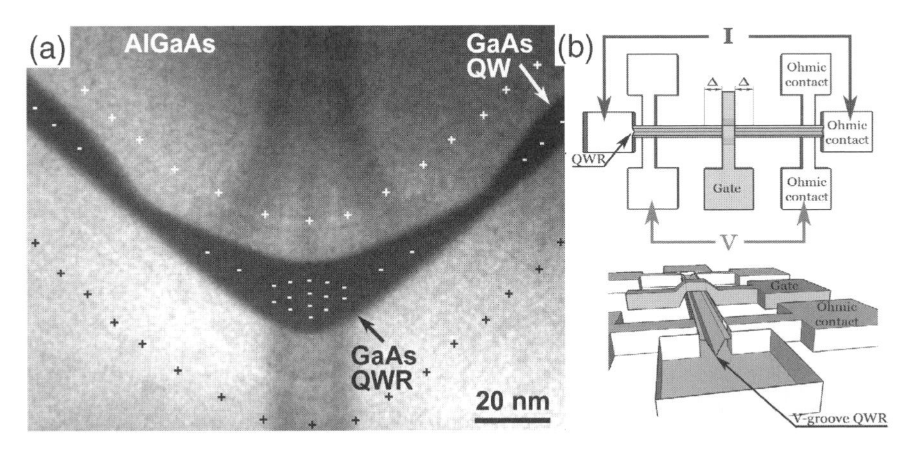

Figure 1. (a) Cross-sectional TEM image of the QWR region, on which the charge distribution due to the doping is schematically shown. (b) Top and perspective schematic views of the device's geometry. The QWR is present at the bottom of the V-groove aligned with the mesa structure.

mK and 4.2 K. The electronic temperature of GaAs 2DEG does not deviate from the bath temperature for $T > 100$ mK, as shown in our previous studies.[18] The data were taken at stabilized temperatures of the bath while the V_G was swept through the entire range. A series resistance of 180 Ω, measured at $V_G = 0$, has been subtracted from all curves. At $T = 4.2$ K, the conductance plateau is smooth with $G = 0.94(2e^2/h)$, indicating that only weak disorder is present in our samples. At lower temperatures, some small undulations of the conductance values appear on the plateau, but its average value is well defined with the standard deviation being much less than the average value (see error bars in Figs. 3 and 4). A similar phenomenon, namely the appearance of such structures at lower temperatures and their disappearance at higher temperatures, was also recently observed in clean cleaved-edged overgrown wires.[19] The variation of the plateau value (approximately 20%) through the wide temperature range (1.5 decades), allows us to make a meaningful comparison of the data to the theories derived in the appropriate limit of weak disorder. Figure 3 shows the measured variation of conductance versus temperature.

Early theories, particularly those of Kane and Fisher[6] (and of Ogata and Fukuyama[7]), proposed that for relatively small barriers (weak disorder, which is assumed to result in relatively small corrections), the conductance of a sufficiently long, single-mode 1D spin-full Luttinger liquid system decreases with temperature in the manner

$$G'(T) = g(2e^2/h)[1 - (T/T_0)^{g-1}] . \tag{1}$$

Here, $g < 1$ is a dimensionless parameter, which is a measure of the strength of the interactions. For repulsive interactions, g obtains roughly from the expression $g^2 = [1 + (U/2E_F)]^{-1}$, where U is the Coulomb interaction energy between neighboring

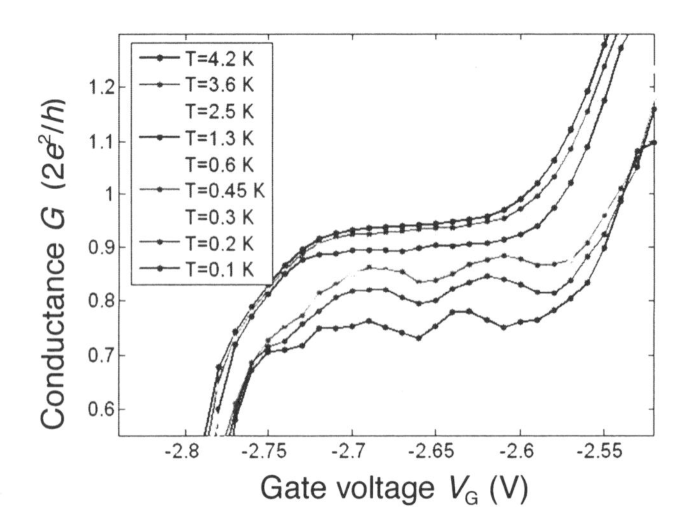

Figure 2. Conductance *vs.* gate voltage V_G for 0.5 μm gate width at various temperatures, after subtraction of series resistance.

electrons. The parameter T_0 in Eq. (1) describes the strength of the backscattering (disorder) in the wire; at $T \sim T_0$, the corrections to $G(T)$ become of order $(2e^2/h)$. Both theories predict a correction of $g(2e^2/h)$ even for ballistic wires at relatively high temperatures. These imply that for sufficiently long wires, one cannot observe values close to $2e^2/h$ in GaAs, since the value of g is expected to be ~0.7 in such wires, as was already pointed out by Tarucha *et al.*[12]

This contradiction was also addressed in detail in several theoretical papers.[8,9,20-22] According to the theory of Maslov,[9] the interaction parameter g of the wire determines the exponent of the temperature variation, whereas the pre-factor g in Eq. (1) should be set to one (for non-interacting reservoirs). Figure 3 shows the curve calculated from this modified equation (dashed line).

A different but numerically equivalent result was derived by Oreg and Finkel'stein.[9] They also demonstrated that for an infinite clean wire, the conductance keeps the universal value $2e^2/h$ per mode, even in the presence of interactions. According to their theory, because of the electric field renormalization by the interactions, the results given by Kane and Fisher[5] of Eq. (1) are modified in the following way:

$$G(T)=2gG'(T)/[(h/e^2)(g-1)G'(T)+2g] \tag{2}$$

As can be easily verified, the leading term in the temperature variation of the conductance of Eq. (2) leads to the same results given by Maslov.[8] As one can see from Fig. 3, an excellent fit is obtained for both theories,[8,9] and we obtain g = 0.64±0.05, as is expected for electrons in GaAs wires. Indeed, this value is consistent with the experiments in Ref. 10, showing g values between 0.66 and 0.82. Moreover, using the Fermi energy $E_F \approx 1.5\pm0.5$ meV (half of level spacing

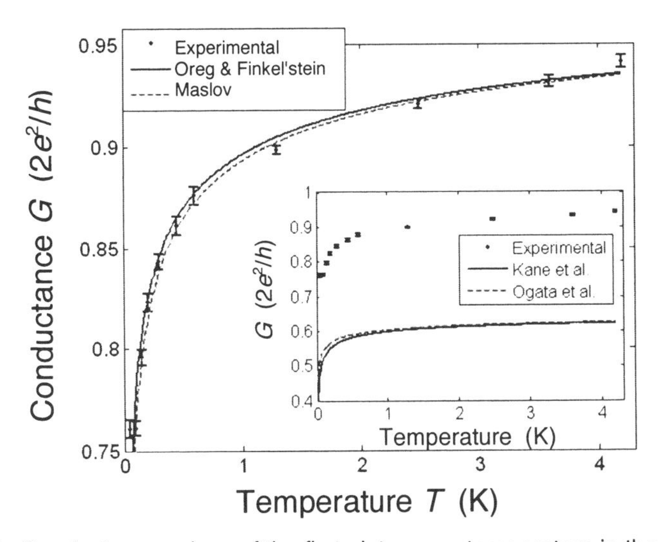

Figure 3. Conductance values of the first plateau *vs.* temperature in the wire of Fig. 2 (points with error bars). Both theoretical expressions are plotted for the same parameters, *e.g.* $g = 0.64$ and $T_0 = 0.7$ mK of Eq. (2).

between 1D subbands estimated in our previous experiments[16]) we calculate the corresponding electron densities at the middle of the plateau, obtaining $n_{1D} = 3.2\pm0.5\times10^5$ cm^{-1}. Substituting the above value for n_{1D} into $U = (e^2/\varepsilon)n_{1D}$ we obtain $U = 3.85\pm0.60$ meV which yields the values of $g = 0.66\pm0.04$, consistent with our fit to the LL model.

Disorder in V-groove QWRs stems mainly from interface roughness brought about by lithography imperfections on the patterned substrate and peculiar faceting taking place during MOVPE on a nonplanar surface.[23] The disorder results in potential fluctuations along the axis of the wire, and manifests itself in localization of excitons and other charge carriers as evidenced in optical spectroscopy studies of these wires.[24] Optical and structural studies indicate the formation of localizing potential wells along the wires with size in the range of several tens of nm.[25] The specific features of the disorder in the QWRs studied here, in terms of depth and size of the localization potential, are expected to vary from sample to sample. In fact, the degree of disorder is represented in our analysis of the temperature dependence of the conductance by the parameter T_0. Repeating the analysis of Fig. 3 for several samples, we observed similar values of g in all the wires characterized by a small amount of disorder ($T_0 < 2$ mK), namely $g = 0.66$. However, other wires with stronger disorder ($T_0 > 2$ mK), showed lower values of g, around $g = 0.5$. Figure 4 summarizes the values of g *vs.* T_0, obtained for our different wires.

The values of the total change in $\Delta G/G$ were calculated for each wire in the 0.1–4.2 K temperature range and are also shown in Fig. 4. Note that there is a complete correspondence between the two indicators for the strength of the disorder, T_0 and $\Delta G/G$.

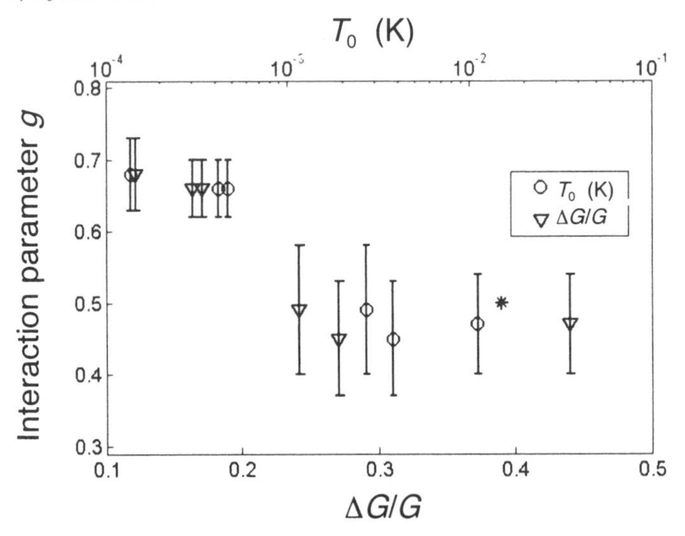

Figure 4. Interaction parameter g vs. disorder parameter T_0 and the values of $\Delta G/G$ (in the temperature range 0.1–4.2 K). The star represents an estimate for the strength of the disorder for the results reported in Ref. 14. The wrong values of $g \approx 0.5$ (at high disorder) are established by using the perturbation formula in a region where it is inapplicable.

The transition between $g = 0.66$ and $g = 0.5$ at $T_0 \approx 2$ mK occurs for $\Delta G/G \approx$ 23%. We believe that above $T_0 \approx 2$ mK the disorder in the wires is strong enough so that the description by perturbation theory is no longer valid. Trying to fit such data with perturbation-theory equations gives inevitably lower (and wrong) values of g. For such wires, one should use other theories, appropriate to stronger disorder due to many impurities,[25] or stronger backscattering[26] in the system. The results of conductance measurements in GaAs wires reported recently by Rother *et al.*[14] also correspond to highly disordered samples, and also give $g = 0.5$. Indeed, analyzing their data, we estimate the value of $T_0 \approx 15$ mK (marked by a star in Fig. 4) and the change in the conductance $\Delta G/G \approx 10\%$ over a small temperature range (1–3 K). These values are even larger than corresponding values for our most disordered sample in the same temperature range.

It is highly unlikely that the observed temperature dependence could be attributed to the contact resistance between the 2DEG and the 1D subbands outside the gated region for the reasons outlined below:

i) if the contact resistance is treated quantum mechanically,[13] as a change of the transmission from the 2DEG to the 1D subbands in the ungated region, we would expect that $\Delta G/G$ would be similar for any number of 1D channels under the gate. However, we observe that $\Delta G_1/G_1$ of the first plateau is much smaller than $\Delta G_2/G_2$ of the second plateau over the same temperature range. This is consistent with the expected result of the Luttinger model where the scattering occurs in the gated region, since the effect of the Coulomb interaction on the transmission depends on the number of 1D subbands. Indeed, from an analysis of higher steps in the conduction depletion curve, used in a smaller temperature range (0.1–0.6 K, where the plateaus are better resolved), we deduce the values $g = 0.55$ and $g = 0.47$ for the second and the third plateaus, respectively, which agrees with the theoretical values of 0.54 and 0.47.[9]

ii) if the decrease of the conductance is considered as an additional contact resistance added in series to the wire (*i.e.* treated classically), than the values of the transmission for each channel at low T at the second plateau would increase with lowering temperature and eventually exceed unity for each channel. Therefore, we conclude that the observed decrease of the conductance is due to the interactions in the LL model.

4. Conclusions

In conclusion, we have measured the temperature dependence of the electrical conductance in single mode quantum wires. We find that our data is consistent with theoretical calculations[8,9] based on the LL model, in the limit of weak disorder in the system. We showed that the use of the perturbative result (namely $G' \propto T^{g-1}$) in order to estimate g, is valid only for wires produced with a *moderate* amount of disorder ($T_0 < 1$ mK).

Acknowledgments

We thank Dganit Meidan for constructive discussions of our results. This research was partially supported by the Israel Science Foundation, founded by the Israeli academy Sciences and Humanities Centers of Excellence Program and by ISF grant 845/04.

References

1. R. Landauer, "Spatial variation of currents and fields due to localized scatterers in metallic conduction," *IBM J. Res. Develop.* **32**, 306 (1988).
2. C. W. J. Beenakker and H. van Houten, *Quantum Transport in Semiconductor Nanostructures*, Vol. 44 in: H. Ehrenreich and D. Turnbull, eds., *Solid State Physics* series, New York: Academic Press, 1991.
3. B. J. van Wees, H. van Houten, C. W. J. Beenakker, *et al.*, "Quantized conductance of point contacts in a two-dimensional electron gas," *Phys. Rev. Lett.* **60**, 848 (1988).
4. D. A. Wharam, T. J. Thornton, R. Newbury, *et al.*, "One-dimensional transport and the quantisation of the ballistic resistance," *J. Phys. C* **21**, L209 (1988).
5. S. Tomonaga, "Remarks on Bloch's method of sound waves applied to many-fermion problems," *Prog. Theor. Phys.* **5**, 544 (1950).
6. C. L. Kane and M. P. A. Fisher, "Transmission through barriers and resonant tunneling in an interacting one-dimensional electron gas," *Phys. Rev. B* **46**, 15233 (1992).
7. M. Ogata and H. Fukuyama, "Collapse of quantized conductance in a dirty Tomonaga-Luttinger liquid," *Phys. Rev. Lett.* **73**, 468 (1994).
8. D. L. Maslov, "Transport through dirty Luttinger liquids connected to reservoirs," *Phys. Rev. B* **52**, R14368 (1995).
9. Y. Oreg and A. M. Finkel'stein, "dc transport in quantum wires," *Phys. Rev. B* **54**, R14265 (1996).
10. O. M. Auslaender, A. Yacoby, R. de Picciotto, K. W. Baldwin, L. N. Pfeiffer, and K.W. West, "Experimental evidence for resonant tunneling in a Luttinger liquid," *Phys. Rev. Lett.* **84**, 1764 (2000).
11. M. Bockrath, D. H. Cobden, J. Lu, *et al.*, "Luttinger-liquid behavior in carbon nanotubes," *Nature* **397**, 598 (1999).
12. C. Tarucha, T. Honda, and T. Saku, "Reduction of quantized conductance at low temperatures observed in 2 to 10 μm-long quantum wires," *Solid State Commun.* **94**, 413 (1995).
13. A. Yacoby, H. L. Stormer, N. S. Wingreen, L. N. Pfeiffer, K. W. Baldwin, and K. W. West, "Nonuniversal conductance quantization in quantum wires," *Phys. Rev. Lett.* **77**, 4612 (1996).
14. M. Rother, W. Wegscheider, M. Bichler, and G. Abstreiter, "Evidence of Luttinger liquid behavior in GaAs/AlGaAs quantum wires," *Physica E* **6**, 551 (2000).

15. F. Gustafsson, F. Reinhardt, G. Biasiol, and E. Kapon, "Low-pressure organometallic chemical vapor deposition of quantum wires on V-grooved substrates," *Appl. Phys. Lett.* **67**, 3673 (1995).
16. D. Kaufman, Y. Berk, B. Dwir, A. Rudra, A. Palevski, and E. Kapon, "Conductance quantization in V-groove quantum wires," *Phys. Rev. B* **59**, R10433 (1999).
17. D. Kaufman, B. Dwir, A. Rudra, I. Utke, A. Palevski, and E. Kapon, "Direct evidence for quantum contact resistance effects in V-groove quantum wires," *Physica E* **7**, 756 (2000).
18. M. Eshkol, E. Eisenberg, M. Karpovski, and A. Palevski, "Dephasing time in a two-dimensional electron Fermi liquid," *Phys. Rev. B* **73**, 115318 (2006).
19. R. de Picciotto, L. N. Pfeiffer, K. W. Baldwin, and K. W. West, "Nonlinear response of a clean one-dimensional wire," *Phys. Rev. Lett.* **92**, 036805 (2004).
20. D. L. Maslov, and M. Stone, "Landauer conductance of Luttinger liquids with leads," *Phys. Rev. B* **52**, R5539 (1995).
21. I. Safi and H. J. Schulz, "Transport in an inhomogeneous interacting one-dimensional system," *Phys. Rev. B* **52,** R17040 (1995).
22. Y. Oreg and A. M. Finkel'stein, "Interedge interaction in the quantum Hall effect," *Phys. Rev. Lett.* **74**, 3668 (1995).
23. F. Lelarge, T. Otterburg, D. Y. Oberli, A. Rudra, and E. Kapon, "Origin of disorder in self-ordered GaAs/AlGaAs quantum wires grown by OMVPE on V-grooved substrate," *J. Crystal Growth* **221**, 551 (2000).
24. F. Vouilloz, D. Y. Oberli, F. Lelarge, B. Dwir, and E. Kapon, "Observation of many-body effects in the excitonic spectra of semiconductor quantum wires," *Solid State Commun.* **108**, 945 (1998).
25. D. Y. Oberli, M.-A. Dupertuis, F. Reinhardt, and E. Kapon, "Effect of disorder on the temperature dependence of radiative lifetimes in V-groove quantum wires," *Phys. Rev. B* **59**, 2910 (1999); T. Otterburg, D. Y. Oberli, M.-A. Dupertuis, *et al.*, "Enhancement of the binding energy of charged excitons in disordered quantum wires," *Phys. Rev. B* **71**, 033301 (2005).
26. I. V. Gornyi, A. D. Mirlin, and D. G. Polyakov, "Dephasing and weak localization in disordered Luttinger liquid," *Phys. Rev. Lett.* **95**, 046404 (2005).
27. P. Fendley, A. W. W. Ludwig, and H. Saleur, "Exact nonequilibrium transport through point contacts in quantum wires and fractional quantum Hall devices," *Phys. Rev. B* **52**, 8934 (1995).

Part II

Biology: We Are All Zoa

2 Biology: We Are All Zoa

Biology in microelectronics? Or, microelectronics in biology? Either way it is an intriguing notion, and very possibly a reality. Indeed, it is intriguing and real enough a subject to comprise a standalone part in this volume.

In many ways the celebrated nanotechnology may be viewed as a natural evolution of human capability from the micro to nano regime that the microelectronics has entered over a decade ago. In this light it is really more of an evolution than a revolution. Biology, on the other hand, is a space where there may be a real chance of a revolution created by the unprecedented access and control at the same size scale as the basic units of life – proteins and biomolecules. Signs of revolution in this space are emerging and can be seen throughout this chapter – direct electronic interface with and read-out from neurons in living human brains; synthesis of viruses as "chemicals with life"; and guided evolution of colonies of photobacteria. By breaking into biology, micro- and nanoelectronics also break out of their traditional realm of information processing and transmission to reach into the equally rewarding but so far under-explored spaces of information acquisition and information execution. Molecular markers, molecular imaging, and molecular diagnostics are examples highlighted in this chapter, not surprisingly aimed above all at applications in healthcare – a sector of the economy that is projected to grow to 25% of the GDP of industrialized countries in the next decade and one that could displace information technology as the dominant sector of economy.

Contributors

2.1 H. van Houten and H. Hofstraat

2.2 A. V. Nurmikko, W. R. Patterson, Y.-K. Song, C. W. Bull, and J. P. Donoghue

2.3 S. Mueller, J. R. Coleman, J. Cello, A. Paul, E. Wimmer, D. Papamichail, and S. Skiena

2.4 R. H. Austin, P. Galajda, and J. Keymer

2.5 C. Weisbuch, A. David, M. Rattier, L. Martinelli, H. Choumane, N. Ha, C. Nelep, A. Chardon, G.-O. Reymond, C. Goutel, G. Cerovic, and H. Benisty

Future Trends in Microelectronics. Edited by Serge Luryi, Jimmy Xu, and Alex Zaslavsky

ISBN 0-471-48

Towards Molecular Medicine

Henk van Houten and Hans Hofstraat
Philips Research, High Tech Campus 34, Eindhoven, The Netherlands

1. Introduction

Historically, the human species has lived with the expectation of dying quite young from violent external factors or from infectious diseases. Considerable progress in world health arose eventually from improvements in living conditions – hygiene, access to safe drinking water, improved quality and variety of nutrition. However, many major infectious diseases, such as tuberculosis or syphilis, were not understood, and could not really be treated. In most cases, the medical profession could only offer palliative care. As Lewis Thomas recalls the years of his training to become a medical doctor in the early 1930's: "... it gradually dawned upon us that we didn't know much that was really useful, that we could do nothing to change the course of the great majority of the diseases we were so busy analyzing, that medicine, for all its façade as a learned profession, was in real life a profoundly ignorant occupation."[1] Medicine only started to be a science with the realization that diseases have their origin in microbiological processes. Since the 1930's, much progress has been made, with the discovery of antibiotics and the development of very effective vaccines.

As a result, life expectancy has increased dramatically, first in Europe and North America, but now also in countries such as China. Unfortunately, the increased life expectancy is going hand in hand with an increase in the number of people affected by chronic or degenerative diseases, which also are at the origin of the current explosion of costs in the healthcare system. Again, treatment for these diseases is typically palliative in nature – mostly there is no cure and at best the progression of the disease can be delayed. In addition, the development and regulatory approval of new drugs is hampered by the fact that the effectiveness and side effects are not the same for all patient groups, leading to very costly trials and late-stage drug withdrawals. Furthermore, bacterial and viral infections still pose very serious threats, ranging from hospital infections to viral pandemics.

Further progress can only be expected from a much more sophisticated understanding and exploitation of molecular biology. It is now well accepted that most diseases have their origin in disturbances of the delicate balance in molecular processes taking place at the cellular and sub-cellular level. This is currently leading to a paradigm shift in medicine: from a focus on dysfunctional organs to an understanding of disease pathways at the cellular and molecular level. The vision is that genetic predisposition testing, early diagnosis using molecular tests, and personalized treatment will transform clinical practice, and lead to improved

 Future Trends in Microelectronics. Edited by Serge Luryi, Jimmy Xu, and Alex Zaslavsky

patient outcomes. Aspects of this vision are referred to as evidence-based medicine, personalized medicine, molecular medicine, or nanomedicine.[2,3]

The promise of molecular medicine is illustrated in Fig. 1: early and faster diagnosis, better prognosis, and tailored therapy with higher efficacy and reduced side effects as compared to the present state-of-the-art, which is based on treatment of the patient on a trial and error basis after serious symptoms have developed.

It is our firm belief that the vision of nanomedicine can only be realized by matching the progress in molecular biology with advances in medical microdevices, nanotechnology and physical instrumentation, and by inventing new ways of dealing with complex data in support of decision-taking, a field referred to as bioinformatics.

2. System biology and biomarkers

The central dogma of molecular biology is that the genetic make-up of the individual, primarily given by hereditary factors, is laid down in the DNA, and subsequently transcribed by RNA into proteins, the molecules that are instrumental in all major biological processes taking place in human cells, tissues and organs.

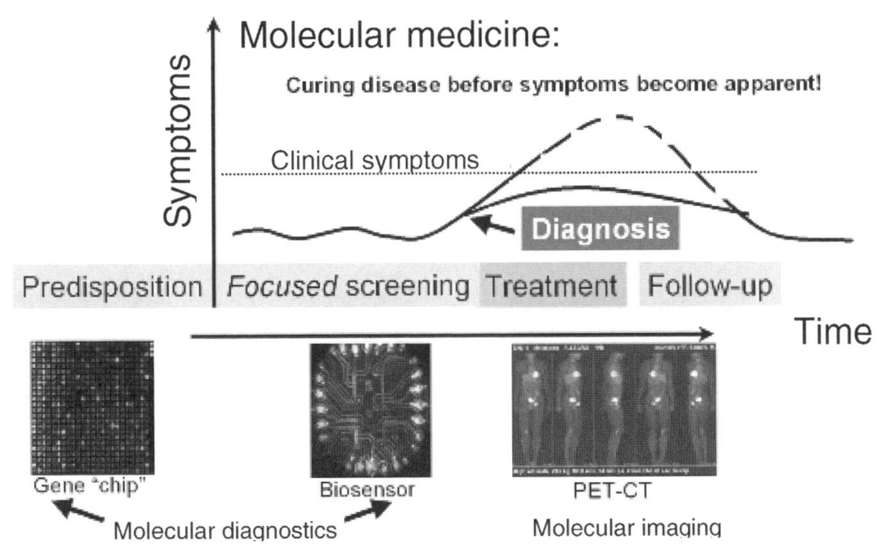

Figure 1. Molecular diagnostics and molecular imaging, coupled to therapy, will change the current practice of health care.

The basis for nanomedicine is the explosive growth in knowledge of the structure of the human genome, and its translation into *functional* elements, the proteins. Advances in technology have led to elucidation of the genetic make-up of many species, including humans as a result of the ambitious human genome project. Similarly, efforts are ongoing to establish RNA patterns (the "transcriptome") and – more challenging still – get insight into the range of proteins present in a human being (the "proteome").

Knowledge of the genome, transcriptome and proteome by itself is not sufficient, however. The next step requires relating the knowledge of the molecular translation cycle to the onset, development and ultimately treatment of disease. This step is still extremely complex and poorly understood. Gaining insight into the functioning of protein signalling, and its impact on cell multiplication, interaction and transformation forms the main challenge of an emerging branch of science called systems biology. In systems biology, the aim is to integrate all relevant genomic, transcriptomic, and proteomic information with the metabolic processes in our cells and organs ("metabolomics") into a consistent model of the intricate biological processes that describe the functioning of the human body. Disturbances of these intrinsic processes are at the origin of disease. External factors, such as nutrition and lifestyle, play an important role as well, as they can have profound effects on the structure of the genome (*e.g.* modification of DNA by methylation, "epigenetics"), and will translate into the proteome. A proper understanding of these environmental factors is required to really get a handle on the origin of most diseases, and to be able to devise effective cures.

In an effort to reach this goal one tries to relate the wealth of information, which is available from patient samples (information from healthy and diseased tissue, generally comprising DNA, RNA, proteins and metabolites), to clinical information. The key challenge is to discover biomarkers, which are observable signatures of relevant disease pathways.

Biomarkers are so important because they are characteristic of a particular disease, and can be used for early diagnosis. Such biomarkers can be discovered in bodily fluids, *e.g.* in blood or serum, so that they may be determined by *in-vitro* diagnostic approaches, but also in tissue or organs, providing handles for targeted contrast agents, which can be visualized *in-vivo* by making use of advanced imaging instrumentation.

Specific biomarkers can be employed for early diagnosis and for monitoring of diseases, but they can also be used to accelerate the process of drug discovery and development: by using biomarkers as "surrogate endpoints" in clinical trials, drug effectiveness (and toxicology, or other side effects) can be detected much earlier than by such conventional metrics as the five-year survival rate. Here the challenge is to link the rich "molecular" information to the relatively scarce patient data, which is furthermore complicated by the inherent biological variability and the environmental factors mentioned before. Bioinformatics plays a central role in coming to grips with this complexity, as illustrated in Fig. 2.

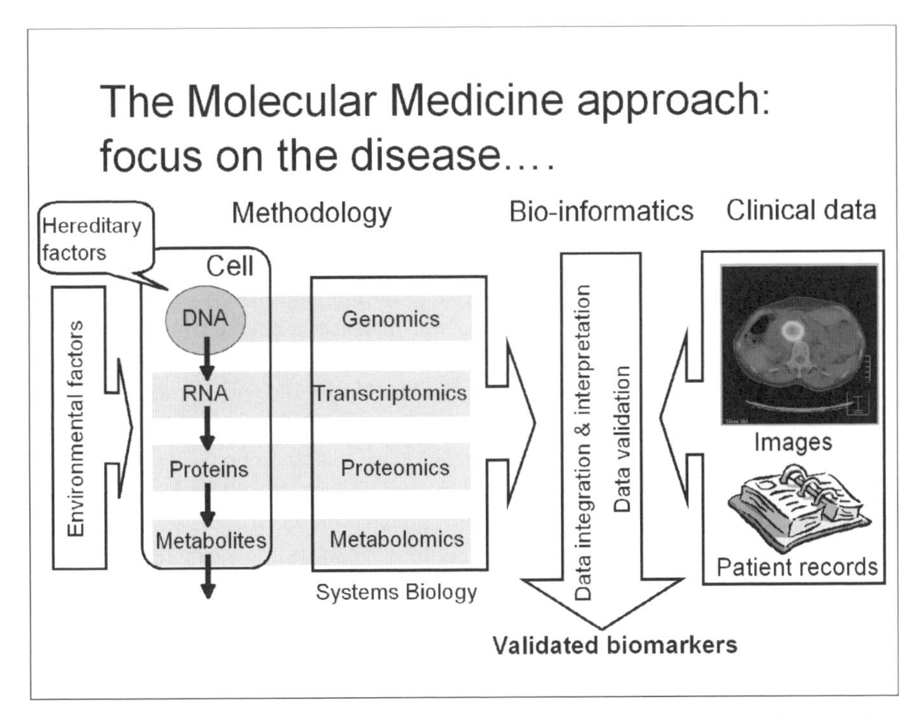

Figure 2. Validated biomarkers are key in the successful introduction of molecular medicine. Their identification requires the interpretation of large and complicated data sets, with the help of bio-informatics tools.

3. Molecular medicine

The major life-threatening and chronic diseases, such as cardiovascular disease, cancer, diabetes, and infectious diseases (tuberculosis, malaria, AIDS) are at least partially genetically determined. The same is true for the major debilitating diseases, such as neuro-degenerative diseases (Alzheimer's, Parkinson's) and autoimmune diseases (rheumatoid arthritis). Early detection of these diseases greatly improves the therapeutic success rate, leading to a prolongation of the healthy and productive lifespan of the individual, and treatment with fewer side effects. In addition, it has a potential cost-containment effect as well: particularly, delaying the onset of debilitating diseases results in a significant reduction of the very high personnel costs associated with nursing the patients.

A more personalized approach of tailored care for every individual will become the standard. Introduction of targeted drugs to block receptors in the membranes of tumor cells, for instance, may result in slowing down tumor cell proliferation or even their elimination. Apart from more effective treatment, some cancer types may well be contained – effectively turning cancer in a manageable "chronic" disease. The first successful targeted drugs have already been introduced. An example is the drug imatinib (Gleevec, by Novartis), developed after the discovery of a chromosome translocation creating a new gene structure, the abl-bcr gene, in chronic myeloid leukemia patients. Gleevec binds specifically to the abl-bcr protein, and can alleviate leukemia in patients for whom other treatments have failed.[4] Other examples of targeted drugs are Herceptin (produced by Genentech/Roche to treat metastatic breast cancer) and the non-Hodgkin's lymphoma drugs Bexxar (GlaxoSmithKline) and Zevalin (BiogenIdec). All these drugs are based on monoclonal antibodies, which bind selectively to tumor cells, and may be equipped with toxic substances to enhance their efficiency (*e.g.* in Bexxar radioactive ^{131}I is present to invoke radio immunotherapy). Such targeted or "smart" drugs can be quite expensive. For instance, a course of treatment with Zevalin and Bexxar amounts to $25–30K per patient for the medication. It is therefore very important for financial reasons to identify the patients who are likely to respond well to the medication prior to the treatment, and this can be done with molecular diagnostic tests.

Traditional medical practice, based on trial-and-error, results both in under-treatment and over-treatment, multiple office visits, the need for drug monitoring, frequent regimen changes, and a staggering number of deaths attributed to adverse drug reactions (more than 100,000 annually in the U.S.A. alone).[2] We expect that molecular medicine will dramatically change the healthcare system, by alleviating many of these problems.

A secondary opportunity may be the application of molecular medicine techniques to accelerate and simplify the drug discovery and development process, driven by collaboration of pharmaceutical and biotech companies on the one hand, and medical technology companies on the other.

Molecular medicine is enabled by two key medical technologies, see Figure 3. *In-vitro* molecular diagnostics is a technique for screening and monitoring to enable early and precise detection of disease. *In-vivo* molecular imaging is a technique to localize, image, and interpret the disease as it takes place in the body. This requires the right combination of advanced imaging equipment with targeted and/or functional contrast agents. Molecular imaging offers unique opportunities for combination with (targeted) therapy, which can be much better planned and monitored with the help of advanced hardware and especially software tools utilizing pharmacodynamic modelling. Typically, molecular diagnostics and molecular imaging will be applied in tandem, with the goal of providing tailored solutions for a wide range of diseases.

4. Molecular diagnostics

In-vitro diagnostic approaches will become indispensable for early diagnosis, for the selection of personalized therapy, and for effective follow-up, after completion of the treatment or to support management of a chronic condition. A distinction should be made between techniques applied for the identification of genomic fingerprints and methods suitable for identification of particular biomarkers.

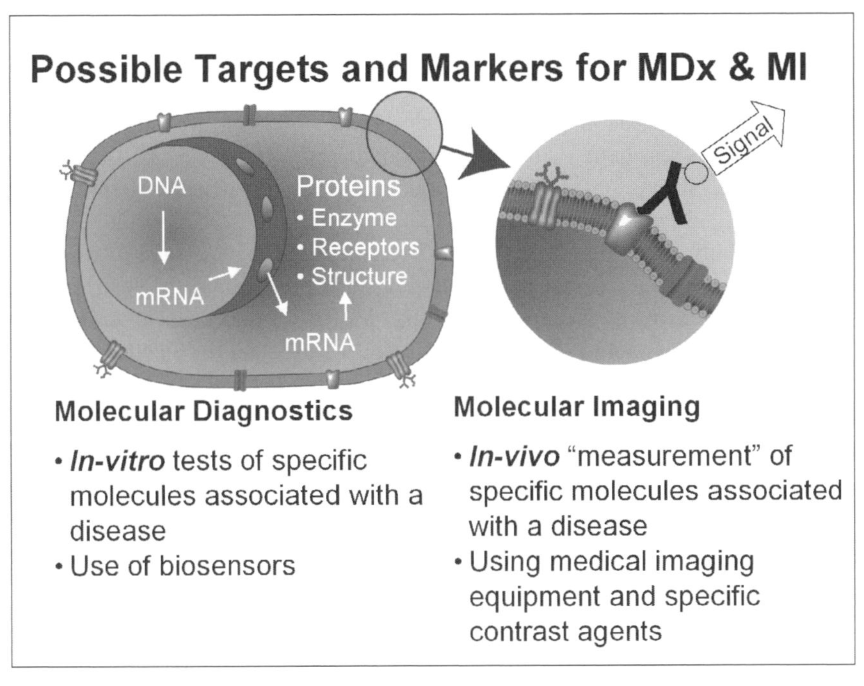

Figure 3. Molecular diagnostics (MDx) and molecular imaging (MI) are the key technologies enabling molecular medicine. In both technologies characteristic and validated biomarkers are needed. The diagram illustrates an antibody with a signalling label, bound selectively to a disease specific molecule expressed at the cell membrane.

Genomic fingerprints thus far have been mostly applied to identify pathogens. In particular, tests are commercially available for the human papilloma virus, for various forms of the human immunodeficiency virus, and for hepatitis B and C. Diagnostic products for infectious diseases therefore dominate the market at present. Detection is predominantly based on amplification of characteristic nucleotide sequences using the polymerase chain reaction (PCR), followed by a hybridization assay.

However, genomic fingerprints are increasingly being utilized to assay the molecular make-up of the host, rather than the pathogen. They are applied to phenotype individuals to identify their predisposition to particular diseases or to tailor individual therapeutic interventions (*e.g.* selection of the appropriate dose of medication on the basis of metabolic characteristics). The resulting "pharmaco-genomic" fingerprints rely on the application of high-density arrays (*e.g.* the GeneChips provided by Affymetrix, or the DNA Microarrays sold by Agilent). Typically, these high-density arrays contain many thousands of different oligonucleotide strings found at different known locations. The presence of complementary oligonucleotides in the sample can be measured optically, through sensitive detection of fluorescent labels; even single mismatches, so-called single nucleotide polymorphisms, can be identified. By careful execution of the measurement protocol, genetic expression profiles highlighting up-regulation or down-regulation of certain parts of DNA or RNA can also be made visible. The observed features can be applied for diagnostic classification, treatment selection and prognostic assessment.[5] In Fig. 1 an image of (part of) a DNA "chip" is shown. Other technologies currently gaining ground, particularly for cancer diagnostics, are *in-situ* hybridization and fluorescent *in-situ* hybridization.

Alternatively, the measurement can be focused on the identification of proteomic biomarkers. Proteomic biomarkers are proteins, such as membrane proteins, triggered by or synthesized in response to disease. Examples are the proteins that signal apoptosis, or programmed cell death, and the enzymes that are released following a stroke or a myocardial infarction. Generally, well-established immunological techniques, such as the widely applied enzyme-linked immunosorbent assay ("ELISA"), are used for protein diagnostics, all based on the application of highly specific antibodies. For many diseases it is necessary to determine a multitude of proteins and, sometimes, additional biomarkers, which requires development of new methodologies. High-throughput analysis of proteins can be applied for the detection of novel drug targets, diagnostic markers, and for the investigation of biological events.[5] Proteomics has the potential of becoming a very powerful tool in modern medicine, but it is still very much under development.

Another kind of biomarker is the presence of a particular kind of pathogen, which can be identified following the approach described above, so that immediately the cause of the infection and the optimal cure can be established.

Essential for the massive introduction of molecular diagnostics is the availability of cheaper and more accessible technologies. Rapid, simple to use, self-contained systems are needed. Miniaturized, integrated, "lab-on-a-chip" tools, based on microfluidic solutions and enabled by advances in micro- and nanotechnology, may serve this need. An example is the work done at Philips Research laboratories on magnetic biosensors, which utilize the highly sensitive magnetic field sensor based on the giant magnetoresistance (GMR) effect in conjunction with paramagnetic nanoparticles as labels.[7]

5. Molecular imaging

Medical imaging modalities have become highly advanced systems, combining high performance image acquisition with sophisticated data and image processing to provide ever increasing quality of information to the medical professional. Molecular imaging (MI) is enabled by such systems, in combination with functional and targeted contrast agents.[8,9]

Two nuclear imaging technologies – single photon emission computed tomography (SPECT) and positron emission tomography (PET) – appear very promising. In SPECT, a tomographic picture is reconstructed from data obtained using an orthogonal set of gamma cameras rotated around the body, to which a radioactive agent has been administered that decays by emitting gamma quanta. In PET, a positron-emitting agent is used instead. On recombination of the positron with an electron, two high-energy photons are emitted in opposite directions, and detected in coincidence. Both imaging modalities have a relatively low resolution, but a very high sensitivity, since no background signals are present. Consequently, SPECT and PET can localize and quantify extremely low concentrations of targets, down to the nanomolar or even picomolar concentration regime. As it intrinsically relies on contrast agents, nuclear imaging offers poor anatomical information, but is strong in functional imaging. An important example is the imaging of increased metabolic rates, related to tumor growth. This is achieved in PET using the modified glucose-based agent SDG, which is taken up by cells as part of the metabolic process.

The combination of PET and SPECT as functional or molecular imaging tools with a complementary imaging modality such as computed tomography (CT), which does provide high-resolution morphological images, leads to very powerful MI tools. Once suitable biomarkers are discovered, it is expected to be relatively easy to develop a targeted contrast agent for PET or for SPECT.

Magnetic resonance imaging (MRI) also offers excellent morphological images, with very good soft tissue imaging capability. In addition, MRI can be used as a functional imaging tool, *e.g.* to measure brain activity. Although MRI is far less sensitive than PET or SPECT, impressive improvements in sensitivity have been achieved in recent years. Thus, it is conceivable that MRI will also develop into a molecular imaging tool. For this purpose, one needs targeted agents present at sub-micromolar concentration. A promising route towards this goal is to use nanoparticles as a scaffold.[10]

An emerging imaging modality is optical imaging. Due to the limited penetration rate of photons in human tissue, the application range is expected to be restricted. A promising application that is being explored by Philips Research in an alliance with Schering is optical mammography.

6. Molecular medicine in the "care cycle"

The full-fledged introduction of molecular medicine is the basis for an integrated approach of tomorrow's healthcare, with the following characteristics:

- earlier detection of disease, by careful screening of people with elevated genetically inherited and/or lifestyle-related risks using highly specific biomarkers;
- better diagnosis for better treatment, based on the individual patient's own biochemistry; and
- targeted and minimally invasive treatment with improved efficacy and fewer side effects.

These advances will be enabled by new technology, and by molecular biology and biomedical science, but they will require important progress in information technology as well.

We are convinced that an optimal way to benefit from nanomedicine requires an approach to healthcare in an integrated fashion, addressing all aspects of the "care cycle", as schematically depicted in Fig. 4. The care cycle approach starts with determination of the individual's predisposition to identified genetically inherited or lifestyle-related risks using molecular diagnostics (MDx). It then moves to focused screening of people at risk, initially employing MDx technologies, but in combination with MI for confirmation, localization and quantification, aiming at early detection of the onset of disease. Subsequently, when needed, individualized therapy is started, guided by treatment planning and monitoring of the therapeutic results with the aid of (molecular) imaging. In addition, imaging techniques can be invoked for minimally invasive treatment, providing more directed surgery and treatment. Finally, post-treatment MDx and MI can be utilized to monitor for recurrence or for active containment of the disease. The proposed approaches, which of course need to be combined with established clinical procedures, lead to an explosive increase of data, both qualitative and quantitative (enabling more objective, "evidence-based" medicine), which makes taking the right decisions more and more complex. Hence, attention needs to be paid to derive transparent information from the rich data sets, in order to support the physician in coming to the correct diagnosis and therapy selection. To this purpose, a clinical decision support system will need to be developed.

The first elements of molecular medicine have already been introduced into clinical practice. As an example, consider the case of breast cancer. Focused screening of women at risk may result in detection of breast cancer at an early stage, when it is still localized, with close to 100% treatment success. Screening for predisposition for breast cancer is offered, for example, by Myriad Genetics. Once breast cancer has been correctly diagnosed, the optimum treatment needs to be selected. The Dutch start-up Agendia, based on the pioneering work of van 't Veer and co-workers, has developed a tool for stratification of breast cancer patients, based on 70 marker genes, allowing for the administering of the best treatment to the individual patient.[11] Genentech has developed an antibody-based targeted drug, which can be applied to cure women with metastasised breast cancer, provided they show over-expression of the Her2/Neu membrane receptor. Vysis has developed a molecular diagnostic test, which can in fact screen for this receptor, to identify the patients who would benefit from the treatment.

Even though individual tests are available, it will take time to introduce molecular medicine throughout the care cycle. For many diseases, no comprehensive insight is available into their origin, and no unambiguous biomarkers have yet been identified. To counter this challenge, a tremendous effort is required, involving advanced academic research, together with contributions from pharmaceutical and biotech companies, and from medical technology companies, which should join forces to realize breakthroughs. At the same time, it is crucial to link the increasing insights in the fundamental biochemistry of disease to clinical observations. In particular, MI can play a crucial role in this translational challenge. Finally, the medical profession is (rightly) conservative; therefore, convincing evidence for the efficacy of the molecular medicine approaches needs to be provided, before they will be accepted. The challenges have been recognized by NIH director Zerhouni, who identifies in his description of the NIH Roadmap the most compelling opportunities in three arenas: new pathways to discoveries, (highly multidisciplinary) research teams of the future, and re-engineering the clinical research enterprise.[12]

We are optimistic, nevertheless, that these challenges can be overcome, and that within the next decades an increasing number of MDx and MI approaches will be introduced, providing molecular medicine-based care cycles for many important diseases.

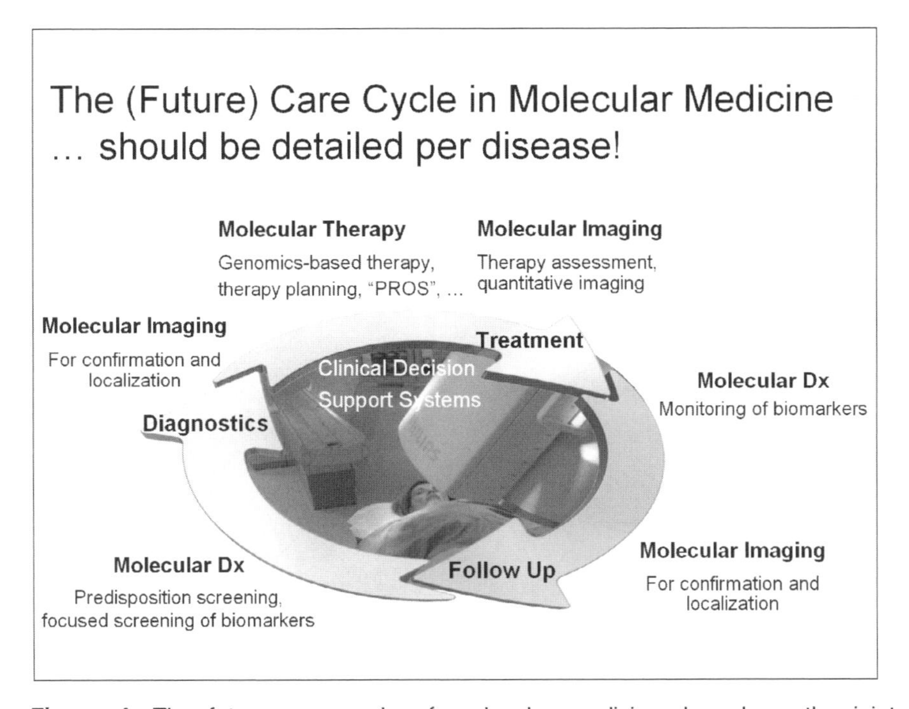

Figure 4. The future care cycle of molecular medicine, based on the joint application of molecular diagnostics (MDx) and molecular imaging (MI) to enable timely and targeted treatment.

References

1. Lewis Thomas, *The Youngest Science: Notes of a Medicine Watcher*, New York: The Viking Press, 1983.
2. G. S. Ginsburg and J. J. McCarthy, "Personalized medicine: Revolutionizing drug discovery and patient care," *Trends Biotechnol.* **19**, 491 (2001).
3. R. I. Pettigrew, C. A. Fee, and K. C. Li, "Changes in the world of biomedical research are moving the field of 'personalized medicine' from concept to reality," *J. Nucl. Med.* **45**, 1427 (2004).
4. B. J. Druker, "Imatinib alone and in combination for chronic myeloid leukemia," *Semin. Hematol.* **40**, 50 (2003).
5. R. Simon, "Using DNA microarrays for diagnostic and prognostic prediction," *Expert Rev. Mol. Diagn.* **3**, 587 (2003).
6. M. Fountoulakis, "Proteomics in drug discovery: Potential and limitations," *Biomed. Health Res.* **55**, 279 (2002).
7. M. Megens and M. W. J. Prins, "Magnetic biochips: A new option for sensitive diagnostics," *J. Magn. Magn. Mat.* **293**, 702 (2005).
8. R. Weissleder and U. Mahmood, "Molecular imaging," *Radiology* **219**, 316 (2001).
9. F. A. Jaffer and R. Weissleder, "Molecular imaging in the clinical arena," *J. Amer. Med. Assoc.* **293**, 855 (2005).
10. S. A. Wickline and G. Lanza, "Nanotechnology for molecular imaging and targeted therapy," *Circulation* **107**, 1092 (2003).
11. L. van 't Veer, H. Dai, M. van de Vijver, *et al.*, "Gene expression and profiling predicts clinical outcome of breast cancer," *Nature* **415**, 530 (2002).
12. E. Zerhouni, "Medicine. The NIH Roadmap," *Science* **302**, 63 (2003).

Interfacing the Brain – With Microelectronics?

Arto V. Nurmikko, William R. Patterson, Y.-K. Song, Christopher W. Bull
Division of Engineering, Brown University, Providence, RI 02912, U.S.A.

John P. Donoghue
Department of Neuroscience, Brown University, Providence, RI 02912, U.S.A.

1. Introduction

While paying appropriate homage to the technological revolution(s) empowered by modern microelectronics, and the continued march towards ever increasing sophistication of nanoscale high-density digital circuits, it maybe occasionally useful to pause and draw a reference to a computer of very different variety, namely the human brain. A product of millions of years of evolution, this non-silicon, biochemical processor features about 10^{15} interconnects amongst its three-dimensional (3D) network of 10^{11} neurons ("neuristors"), while consuming about 10 W of power. Approximately 10^{16} synaptic events per second occur within the sublime architecture of our "cortical computer", whose detailed design and operational principles define the main quest of modern neuroscience.

Science fiction has frequently visited the question of direct communication with the brain, including the prospects of interfacing the brain with external machines and extrasensory communication such as telepathy. Peeling away many such ideas of "brain alchemy", in addition to their ethics, there is a nonetheless a serious and compelling medical rationale to pursue direct retrieval of command and control signals from the brain. Although difficult to quantify and express by a single number, there are millions of individuals who suffer from serious neurological illnesses, ranging from Parkinson's disease to complete paralysis. Spinal cord injuries and degenerative diseases, such as ALS in particular, involve very large populations of people whose quality of life is much impaired, even if the brain is cognitively functional. Enabling such individuals to translate their command thoughts from the brain to direct activation of external assistive devices offers a potential intersection with modern technologies of significant societal and health care value.

In this chapter, we outline some of the first steps in this new area of neuroengineering and neurotechnology, focusing on specifics examples of contemporary progress. We first review very recent work, where direct extraction of signals from the human brain, preceded by vast amount of animal work, is showing striking early promise for enabling severely paralyzed individuals to interface with their external surroundings. We then consider the role of modern

Future Trends in Microelectronics. Edited by Serge Luryi, Jimmy Xu, and Alex Zaslavsky

ISBN 0-471-48

(and future) microelectronics in creating powerful platforms for future neuro-technologies featuring brain- and body-implantable neural prosthetic microsystems.

2. "Thought-to-action" using brain-implanted cortical neurosensor arrays

Within the enormously complex human brain, a highly distributed and hierarchical biological computer, modern neuroscience has made much progress in identifying certain areas of the cortex where direct access to and extraction of electrical signal impulses at the single neuron level can provide information about the commands which enable *e.g.* our motor functions (arm and leg movement).[1] For example, the so-called arm area can be identified within the primary motor cortex, as indicated in Fig. 1, which also shows the typical local neuronal architecture. Since all current noninvasive (*i.e.* external to the skull) brain "imaging" techniques from EEG to MEG to fMRI lack the spatial resolution anywhere near the single neural cell level, invasive insertion of microelectrodes to an appropriate depth is required to record the ~ms duration biphasic action potentials (or "spikes") from single neurons, on which neural coding is mainly based – see Fig. 1. Two-dimensional arrays of up to 100 of such needle-like surgically implanted electrodes with μm size tips and inter-electrode separation on the 100 μm scale (to roughly match the neuronal density) have been employed usefully in several animal "models" to develop understanding of the neural code that drives *e.g.* the arm motion in a monkey.[2] The decoding approaches involve correlating the rates of spike train activity recorded across the microelectrode array with the observed motion of the monkey's arm *e.g.* in the *x*-*y* plane, while the monkey uses a joystick or mouse to move a cursor on a computer screen.

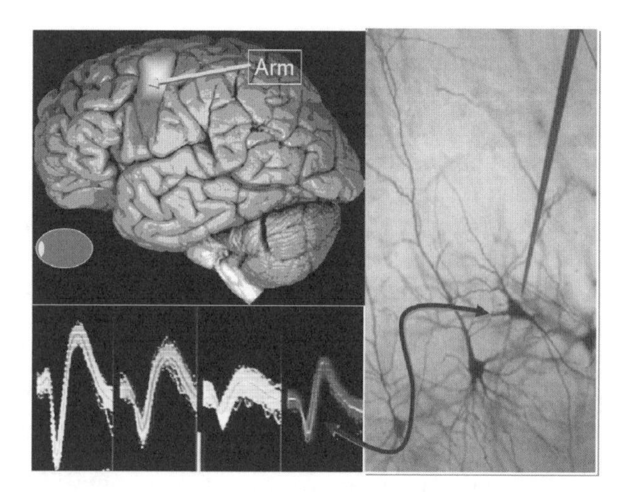

Figure 1. Location of the arm area in the cortex (upper left); typical action potentials of neural spikes (lower left); local neural landscape with a single needle-like microelectrode in the vicinity of a neuron (right). Cell body size ~20 μm.

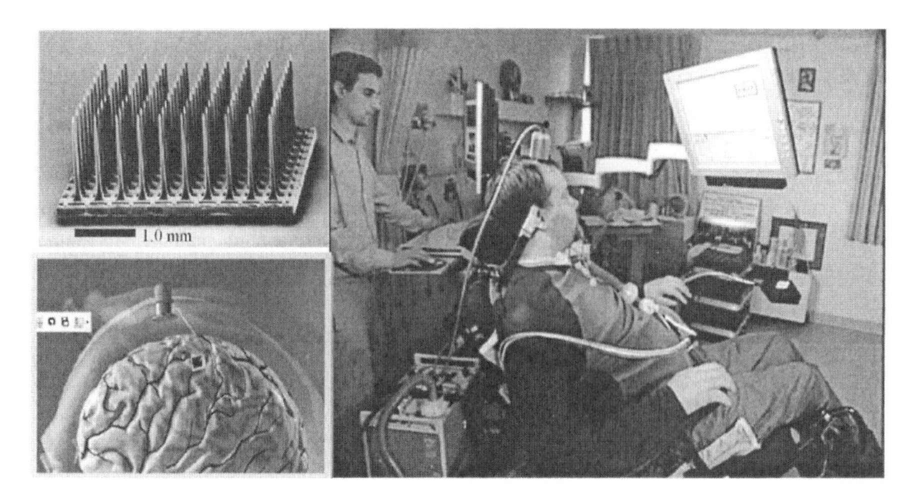

Figure 2. A silicon-based cortical microelectrode array (upper left); schematic of insert of the array and percutaneous connection to a skull-mounted pedestal connector (lower left); and a patient in a human clinical trial (courtesy of Cyberkinetics Inc).

A contemporary version of this brain recording platform, connecting the cortical neuroprobe to the signal processing electronics systems, summarized in Fig. 2 has been recently used in the very first human pilot trials involving chronic implants inserted for up to two years in severely paralyzed patients by a team combining expertise from Brown University, Cyberkinetics Inc., and Harvard Medical School.[3] The Cyberkinetics Braingate™ system employs a 10x10 element silicon-based multielectrode neural array[4] encased in silicone, with protruding platinized tips in physical contact with the neural tissue (Pt is chosen for impedance "matching" to the neural "electrolyte"). A 100-wire bundle (Au, 25 μm diameter) conducts the ~10–50 μV amplitude neural signals through a skull-mounted titanium-based percutaneous connector to an electronics instrumentation package which features low noise preamplifiers for each channel, signal multiplexing, A/D conversion, spike "sorting" and other signal processing tasks. Finally, the neural signals are decoded by specific algorithm approaches and, once deciphered, are connected to an external device, such as an electronic mouse, for exploring direct cortical control by the brain of a cursor on a computer screen or an artificial prosthetic device. While the process of surgically implanting the multielectrode array places individual recording tips randomly into the neural cell tissue (within a depth of about 1 mm from the top of the cortical surface), the array geometry and dimensions are such that in general there is a high probability for a given electrode registering neural impulses from a single nearby neuron (the practical effective sampling radius of each tip is on the order of 150 μm). The occasional pick up from two nearby neurons can generally be discriminated and separated from their time-amplitude stamps. Prior animal studies in monkeys, involving the removal and reimplantation of the "bed-of-needles" arrays,[5] have shown that these neuroprobes in the primary motor cortex create generally only modest tissue damage over time spans of years. Likewise, long-term tissue-electrode interactions (from immune reactions) appear to be moderate in terms of decreasing the recording performance of the array and impairment of neural circuitry. Figure 3 shows a typical multichannel recording from a monkey with most of the 100 channels in the 10x10 microelectrode array picking up useful biphasic action potential impulses of ~ms duration.

Returning back to Fig. 2, which shows a photograph of a human tetraplegic patient in the clinical test setting, we see a patient with an implanted multielectrode array connected to a head-mounted module that mainly houses preamplifiers and multiplexing circuitry, subsequently cabled to multiple electronic system modules. As summarized in Ref. 3, this clinical trial, conducted over nearly one year (and currently underway with several other paralyzed patients suffering from major impairment of motor functions) has enabled the patient, whose spinal cord is severed at the neck, to control a cursor on a computer screen for communication activities such as reading email, typing messages, drawing elementary freeform shapes, and operating a open-close prosthetic hand. Other assistive devices amenable to direct brain control that are being developed and tested include a wheelchair and a robotic arm. It is the culmination of much research and development of brain recording devices that has not only led to the these striking results in the first human trials, but is now motivating attempts to microminiaturize and enhance the performance of the neuroprobe systems by turning to advanced microelectronics.

3. Application of microelectronics to brain interfaces in neuroprosthetics

The progress in the early development of human neuroprosthetic approaches, summarized above, offers a tantalizing glimpse of a future where more complex

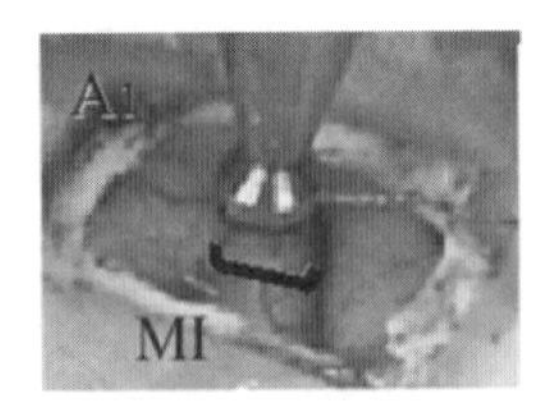

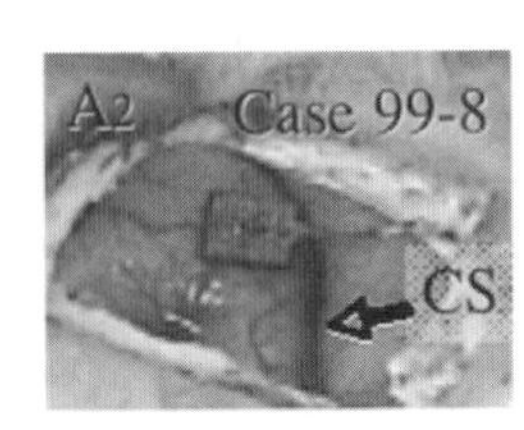

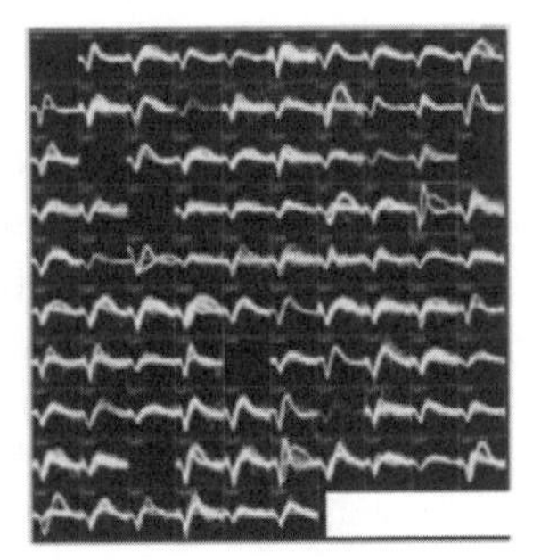

Figure 3. Images of the cortical array being implanted into a monkey's brain by a pneumatic inserter (left) and "snapshot" neural recordings after repeat implants from the motor cortex, showing activity at significant majority of the 10x10 channels, and displayed to reflect the rows and columns of the multielectrode array.

tasks can be accomplished by neurologically impaired persons via direct retrieval of command and control signals from the brain. As noted, the present approaches to chronic multichannel cortical recording involve a passive (*i.e.* unpowered) implanted neural probe arrays with a percutaneous and often bulky and fragile cabling to electronic processors outside the body. The integration of a significant portion or all of the electronics onto a body-implantable platform is highly desirable for future portable and wearable prosthetics, inclusive of on-board telemetry to transmit signals to internal body sites (*e.g.* abdominal cavity) or external remote processors and other assistive technologies.

Such an implantable cortical microsystem presents a multifaceted technical challenge, which includes the development and integration of ultralow-power microelectronic ICs to the neuroprobe recording platform, approaches to (broadband) on-board telemetry, either by optical (IR) or radiofrequency components, as well as the means to deliver power for the active components. The choice of an optical telemetry link, either free-space through the skin (from the top of head) or by fiber optical guides to less anatomically and physiologically demanding sites in the body is very attractive, as the expected data rates for more complex neural activity to be sampled in the future reach well beyond 1 GB/s. Finally, encapsulation and biocompatibility of such an electrically active multielement implants, in conjunction with surgical and other clinical considerations, are challenges of fundamental importance for future human applications. Here we summarize efforts to develop one approach to an implantable cortical microsystem, presently underway in the authors' laboratories, and shown schematically as a block diagram in Fig. 4. By employing the silicon-based microelectrode array as a recording platform, we have completed a full 16-channel microsystem, and experimentally tested key elements, including *in-vivo* evaluation in rats.

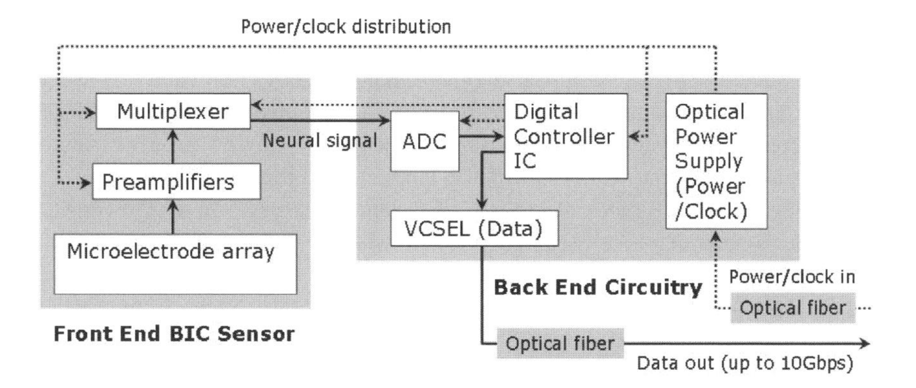

Figure 4. Block diagram for the implantable microsystem discussed in the text, enumerating the key micro- and optoelectronic on-chip circuit elements.

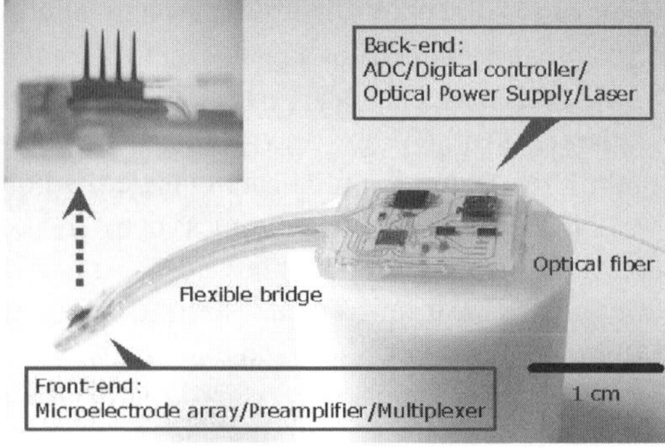

Figure 5. Photograph of a 4x4 element microelectrode array flip-chip bonded to a VLSI CMOS chip as an integrated cortical neuroprobe (left) and the full implantable microsystem fabricated on a flexible LCP substrate and encapsulated by silicone (right). The "back-end" panel houses A/D and command/control CMOS chips, together with a photovoltaic converter for incoming power/clock, and a VCSEL for outgoing IR digitized data stream, both by using an optical fiber.

The full microsystem design discussed here is based on specific device layout geometry and architecture, aimed firstly at non-human primates (monkeys), and is composed of analog and digital microelectronic components as well as infrared digital optical telemetry. The microsystem is powered by a photovoltaic "optical power supply" and has a soft and flexible encapsulation. Its design emphasizes scalability to increasingly larger amounts of neural information transmitted from the cortex to assistive technologies, aiming ultimately at human neuroprosthetics.

From a purely engineering viewpoint, the concept of using the cortical microelectrode array as a platform onto which *all* the microelectronic and telemetric components are directly integrated as a single monolithic cortical implant module might at first appear to represent an attractive approach. However, after weighing a number of neurophysiological, biomedical, telemetric, and surgical considerations, we have chosen to pursue a spatially more distributed microsystem architecture that is shown as a fabricated prototype in Fig. 5.

The layout in the figure reflects our "dual-panel" microsystem design concept, composed of two electrically interconnected islands that are landscaped on a common flexible liquid crystal polymer (LCP) substrate. The "front-end" panel of the dual-panel system is directly implanted into the cortex. This front panel is flexibly connected to the back panel which threads through a sealable "burr" hole in the skull and resides between the skull and the skin. It houses the cortical microelectrode array (in the present prototype version, a 4x4 unit), directly flip-chip bonded onto an ultra-low power analog CMOS chip.[6] The custom designed and fabricated CMOS IC includes low power preamplifier and multiplexing circuitry – see Refs. 6 and 7 for details and performance. A custom hybrid flip-chip bonding technique was developed at Brown to produce the functional, encapsulated microminiaturized front-end neuroprobe device whose separate recording characteristics have been reported in detail.[7] The front end has been implanted in a rat, to show its viability in recording brain signals from the somatosensory cortex.[8]

A photograph of the fabricated implantable microsystem of Fig. 5 shows some of the details of the full dual-panel system layout with its micro- and optoelectronic component functional grouping, according the schematic of Fig. 4. The multiplexed analog neural signals are routed from the front panel to the peripheral circuits of the second "back-end" panel of the implant. The back end circuitry integrates a low-power A/D converter, a digital control-and-command chip, an optional microcrystal photovoltaic energy converter, and an infrared (IR) data-out transmission link employing a very low-threshold IR vertical cavity surface emitting semiconductor laser (VCSEL). The concept of an optoelectronic data link enables considerable system flexibility for future neural prosthesis applications, including very large bandwidth capability. Extraction of the digitized IR neural signals from the neural interface unit can be performed either via a free-space beam from the laser emerging directly through the skin, which is sufficiently transparent in the near infrared, or via a fiber optic strand threaded subcutaneously to other information-linking sites in the body (*e.g.* in the thoracic unit). Likewise, our platform enables different options for power delivery to the active components of the implant. This implementation employs a photovoltaic GaAs/AlGaAs three tandem cell photovoltaic energy converter (>50% efficiency) as the "optical power supply" of Fig. 5, but the system can be adapted to an inductively coupled RF link which we have also demonstrated in the laboratory.

To summarize the rationale for the architecture and design of our implantable, compact dual platform neural interface, we note that it is based on the convergence of engineering considerations, input from neurosurgeons and neurologists, all leading to a number of practical electronic and biomedical device choices, compromises, and solutions, with the ultimate goal of human implants providing the project motivation. Important factors include practical surgical considerations, minimizing the heat transferred by the unit directly to the cortex (temperature rise below 1°C based on thermal modeling scaled up to a full 10x10 array, *i.e.* a 100 channel system), mechanical stability against stress on the tether from movement of brain, the microsystem's scalability to ever higher performance in terms of its signal bandwidth, its convenient modularity, adaptability to different wireless telemetries and powering (infrared, either free space or via optical fiber, RF), adaptability to multiple encapsulation strategies, and so on.

A performance demonstration result of the full 16 channel microsystem is shown in Fig. 6, in a benchtop evaluation test, where the microelectrode array has been excited by a series of "pseudospikes" (or artificial neurons) by applying a bipolar transient voltage through a wire to a saline solution (ACSF, which mimics closely the electrolyte background in the brain) into which the microsystem unit was immersed (labeled "electrical" in the left panel of Fig. 6).[8] After detection and reconstruction of the received IR optical data stream transmitted by the full microsystem into analog form (labeled "all-optical system"), a typical series of pseudospikes at high fidelity and good signal-to-noise ration were obtained as shown. (The upper trace is acquired from electrically accessing the front end of the system only). The signals acquired from the all-optical system in the lower trace shows almost the same preamplifier gain of 42 dB as the front-end microelectronic system, with comparable input noise characteristics. In the panel on the right of Fig. 6, a detail of the digital optical data stream trace is compared with the digital sampling rate clock trace on a microsecond time scale. The pulse coded modulation (PCM) data at the photoreceiver demonstrates the robust fiber optic data link in the present microsystem. The lower trace originates from the sinusoidal waveform applied to the external diode laser power source, current modulated to generate an infrared optical input carrying a 15.36 MHz clock signal.

At the time of writing, the microsystem described above is about enter animal testing phase in monkeys. A proof-of-concept 16 channel version prototype of this microsystem has been partially implanted in a rat animal model and proven to be functional on benchtop experiments, as summarized above. Many developmental challenges remain, nonetheless. These include long term reliability of encapsulation, development of practical and safe surgical procedures, and *in-vivo* microsystem trouble-shooting and maintenance strategies.

4. Future application prospects for brain interface microsystems

While brain implantable biomedical devices, such as the microsystem outlined above as well as others,[9-11] are still at early development stages, with many

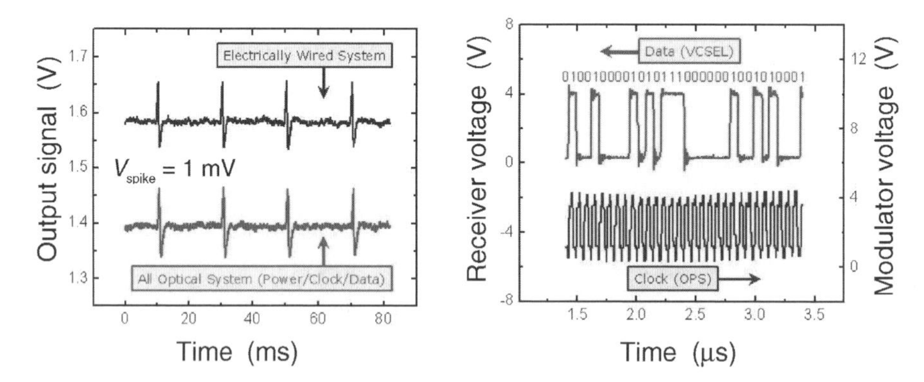

Figure 6. (a) Comparison of an electrically recorded train of "pseudospikes", generated in a saline solution, with those retrieved from the digital optical data stream transmitted from the full microsystem by the IR optical fiber; (b) upper trace shows a piece of the digitized outgoing optical data train while lower trace shows the incoming clock signal retrieved by photovoltaic conversion of modulated IR laser light entering the microsystem via the fiber.

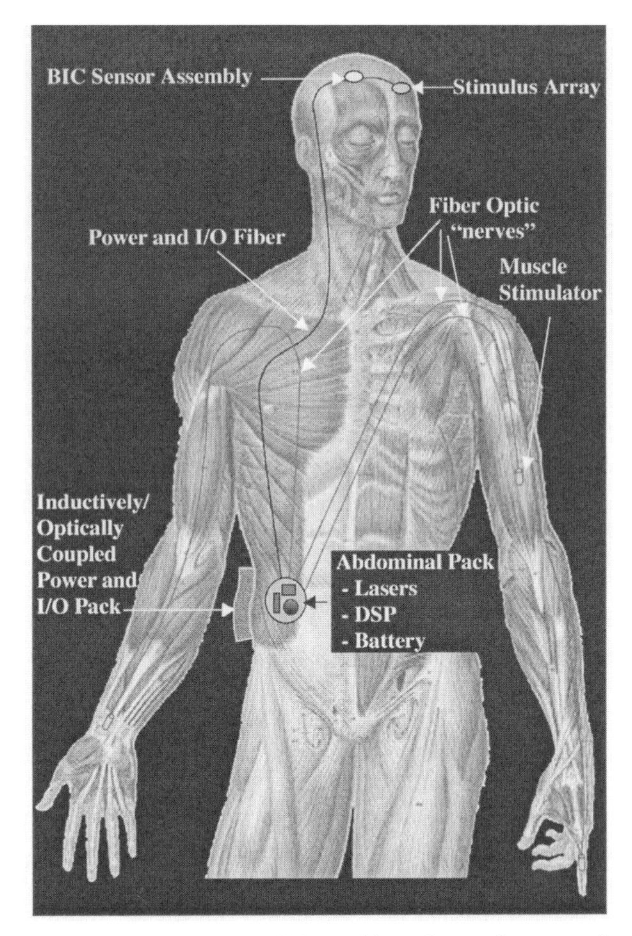

Figure 7. Futuristic view of an optical fiber wiring of neural communication system where cortical recording and stimulation are combined with peripheral nerve stimulation and sensing, all part of a closed-loop interactive prosthetic system.

technical challenges to be solved, it is important to consider the longer-term impact and prospects for an implanted neural prosthetic system, as an extension of the technology currently being advanced. A visionary/imaginary example of a fully "closed-loop" design augmenting the human nervous system is suggested schematically in Fig. 7, synergizing three principal performance functionalities: (a) a cortical recording microsystem (BIC sensor), capable of simultaneously addressing several cortical sites; (b) the ability to stimulate both the cortex, as well as peripheral nerves (such as muscle nerves in an arm of a leg); and (c) a very wide bandwidth "artificial" neural signal distribution network.

The cartoon of Fig. 7 suggests a scenario, where (biocompatible) optical fibers – analogous to optical telecommunication systems, – provide the information pathway for such an artificial nervous system. We note that the demand for processing large numbers of single neurons is already substantial at the present early stages of interfacing the brain, and is expected to increase significantly as the field of neuromotor prosthesis develops to mimic increasingly complex neurally driven tasks. Neural stimulation of the cortex will become essential in any closed-loop system, where feedback and sensory inputs from prosthetic devices or intact limbs are returned to the brain. For example, one would seek to evoke movement response and correlate this with spatiotemporal stimulation patterns.

As noted above, individual action potentials have duration of ~1 ms. Hence neural recording data stream from 100 electrodes at a sampling rate of only 10 kHz with a resolution of 10 bits (which is ultimately not adequate), requires a baud rate of 2 Gb/s. In the futuristic scenario of Fig. 7, the demands of implementing a closed-loop system will ultimately push ultralow-power and low-noise CMOS custom designed microelectronics to the limits, especially for analog circuits. The corresponding optoelectronic components would also need to provide ultrahigh speeds. In the scheme of Fig. 7, a cortical microsystem-on-chip is optically interconnected by a fiber to an advanced signal processing and command-control center module placed in the chest or abdominal cavity (akin to the heart pacemaker). Glass or plastic optical fibers are a lightweight, biocompatible, flexible means of routing neural signals within the body, without being subject to electromagnetic interference; they can carry very large amounts of information along a single strand, in contrast to metal wiring. This will drastically reduce the wiring burden and facilitate configurations where sensors and nerve stimulators in the limbs are also connected to a network by the optical neural information highway.

In terms of peripheral nerve stimulation, recent advances in functional electrical stimulation (FES) of muscle nerves by using electrical impulses show significant promise for restoring voluntary movement in patients with paralysis or other severe motor impairments.[12] Current approaches for implantable FES systems involve multisite stimulation, often with extensive wiring, posing research issues related to the physical size, power and signal delivery of the wiring, in conjunction with surgical and safety challenges. We note that by exploiting high-efficiency epitaxially-grown solar-cell type microdevices, efficient conversion of light to electrical stimulus (and system power) can be achieved. To explore a different means for delivering the stimulus to a distant muscle nerve site, we have recently elicited *in-vitro* FES response using a high-efficiency microcrystalline photovoltaic device as a neurostimulator, integrated with a biocompatible glass optical fiber that forms a lossless, interference-free lightwave conduit for signal and energy transport. As a proof-of-concept demonstration, the sciatic nerve of a frog was stimulated by the microcrystalline device connected to a multimode optical fiber (core diameter of 62.5 μm). This set-up converts optical activation pulses (~100 μs) from an infrared semiconductor laser source (at 852 nm wavelength) into an FES signal.[13]

5. Summary

In this overview, we have attempted to give the reader an impression of the current intersection of microelectronics and brain science, in the quest for hybrid device platforms to provide a direct electronic interface with the brain. Only early steps in this direction have been taken, in what is a major technological and biomedical challenge that will have far-reaching societal impact in the future.

We have focused specifically on an approach for retrieving neural signals from the brain through a contemporary implantable microsystems, where the performance demands on the micro- and optoelectronics devices at the moment are modest by *e.g.* current computer industry standards (though packaging of brain implantable devices is highly nontrivial). However, if and when the integrated neuroprobe platform discussed above, or related approaches, will be shown to be practical in a human patient setting, it is clear that this emerging neurotechnology will demand the best that semiconductor technology can offer. Truly far-reaching ideas, such as implementation and/or integration of silicon "brain" components, might then also become more realistic.[14]

Acknowledgments

The authors acknowledge the participation of many colleagues and students in the work described above, including Leigh Hochberg, Mijail Serruya, Selim Suner, Matt Fellows, John Simeral, Kristina Davitt, Paula Petrica, Joanna Zhang, and Heng Xu. We especially thank Dr. Eric Eisenstadt for his role in catalyzing this work. Research supported by ONR, DARPA, and NIH.

References

1. Among many general brain science textbooks see *e.g.* E. R. Kandel, J. H. Schwartz, and T. M. Jessel, *Principles of Neural Science*, 4th ed., New York: McGraw-Hill Medical, 2000.
2. M. D. Serruya, N. G. Hatsopoulos, L. Paninski, M. R. Fellows, and J. P. Donoghue, "Instant neural control of a movement signal," *Nature* **416**, 141 (2002); D. M. Taylor, S. I. Tillery, and A. B. Schwartz, "Information conveyed through brain-control: Cursor versus robot," *IEEE Trans. Neural Syst. Rehabil. Eng.* **11**, 195 (2003).
3. L. R. Hochberg, M. D. Serruya, G. M. Friehs, *etc.*, "Neuronal ensemble control of prosthetic devices by a human with tetraplegia," *Nature* **442**, 164 (2006); the website includes video clips.
4. K. E. Jones, P. K. Campbell, and R. A. Normann, "A glass/silicon composite intracortical electrode array," Ann. Biomed. Eng. **20**, 423 (1992).
5. S. Suner, M. R. Fellows, C. Vargas-Irwin, K. Nakata, and J. P. Donoghue, "Reliability of signals from chronically implanted, silicon-based electrode

array in non-human primate primary motor cortex," *IEEE Trans Neural Syst. Rehabil. Eng.* **13**, 524 (2005).

6. W. R. Patterson, Y.-K. Song, C. W. Bull, *et al.*, "A microelectrode/micro-electronic hybrid device for brain implantable neuroprosthesis applications," *IEEE Trans. Biomed. Eng.* **51**, 1845 (2004).
7. Y.-K. Song, W. R. Patterson, C. W. Bull, *et al.*, "Development of a chipscale integrated microelectrode/microelectronic device for brain implantable neuro-engineering applications," *IEEE Trans. Neural Rehabil. Eng.* **13**, 220 (2005).
8. Y.-K. Song, W. R. Patterson, C. W. Bull, *et al.*, "Development of brain implantable microsystems with infrared optical telemetry for neuromotor prosthesis," submitted to *IEEE Trans. Biomed. Eng.* (2006).
9. M. Mojarradi, D. Binkley, B. Blalock, R. Andersen, N. Uslhoefer, T. Johnson, and L. Del Castillo, "A miniaturized neuroprosthesis suitable for implantation into the brain," *IEEE Trans. Neural Syst. Rehabil. Eng.* **11**, 38 (2003).
10. I. Obeid, J. Morizio, K. Moxon, M. A. Nicolelis, and P. D. Wolf, "Two multichannel integrated circuits for neural recording and signal processing," *IEEE Trans. Biomed. Eng.* **50**, 255 (2003).
11. R. Harrison, P. Watkins, R. Kier, R. Lovejoy, D. Black, R. Normann, and F. Solzbacher, "A low-power integrated circuit for a wireless 100-electrode neural recording system," *Tech. Dig. ISSCC* (2006), pp. 2258–67.
12. W. M. Grill and R. F. Kirsch, "Neuroprosthetic applications of electrical stimulation," *Assist. Technol.* **12**, 6 (2000); B. P. Heilman, M. L. Audu, R. F. Kirsch, and R. J. Triolo, "Selection of an optimal muscle set for a 16-channel standing neuroprosthesis using a human musculoskeletal model," *J. Rehabil. Res. Dev.* **43**, 273 (2006).
13. Y.-K. Song, J. Stein, W. R. Patterson, *et al.*, "A microscale photovoltaic neurostimulator for fiber optic delivery of functional electrical stimulation", to appear in *J. Neuroeng.* (2007).
14. T. W Berger, M. Baudry, R. D. Brinton, *et al.*, "Brain-implantable biomimetic electronics as the next era in neural prosthetics", *Proc. IEEE* **89**, 993 (2001).

Synthetic Biology: Synthesis and Modification of a Chemical Called Poliovirus

S. Mueller, J. R. Coleman, J. Cello, A. Paul, and E. Wimmer
School of Medicine, SUNY–Stony Brook, Stony Brook, NY 11974, U.S.A.

D. Papamichail and S. Skiena
Dept. of Computer Science, SUNY–Stony Brook, Stony Brook, NY 11974, U.S.A.

1. Introduction

Synthetic biology is a newly emerging scientific discipline, encompassing knowledge of different disciplines such as engineering, physics, chemistry, computer sciences, mathematics, and biology.[1] Synthetic biology aims to create novel biological systems with functions that do not exist in nature. There seem infinite possibilities of constructing unique derivatives of existing organisms (bacteria, yeast, viruses). Apart from designing novel building blocks for engineering biological systems, a fundamental requirement in synthetic biology is the ability of large-scale DNA synthesis and DNA sequencing.

Viruses can be described in chemical terms; the empirical formula of the organic matter of poliovirus being[2]

$$C_{332,652}H_{492,388}N_{98,245}O_{131,196}P_{7,501}S_{2,340}\ .$$

There might be little practical use in describing poliovirus by an empirical formula, but it persuasively portrays the virus as a chemical. Placing these atoms into order, particles of high symmetry emerge,[3] see Fig. 1. These particles can be purified and crystallized.

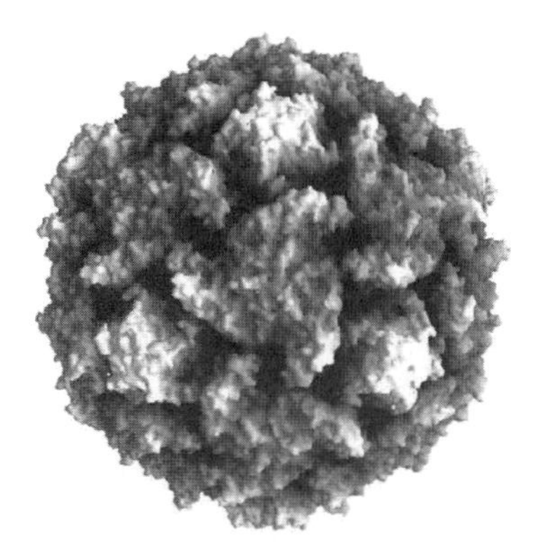

Figure 1. Computer model of poliovirus, generated from x-ray crystallographic data.[3]

Future Trends in Microelectronics. Edited by Serge Luryi, Jimmy Xu, and Alex Zaslavsky
ISBN 0-471-48

The structure of the particle and the strategy for proliferation are encoded in the viral genome, which is a single stranded nucleic acid (RNA) of about 7,500 nucleotides – see Fig. 2. We have determined the sequence of this viral genome in 1981,[4] which was a key to understanding aspects of the molecular biology of poliovirus proliferation.

Whereas the poliovirion is a beautiful molecular entity, it has frightening properties as a human pathogen. Infecting by the oral-fecal route, the virus

UUAAAACAGCUCUGGGGUUGUACCCACCCCAGAGGCCCACGUGGCGGCUAGUACUCCGGUAUUGCGGUACCCUUGUACGCCUGUUUUAUACUCCCUUCCCG
UAACUUAGACGCACAAAACCAAGUUCAAUAGAAGGGGGUACAAACCAGUACCACCACGAACAAGCACUUCUGUUUCCCCGGUGAUGUCGUAUAGACUGCUU
GCGUGGUUGAAAGCGACGGAUCCGUUAUCCGCUUAUGUACUUCGAGAAGCCCAGUACCACCUCGGAAUCUUCGAUGCGUUGCGCUCAGCACUCAACCCCAG
AGUGUAGCUUAGGCUGAUGAGUCUGGACAUCCCUCACCGGUGACGGUGGUCCAGGCUGCGUUGGCGGCCUACCUAUGGCUAACGCCAUGGGACGCUAGUUG
UGAACAAGGUGUGAAGAGCCUAUUGAGCUACAUAAGAAUCCUCCGGCCCCUGAAUGCGGCUAAUCCCAACCUCGGAGCAGGUGGUCACAAACCAGUGAUUG
GCCUGUCGUAACGCGCAAGUCCGUGGCGGAACCGACUACUUUGGGUGUCCGUGUUUCCUUUUAUUUUAUUGUGGCUGCUUAUGGUGACAAUCACAGAUUGU
UAUCAUAAAGCGAAUUGGAUUGGCCAUCCGGUGAAAGUGAGACUCAUUAUCUAUCUGUUUGCUGGAUCCGCUCCAUUGAGUGUGUUUACUCUAAGUACAAU
UUCAACAGUUAUUUCAAUCAGACAAUUGUAUCAUAAUGGGUGCUCAGGUUUCAUCACAGAAAGUGGGCGCACAUGAAAACUCAAAUAGAGCGUAUGGUGGU
UCUACCAUUAAUUACACCACCAUUAAUUAUUAUAGAGAUUCAGCUAGUAACGCGGCUUCGAAACAGGACUUCUCUCAAGACCCUUCCAAGUUCACCGAGCC
CAUCAAGGAUGUCCUGAUAAAAACAGCCCCAAUGCUAAACUCGCCAAACAUAGAGGCUUGCGGGUAUAGCGAUAGAGUACUGCAAUUAACACUGGGAAACU
CCACUAUAACCACACAGGAGGCGGCUAAUUCAGUAGUCGCUUAUGGGCGUUGGCCUGAAUAUCUGAGGGACAGCGAAGCCAAUCCAGUGGACCAGCCGACA
GAACCAGACGUCGCUGCAUGCAGGUUUUAUACGCUAGACACCGUGUCUUGGACGAAAGAGUCGCGAGGGUGGUGGUGGAAGUUGCCUGAUGCACUGAGGGA
CAUGGGACUCUUUGGGCAAAAUAUGUACUACCACUACCUAGGUAGGUCCGGGUACACCGUGCAUGUACAGUGUAACGCCUCCAAAUUCCACCAGGGGGCAC
UAGGGGUAUUCGCCGUACCAGAGAUGUGUCUGGCCGGGGAUAGCAACACCACUACCAUGCACACCAGCUAUCAAAAUGCCAAUCCUGGCGAGAAAGGAGGC
ACUUUCACGGGUACGUUCACUCCUGACAACAACCAGACAUCACCUGCCCGCAGGUUCUGCCCGGUGGAUUACCUCCUUGGAAAUGGCACGUUGUUGGGGAA
UGCCUUUGUGUUCCCGCACCAGAUAAUAAACCUACGGACCAACAACUGUGCUACACUGGUACUCCCUUACGUGAACUCCCUCUCGAUAGAUAGUAUGGUAA
AGCACAAUAAUUGGGGAAUUGCAAUAUUACCAUUGGCCCCAUUAAAUUUUGCUAGUGAGUCCUCCCCAGAGAUUCCAAUCACCUUGACCAUAGCCCCUAUG
UGCUGUGAGUUCAAUGGAUUAAGAAACAUCACCCUGCCACGCUUACAGGGCCUGCCGGUCAUGAACACCCCUGGUAGCAAUCAAUAUCUUACUGCAGACAA
CUUCCAGUCACCGUGUGCGCUGCCUGAAUUUGAUGUGACCCCACCUAUUGACAUACCCGGUGAAGUAAAGAACAUGAUGGAAUUGGCAGAAAUCGACACCA
UGAUUCCCUUUGACUUAAGUGCCACAAAAAAGAACACCAUGGAAAUGUAUAGGGUUCGGUUAAGUGACAAACCACAUACAGACGAUCCCAUACUCUGCCUG
UCACUCUCUCCAGCUUCAGAUCCUAGGUUGUCACAUACUAUGCUUGGAGAAAUCCUAAAUUACUACACACACUGGGCAGGAUCCCUGAAGUUCACGUUUCU
GUUCUGUGGAUCCAUGAUGGCAACUGGCAAACUGUUGGUGUCAUACGCGCCUCCUGGAGCCGACCCACCAAAGAAGCGUAAGGAGGCGAUGUUGGGAACAC
AUGUGAUCUGGGACAUAGGACUGCAGUCCUCAUGUACUAUGGUAGUGCCAUGGAUUAGCAACACCACGUAUCGGCAAACCAUAGAUGAUAGUUUCACCGAA
GGCGGAUACAUCAGCGUCUUCUACCAAACUAGAAUAGUCGUCCCUCUUUCGACACCCAGAGAGAUGGACAUCCUUGGUUUUGUGUCAGCGUGUAAUGACUU
CAGCGUGCGCUUGUUGCGAGAUACCACACAUAUAGAGCAAAAAGCGCUAGCACAGGGGUUAGGUCAGAUGCUUGAAAGCAUGAUUGACAACACAGUCCGUG
AAACGGUGGGGGCGGCAACAUCUAGAGACGCUCUCCCAAACACUGAAGCCAGUGGACCAACACACUCCAAGGAAAUUCCGGCACUCACCGCAGUGGAAACU
GGGGCCACAAAUCCACUAGUCCCUUCUGAUACAGUGCAAACCAGACAUGUUGUACAACAUAGGUCAAGGUCAGAGUCUAGCAUAGAGUCUUUCUUCGCGCG
GGGUGCAUGCGUGACCAUUAUGACCGUGGAUAACCCAGCUUCCACCACGAAUAAGGAUAAGCUAUUUGCAGUGUGGAAGAUCACUUAUAAAGAUACUGUCC
AGUUACGGAGGAAAUUGGAGUUCUUCACCUAUUCUAGAUUUGAUAUGGAACUUACCUUUGUGGUUACUGCAAAUUUCACUGAGACUAACAAUGGGCAUGCC
UUAAAUCAAGUGUACCAAAUUAUGUACGUACCACCAGGCGCUCCAGUGCCCGAGAAAUGGGACGACUACACAUGGCAAACCUCAUCAAAUCCAUCAAUCUU
UUACACCUACGGAACAGCUCCAGCCCGGAUCUCGGUACCGUAUGUUGGUAUUUCGAACGCCUAUUCACACUUUUACGACGGUUUUUCCAAAGUACCACUGA
AGGACCAGUCGGCAGCACUAGGUGACUCCCUUUAUGGUGCAGCAUCUCUAAAUGACUUCGGUAUUUUGGCUGUUAGAGUAGUCAAUGAUCACAACCCGACC
AAGGUCACCUCCAAAAUCAGAGUGUAUCUAAAACCCAAACACAUCAGAGUCUGGUGCCCGCGUCCACCGAGGGCAGUGGCGUACUACGGCCCUGGAGUGGA
UUACAAGGAUGGUACGCUUACACCCCUCUCCACCAAGGAUCUGACCACAUAUGGAUUCGGACACCAAAACAAAGCGGUGUACACUGCAGGUUACAAAAUUU
GCAACUACCACUUGGCCACUCAGGAUGAUUUGCAAAACGCAGUGAACGUCAUGUGGAGUAGAGACCUCUUAGUCACAGAAUCAAGAGCCCAGGGCACCGAU
UCAAUCGCAAGGUGCAAUUGCAACGCAGGGGUGUACUACUGCGAGUCUAGAAGGAAAUACUACCCAGUAUCCUUCGUUGGCCCAACGUUCCAGUACAUGGA
GGCUAAUAACUAUUACCCAGCUAGGUACCAGUCCCAUAUGCUCAUUGGCCAUGGAUUCGCAUCUCCAGGGGAUUGUGGUGGCAUACUCAGAUGUCACCACG
GGGUGAUAGGGAUCAUUACUGCUGGUGGAGAAGGGUUGGUUGCAUUUUCAGACAUUAGAGACUUGUAUGCCUACGAAGAAGAAGCCAUGGAACAAGGCAUC
ACCAAUUACAUAGAGUCACUUGGGGCCGCAUUUGGAAGUGGAUUUACUCAGCAGAUUAGCGACAAAAUAACAGAGUUGACCAAUAUGGUGACCAGUACCAU
CACUGAAAAGCUACUUAAGAACUUGAUCAAGAUCAUAUCCUCACUAGUUAUUAUAACUAGGAACUAUGAAGACACCACAACAGUGCUCGCUACCCUGGCCC
UUCUUGGGUGUGAUGCUUCACCAUGGCAGUGGCUUAGAAAGAAAGCAUGCGAUGUUCUGGAGAUACCUUAUGUCAUCAAGCAAGGUGACAGUUGGUUGAAG
AAGUUUACUGAAGCAUGCAACGCAGCUAAGGGCCUGGAGUGGGUGUCAAACAAAAUCUCAAAAUUCAUUGAUUGGCUCAAGGAGAAAAUUAUCCCACAAGC
UAGAGAUAAGUUGGAAUUUGUAACAAAACUUAGACAACUAGAAAUGCUGGAAAACCAAAUCUCAACUAUACACCAAUCAUGCCCUAGUCAGGAACACCAGG
AAAUUCUAUUCAAUAAUGUCAGAGUGGUUAUCCAUCCGGUCUAAGAGGUUUGCCCCUCUUUACGCAGUGGAAGCCAAAAGAAUACAGAAACUAGAGCAUACU
AUUAACAACUACAUACAGUUCAAGAGCAAACACCGUAUUGAACCAGUAUGUUUGCUAGUACAUGGCAGCCCCGGAACAGGUAAAUCUGUAGCAACCACCU
GAUUGCUAGAGCCAUAGCUGAAAGAGAAAACACGUCCACGUACUCGCUACCCCCGGAUCCAUCACACUUCGACGGAUACAAACAACAGGGAGUGGUGAUUA
UGGACGACCUGAAUCAAAACCCAGAUGGUGCGGACAUGAAGCUGUUCUGUCAGAUGGUAUCAACAGUGGAGUUUAUACCACCCAUGGCUUCCCUGGAGGAG
AAAGGAAUCCUGUUUACUUCAAAUUACGUUCUAGCAUCCACAAACUCAAGCAGAAUUUCCCCCCCACUGUGGCACACAGUGAUGCAUUAGCCAGGCGCUU
UGCGUUCGACAUGGACAUUCAGGUCAUGAAUGAGUAUUCUAGAGAUGGGAAAUUGAACAUGGCCAUGGCUACUGAAAUGUGUAAGAACUGUCACCAACCAG
CAAACUUUAAGAGAUGCUGUCCUUUAGUGUGUGGUAAGGCAAUUCAAUUAAUGGACAAAUCUUCCAGAGUUAGAUACAGUAUUGACCAGAUCACUACAAUG
AUUAUCAAUGAGAGAAACAGAAGAUCCAACAUUGGCAAUUGUAUGGAGGCUUUGUUUCAAGGACCACUCCAGUAUAAAGACUUGAAAAUUGACAUCAAGAC
GAGUCCCCCUCCUGAAUGUAUCAAUGACUUGCUCCAAGCAGUUGACUCCCAGGAGGUGAGAGAUUACUGUGAGAAGAAGGGUUGGAUAGUCAACAUCACCA
GCCAGGUUCAAACAGAAAGGAACAUCAACAGGGCAAUGACAAUUCUACAAGCGGUGACAACCUUCGCCGCAGUGGCUGGAGUUGUCUAUGUCAUGUAUAAA
CUGUUUGCUGGACACCAGGGAGCAUACACUGGUUUACCAAACAAAAAACCCAACGUGCCCACCAUUCGGACAGCAAAGGUACAAGGACCAGGGUUCGAUUA
CGCAGUGGCUAUGGCUAAAAGAAACAUUGUUACAGCAACUACUAGCAAGGGAGAGUUCACUAUGUUAGGAGUCCACGACAACGUGGCUAUUUUACCAACCC
ACGCUUCACCUGGUGAAAGCAUUGUGAUCGAUGGCAAAGAAGUGGAGAUCUUGGAUGCCAAAGCGCUCGAAGAUCAAGCAGGAACCAAUCUUGAAAUCACU
AUAAUCACUCUAAAGAGAAAUGAAAAGUUCAGAGACAUUAGACCACAUAUACCUACUCAAAUCACUGAGACAAAUGAUGGAGUCUUGAUCGUGAACACUAG
CAAGUACCCCAAUAUGUAUGUUCCUGUCGGUGCUGUGACUGAACAGGGAUAUCUAAAUCUCGGUGGGCGCCAAACUGCUCGUACUCUAAUGUACAACUUUC
CAACCAGAGCAGGACAGUGUGGUGGAGUCAUCACAUGUACUGGGAAAGUCAUCGGGAUGCAUGUUGGUGGGAACGGUUCACACGGGUUUGCAGCGGCCCUG
AAGCGAUCAUACUUCACUCAGAGUCAAGGUGAAAUCCAGUGGAUGAGACCUUCGAAGGAAGUGGGAUAUCCAAUCAUAAAUGCCCCGUCCAAAACCAAGCU
UGAACCCAGUGCUUUCCACUAUGUGUUUGAAGGGGUGAAGGAACCAGCAGUCCUCACUAAAAACGAUCCCAGGCUUAAGACAGACUUUGAGGAGGCAAUUU
UCUCCAAGUACGUGGGUAACAAAAUUACUGAAGUGGAUGAGUACAUGAAAGAGGCAGUAGACCACUAUGCUGGCCAGCUCAUGUCACUAGACAUCAACACA
GAACAAAUGUGCUUGGAGGAUGCCAUGUAUGGCACUGAUGGUCUAGAAGCACUUGAUUUGUCCACCAGUGCUGGCUACCCUUAUGUAGCAAUGGGAAAGAA
GAAGAGAGACAUCUUGAACAAACAAACCAGAGACACUAAGGAAAUGCAAAAACUGCUCGACACAUAUGGAAUCAACCUCCCACUGGUGACUUAUGUAAAGG
AUGAACUUAGAUCCAAAACAAAGGUUGAGCAGGGGAAAUCCAGAUUAAUUGAAGCUUCUAGUUUGAAUGACUCAGUGGCAAUGAGAAUGGCUUUUGGGAAC
CUAUAUGCUGCUUUUCACAAAAACCCAGGAGUGAUAACAGGUUCAGCAGUGGGGUGCGAUCCAGAUUUGUUUUGGAGCAAAAUUCCGGUAUUGAUGGAAGA
GAAGCUGUUUGCUUUUGACUACACAGGGUAUGAUGCAUCUCUCAGCCCUGCUUGGUUCGAGGCACUAAAGAUGGUGCUUGAGAAAAUCGGAUUCGGAGACA
GAGUUGACUACAUCGACUACCUAAACCACUCACACCACCUGUACAAGAAUAAAACAUACUGUGUCAAGGGCGGUAUGCCAUCUGGCUGCUCAGGCACUUCA
AUUUUUAACUCAAUGAUUAACAACUUGAUUAUCAGGACACUCUUACUGAAAACCUACAAGGGCAUAGAUUUAGACCACCUAAAAAUGAUUGCCUAUGGUGA
UGAUGUAAUUGCUUCCUACCCCCAUGAAGUUGACGCUAGUCUCCUAGCCCAAUCAGGAAAAGACUAUGGACUAACUAUGACUCCAGCUGACAAAUCAGCUA
CAUUUGAAACAGUCACAUGGGAGAAUGUAACAUUCUUGAAGAGAUUCUUCAGGGCAGACGAGAAAUACCCAUUUCUUAUUCAUCCAGUAAUGCCAAUGAAG
GAAAUUCAUGAAUCAAUUAGAUGGACUAAAGAUCCUAGGAACACUCAGGAUCACGUUCGCUCUCUGUGCCUUUUAGCUUGGCACAAUGGCGAAGAAGAAUA
UAACAAAUUCCUAGCUAAAAUCAGGAGUGUGCCAAUUGGAAGAGCUUUAUUGCUCCCAGAGUACUCAACAUUGUACCGCCGUUGGCUUGACUCAUUUUAGU
AACCCUACCUCAGUCGAAUUGGAUUGGGUCAUACUGUUGUAGGGGUAAAUUUUUCUUUAAUUCGGAGGAAAAAAAAAAAAAAAAAAAAAAAAAAAAAAAA
AAAAA

Figure 2. Nucleotide sequence of the poliovirus genome.[4]

proliferates very efficiently in the gastrointestinal tract. Rarely, the virus finds itsway into the central nervous system (CNS), where it destroys with cunning specificity motor neurons that control muscle movement. This results in irreversible paralysis and sometimes death, a disease called poliomyelitis.[5] The virus has caused horrific epidemics in the first part of last century until two excellent vaccines were developed. The killed poliovirus vaccine (IPV) by Jonas Salk, and the live oral vaccine (OPV) by Albert Sabin now effectively control the disease.[5] The fact that humans are the only reservoir for the poliovirus and the efficacy of the vaccines has prompted the World Health Organization to attempt eradicating the agent from the globe. Although great progress has been made over a period of eighteen years, poliovirus has shown great resilience and a successful end to the eradication campaign is not yet in sight.

In the early 1990's, the intriguing dual nature of poliovirus as an inanimate entity, as well as a replicating organism, has led us to ask the question whether the virus can be synthesized in the test tube. To be sure, nucleic acid synthesis by then was already developed to a stage where the DNA of individual genes could be assembled from their known sequence. But the synthesis of a replicating entity, requiring the assembly of nucleic acid much larger than had been reported before, was unprecedented. It provided the *proof of principle* that viruses can be considered to be chemicals, which can be regenerated in the absence of a natural template outside living cells.[6] Moreover, we were probably amongst the first to realize that the advances in nucleic acid synthesis could pose a threat if misused for the synthesis of bioterrorist agents. The Defense Advanced Research Project Agency (DARPA) had come to the same conclusion and supported the synthesis. It was meant as a "wake up call" to pinpoint the dual nature of advances in biotechnology[6].

We are now studying the possibility of generating via whole genome synthesis novel polioviruses whose ability to proliferate in the CNS is debilitated, whereas its efficiency to replicate in tissue culture cells remains largely unchanged. The basis of engineering new polioviruses is altering codon usage of the viral mRNA.[7]

2. The synthesis of poliovirus

At the present time, RNA of the size of the poliovirus genome cannot be synthesized. The DNA-complement of the RNA, however, can be assembled from small oligonucleotides, available from biotech companies. As shown in Fig. 3, the assembly of DNA fragments proceeds via the double-stranded form, allowing subsequent transcription of the DNA with a RNA transcriptase[8] into synthetic, infectious poliovirus genomic RNA.[8] Incubation of the RNA in a cell-free extract leads to the translation and replication of the genome until sufficient quantities of components are synthesized for the spontaneous assembly of virions to occur.[2]

The identity of the virus was proven by a variety of biological and serological methods. For example, the synthetic virus was shown to produce plaques on

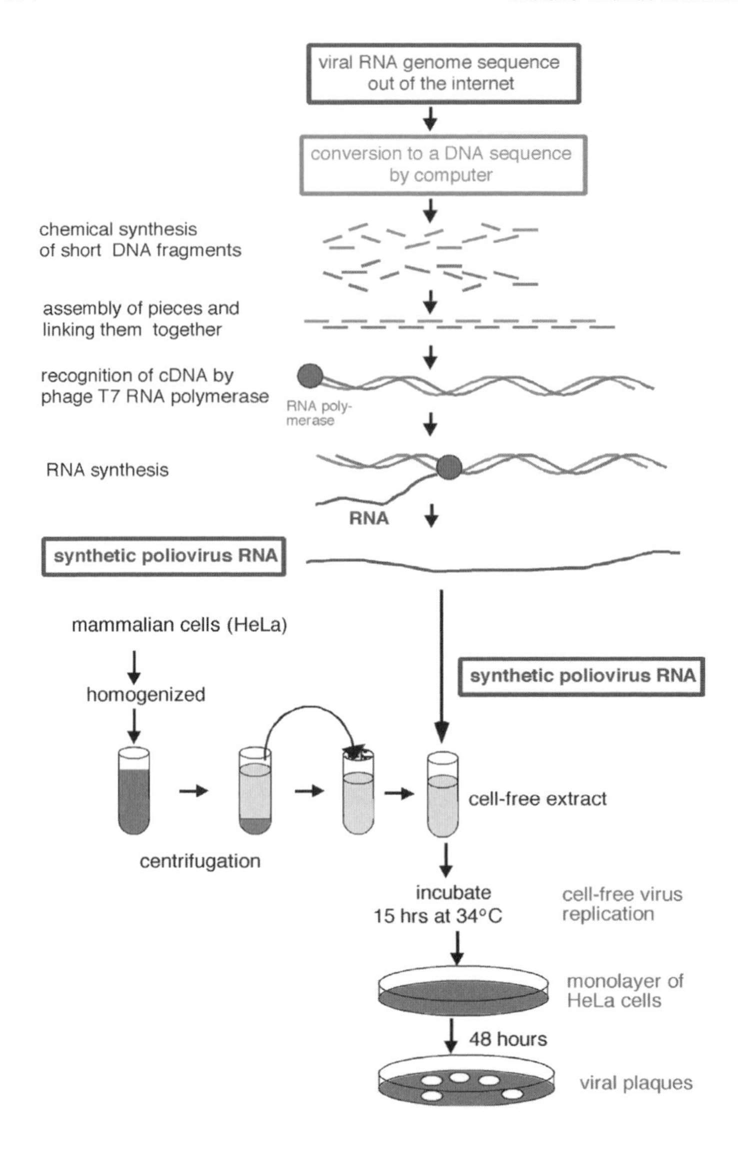

Figure 3. Synthesis of poliovirus in the absence of a natural template.[6] The sequence derived from the Internet[4] was converted to a DNA sequence and divided into small oligonucleotides. These were obtained from a biotech company and assembled to double stranded DNA. The DNA was enzymatically transcribed into infectious synthetic poliovirus RNA[8] that, in turn, was used to generate poliovirus in an extract of human cells.[2] The newly synthesized virus grew on a layer of human cells, forming plaques. Reproduced with permission from Mueller *et al.*[10]

monolayers of human cells that could be prevented by anti sera specific to the virus.[6] Moreover, the virus produced disease symptoms in experimental animals (transgenic mice) indistinguishable from those of wild type poliovirus.[6] However, because the synthetic virus was genetically marked through the intentional introduction of 27 nucleotide changes, the virus proved "attenuated" (less virulent), which we later showed to be due to the exchange of a single nucleotide within the 7500 nucleotide long genome.[9]

3. The response to the poliovirus synthesis

The synthesis of a replicating biological entity in the absence of the natural template was without precedence at the time of publication and provoked unusually strong and conflicting responses. As summarized in Table 1, it was praised as landmark in biology, perceived as a challenge to divine power, scorned as a stunt, considered to wreck poliovirus eradication, and condemned as dangerous to national security.

A major reason for the intense discussions surrounding the publication[6] by Cello *et al.* in 2002 was the timing of publication. Following the terrorist assault

SIX CATEGORIES OF THE RESPONSES TO THE POLIOVIRUS SYNTHESIS
1. Landmark in biology
2. Ethical issues: *Has life been created in test tube? It the synthesis a challenge to divine power?*
3. The value of the experiment: *Did the predictability of the outcome render the experiment a stunt?*
4. Synthesis of poliovirus and global eradication: *Does the regeneration of poliovirus render global eradication of the virus impossible?*
5. Shock and fear in relation to bioterrorism: *Is the poliovirus synthesis a blueprint for bioterrorists?*
6. Publication of the synthesis of poliovirus and national security: *Should biological research and its publication be censored?*

Table 1. Responses to the poliovirus synthesis.

of September 11, 2001, and the anthrax bioterrorist attack in 2001–2002, the publication reached a highly sensitized public, particularly in the U.S.A. Ethical considerations, *e.g.* whether viruses are chemicals amenable to synthesis and modification or living entities, were less of an issue. Instead, the synthesis was considered almost exclusively in the context of bioterrorism. It was concluded, correctly, that bioterrorist agents, such as smallpox virus, could in the future be synthesized from information available on the Internet. Indeed, a discussion of this possible threat was included into the original draft of our Cello *et al.* manuscript submitted to *Science* in 2002.[6] Regrettably, the editors of *Science* eliminated a passage referring to possible dual use of modern technology from our original manuscript. We emphasize that the threat of synthesizing dangerous bioterrorist agents for malicious intent did not arise from the publication of the poliovirus synthesis. This threat is intrinsic to the recent advances in biotechnology.[6,11] However, we believe that it is important to bring these issues to the attention of the public, as we disagree with suggestions that the poliovirus synthesis should have been kept a secret. A detailed discussion of the responses to the poliovirus synthesis listed in Table 1 can be found in Ref. 11.

4. Application of the strategy of whole genome synthesis to viruses

Synthetic biology strives to generate novel biological systems that do not exist in nature, mainly with medical or commercial applications in mind. The poliovirus synthesis as described above does not fall into this category since it resulted in poliovirus with nearly identical phenotypes as the model wild-type virus (as mentioned, only the pathogenic phenotype of the synthetic virus in transgenic mice was different because of nucleotide changes introduced intentionally as genetic markers[6,9]).

Recently, we have generated novel polioviruses with vastly different phenotypes when compared to wild-type poliovirus.[7] This was done by chemical/biochemical synthesis of very large segments of the viral genome. In these segments (~3,000 nucleotides), synonymous codon usage was either deoptimized or the position of synonymous codons was randomized. This strategy is based on the fact that all amino acids but two are encoded by more than one codon (there are 20 amino acids and 64 codons). However, the frequency of synonymous codon use for each amino acid is unequal and, overall, the frequencies have co-evolved with the cell's translation machinery to avoid excessive use of suboptimal codons. Changing synonymous codons altogether or randomizing them can unbalance the synthesis of proteins without changing the amino acid sequence of the protein. Hence, the product of the translational machinery remains the same while the efficiency of protein synthesis may be vastly changed. Thus, regardless of the changes introduced into the genome, a virus synthesized in the infected cell will have the same structure and will encode the same replication proteins as the virus with the wild-type nucleotide sequence. This means that the virus variant with the altered genome will infect and replicate in the same cells as the wild-type virus, but it may be significantly disadvantaged when it comes to proliferation. Such a variant of a human pathogenic virus may enter the host by its normal route, replicate poorly, but still allow the host to mount a sufficiently strong immune response to clear the infection and develop lasting immunity. In other words, a human virus with altered codon usage or codon pairs could possibly serve as vaccine.

We have made use of the two possibilities to change codons in the (messenger RNA) template for translation without changing the sequence of the translation product.[7] One is to shuffle existing codons to maximize their positional changes while codon usage (the overall frequency of synonymous codons in the viral mRNA) is maintained. The extent of codon "shuffling", which we have applied, did not exert any influence on poliovirus proliferation, regardless of whether the assays were conducted in tissue culture cells or in poliovirus-sensitive transgenic mice.[7] On the other hand, changes in the nature of codons (exchanges of commonly used codons to codons that are rarely used) had a profound effect on viral infectivity:[7] the ratio of particles to infectious units (the measure of how many particles are necessary to achieve productive infection of one cell) was changed from 100:1 to 100,000:1. However, once a successful infection was achieved, the virus managed to produce per cell as many particles as wild-type virus.[7] Experimental evidence clearly showed that the reason for the altered particle to infectious units ratio was due to a translational handicap early in the infectious cycle.[7]

By fine-tuning the relationship of translation to pathogenesis, a virus variant may be developed with properties useful for vaccine application. Viral vaccines generated by large-scale genome alteration have the advantage over conventional live virus vaccines in that, due to the large-scale codon alterations (changes in the nucleotide sequence of the viral mRNA), "reversion" of the genetic information from attenuation to virulence in vaccine recipients is all but impossible.

5. Conclusion

The synthesis of poliovirus in the absence of natural template was a landmark experiment without precedence. Although it had been predicted prior to its publication, the experiment has caused unusual responses ranging from ethical to political issues. It was largely overlooked that a virus was regenerated ("recreated") outside living cells from written information in the public domain, a new reality that reduced viruses to chemicals. Since viruses assume properties of living entities upon entering the cell – heredity, genetic variation, selection towards fitness, evolution into different species, and even some form of sex (exchange of genetic material with closely related viruses through recombination) – we consider viruses as "chemicals with a life cycle".

Whole scale genome synthesis of viruses can now be easily realized with rapidly developing new technologies of DNA synthesis.[12] We believe that there is great promise in using this strategy to developing novel viral derivatives useful in medicine. Fully exploiting the potential of genome-level synthesis will require a new generation of sequence design tools. We have developed algorithms for designing optimal synthetic coding sequences under a variety of criteria, including avoiding undesired subsequences[13] and RNA secondary structure[14] or interleaving in overlapping genes on alternate reading frames (B. Wang, D. Papamichail, S. Mueller, and S. Skiena, unpublished results). Undoubtedly, the new technologies could be misused for malicious intent from which only powerful and open research can protect us.[11] The experiments that we have described in this chapter[7] are but the beginning of exploring the vast space of genetic information for the generation of novel kinds of vaccines. This, indeed, connects our work with Synthetic Biology.

Acknowledgements

This work was supported in part by grants from the National Institutes of Health and the Defence Advanced Research Project Agency (to E. W.) and from the National Science Foundation (to S. S.).

References

1. D. Endy, "Foundations for engineering biology," *Nature* **438**, 449 (2005).
2. A. Molla, A. V. Paul, and E. Wimmer, "Cell-free, *de novo* synthesis of poliovirus," *Science* **254**, 1647 (1991).
3. J. M. Hogle, M. Chow, and D. J. Filman, "Three-dimensional structure of poliovirus at 2.9 Å resolution," *Science* **229**, 1358 (1985).
4. N. Kitamura, B. L. Semler, P. G. Rothberg, *et al.*, "Primary structure, gene organization and polypeptide expression of poliovirus RNA," *Nature* **291**, 547 (1981).
5. S. Mueller, E. Wimmer, and J. Cello, "Poliovirus and poliomyelitis: A tale of guts, brains, and an accidental event," *Virus Research* **111**, 175 (2005).
6. J. Cello, A. Paul, and E. Wimmer, "Chemical synthesis of poliovirus cDNA: Generation of infectious virus in the absence of natural template," *Science* **297**, 1016 (2002).
7. S. Mueller, D. Papamichail, R. J. Coleman, S. Skiena, and E. Wimmer, "Reduction of the rate of poliovirus protein synthesis through large scale codon deoptimization causes attenuation of viral virulence by lowering specific infectivity," *J. Virology* **80**, 9687 (2006).
8. S. van der Werf, J. Bradley, E. Wimmer, F. W. Studier, and J. J. Dunn, "Synthesis of infectious poliovirus RNA by purified T7 RNA polymerase," *Proc. Natl. Acad. Sci. U.S.A.* **78**, 2330 (1986).
9. N. DeJesus, N., D. Franco, A. Paul, E. Wimmer, and J. Cello, "Mutation of a single conserved nucleotide between the cloverleaf and internal ribosome entry site attenuates poliovirus neurovirulence," *J. Virology* **79**, 14235 (2005).

10. S. Mueller, P. Jiang, E. Rieder, *et al.*, "Pathogenesis and prevention of poliomyelitis and the chemical synthesis of poliovirus," *Nova Acta Leopoldina* NF 92, Nr. 344 (2005), pp. 35–43.
11. E. Wimmer, "The test–tube synthesis of a chemical called poliovirus. The simple synthesis of a virus has far reaching societal implications," EMBO Reports 7:S3-S9 (2006).
12. J. Tian, H. Gong, N. Sheng, X. Zhou, E. Gulari, X. Gao, and G. Church, "Accurate multiplex gene synthesis from programmable DNA microchips," *Nature* **432**, 1050 (2004).
13. S. Skiena, "Designing better phages," *Bioinformatics* **17**, 253 (2001).
14. B. Cohen and S. Skiena, "Natural selection and algorithmic design of mRNA," *J. Comput. Biol.* **10**, 419 (2003).

Guided Evolution in Interacting Microchemostat Arrays for Optimization of Photobacterial Hydrogen Production

R. H. Austin, P. Galajda, and J. Keymer
Dept. of Physics, Princeton University, Princeton, NJ 08540, U.S.A.

1. Introduction

Microorganisms are incredibly sophisticated chemical factories that can transform, with great precision and efficiency, raw chemical energy (food) into the most directed and precise of products necessary for the organism's survival in the particular ecological niche it finds itself. Microorganisms do not always need to eat food to survive: some have the ability to use not chemicals but light as an energy source. Although biological chemistry has attempted to deconstruct these remarkable factories with much success, and this is a worthy goal, a more opportunistic approach is simply to guide the intact factory in the cell through directed evolution towards an externally defined goal, that goal being the most efficient production of hydrogen gas (H_2) by the organism.

Ordinarily evolution experiments are done in chemostats that maintain a constant inflow of nutrients and outflow of waste in a stirred vessel containing a colony of bacteria or similar microorganism in solution. We wish to take the present chemostat design dramatically further by using our microfabrication technology to construct interacting arrays of chemostats, and then use our ability to sense the level of H_2 production within each microchemostat to punish colonies that have low H_2 gas production and reward colonies that have high H_2 production. As part of this work in competing microchemostat arrays, we will also conduct research in the basic dynamics of how populations evolving in the face of limited resources compete with each other, a subject of fundamental importance in biology at many size scales, from microbes to mankind.

We are using our expertise in microfabrication in conjunction with the ongoing work on microbiological H_2 production at Princeton to develop a new class of interacting microchemostat arrays (IMCAs) that will direct the evolution of single-cell cyanobacteria. An IMCA consists of a series of small microfabricated microchemostat squares, which have small channels of 10 μm diameter or smaller, to allow species in adjacent microchemostats to exchange with each other in a slow way. As a necessary part of this work, we also explore the evolution of competing species at a new level of complexity within a small world of topological design. This part of the work will probe, at a very general level, how species compete with each other both cooperatively and as antagonists.

 Future Trends in Microelectronics. Edited by Serge Luryi, Jimmy Xu, and Alex Zaslavsky

The reasons for maximizing the efficiency of photobiological hydrogen gas production are abundantly clear: i) fossil fuels are polluting our air, land and water; ii) carbon dioxide gas and related gas accumulation due to fossil fuel combustion will at some point in time start to affect the climate of the earth; iii) there are major technological problems with fusion power that remain to be overcome, nuclear (fission) power faces many political and security issues, and alternate sources, such as biomass production and solar power via semiconductor technologies, have substantial economic problems that do not scale well to replacing fossil fuels. At some point in time, we will find ourselves in a difficult position unless we begin to develop – as soon as possible – fuel-producing technologies with nearly zero environmental impact that can scale to a secure world consisting of all nations rich and poor.[1]

An H_2 technology has a major advantage in that much of the fossil fuel technology infrastructure could be relatively easily converted to a hydrogen gas fuel stock, and of course hydrogen gas burns extremely cleanly, yielding only water. Unfortunately hydrogen is not a "primary fuel" in the technical sense of the word. There is no free hydrogen and it takes chemical energy to make hydrogen from other sources. It is not so easy to generate hydrogen gas from organic food stocks using conventional organic chemistry technologies without consuming more free energy than is produced. Fortunately, however, there is a very large fusion reactor in the sky called the Sun, which produces about 500 W/m^2 of photon energy flux at the surface of the earth, and biology has long ago learned to produce chemical energy from light. In particular, some organisms have developed a photochemistry that can generate H_2. Hydrogen gas production from photobiology has an enormous advantage over H_2 production from fermentation in that the intermediate step of biomass production is eliminated. There are three classes of microorganisms that can generate H_2 from light:[2] cyanobacteria, single-celled algae and photosynthetic algae. Of the three, photosynthetic bacteria do not directly oxidize water, but use a combination of photosynthesis and biomass. Cyanobacteria and certain algae can, however, use light to oxidize water and generate H_2 directly from water. This chapter concentrates on using a new IMCA to study directed evolution of H_2 production by microorganisms. Both algae (which are complex eukaryotic organisms) and photobacteria (somewhat simpler prokaryotic organisms) can produce H_2 from sunlight and can in principle be evolved with our device. However, since this is "pioneering" work, we will work with bacteria and not the more complex eukaryotic algae, so that we can effectively harness extensive molecular biology technology that has been developed for prokaryotes, particularly *E. coli*. Perhaps what we learn here can be applied to eukaryotes, such as the green algae.

There exist two classes of bacteria that use light to produce oxygen: photosynthetic bacteria such as *Rhodopseudomonas viridis* and the cyanobacteria, such as single cell cyanobacterium *Synechocystis* sp. PCC 6803 or the single cell algae *Chlamydomonas reinhardtii*. Both of these cyanobacteria are single-celled and can generate H_2. Of the two, *Synechocystis* sp. PCC 6803 is of greater interest for directed evolution because it is the first photosynthetic organism to be completely sequenced and this is a huge advantage for any genomic work. However, *Chlamydomonas reinhardtii* is also interesting because it is unicellular, grows quickly, forms colonies on plates and is easy to transform.

2. Basic interacting microchemostat technology

We are using our expertise in microfabrication to develop a new form of microchemostat array that will take advantage of three aspects of microlithographic technology to direct the evolution of cyanobacteria to optimize H_2 gas production. The key idea here, due to J. Keymer, is the IMCA. There are several unique features of this device, listed below.

- Because only a few thousand, rather than 10^{10} cells are in each microchemostat, it is far easier for genetic fluctuations to compete with the dominant species. Since our aim is the rapid and directed evolution of H_2 production, it is important that mutations that produce higher levels of hydrogen gas be quickly identified and rewarded, even if they are slower growing or have other non-competitive properties.
- The microchemostats will be optically thin (on the order of 20 μm to 400 μm thick) so that light can uniformly illuminate the entire depth of the chemostat.
- The transparent nature of the microchemostat will allow us to make sensitive colorimetric measurements of H_2 gas remotely in each microchemostat, as we will discuss below.
- Finally, each microchemostat will be "weakly coupled" via 10 μm channels with neighboring chemostats, so that we can both punish poorly performing sub-species in a microchemostat and allow highly performing bacteria to move into one where the species is weak, as part of the reward process.

Arrays of microchemostats offer a fundamental advantage in that the small volume and consequently small numbers of bacteria in each space allow for fluctuations in the genotype within a microchemostat to compete efficiently with the established colony, avoiding the tyranny of numbers if a mutated strain would have to compete with a vast number of dominant bacteria. The "optimum" volume each microchemostat should have in order to accelerate evolution is not clear at this point, but this is a variable completely under our control and will be varied in conjunction with both theoretical analysis and experiments. In order to develop the basic technology, the initial experiments used an even simpler design than a 2D array. Instead, we used a line of coupled microchemostats, fed by 100 nm deep nanofeed channels capable of introducing and removing chemicals without removing the bacteria, since the bacteria cannot penetrate a 100 nm wide gap. Transverse to this line of microchemostats is an array of nanochannels that are etched down 100 nm but are a full 100 μm wide. This transverse set of nanochannels allows us to: i) confine the bacteria to the coupled array of chemostats; ii) feed the bacteria as in a standard chemostat; iii) poise the bacteria in the correct redox and solvent oxygen state so that hydrogen gas production occurs in the dark; iv) sample the output of the microchemostat arrays for hydrogen gas production on an individual scale, but spatially separating the probing chemistry from the colony itself to avoid interference. Figure 1 shows the basic scheme of the transverse flow microchemostats.

The reward/punishment cycle can be carried out in a discrete manner based upon the following cycle: light based photosynthesis, dark anaerobic production of hydrogen, analysis of the H_2 gas production during the dark anaerobic phase, and repetition of the cycle with light and food delivery to a particular microchemostat. Those colonies that have been determined to produce H_2 at a high rate during the anaerobic phase will be rewarded with more light at the wavelength where H_2 photoproduction is highest, while those colonies that are poor H_2 producers will be both punished and urged to evolve: we will starve them for actinic light and increase mutagenic UV light to increase the mutation rate. Since the chemostats are interconnected, rapidly growing colonies that are good H_2 producers will be allowed to gain more space in addition to being rewarded with more light.

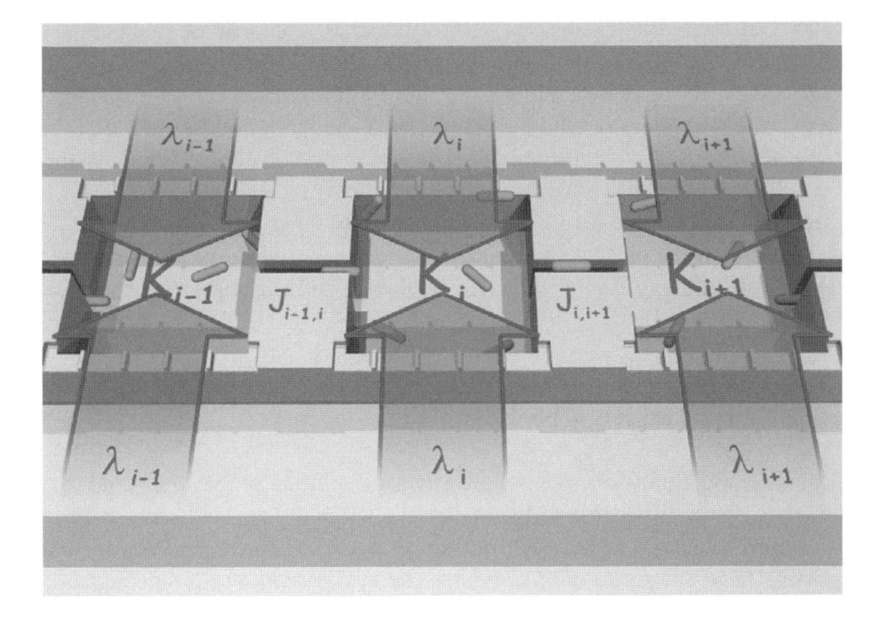

Figure 1. Cartoon of the coupled chemostats. Each chemostat is characterized by a carrying capacity K_i, a measure of how many organisms can live there, and a diffusional food flux with coefficient λ_i. There is also a flux of organisms between adjacent microchemostats which is denoted by $J_{i,i+1}$ and $J_{i,i-1}$. The microchemostats and the feed/sensor channels are deep etched (10–100 μm), while the broad shallow etch feeds that couple the feed and sensor channels to the microchemostat are 100 nm deep and too shallow for bacteria penetration.

Another possible reward is the delivery of food stocks to good H_2 producers and punishment via either starvation or the delivery of toxins (or mutagenic chemicals) to poorly producing bacteria. This delivery of fluids to specific chambers is easily implemented in the 1D array of Fig. 1, since the transverse 100 nm deep channels will have strictly laminar flow profiles that confine fluids within single chambers across the width, while the connection channels are long enough (hundreds of μm) that cross-diffusion is not a problem on the hour time scale.

Remote detection of the local H_2 gas concentration during the anaerobic hydrogen-producing phase will be critical in a directed evolution experiment. Fortunately there is a fairly simply way to do this and it involves use of the same enzyme developed by these remarkable microorganisms. The soluble hydrogenase of the aerobic protobacterium *Ralstonia eutropha* can catalyze the reduction of the colorimetric redox dye benzyl viologen (BV). If the hydrogenase is either present in the input fluid flow to the microchemostat or immobilized on the surface, it will in the presence of H_2 gas catalyze the oxidation of H_2 and reduce colorless redox dye benzyl viologen BV_{2+} to the red light absorbing BV^+.[4] A three-color chip CCD camera imaging the array, with each pixel mapped to a microchemostat, can determine the level of BV^+ being produced by the cyanobacteria in each microchemostat and hence the H_2 production of that particular microstrain. Decisions of reward and punishment for a population in a given microchemostat can be made based on a combination of growth rate and H_2 production.

Of course, it could well happen that the BV present in the same chamber as the bacteria could interfere with the assay. Fortunately, in the case of the 1D array this can be mitigated by confining the H_2 assay to one of the feed stock channels, and use the laminar flow we discussed above to bring the effluent from the bacteria microchemostats into the feed stock channel and do the assays remotely. No doubt some redesign of the present chip will be in order but we believe the basic concept of compartmentalizing the bacteria microcolonies and the assay regions is easily done via micro/nanofluidics.

3. Fundamental aspects of guided evolution

It is important to emphasize that the growth and evolution of even bacterial species is a complex process because of the highly evolved biological networks in bacteria that enable bacteria to compete or cooperate with other strains for limited resources. In our case, we will be guiding the evolution of the competing species, but we will by necessity be working within the framework of competing species in stationary growth conditions. Some of the relevant concepts are discussed below.

- *The Prisoner's dilemma*

The Prisoners dilemma[5] is a game that comes from game theory. A succinct description of this game might run as follows:[6]

"The game got its name from the following situation: imagine two criminals arrested under the suspicion of having committed a crime together. However, the police do not have sufficient proof in order to have them convicted. The two prisoners are isolated from each other, and the police visit each of them and offer a deal: the one who offers evidence against the other one will be freed. If none of them accepts the offer, they are in fact cooperating against the police, and both of them will get only a small punishment because of lack of proof. They both gain. However, if one of them betrays the other one, by confessing to the police, the defector will gain more, since he is freed; the one who remained silent, on the other hand, will receive the full punishment, since he did not help the police, and there is sufficient proof. If both betray, both will be punished, but less severely than if they had refused to talk. The dilemma resides in the fact that each prisoner has a choice between only two options, but cannot make a good decision without knowing what the other one will do."

Since the Prisoner's dilemma (PD) is widely used as a metaphor for the evolution of cooperation, we are interested on using bacterial genetics to implement the iterative (and spatial) PD of bacterial societies on our metapopulations biochips. Ultraviolet-laser disturbance will be used to study the game (and its evolutionary strategies) in different dynamic landscapes of opportunity patches. In the standard formal form, the PD game goes like this:[7] there are two players which can choose (independently but simultaneously) to cooperate, C, or defect, D, in any one encounter. If both players cooperate (C), they get a payoff of magnitude R (a "reward"); if one defects and the other cooperates, D gets the biggest payoff T (the "temptation"), while C gets the smallest payoff S. If both defect, both get P. The payoff matrix can be written as:

$$\begin{pmatrix} R & S \\ T & P \end{pmatrix} . \tag{1}$$

The most interesting situation occurs when the elements of the payoff matrix (1) satisfy the following inequality:

$$T > R > P > S . \tag{2}$$

The paradox embedded in the PD relationship of Eq. (2) is that strategy D is unbeatable at any one round of the game. However, if both players play it iteratively, both end up with less total payoff than if had they cooperated. This paradox is a problem not just for humans but also for *E. coli*[8] (and our cyanobacteria evolving in the IMCA structures). As resources change during growth and cell density increases, organisms like *E. coli* change their gene expression and enter what is called the GASP phase, where GASP is an acronym for the growth advantage in stationary phase phenotype – see Fig. 2. Much like us, *E. coli* organisms are victims of the tragedy of the commons.[9] In closed cultured systems (test tubes) after the mid-log phase, resources are not as available and toxic waste is building up. At this moment, the cells signal one another (*i.e.* by secreting auto-inducer molecules like AI_2)[10] and decide to cooperate and stop replicating. This is the onset of the "stationary phase", and the standard condition in the IMCA as we evolve the organisms. Stationary phase *E. coli* cells express the gene rpoS and produce the sigma factor σ^S (a genetic switch) responsible for the relevant gene regulation needed to express stationary phase programs.

Under this "phenotipic state", wild-type organisms "cooperate" (by not dividing) in order to save resources. After a while in the stationary phase (middle part of the temporal axis in Fig. 2), "GASP mutants" appear in the population (underlying curves in Fig. 2) and are constantly replaced. Thus, when we look carefully, we note that the nominally stationary phase is actually very dynamic. The older we allow the culture to get, the more mutants appear, until a point diversity starts accumulating and highly diverse and polymorphic assemblages start to develop. This is the onset of "stationary phase" in the IMCA under which we will be working.

Generally, GASP mutants carry mutations on the rpoS gene.[11] In particular, the GASP phenotype of the early mutants that take over stationary phase cultures of wild-type cells is thought to arise because they carry the $rpoS_{819}$ allele. The $rpoS_{819}$ allele (the cheating gene?) has a 46 base pair duplication of part of the original sequence of the wild-type allele ($rpoS_{wt}$), which allows easy identification of PCR fragments by running agarose gels. More information on the gene and its products can be easily obtained from Ecocyc (*http://www.ecocyc.org*) by searching for rpoS.

- *Rewards and punishments*

Extinction of local populations, and the colonization of new suitable habitat patches (metapopulation dynamics) as well as the regime of patch creation and destruction (patch dynamics) are the basic components of the evolutionary process shaping natural populations – "the zoology of archipelagoes", as Darwin liked to

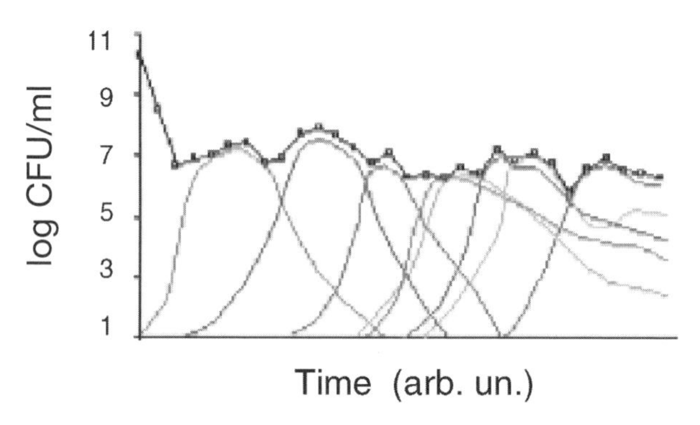

Figure 2. GASP stationary phase dynamics, plotting colony-forming units (CFU)/ml *vs.* time. Stationary phase really is a very dynamic regime, with different GASP mutants successively replacing each other.

put it. The theories of island biogeography[12] and metapopulation biology[13] are the modern legacy of Darwin's vision about the role of environmental patchiness in structuring ecological niches. At a more abstract level, the basic idea of this project is to use metapopulation and patch (landscape) dynamics in conjunction with fluorescence measurements to build fitness landscapes and evolve populations of microorganisms in microfluidic devices. The more fundamental goal of this experiment is the manipulation of a dynamic landscape of opportunity patches, as described by Keymer and collaborators.[14] In our UV-landscaping machine, chambers under UV radiation correspond to "destroyed patches" (state −1) and UV-free chambers correspond to "suitable patches" (state 0), as shown in Fig. 3. Patch dynamics can be easily implemented thanks to the lethal action of UV laser beams upon microorganisms. By adding a chamber to the laser "hit list" we can drive a local population (located at x) to extinction by habitat destruction:

$$1_X \rightarrow -1_X . \tag{3}$$

Habitat destruction can also affect empty patches:

$$0_X \rightarrow -1_X . \tag{4}$$

By removing a chamber from the laser's "hit list" we can "create" a suitable habitat patch:

$$-1_X \rightarrow 0_X , \tag{5}$$

which later can be colonized by propagates coming from nearby populations (chambers). Unlike colonization-extinction reactions that are evolved by the organisms in the culture, these new chamber reactions are generated by laser irradiation (under our control). Using the laser, we implement a regime of landscape dynamics and guide the organisms to evolve in ways we desire.

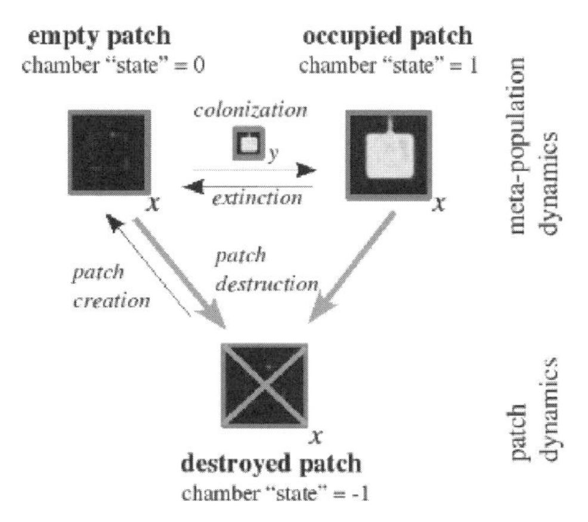

Figure 3. The use of lethal UV laser radiation to implement a model of meta-populations in dynamic landscapes. Monte Carlo sampling of chambers and the UV laser create a dynamic landscape of habitat patches.

4. Preliminary work

- *Initial work on population dynamics*

We have pioneered the study of the colonies of microorganisms in microfabricated spaces. The initial research involved studying how bacteria colonize and communicate with each other in microfabricated environments designed to mimic the complex world that microorganisms inhabit in the real world, as opposed to an agar plate. We found, as was anticipated by many biological studies of course, that even *E. coli* had a complex but understandable signaling pathway that, under conditions of metabolic stress, drove the organisms to form a microcolony in the smallest enclosed volume they could find. The fact that we could simulate this behavior mathematically, using a coupled set of physically realistic equations governing bacterial density and chemotactic response, gave us faith in the general power of this approach.

Figure 4 shows the basic idea behind these original experiments.[15] We studied the behavior of both *E. coli* and *V. harveyi* in these structures. We were able to observe the metabolic stress-induced clustering of the bacteria together into compact colonies, as well as simulate these population movements.[16]

We are now working towards performing similar experiments in a 2D array of microchemostats, shown schematically in Fig. 5. We have carried out initial culturing of green-fluorescent expressing *E. coli* in this array. Each array element is a 100x100x20 μm voxel, linked to its neighbors by a 10 μm wide channel. At present, we have a simple chemostat design where nutrients are feed into one end of the array and waste is removed out the other end, with a syringe pump

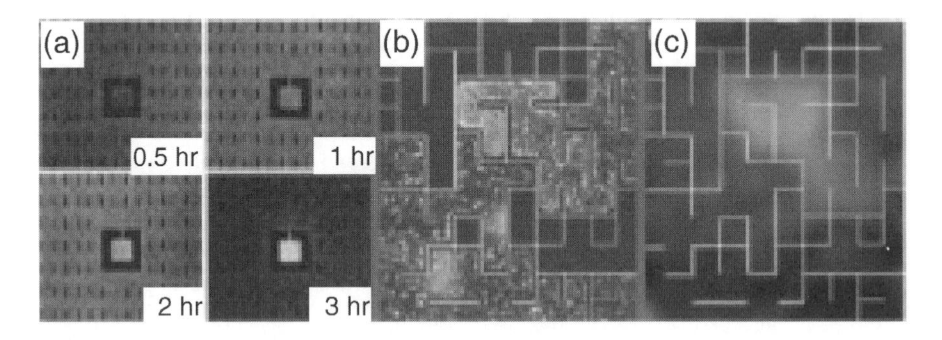

Figure 4. *E. coli* and *V. harveyi* accumulation and quorum sensing. (a) Epifluorescence images of GFP-labeled *E. coli* as they accumulate into a central 250x250 μm enclosure via a 40 μm wide channel through 100 μm wide walls. After 3 hours the density of cells is more than seven times greater inside than outside. The rectangles are silicone pillars that support the roof of the chamber. (b) Dark-field image of *V. harveyi* after 8 hours in the maze. Lines corresponding to the walls of the maze are overlaid for clarity. (c) Photon-counting image of the intrinsic luminescence, indicating active quorum sensing.

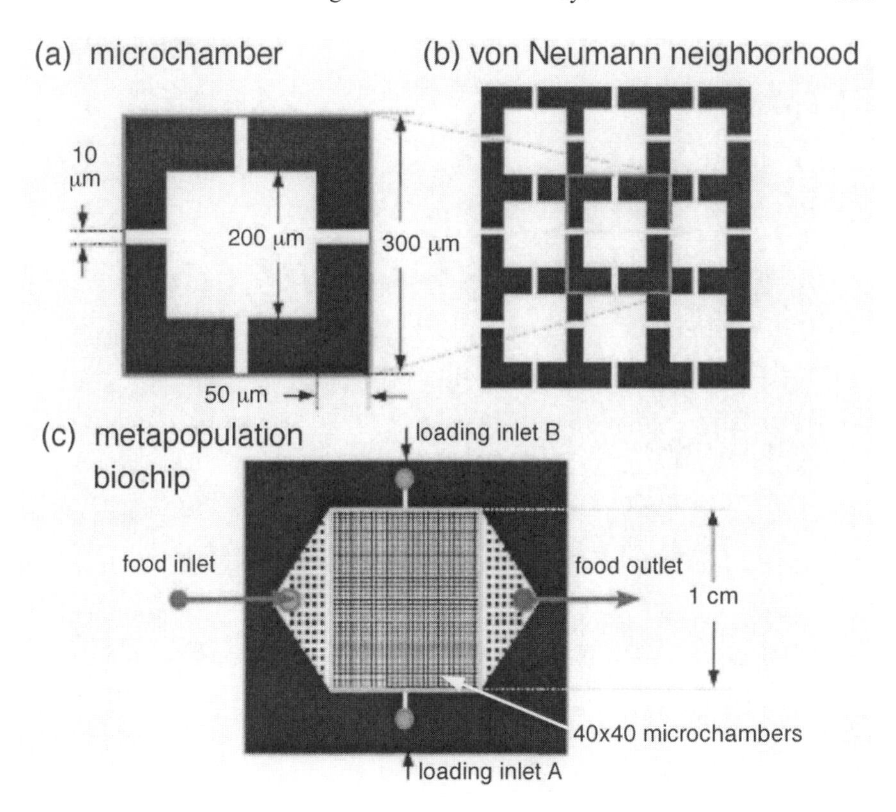

Figure 5. Basic design of a metapopulation biochip: (a) single microchamber; (b) a "von Neumann" neighborhood used to introduce local connectivity (limited dispersal) to four nearest-neighbor chambers; (c) main chamber at the center of the biochip. The food inlet (left) and outlet (right) used for flowing culture media are interfaced to a syringe pump via o-rings. The main chamber is embedded into a transition (field of posts) area that distributes the flow in the vertical direction. Two inlets at top and bottom are for loading organisms.

delivering metered amounts of fluid. The basic design of a layout of 6 arrays, with each array containing 10^4 microchemostats, is shown in Fig. 5. *E. coli* grow vigorously in these devices, as can be seen in the epifluorescence image of Fig. 6, showing a section of the array 12 hours after the array was inoculated with a low density of GFP-expressing *E. coli* (here the bacteria have reached stationary levels of population density). We have also shown that pulsed UV light at 337 nm from a N_2 laser can be used to kill bacteria within an individual array, as expected.[17,18]

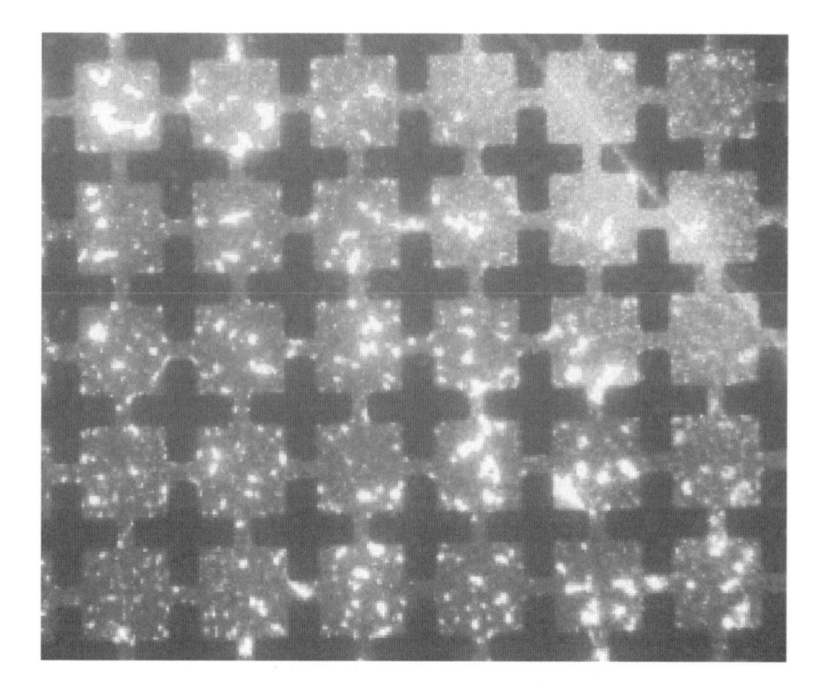

Figure 6. GFP-expressing *E. coli* growing in 100x100 μm microchemostats.

- *Spatial evolutionary games: E. coli metapopulations in dynamic landscapes*

We are interested in understanding how different strains solve the spatio-temporal allocation of local fitness and apply it to the H_2 evolution experiment. We plan to use strains which are chemotactic as opposed to non-chemotactic and quorum-sensors as opposed to non-quorum-sensors in order to construct cheaters and cooperators for the PD problem. Armed with these strains we hope we will be able to study game dynamics and the relationship between genetic architectures and spatiotemporal evolutionary stable strategies evolved by populations growing on our biochips.

We have begun a series of experiments to test the 1D technology that will be key to optimizing H_2 production in the cyanobacteria, as we have discussed earlier. These initial experiments with the 1D array are very simple. They begin with the inoculation of the 1D array with a GFP-expressing strain of bacteria at a very low density of about 10 bacteria per microchamber. The bacteria can pass individually between the chambers. To our surprise, we have found that the population dynamics even in such a systems are astonishingly complex, with different levels of movement of the bacteria between the chambers in both wave-like and chaotic manner. Figure 7 gives a brief taste of the population dynamics that we are just beginning to explore.

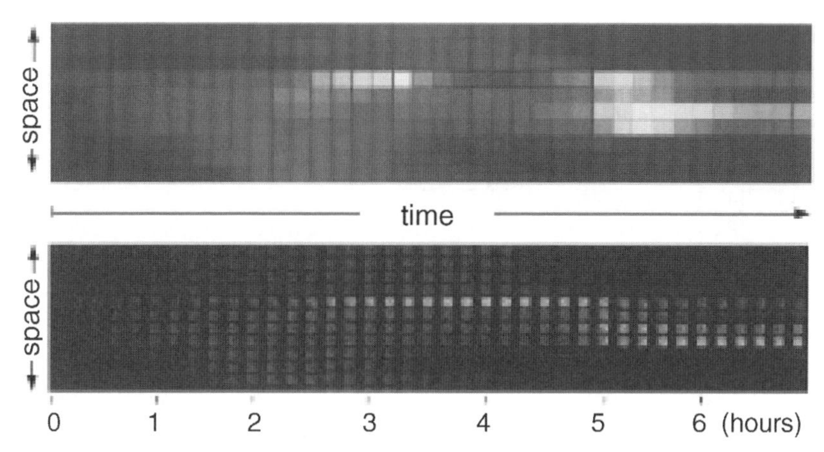

Figure 7. Time-sequence images of bacterial populations in a coupled set of linear microchemostats. The lower image shows bacterial densities for 50 coupled microchemostats, the upper image shows the densities in grayscale. As bacteria grow, they jump between different microchemostats in a collective manner.

5. Conclusions

Evolution itself is strongly influenced by topology and metapopulation dynamics, as Darwin recognized in his revolutionary *Origin of the Species*. Indeed, island metapopulation dynamics was central to his experimental observations. The key idea for Darwin was that small, interconnected metapopulations evolve in ways that can be analyzed because neighboring populations are always different but related to one another. By working at the interface between landscape ecology and nanobiotechnology, we believe the technology presented here can implement ecological forces in microfluidic devices and use these forces to study the interplay between the spatiotemporal structure of a habitat and the organismal biology of life-cycle propagation. We believe that one possible result of this guided evolution might be the optimization of H_2 gas production by photosynthetic organisms.

Acknowledgments

We thank Ted Cox, Miguel Gaspar, Juliana Malinverni, Sungsu Park, Pascal Silberzan, Tom Silhavy, and Ned Wingreen for comments, discussions and support, and Peter Wolanin for supplying us with the *E. coli* GFP construct. This work was supported by the AFOSR (FA9550-05-01-0365), NIH (HG01506), NSF Nanobiology Technology Center (BSCECS9876771), and Cornell NanoScale Science and Technology Facility (NSF Grant ECS9731293).

References

1. National Research Council Committee on Alternatives and Strategies for Future Hydrogen Production and Use, *The Hydrogen Economy: Opportunities, Costs, Barriers, and R&D Needs*, Washington, DC: The National Academies Press, 2004.
2. G. A. Peschek, W. Loeffelhardt, and G. Schmetterer, eds., *The Phototrophic Prokaryotes*, New York: Kluwer Academic, 1983, p. 836.
3. B. J. Lutz, Z. H. Fan, T. Burgdorf, and B. Friedrich, "Hydrogen sensing by enzyme-catalyzed electrochemical detection," *Anal. Chem.* **77**, 4969 (2005).
4. We note here that Prof. Hugo Fan of the University of Florida suggested this redox sensor to us.
5. F. Heylighen, "Evolution, selfishness and cooperation," *J. Ideas* **2**, 70 (1992).
6. See *http://pespmc1.vub.ac.be/prisdil.html*
7. M. Nowak and R. May, "Evolutionary games and spatial chaos," *Nature* **359**, 826 (1992).
8. M. Vulic and R. Kolter, "Evolutionary cheating in *Escherichia coli* stationary phase cultures," *Genetics* **158**, 519 (2001).
9. G. Hardin, "The tragedy of the commons," *Science* **162**, 1243 (1968).
10. M. G. Surette and B. L. Bassler, "Quorum sensing in *Escherichia coli* and *Salmonella typhimurium*," *Proc. Nat. Acad. Sci. USA* **95**, 7046 (1998).
11. M. M. Zambrano and R. Kolter, "GASPing for life in stationary phase," *Cell* **86**, 181 (1996).
12. R. H. McArthur and E. O. Wilson, *The Theory of Island Biogeography*, Princeton, NJ: Princeton University Press, 1967.
13. I. A. Hanski and M. E. Gilpin, *Metapopulation Biology: Ecology, Genetics, Evolution*, San Diego, CA: Academic Press, 1997.
14. J. E. Keymer, P. A. Marquet, J. X. Velasco-Hernandez, and S. A. Levin, "Extinction thresholds and metapopulation persistence in dynamic landscapes," *Amer. Naturalists* **156**, 478 (2000).
15. S. Park, P. M. Wolanin, E. A. Yuzbashyan, P. Silberzan, J. B. Stock, and R. H. Austin, "Motion to form a quorum," *Science* **301**, 188 (2003).
16. S. Park, P. M. Wolanin, E. A. Yuzbashyan, *et al.*, "Influence of topology on bacterial social interaction," *Proc. Nat. Acad. Sci. USA* **100**,13910 (2003).
17. A. Tiphlova, T. I. Karu, and N. R. Furzikov, "Lethal and mutagenic action of XeCl laser radiation on *Escherichia coli*," *Lasers Life Sci.* **2**, 155 (1998).
18. L. Xue, Y. Zhang, T. Zhang, L. An, and X. Wang, "Effects of enhanced ultraviolet-B radiation on algae and cyanobacteria," *Critical Rev. Microbiol.* **31**, 79 (2005).

Improvements in Light Emitters by Controlling Spontaneous Emission: From LEDs to Biochips

C. Weisbuch
Ecole Polytechnique, 91120 Palaiseau, France and
Materials Department, UC–Santa Barbara, Santa Barbara, CA 93106, U.S.A.

A. David
Materials Department, UC–Santa Barbara, Santa Barbara, CA 93106, U.S.A.

M. Rattier, L. Martinelli, H. Choumane, N. Ha, C. Nelep, A. Chardon, G.-O. Reymond, C. Goutel, and G. Cerovic
Genewave, bât. XTEC, Ecole Polytechnique, 91128 Palaiseau Cedex, France

H. Benisty
Labo. Charles Fabry de l'Institut d'Optique, CNRS, 91127 Palaiseau, France

1. Introduction

The need to control electron and photon motion in solids, which would in particular enable better optoelectronic devices, has driven the field of light–matter interactions for the past thirty years.

Semiconductors, although uniquely useful for electrically pumped light emission, are hampered in this application by their high $n \geq 2.5$ refractive index. At a planar face of a standard light-emitting diode (LED), extraction efficiency is limited to only $\eta = 2$–4% by the small fraction of internal emission that impinges at angles smaller than the critical angle, while substrate absorption usually consumes the remaining 96–98% of the internally reflected photons.[1,2] One possible solution to this drawback is to have light make multiple attempts to escape by impinging on the semiconductor-air interface at different angles. This is the solution offered by *geometrical* or *ray optics*, relying on shaped devices or disordered interfaces. A more complete control relies on *wave optics* and optical mode quantization, which can be achieved in a number of ways, through microcavities of varied photonic dimensionalities (including photonic crystals), or simpler interference systems.

In this chapter we will describe some recent results on microcavity and photonic crystal LEDs. It will be shown that in order to obtain a significant improvement in extraction efficiency, the structures must be fully designed, to control both in-plane and vertical modes of spontaneous emission.

The concepts used for semiconductor devices can be fruitfully applied to biological systems. Like LEDs, fluorescent DNA or protein microarrays[3] suffer from poor luminescence extraction efficiency. Fluorescent spots originating from

Future Trends in Microelectronics. Edited by Serge Luryi, Jimmy Xu, and Alex Zaslavsky

ISBN 0-471-48

the spatially selective attachment of fluorescent species to glass surfaces mostly emit into the substrate; the remaining light is not efficiently collected for detection. We will describe amplifying substrates and integrated CCD/biochips that provide greatly enhanced fluorescence collection, translating into economies of biological material, improved detection of genes with low expression, real-time measurements of hybridization, all achieved in high-functionality integrated systems.

2. Microcavity LEDs

- *A simple model of microcavity emission*

Placing an emitter between the two mirrors comprising a microcavity leads to emission in those directions for which constructive interference occurs. A simple prediction for how much light can be extracted from a microcavity follows from a mode-counting argument illustrated in Fig. 1. It can be shown that each cavity mode, either extracted or guided, carries the same emission intensity. Then, the extraction efficiency η is the ratio of the number of modes in the extraction window (limited by the critical angle) to the total number of modes m_C. If there is only one mode in the escape window, the extraction is therefore simply given by $\eta = 1/m_C$.

More exact calculation methods[4] must be used to optimize microcavity (MC) LED performance. However, the ultimate efficiency that can be expected in microcavity LEDs in the GaN material system is quite limited: with standard materials, the best achievable extraction efficiency would be in the 40% range (into epoxy shaped domes). In addition, this optimized structure would require fabrication precision in the thickness of the cavity and in the positioning of the emitting QWs that appear technologically impossible.[5] Therefore, MC LEDs are not competitive with existing solutions based on geometrical optics. The way to further progress in LEDs based on wave optics is to use photonic crystals to extract the guided light remaining in MC LEDs.

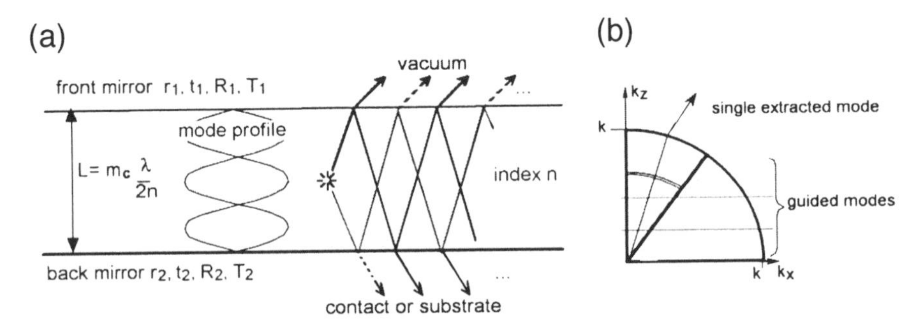

Figure 1. (a) Schematics of light extraction in a microcavity LED cavity with a resonant mode profile. A source emits two series of resonant waves, as shown. (b) Mode counting: for enhanced extraction efficiency, one mode is extracted while only a few others are guided.

3. Photonic crystal LEDs

A number of photonic crystal (PC) schemes have been proposed to extract light from LEDs. As illustrated in Fig. 2, three are based on the forbidden bandgap of PCs, whereas two other schemes are based on PC diffraction properties.

Photonic crystals can act as mirrors in their forbidden band to reinject the waveguided light into the active region, where it can undergo photon recycling as in Fig. 2(a). This can only be efficient in those cases where the internal quantum efficiency is very near unity, leading to a good improvement by photon recycling. Defects in PCs can define microcavities such that no in-plane modes exist within the emission spectrum, making the cavity between the PC mirrors anti-resonant – see Fig. 2(b). Then light can only escape through vertical optical modes. Besides the difficulty of achieving such tiny structures with the desired outgoing optical modes, this scheme would yield devices with minute active volumes and hence very low emitting powers, unless many such devices are operated in parallel. It

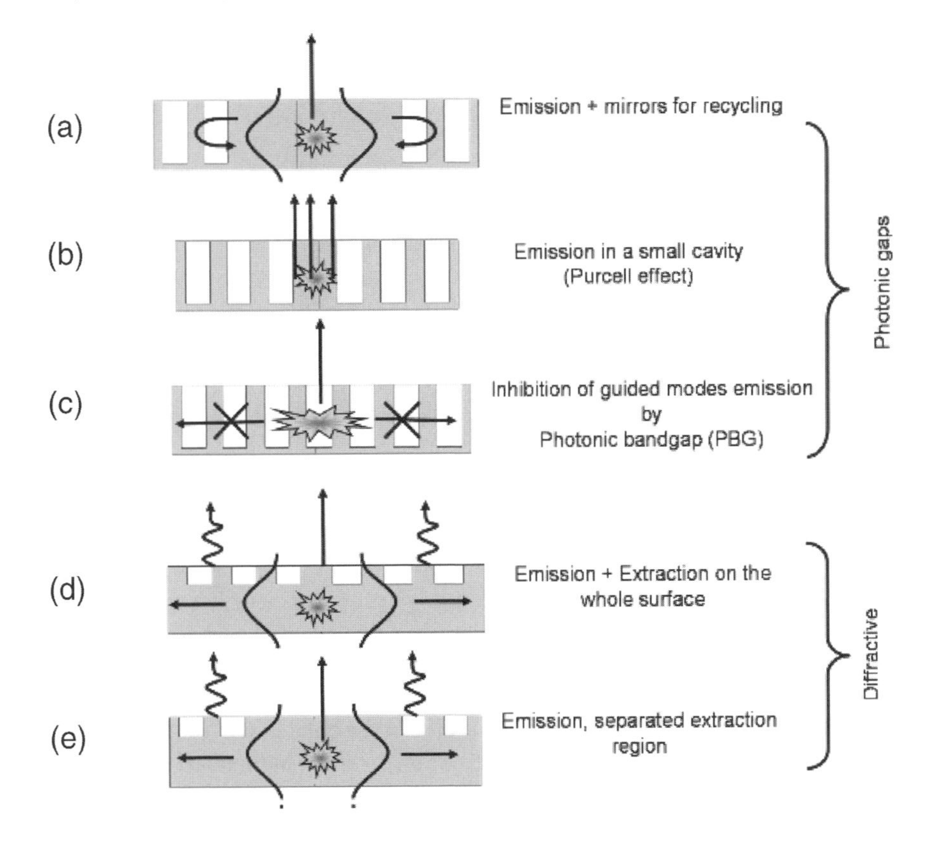

Figure 2. Schematics of light extraction schemes by photonic crystals: (a) recycling of guided modes; (b) extreme confinement; (c) photonic band gap forbidding guided modes; (d) extraction by photonic crystals in emitting area; (e) same but only around emitting area.

should be noted, however, that such structures are the basis of single photon emitters or of strongly coupled systems for the light-matter interaction. Another approach relies on modifying the emission region with a deep-etched pattern, so as to forbid propagation of in-plane guided modes, as in Fig. 2(c). This forces the emitted light to be redirected out of plane through the leakage of guided modes.[6] However, this approach should lead to poor internal quantum efficiencies at room temperature, given that the structure is etched through the active layer. A shallow etched grating near second order diffraction does not prohibit emission into the guided modes, but diffracts them towards radiating modes – see Figs. 2(d, e). This appears to be the most promising route towards improved light extraction in low-cost LEDs. Note that PCs can either be used to extract guided modes outside the excited region, or from the excited region. The exact choice will depend on fabrication parameters, such as the etch damage to the active region.

A PC-based light extractor should have the following properties: i) any light impinging on the PC should be scattered towards air, within the device length, which implies an optimized diffractive structure; ii) no light should escape in the plane, either in reflection or transmission; iii) no light should scatter to the substrate (however this condition can be alleviated by using a mirror below the substrate); iv) the crystal extraction efficiency should be insensitive to the incident angle.

In order to study the properties of the preferred PC diffractive extraction schemes, we use angle-resolved spectroscopy of photoluminescence (PL) or electroluminescence (EL).[7] Figure 3 represents the result of such a PL experiment for a simple structure, consisting of a thick (2.5 μm) buffer layer of undoped GaN on top of a sapphire substrate with a submicron GaN active layer comprising three GaInN QWs, and a triangular, shallow-etched PC structure (e-beam lithography period is a = 200 nm, reactive ion etch depth of 120 nm). The PC is located outside the excited region. This spectrum can be transformed into a bandstructure of the PC by noting that light is emitted due to GaN waveguided modes being diffracted by the PC. Therefore, intensity at a given angle corresponds to a diffracted lightwave inside the structure. The outside angle of the emitted beam makes it possible to determine the in-plane wavevector inside the photonic crystal mode due to in-plane wavevector conservation when light escapes the structure. One can thus replot Fig. 3 as a dispersion plot of wavelength (or, better, the dimensionless parameter a/λ) *vs.* in-plane wavevector, see Fig. 4.

The many lines observed correspond to the many waveguided modes of the thick GaN layer.[7] Simulation of such a crystal, however, indicates that many modes are not observed in the experiment: they correspond to the low-order modes of the waveguide that are mainly localized near its center and interact very weakly with the PC (*i.e.* they have an extraction length much longer than the size of the sample, here 160x320 μm).

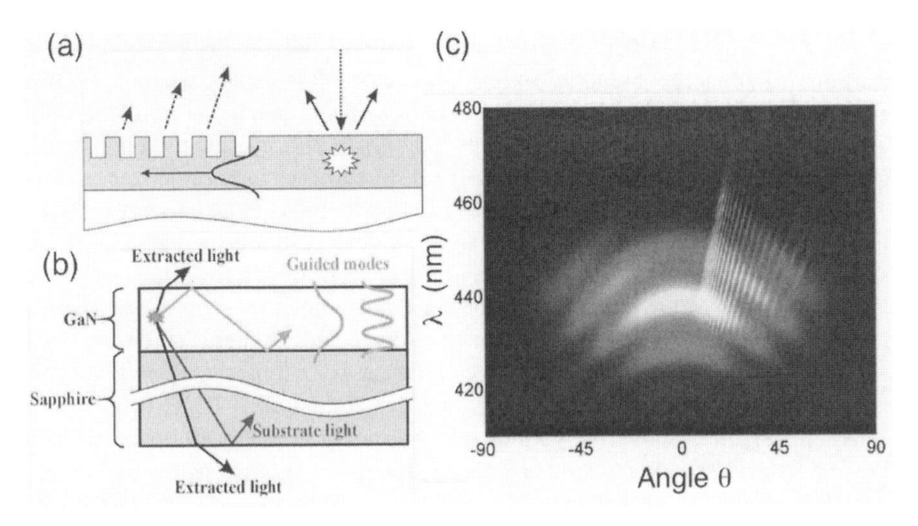

Figure 3. Angle-resolved PL spectra of the photonic crystal extractor structure: (a) measurement schematic; (b) various modes in the structure; (c) angle-resolved PL color plot. The large features in (c) are the vertical resonances, the fine lines are signatures of guided modes extracted by the photonic crystal.

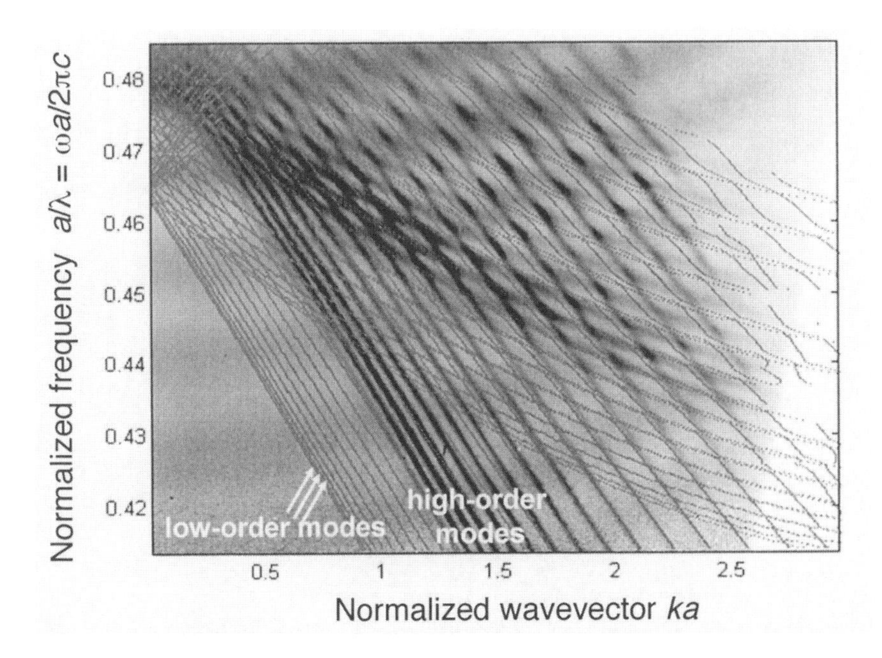

Figure 4. Experimental (heavy lines) and calculated (light lines) dispersion curves of the PC as deduced from the data shown in Fig. 3(c).

The situation is therefore quite clear. Without any additional measures, the fraction of light accessible to PC extraction in an LED structure is limited by two effects: part of the light is lost into the substrate and another part of the light emitted into waveguided modes interacts too weakly with the PC to be extracted. To improve on this situation, the LED vertical structure has to be optimized.

One way to solve the issue of unextracted waveguided modes is to use a low-index confining layer below the active layer.[8] In that case, waveguided modes are split into three types: i) a confined mode localized in the active region above the low index layer; ii) low-order modes localized below the low-index layer; and iii) high-order modes delocalized throughout the waveguide. Modes i) and iii) interact strongly with the PC and they are efficiently extracted, but modes ii) are not. This is of little concern, however, as such modes do not carry QW spontaneous emission. Figure 5 shows the various relevant modes in a regular structure and a low-index layer-containing structure designed to direct spontaneous emission into well-defined modes. For comparison, the observed modes in the PC dispersion diagram as revealed by angle-resolved PL are also shown. One indeed observes very intense new modes corresponding to the concentration of all previously unextracted low-order modes into well-extracted "cap-layer modes". Of course, this still leaves the light emitted into the substrate modes unavailable for extraction. One way to have complete extraction of spontaneous emission is to design structures without a substrate and with sufficiently thin active layers for the waveguided modes to interact strongly with the PC extractor. This is possible in

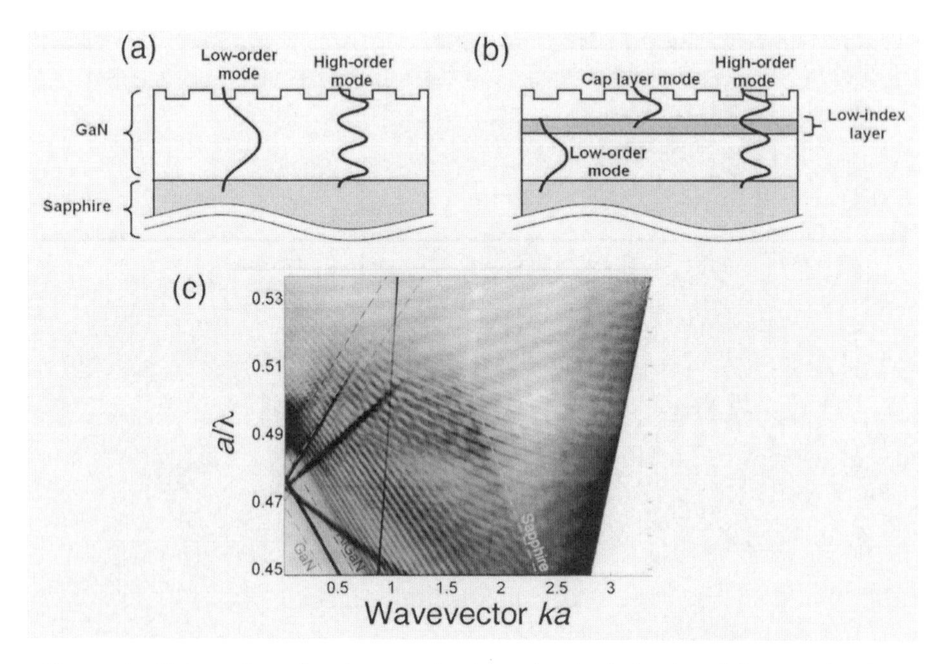

Figure 5. Schematics of various modes in a standard structure (a) *vs.* a structure designed for the control of spontaneous emission modes (b). Such structures give rise to a strong guided mode that is well extracted by the PC (c) (see Fig. 4).

principle with thinned down structures transferred by laser lift-off, although fabrication with the required yield and precision remains a major challenge.[9]

All told, with the various designs outlined above, PC LEDs promise extraction efficiencies in the 70–90% range, depending on the level of fabrication difficulty and yield. Compared with the other solutions based on geometrical optics that may achieve similar efficiencies, PC-based devices bring two major additional advantages: high directionality and brightness, and manufacturability in a fully planar process.

4. Fluorescent biochips I: Amplifying slides and associated reading systems

A high detection sensitivity is extremely important in microarray biochips for analysis of gene expression patterns, patient genotypes, drug metabolism, *etc.*, especially when RNA or c-DNA is involved. While many detection schemes exist (based on electrochemistry, surface plasmon resonance, ellipsometry, *etc.*), fluorescence is a preferred detection technique because of its high intrinsic sensitivity, excellent selectivity and specificity. The drawbacks associated with fluorescence detection are related to inadequate integration (separate hybridization and measuring set-ups) and portability, in particular when considering the new bioanalytical approaches based on integrated lab-on-chips. We describe below advances in these aspects of sensitivity, integration and portability.

A typical microarray biochip[3] consists of thousands of spots of known DNA molecular *probes* immobilized at known locations on a microscope glass slide functionalized with adequate surface attachment chemistry. Dye-tagged DNA *targets* originating from a sample of interest are specifically hybridized *in-situ* onto those probes, achieving a spatially-resolved molecular recognition. Quite usually, two dyes are used (*e.g.* green cyanine 3 (Cy3) for the DNA sample of interest and red cyanine 5 (Cy5) for a control target DNA sample serving as a reference[3]). Excess target material, *e.g.* resulting from nonspecific adsorption, is washed away before fluorescence mapping in a scanner. The scanner detects which probe spots have been hybridized, leading to identification of the target molecule present in the analyzed sample. The typical sensitivity of existing scanners for a glass slide is in the range of 10 fluorophores per μm^2, the typical pixel size being 10 μm.

Fluorescence-based microarrays discussed here enable very compact assays, a typical slide containing up to tens of thousands of spots. The density of fluorophore molecules grafted on DNA strands remains small (one per oligo-DNA) to limit fluorescence quenching. This makes the detection of genes with low expression levels difficult due to the weak signal (gene expression in a cell is measured by the amount of m-RNA generated by a given gene, that RNA being reverse-transcripted into specific DNA strands acting as targets to allow for their specific attachment to c-DNA or oligo probes, followed by scanner detection). It is therefore highly desirable to obtain higher luminous efficiency per hybridization event. An additional advantage of higher efficiency is the possible reduction in the quantity of rare and expensive biological materials used as probes or targets.

We first describe an enhancement technique for fluorescent signals capitalizing on simple interference effects occurring close to a mirror coating the microarray surface underneath the probes, see Fig. 6. It is also applicable to other luminescence-based assays (chemo and electro-chemo-luminescence, not discussed here), and resembles very much the microcavity approaches used for LEDs.

Let us first remark that the use of a simple glass substrate leads to a very poor luminous efficiency. First, the opposite phase of the reflected beam compared to the incident one means that only 64% of the incoming intensity excites the fluorophores. Second, as can be expected for emission on a high-index substrate, most of the emitted light (typically two thirds) is emitted into the substrate, as shown in Fig. 7(a). By using a DBR stack on a glass substrate, one observes an enhancement of light emitted towards air, in addition to a strong directionality, making it possible to use optics with limited aperture while retaining a good collection efficiency – see Fig. 7(b). The resulting increase in light emission, reaching a factor of approximately 16 (enhancements by a factor of 4 for both excitation and emission intensities, see Fig. 6) is confirmed in experiments.[10]

- *Imaging reader systems*

Amplifying slides have an important property regarding background slide fluorescence: due to the optical isolation provided by the mirror layers, the portion of the slide below the mirror does not produce any light that reaches the detection system. Therefore, the background fluorescence due to the slide material is only limited to the volume located above the mirror. Instead of having to spatially filter the fluorescence light through the use of a confocal microscope and mapping the entire surface by a high-precision scanning mechanism, large surface fluorescence images can be obtained using standard high-sensitivity cameras with a 15x25 mm field of view, thus reducing the imaging of the full microscope slide to five fields. It can be shown that sensitivities similar to photomultiplier-based scanning systems

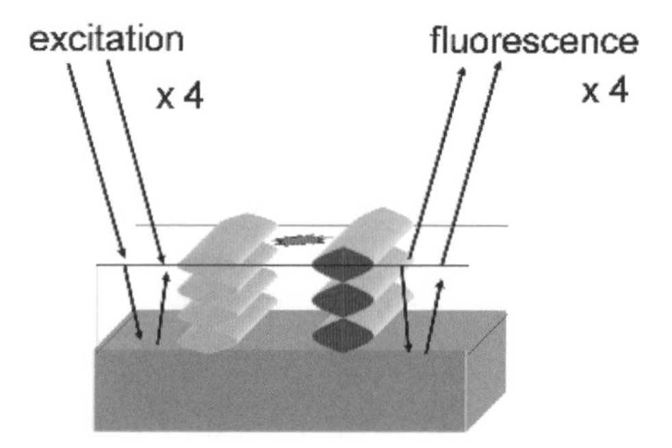

Figure 6. Amplification of excitation intensity (lighter lobes) and far-field fluorescence emission intensity (darker lobes) of a fluorophore adequately spaced from a perfect mirror. Each two-wave interference leads to a x4 gain in intensity, hence a x16 factor overall.

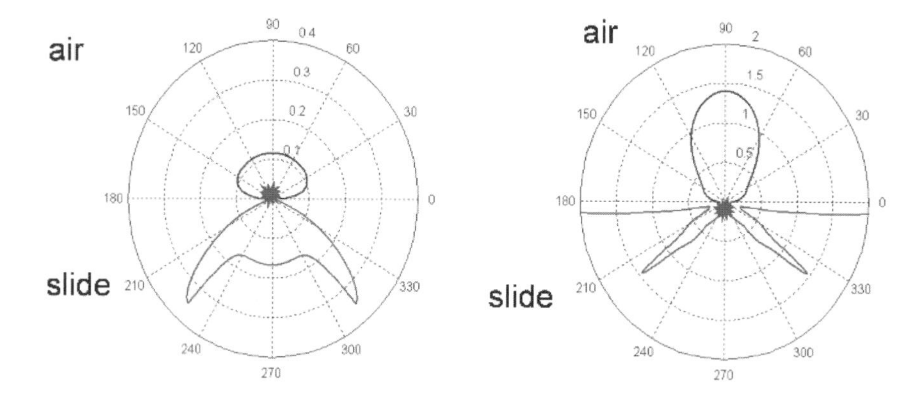

Figure 7. Polar diagram of emission intensities for fluorophores placed on a bare glass slide (left) and on an amplifying substrate consisting in a four period TiO_2/SiO_2 DBR stack (right).

can be obtained, with parallel detection allowing a gain in speed by at least a factor of five. Further, the use of CMOS or CCD imagers allows the easy extension of useful wavelengths to the infrared, where the background fluorescence is smaller.

- *Real-time imaging systems*

Once parallel imaging of a large microarray surface is accessible, one can image the hybridization process in real time. Compared to other real-time schemes, such as those based on surface plasmon resonance, this technique brings the ultra high sensitivity of fluorescence. Figure 8 shows results of such real-time measurements. It is remarkable that spots are observed at very short times, although there is still a large concentration of unhybridized fluorescent targets in the solution. This is due to the fact that bound fluorophores undergo the interference-based amplification due to their proximity to the mirror while dilute molecules do not. There is also a marked increase in the concentration of bound fluorophores, providing contrast against the dilute background.

The uses of such a time-resolved fluorescence system are multiple: it allows for the rapid discrimination of a particular recognition event, a highly demanded task in biodetection alert or in fast diagnostics systems. It also allows for systematic studies of microarray protocols, such as quantitative effects of buffer or washing solutions on binding efficiencies, background fluorescence, *etc*. The time resolution of binding and dissociation adds another dimension to fluorescence measurements, allowing the detailed study of binding selectivity, which is of very high importance in protein studies for drug activity selection. One can also use temperature ramps to observe continuously selective binding or dissociation, a very useful function, for instance, in single-nucleotide polymorphism studies.

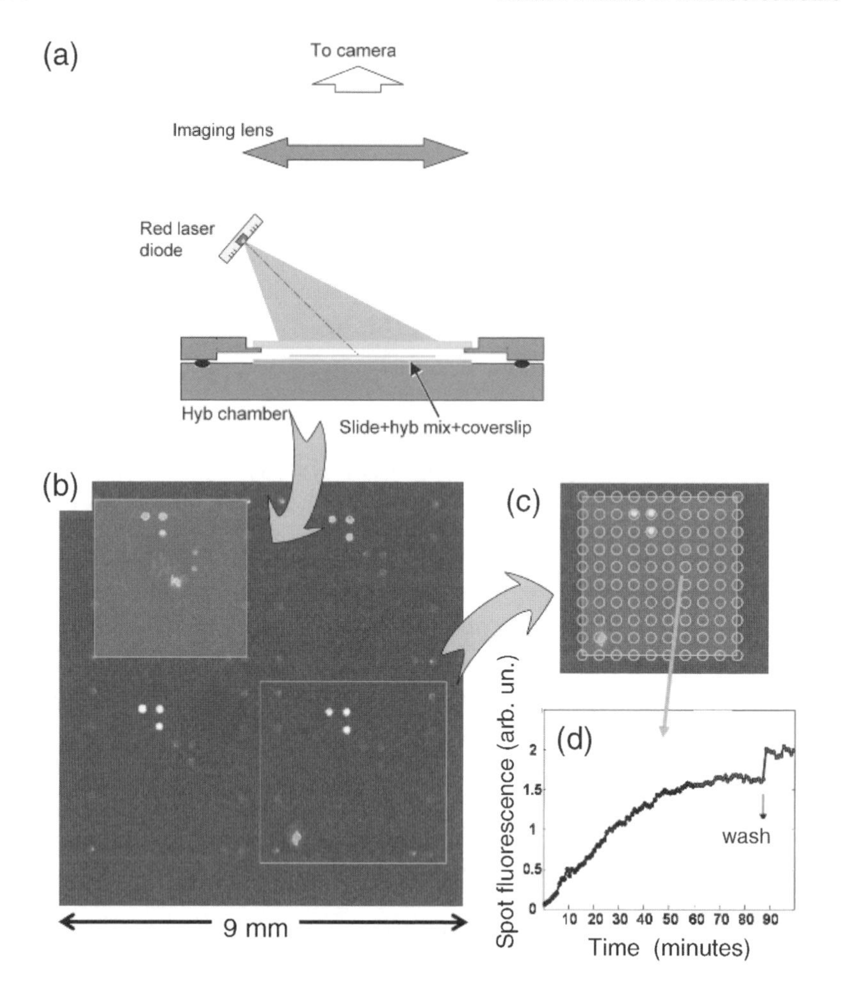

Figure 8. Real-time imaging of a microarray slide during hybridization. (a) Set-up using a fluorescence-enhancing slide; (b) fluorescence images at t = 16 minutes on Genewave's AmpliSlide™ after background subtraction; (c) grid for analysis superimposed on bottom right block (L-shaped triplets within each block show PCR products hybridization, the spot intensities match their initial concentration in the mix); (d) spot fluorescence *vs.* time for the weaker spots of (c). Signal minus background increases linearly and then saturates at t ~ 60 minutes; washing, at 90 minutes, decreases significantly the background causing an apparent rise in signal.

Of course, one still observes significant background fluorescence due to the unbound fluorophores of the solution. It can be suppressed by washing the hybridization chamber, yielding images with a very high contrast ratio – see Fig. 8(d). Other schemes can increase the bound-to-free fluorophore contrast during hybridization, such as the use of evanescent wave excitation or spatially modulated amplification slides. The former scheme relies on the fact that exciting with a waveguided light beam will only excite those molecules within the evanescent part of the wave, *i.e.* those attached to the surface, together with only a very small fraction of those in the solution.[11] Several ways of producing such waveguided modes are possible, such as coupling in a laser beam from the outside or using fluorescent species embedded in the waveguide. The spatial modulation scheme relies on an in-plane structuring of the amplifying slide, such that the amplification coefficient is spatially modulated between zero and maximum gain, preferably on a smaller scale than the spot size (one way to do it is to modulate the top layer thickness to obtain constructive or destructive interference). In that case, while the background fluorescence of the solution is constant over the slide, the signal originating from bound species is spatially modulated, so its demodulation and detection can recover a signal with enhanced dynamic range.

- *The future*

There are other ways to further improve the sensitivity of microarray slides. One avenue is to improve the signal to noise per hybridization event. One can try to increase the fluorescence per event (by using fluorophores with larger excitation cross sections and higher fluorescence yield, such as quantum dots, molecular clusters, dendrimers, *etc.*) or suppress the fluorescence background (evanescent wave excitation, better molecular recognition schemes, such as molecular beacons or hairpin probes leading to less non-specific binding, *etc.*). Figure 9 shows a scheme where one uses a slide with a metal mask: in the imaging mode, the surface and volume generating spurious signals can be made extremely small by going with a small hole size. Further refinements can include other local techniques, like electrophoresis, to drive DNA to detection regions, opening the way to the detection of a single molecule hybridization.

It has been shown[12] that deep-submicron holes actually enhance fluorescence due to surface plasmon enhancement. Such metal-masked system would achieve spatial filtering of background fluorescence in the object plane. The equivalent operation in the image plane, which is usually done in a confocal microscope on a single spot, is more difficult because it requires exact alignment of the filtering mask with the image.

5. Fluorescent biochips II: Integrated microarray and imaging systems

Contrary to the approach of increasing the efficiency fluorescence systems by diminishing emission towards the substrate, another approach relies on the fact that most of the light generated at an interface is emitted into the higher index material. A remarkable opportunity arises when using an imaging semiconductor integrated circuit, CCD or CMOS-based, as the substrate for the microarray. Then, the collection efficiency can be as high as 60-70% (all the light that is emitted downwards–see Fig. 7(a)) and no optical imaging of the spot is required, provided the spots are located close enough to the pixels: in a typical geometry, spots are 100 μm wide, distance between spot and pixel is a few microns, and pixel size is

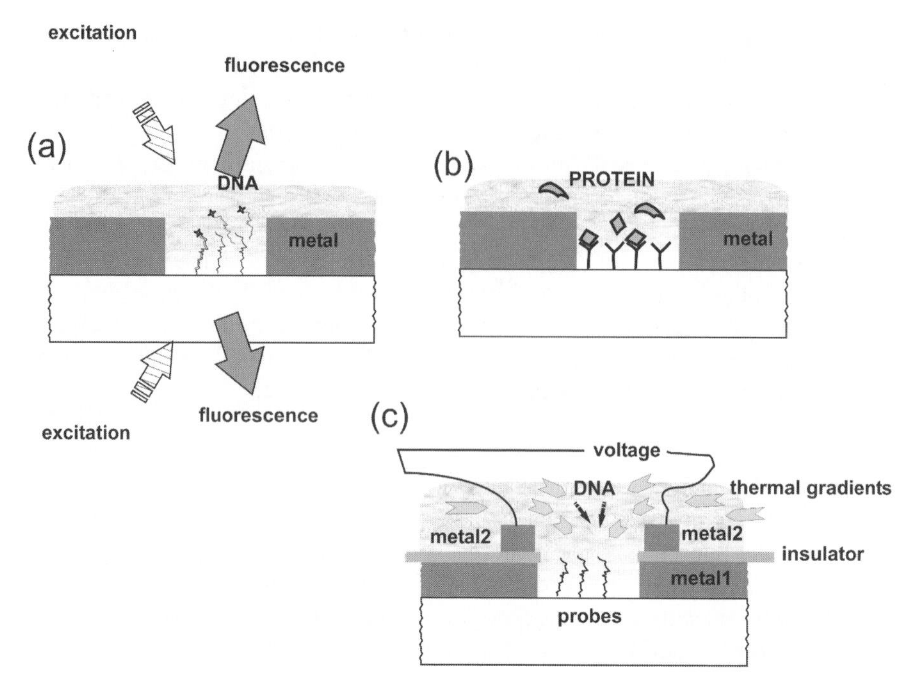

Figure 9. Hybridization detection can take advantage of localization by apertures in metal masks (a) for DNA detection by fluorescent labels; (b) for protein detection, with fluorescent tags or without; (c) with extra means (electrophoresis, thermophoresis, *etc.*), one can furthermore drive the DNA targets of other biomolecular targets towards the relatively tiny detection regions, increasing dramatically the rate of capture.

5–15 μm. In that case, fluorescence from a given spot illuminates a few dozen pixels. Direct imaging of the spot is enabled by the proximity of the spot to the pixel and by the very small (undetectable) crosstalk between spots. Figure 10 shows the images obtained for a CCD imager directly spotted with fluorescent QDs (#QD800 dots from QD Corporation) at a concentration of 0.5 μM, observing the spots either directly from the CCD or imaging the spots on a similar CCD with wide aperture optics ($NA = 0.5$). The measured signal intensities differ by a factor of 30, as expected from the differences in collection efficiencies between the two geometries (> 60% for the integrated detection, ~2% for the imaging detection).

Of course this detection is made possible by the use of a high rejection, thin-film, wavelength-selective filter intercalated between the spots and the imager. Such a filter has to be integrated without a break in refractive index (*i.e.* without an air gap) in order to have the high collection efficiency due to the emission into a high index material. It also requires, as an order of magnitude calculation involving target fluorophore absorption cross sections easily shows, a wavelength-selective rejection ratio above 10^6 at all incidence angles.

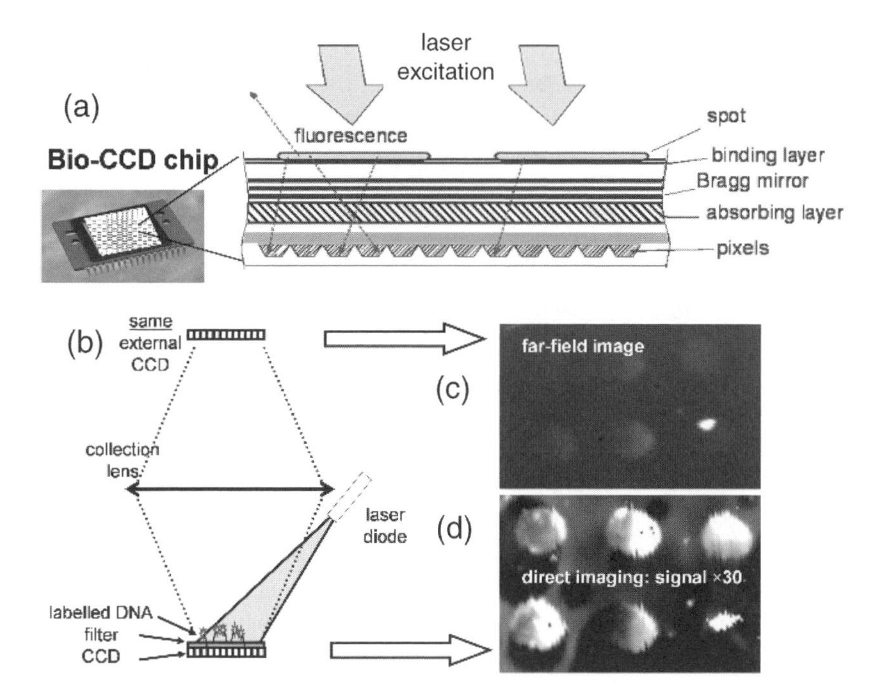

Figure 10. Detection of fluorescent molecules directly spotted on a CCD camera through direct imaging on that camera: (a) principle of operation; (b) comparison with ordinary far-field fluorescence setup, using the same CCD; (c) far field image; (d) direct image, found to be 30x enhanced, i.e. harvesting 30 times more photons.

This cannot be achieved by a multilayer filter, but rather by using transition-metal dyes incorporated in a spin-on sol-gel matrix. The intrinsic fluorescence of this filter must low enough to observe the target fluorophore emission. The filter can either be deposited on the imager wafer or die, or be fabricated on a sacrificial substrate and bonded after fabrication on the imager, before or after chemical functionalization for probe attachment and/or probe spotting.

- *The future*

Fabricating integrated hybridization-readout systems in standard silicon technology would integrate more functions in the device: the entire control and communication system can be integrated, with a USB interface to a computer. Full integration, with built-in biological tests, quality controls, and diagnostic software, will allow standalone biochip operation. In addition, more operations could be added thanks to the many opportunities provided by silicon technology, such as built-in electrodes for the fabrication of probes or high-efficiency hybridization thanks to electro-kinetic control of DNA or protein fragment motion. Further, the usual lab-on-chip functions can also be integrated. A schematic picture of such an integrated system is shown in Fig. 11.

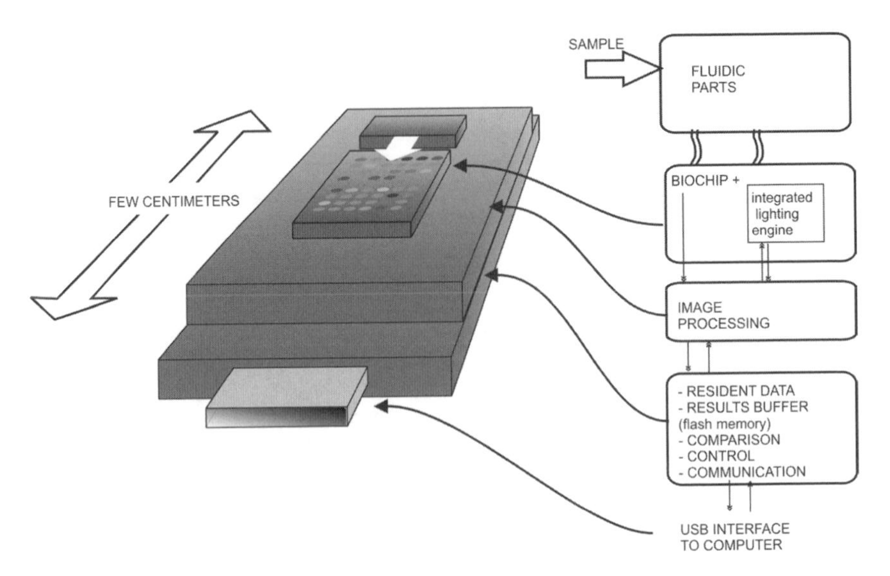

Figure 11. Portable, sensitive and versatile biological detection system, integrated on a USB interface, focusing on the optical and data architecture (fluidic parts not shown). The light engine is integrated (side excitation) and the collected data from the array imager are partly treated on-board, and stored in a flash memory, together with the spotting information. This ensures perfect traceability and allows first-order analysis by an on-board processor. Full analysis is possible after transfer of raw or pre-treated data to a general-purpose computer.

6. Conclusions

High-efficiency emission collection strategies are central to both LED and microarray areas. For LEDs, while solutions appear to exist based on PC light extractors, industrial implementation faces stiff competition from schemes based on geometrical optics. Success will be more due to advantages in brightness and manufacturing performance than in raw extraction efficiency numbers.

For fluorescent microarrays, increased fluorescence collection efficiency not only can improve significantly the performance of biochips, but also allows new biodetection approaches. The integration of the hybridization and detection systems opens the way to fully integrated microarray bioanalysis systems with full automation: there have been many studies for the integration of the sample preparation (extraction and purification of DNA, RNA or proteins; biological amplification through PCR; hybridization) in the so-called lab-on-a-chip or micro total analysis systems. If these steps can be integrated with the detection systems, as described in this paper, they would lead to a fully integrated fluorescence detection scheme, which would be the most sensitive of all once proper light harvesting schemes are implemented.

Acknowledgments

The work on LEDs was supported in part by UC–Santa Barbara's solid state lighting and display consortium (SSLDC), while the work on microarrays was supported in part by ANVAR, DGA and the E.U. project INDIGO.

References

1. W. N. Carr, "Photometric figures of merit for semiconductor luminescent sources operating in spontaneous mode," in: S. M. Sze, ed., *Semiconductor Devices: Pioneering Papers*, London: World Scientific, 1991, pp. 919–37.
2. M. G. Craford, "Overview of device issues in high-brightness light-emitting diodes," in: G. B. Stringfellow and M. G. Craford, eds., *High-Brightness Light-Emitting Diodes*, San Diego: Academic Press, 1997, pp. 47–64.
3. M. Schena, *Microarray Analysis*, Hoboken, NJ: Wiley-Liss, 2003.
4. H. Benisty, H. De Neve, and C. Weisbuch, "Impact of planar microcavity effects on light extraction—Part I: Basic concepts and analytical trends," *IEEE J. Quantum Electronics* **34**, 1612 (1998); "Impact of planar microcavity effects on light extraction—Part II: Selected exact simulations and role of photon recycling," *IEEE J. Quantum Electronics* **34**, 1632 (1998).
5. C. Weisbuch, A. David, T. Fujii, *et al.*, "Recent results and latest views on microcavity LEDs," *Proc. SPIE* **5366**, 1 (2004).
6. M. Fujita, S. Takahashi, Y. Tanaka, T. Asano, and S. Noda, "Simultaneous inhibition and redistribution of spontaneous light emission in photonic crystals," *Science* **308**, 1296 (2005).
7. A. David, C. Meier, R. Sharma, *et al.*, "Photonic bands in two-dimensionally patterned multimode GaN waveguides for light extraction," *Appl. Phys. Lett.* **87**, 101107 (2005).
8. A. David, T. Fujii, R. Sharma, *et al.*, "Photonic-crystal GaN light-emitting diodes with tailored guided modes distribution," *Appl. Phys. Lett.* **88**, 061124 (2006).
9. A. David, T. Fujii, B. Moran, S. Nakamura, S. P. DenBaars, C. Weisbuch, and H. Benisty, "Photonic crystal laser lift-off GaN light-emitting diodes," *Appl. Phys. Lett.* **88**, 133514 (2006).
10. H. Choumane, N. Ha, C. Nelep, *et al.*, "Double interference fluorescence enhancement from reflective slides: Application to bicolor microarrays," *Appl. Phys. Lett.* **87**, 031102 (2005).
11. M. Pawlak, E. Grell, E. Schick, D. Anselmetti, and M. Ehrat, "Functional immobilization of biomembrane fragments on planar waveguides for the investigation of side-directed ligand binding by surface-confined fluorescence," *Faraday Discuss.* **111**, 273 (1998).
12. H. Rigneault, J. Capoulade, J. Dintinger, *et al.*, "Enhancement of single-molecule fluorescence detection in subwavelength apertures," *Phys. Rev. Lett.* **95**, 117401 (2005).

Part III

Electronics: Challenges and Solutions

3 Electronics: Challenges and Solutions

The current state of nanoelectronics is both pleasing and concerning. On the one hand, 60 nm CMOS is already reaching volume production and 40 nm technology is set for the next generation. But en route to 32 nm, lithography is at a crossroad. Irrespective of the eventual solution, CMOS scaling, unabated for the past 50 years, simply cannot continue forever, and at the present rate it would reach sub-10 nm critical dimensions in 2 or 3 more generations.

What lies ahead in the post CMOS-scaling world? What prospects do we see in the likely increasing dissociation of nanomanufacturing from ICs? Are there fundamental physical limits? Are we looking at the right problem? These are some of the big issues being addressed by captains of industry and other electronics researchers in this part. While research continues to be invested in finding solutions to the scaling problems, new materials, new concepts, and new transport mechanisms are being developed and promise to open up new opportunities for future advances in electronics.

Contributors

3.1 M. R. Pinto
3.2 J. P. H. Benschop
3.3 M. Brillouët
3.4 O. Engström
3.5 Y. Nishi
3.6 P. M. Solomon
3.7 M. J. Kelly
3.8 F. Arnaud d'Avitaya, V. Safarov, and A. Filipe
3.9 J. Murota, M. Sakuraba, and B. Tillack
3.10 M. Banu and V. Prodanov
3.11 K. Akarvardar, S. Cristoloveanu, and P. Gentil
3.12 E. Sangiorgi, S. Eminente, C. Fiegna, P. Palestri, D. Esseni, and L. Selmi
3.13 M. I. Dyakonov and M. S. Shur
3.14 C. Dais, P. Käser, H. Solak, Y. Ekinci, E. Deckhardt, E. Müller, D. Grützmacher, J. Stangl, T. Suzuki, T. Fromherz, and G. Bauer
3.15 D. P. Wang, F. Y. Biga, A. Zaslavsky, and G. P. Crawford

Future Trends in Microelectronics. Edited by Serge Luryi, Jimmy Xu, and Alex Zaslavsky
ISBN 0-471-48 © 2007 John Wiley & Sons, Inc.

Nanomanufacturing Technology: Exa-Units at Nano-Dollars

Mark R. Pinto
Applied Materials Inc., Santa Clara, CA 95054, U.S.A.

1. Introduction

What has become known as nanotechnology – "the understanding and control of materials at the sub-100nm level" – holds enormous promise for a range of industries and everyday applications.[1] Whether measured by number of research publications (which outstripped those on silicon integrated circuits in the year 2000),[2] the amount of government-sponsored research funding (over $4B worldwide in 2005),[3] or even the number of web hits for "nanocenter" (over 50,000 via Google™ in September of 2006), there is an ever-growing amount of R&D activity in nanotechnology, presumably with the expectation of its funders that there will be substantial commercial impact. In fact, the U.S. National Science Foundation has estimated that nanotechnology will provide U$1 trillion in revenue by the year 2015, led by applications in materials, electronics and pharmaceuticals.[1]

However, outside of integrated circuits (ICs) and some popularized material coatings (*e.g.* cosmetics and stain-resistant textiles), the pathway to economic impact for much of current nanotechnology research has thus far proved elusive. Potential barriers include not only lab feasibility of the basic concepts, but also how nanostructures will be manufactured in volume. The predictable placement of both ends of a carbon nanotube (CNT) in many envisioned applications typifies this challenge. In order to realize the pervasive scope of nanotechnology-based products, it is therefore essential to develop "nanomanufacturing technologies" that deliver required scale, cost, reproducibility and reliability. In many instances, the conceptualization of the manufacturing method may be substantially different from and of equal importance to the underlying device idea – there is perhaps no better example of this than the IC itself, where both Kilby (first realization) and Noyce (manufacturable process) are recognized as the primary inventors.

As both the largest single nanotechnology end-market and as an established example of the importance of nanomanufacturing technology, we begin this chapter by looking at the current state of nanoelectronics – *i.e.* ICs with critical features at or below 100 nm. We will then examine nanomanufacturing beyond ICs, by focusing on displays and photovoltaics, areas that are poised to have a major economic impact in the near future.

 Future Trends in Microelectronics. Edited by Serge Luryi, Jimmy Xu, and Alex Zaslavsky

2. Nanoelectronics: Nanotechnology and more

The IC industry has continued pushing component count per die and scaling constantly for now over 40 years since Moore published his seminal projection in 1965.[4] In fact, in 2003 the industry began shipping the first volume products in 90 nm technology – a full 4 years ahead of the first industry-wide semiconductor roadmap published only 9 years earlier[5] – thereby transitioning the industry from the microelectronics to the nanoelectronics era. For the industry to have invested so heavily to achieve this milestone, there have been major benefits to scaling – product form factor, performance, power per computation, reliability – but none has been as important as the reduction in cost per transistor or bit, the foundation of Moore's prediction. Figure 1 shows the number of transistors produced yearly, as well as the average cost per transistor from 1968–2002, as updated by Moore in 2003.[6] Recent estimates for the price per memory bit are on the order of 1 billionth of a dollar, *i.e.* 1 nano-dollar.

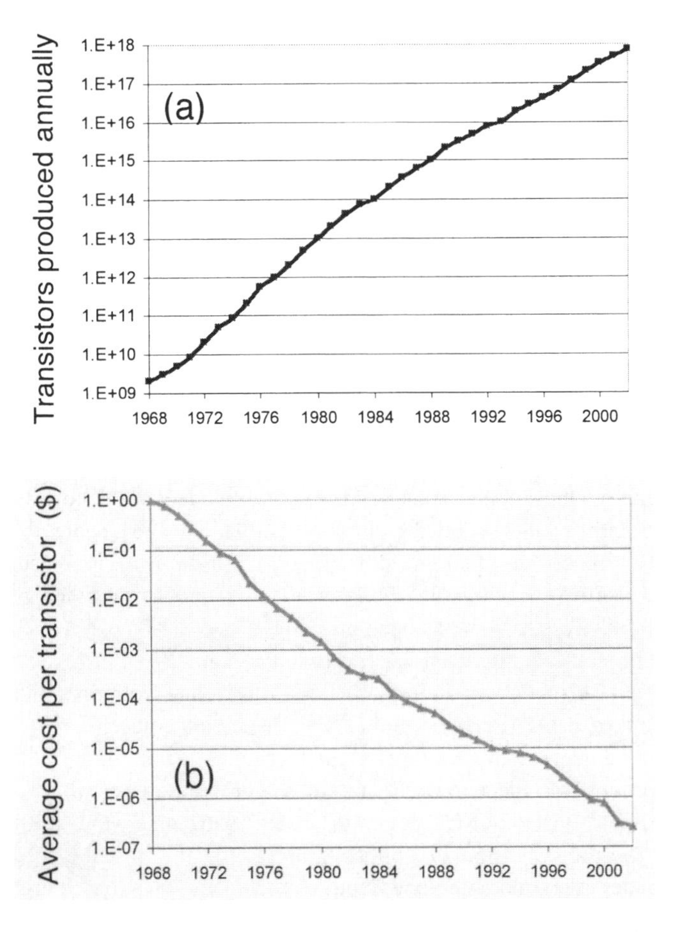

Figure 1. Historical transistor total annual production (a) and average cost (b).[6]

This reduction has followed the traditional learning curve, where increasing scale and units accompanies the decreasing cost – for the IC, the cost per transistor has fallen by ~28% for every doubling of production volume. The industry is now producing well in excess of 10^{18} transistors ("exa"-transistors) per year, and current forecasts predict that there will be over 3 exa-bits of NAND memory alone produced in 2006. The primary driver of the cost per transistor (or bit) reduction is dimensional scaling – "cramming more components onto ICs" as Moore described. Additional factors have included increased wafer sizes and other innovations, such as increased process throughput. As long as yield loss does not increase at a correspondingly higher rate, the economics of transistor production improve dramatically. And as long as there are new applications that will use more of these cheaper transistors or bits, the IC market will grow.

Consumption of transistors has continued unabated in the past 50 years, although the end applications have shifted. Whereas previously corporate consumption dominated, by 2004 more semiconductor revenue was derived from consumer applications. This change – brought about by price elasticity for cell phones, MP3 players, DVDs and so on – further emphasizes the importance of cost per transistor as the primary driving function for scaling. Consumer consumption of semiconductors will continue to grow faster than any other segment, and costs must continue to fall in order for products to hit the cost points that will drive demand in emerging economies. As evidence of the consistency of scaling, cost and consumer demand, Applied Materials saw orders for <100 nm process tools reach 85% in terms of total dollars for the quarter ending in July of 2006.

With 65 nm technology reaching volume production in 2006, the technical challenges to continued scaling have become greater, requiring ever-increasing R&D investment. Looking first at the overall landscape, comparisons of recent industry roadmaps (see Ref. 7) clearly demonstrate that the so-called "red brick wall" – the date by which the anticipated component specifications will outstrip currently available industrial solutions – has gotten closer to the present. To circumvent many of these problems, IC technologists have turned to new materials, and the number of these under consideration has exploded since the 180 nm node. These include cobalt and nickel for contacts, ruthenium and copper for vias and interconnects, silicon germanium and silicon carbide for strain engineering, hafnium and lanthanum for gate dielectrics and chalcogenides for nonvolatile memory. NAND flash, which has been growing at over 120% per year for the past 10 years in terms of bits produced, has become the main driver of IC density. And logic design has become more and more difficult to isolate from manufacturability, requiring changes to EDA, process control and inspection equipment.

In the remainder of this section, nanoelectronics technology challenges and process directions will be reviewed in four major areas: patterning, transistors, interconnect, and inspection. Beginning with patterning, the central driver continues to be the extension of optical lithography, now moving to immersion lenses (for better index matching and depth of focus) and heavily feature-enhanced masks rather than towards pushing the radiation wavelength beyond 193 nm – these topics are covered in detail elsewhere in this volume. Most exposure steps are followed by a plasma etch, where now in addition to smaller features there are additional challenges of new materials and higher aspect ratios. Furthermore more steps have been added prior to exposure to assure highest fidelity patterns, including wafer planarization (see below) and pattern-enhancing CVD films deposited between the resist and the substrate. It currently appears that economic implementations of lower wavelength radiation are a ways off, so emphasis will continue on pushing 193 nm to its absolute limits through further optimization of immersion optics, dual patterning and resolution enhancement masks and processes.

(a)

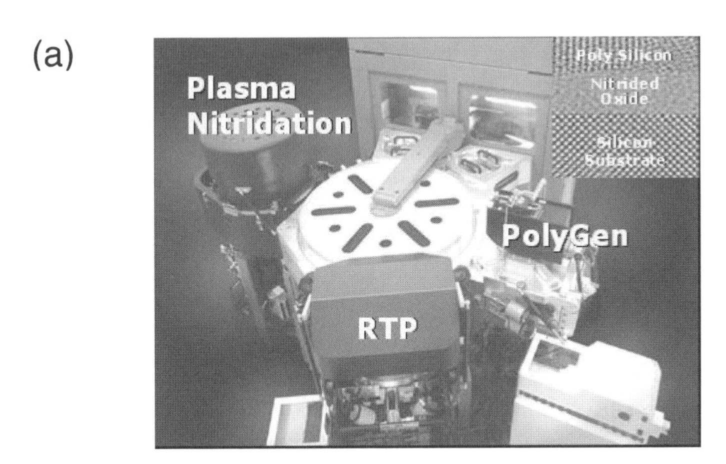

(b)

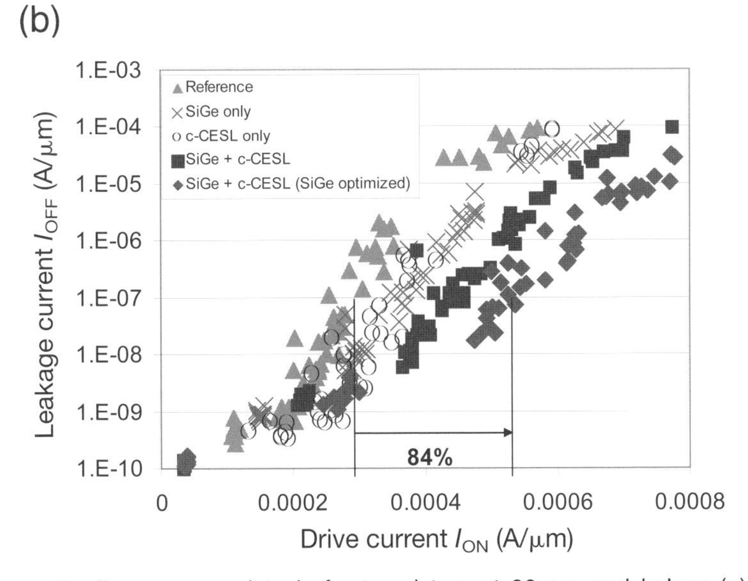

Figure 2. Processes and tools for transistors at 90 nm and below: (a) Applied Materials Centura® DPN gate stack cluster tool with TEM cross section of sub-1nm EOT dielectric in the inset; (b) drive current I_{ON} vs. leakage current I_{OFF} using different strain enhancement techniques.[9]

The major challenge in transistor scaling arises from the rapidly increasing standby power caused by balancing adequate current drive with scaled power supply voltages and nonideal subthreshold turnoff. As in prior generations, the industry continues to focus on both gate stack and junction scaling to manage this challenge. Lately, these approaches have been combined with increased effort in design techniques to avoid unnecessary power dissipation wherever possible. As in the past, polysilicon remains the production gate electrode of choice. However, the gate dielectric has evolved to extremely well-controlled processes in cluster tools like that pictured in Fig. 2(a) that manage material interfaces and incorporate nitrogen into thin SiO_2 films to increase capacitance while avoiding degradation through mechanisms like NBTI.[8] Effective oxide thickness of < 1 nm can now be achieved. Advanced research continues on a full replacement of SiO_2 based dielectrics with higher-κ materials and the gate electrode with suitable metal gates (at low enough process cost). Because of damage and uniformity advantages, shallow junctions are now formed almost exclusively by single wafer implantation systems and rapid lamp or laser anneals to preserve junction depths of < 10 nm and limit diffusion to < 25 Å, with plasma emersion being used in most advanced R&D labs.

A major change that happened around the 180 nm node was the introduction of strain in the MOSFET to achieve higher drive currents. This was first accomplished by inclusion of materials intended to assist contact formation, but below 130 nm substantial efforts have been made to maximize this effect for both NMOS and PMOS devices. Mechanisms being used include silicon nitride overlayers, selective epi in the MOSFET source/drain, stress inducing films in the isolation structure between transistors, and biaxially strained substrates. Figure 2(b) shows a recent result where selective SiGe in a PMOS source/drain was combined with a compressive silicon nitride overlayer to achieve an 85% increase in drive current at a fixed I_{OFF}.[9] While initially used for high-speed applications like server microprocessors, the magnitude of the effect and the option of trading off leakage for performance (for instance, in Fig. 2(b) one can achieve 100x improvement in I_{OFF} at a fixed drive current) have promoted the use of strain more universally.

Due to higher transistor area packing, interconnect must also be scaled to achieve circuit density improvements. However, the potential increase in parasitic impedances together with the added wiring due to nonshrinking (or even slightly growing) average die sizes has led to the evolution of the interconnect structure from generation to generation. Typically, this involves gradual reductions in the dielectric constant of the densest interlayer dielectrics, together with the addition of more metal layers – with different layers using different thicknesses and minimum pitches depending on the maximum reach of each layer (local, semi-global, global), as in Fig. 3(a). Key challenges for manufacturing include new materials, smaller features, more control (thickness, planarity, edge roughness), and thinner films, as well as higher equipment and process productivity to overcome the use of more interconnect layers.

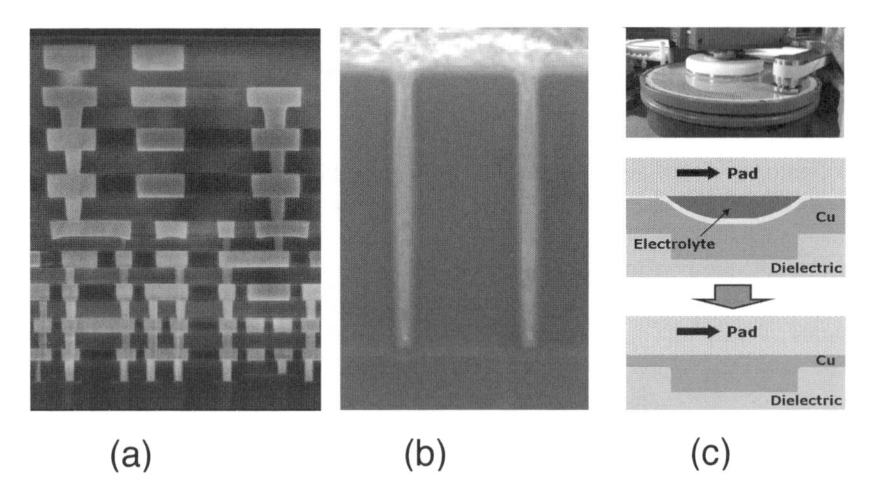

Figure 3. Nanoelectronic interconnects: (a) typical multilevel cross section using different thicknesses and pitches to manage interconnect delay (source: ICE); (b) 45 nm high aspect ratio contacts filled with PVD liners; (c) eCMP process for improved Cu planarization.

The basic damascene process architecture for logic interconnect has not been dramatically altered since the change to copper at 130 nm – trenches and vias are etched into dielectrics and then filled with liners followed by copper and then planarized. Key challenges include creating lower-κ materials that maintain enough mechanical strength for polishing and packaging, filling small and deep contacts/vias with thin liners (before electroplating copper), and reproducible low-cost dielectric and metal planarization. State of the art low-κ dielectrics are now available at $\kappa < 2.5$ by creating well-distributed pores of < 0.1 nm radius through curing carbon-doped CVD SiO_2 films. Thin barrier films (approaching 10 Å) are still typically deposited into contact/via holes by physical vapor deposition because of its high throughput and reliability (see Fig. 3(b) – however the complexity of the tools, materials and processes has continued to increase both for scaling and to increase tool productivity. Innovations in planarization have also been required to achieve required flatness, control and cost. Figure 3(c) shows a recent innovation in copper planarization where electropolishing is combined with the more traditional CMP to accommodate fragile lower-κ dielectrics, reduce design variability, and lower the cost of consumable materials.[10]

Finally, continuous improvements in inspection and metrology are required to be able to control a complicated process to the extraordinary precision and defect levels that produce adequate yield. Defect inspection and review must not only reach higher resolution to find failures in smaller structures, but also must do so at higher throughput in order to cost-effectively examine more pixels – see Fig. 4. Because yields have become more design-dependent (or equivalently, because simple design rules do not scale), there is a growing need to add more connectivity between fab measurement tools and design databases to quickly analyze and correct yield-sensitive layouts. And in more and more process tools, integrated metrology and automated control is required to meet stricter variability tolerances that avoid exploding component statistics at smaller dimensions.

In each of the above areas, the nanoelectronics industry is clearly practicing the control of materials at 100 nm and below – whether by top-down or bottom-up processes. Most essential MOS device properties, like inversion layers, strain enhancement or NAND programming also depend on <100 nm size effects. So there is no question that sub-100 nm CMOS is nanotechnology – but it is also much more. Nanoelectronics requires repeatability and control in order to produce billions of working components on a single chip. Moreover, IC components need to be produced at ever lower costs to continue growing the market. There is perhaps no better example of what nanomanufacturing technology has enabled than the MP3 player – a 4GB Apple iPod Nano using NAND flash can store up to ~1000 songs at a price point of ~$200, creating enormous global demand. If envisioned (and implementable) in 1975, at the 1975 price per transistor of Fig. 1(b), the same iPod would have cost $1B and generated little demand. This is the power of nanomanufacturing technology.

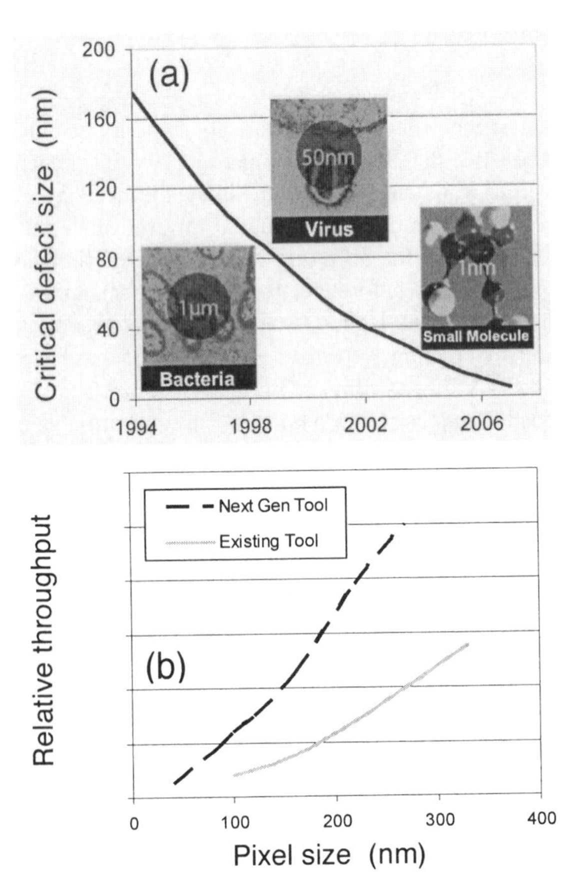

Figure 4. Inspection requirements: smaller defects (a) and higher throughput (b).

3. Nanomanufacturing technology beyond ICs: Flat panel displays

Like the iPod nano, the solar LCD calculator of the late 1970's represents a prototypical, mobile electronic machine. Both include functional elements for energy conversion, energy storage, user input, data processing, data storage and user output, as well as a flat display. As machines like this are an interesting model for future and even more pervasive systems, it is instructive to look at the possible evolution of the contents and the implications that nanomanufacturing may have, particularly for the flat panel display.

While numerical LCD displays have been available for 30 years, the move to full color, higher resolution flat panels began with the laptop PC, which led to desktop monitors and more recently embedded monitors, cell phones/PDAs and TVs. Viewer preferences for direct view, as opposed to projection, displays together with production cost reduction has led to the wide adoption of TFT-driven active matrix LCD displays as the most popular implementation. Plasma displays are running as a distant second and only for very large panels (now considered to be over 52"), where TFT-LCDs have not (yet) reached economical costs.

The primary driver for flat panel display adoption is cost per area. While ICs have a cost/area component, the technical priority has been to increase the transistors/area using scaling while containing growth of cost/area – an approach somewhat unique to information processing. Display pixels have not been getting smaller, and growth in pixels has not been dramatic in going even to full HDTV. So instead display roadmaps target decreasing cost/area with evolutionary improvements in film quality and control – by increasing substrate size while maintaining equipment productivity.

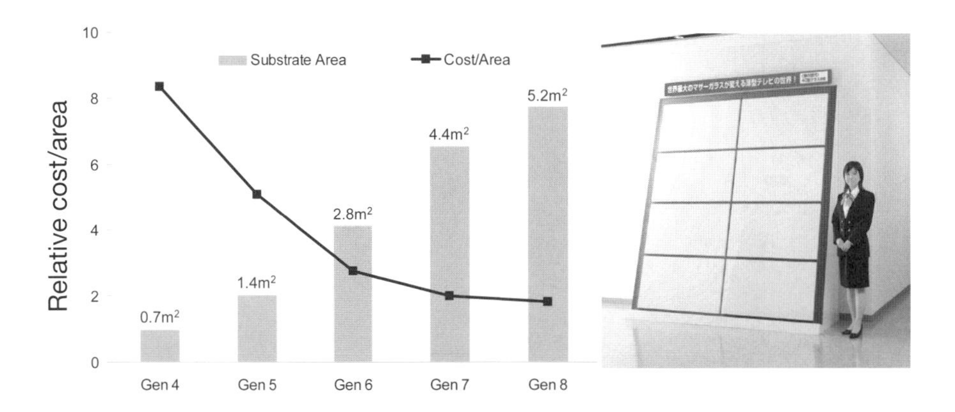

Figure 5. LCD flat panel production: (left) relative PECVD deposition cost/area reduction with glass size; (right) photograph of Gen 8 glass substrate used at Sharp Kameyama fab (courtesy Sharp Corporation).

Figure 5 shows the deposition process cost/area reduction of > 4x as glass substrates have been scaled upwards from < 1 m^2 (Gen 2) to > 5 m^2 (Gen 8), corresponding to 25% per year growth in size for 14 years. Also shown is a picture of the largest LCD TFT flat panel substrate currently in production at the Sharp Gen 8 fab in Kameyama, Japan, which can produce six 52" TV panels on one sheet of mother glass. In order to achieve the cost reductions, the deposition tools not only have to work over larger areas but they also need to maintain the same level of film properties and control (~50 nm uniformity) as well as throughput in excess of 50 substrates/hour – enough glass area to cover a full-size international football pitch every day. In 2005 alone, even before the Gen 8 introduction, total fabricated TFT LCD display area grew by 82% over 2004 to an astounding 23 km^2.[11]

Figure 6. Large area deposition equipment for LCD flat panel displays: (top) PECVD cluster tool for the TFT array; (bottom) in-line PVD tool for color filters.

As a result of this progress, the cost of LCD TFT monitors and now HDTVs have dropped dramatically. At the time of writing, a first-tier name brand 40" LCD HDTV can be found on the internet for $1,500, compared to over $5,000 only two years ago. Today, LCD flat panel monitors have completely taken over the PC market and are expected to ship in over 40 million TVs or about 20% of the total TV market, a factor of four more units than the nearest flat panel alternative (plasma). At the current pace, it is expected that LCD TVs will pass CRTs in unit shipments in 2009.

Early speculation about nanotechnology made much of hypothetical "nanomachines" – miniature machines that would self-assemble large volumes of nanomaterials.[12] In fact, the machines that embody the nanomanufacturing technology for displays are of a size proportional to their large substrates as shown in Fig. 6. Other applications that are also driven primarily by area are also likely to be best served by similar types of large area equipment. In the next section we examine one such area – low cost photovoltaics.

4. Nanomanufacturing technology beyond ICs: Photovoltaics

Like the logic and memory chips (cost/transistor) and the display (cost/area) in the calculator and the iPod Nano, photovoltaic (PV) conversion is also driven by a unit cost – in this case cost/W-peak of electricity generated. And similar to ICs and flat panel displays, photovoltaic technology has followed a learning curve, albeit a slightly less steep ~18% cost reduction for every doubling of volume, as shown in Fig. 7(a).

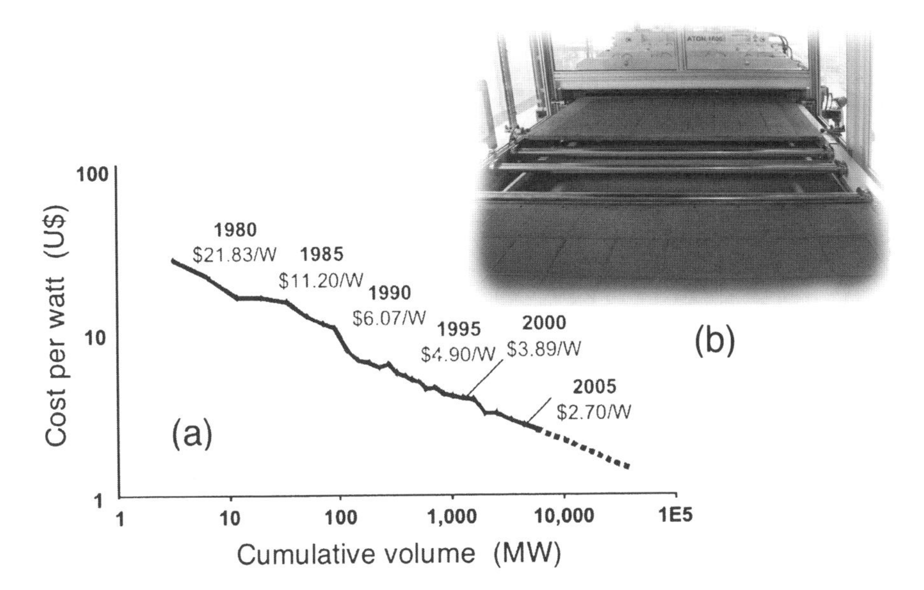

Figure 7. Photovoltaic module manufacturing learning curve (a) and a photograph of Applied Materials ATON™ silicon nitride PVD system for photovoltaics (b).

Photovoltaics were originally used for off-grid applications, like mountaintop receivers and satellites where there was no other convenient source of power. Today, PV accounts for a very small fraction (< 0.1%) of total electricity usage worldwide. Although it has the advantage of being clean and renewable, PV will only gain adoption if its cost can be near competitive with retail electricity rates. Residential subsidies in Japan and Germany over the past 5–10 years have helped stimulate adoption, and at current costs, PV is at the threshold becoming competitive with peak rates in Japan. And unlike ICs and displays, demand trigger points actually increase with time (in favor of earlier PV adoption) because of the increasing costs of traditional energy sources.

Both scaling and cost/W-peak drive the PV learning curve – without scaling, neither cost reduction nor volume demand will be met. In 1980, the ~$20/W module cost was achieved in the first 1 MW/year PV fab, but it took 20 years to reach 10 MW/year and ~$4/W module cost. Future PV fabs need to scale to a GW/year in capacity – or total silicon area output of ~200x the largest current 300 mm IC fab. This kind of scaling will require new factory concepts, not dissimilar to those developed for the display industry. Figure 7(b) shows a recently introduced silicon nitride deposition system for crystalline silicon PV cells using an adapted in-line platform from FPD/glass coating. A single such tool can produce more than 3000 156 mm wafers per hour or more than 50 MW/year.

Turning to the methods to reduce cost/W, it is helpful to decompose the driving function as:

$$\text{Cost/Watt} = (\text{Cost/Area}) \, / \, (\text{Watts/Area}) \, .$$

Like displays, PV cost/area can be driven by large area nanomanufacturing equipment like that shown in Fig. 7(b) for wafers or Fig. 6(a) for thin film absorber layers on glass. Watts per area corresponds directly to conversion efficiency, which can be driven by improvements to materials, device structures and even better manufacturing uniformity. Figure 8 shows an example of material innovation where a tandem amorphous-Si/microcrystalline-Si (a-Si/μc-Si) junction is used to capture more of the photon spectrum than just a single a-Si thin film cell, thereby improving module efficiency. It is critical, however, that the improvements in efficiency not come at an overcompensating increase in cost/area; otherwise overall cost/W will not be reduced. This leads to implementation in sophisticated, large area PECVD tools like that in Fig. 6(a), where large panels as in Fig. 8 can be produced, so that module production costs can be driven down to approximately $1/W.

Nanomanufacturing technology can also have a substantial effect on other energy-related applications, including fuel cells, thin film batteries and even energy-efficient lighting. As with PVs, these devices have been shown to improve through nanoengineered films and structures, but will only become pervasive through controlled, high-volume manufacturing. Like displays and PVs, the main driver for these applications is typically cost/area. Hence the focus on tools and processes that efficiently produce large areas.

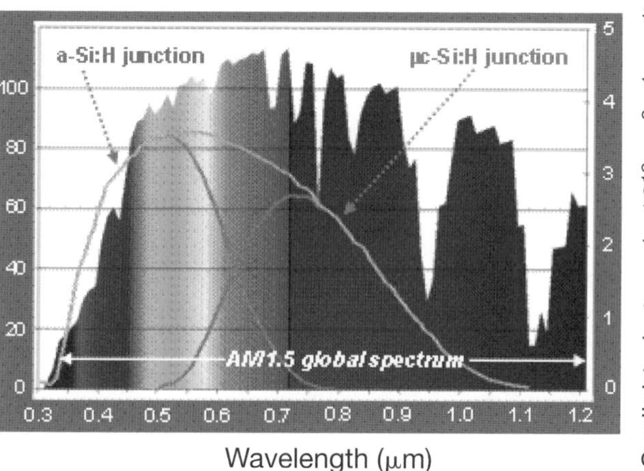

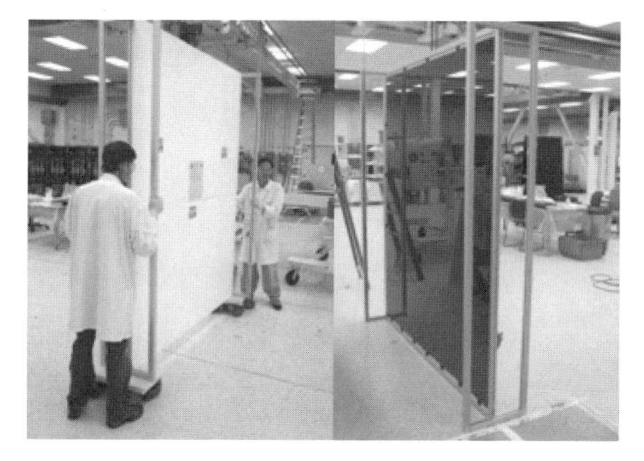

Figure 8. Thin film silicon solar cells: (top) spectral response of a-Si/µc-Si tandem junction; (bottom) Gen 7 panel produced by a large area PECVD tool like that in Fig. 6(a). Note the dark gray color of the panel, indicating the absorber is not pure a-Si.

5. Summary and conclusions

In conclusion, innovations in nanomanufacturing technology – not just the underlying nanotechnology concept – are essential for high volume applications and not just an afterthought. The evolution of IC technology into the nanoelectronics era has been enabled by nanomanufacturing technologies that go beyond top-down add/remove steps. Continuous scaling of IC dimensions and component integration has been driven primarily by reducing the cost/transistor and cost/bit, which has become ever more important as consumer consumption of semiconductors has become dominant over corporate use. No showstoppers are seen to continuing scaling through at least the 32 nm node, although complexity of structures and materials will continue to increase. Nanomanufacturing technology will enable a number of other markets. For example, the flat panel display market has been revolutionized through tremendous cost/area reduction of LCD TFT panels. Significant impact in future is expected for photovoltaics and other energy-related applications.

Returning to the concept of model electronic machines, many have suggested the emergence of "pervasive intelligence", where electronics of various kinds will be embedded throughout our environment. Potentially aided both in cost and form factor by implementations on plastic substrates, miniature machines could, for instance, be embedded within consumer packaging to interact with a refrigerator or as an added feature to attract a buyer. Such applications could drive trillions of machines, but the required cost must be equivalently low. Continuing the current pace of progress though gives some hope – extrapolating IC per bit costs and display per area costs suggests a 4G iPod nano could be included in/on a box of cereal in 15 years for ~5¢ – not including the cost of the recorded content, which unfortunately we have not found a way to scale.

References

1. U.S. National Nanotechnology Initiative, see *http://www.nano.gov*
2. D. Eaglesham, "The nano age?" *MRS Bull.* **30**, 360 (2005).
3. M. Roco, "International perspective on government nanotechnology funding in 2005," *J. Nanoparticle Res.* **7**, 707 (2005).
4. G. Moore, "Cramming more components onto integrated circuits," *Electronics* **38**, 114 (1965).
5. SIA National Technology Roadmap for Semiconductors, 1994.
6. G. Moore, "No exponential is forever: but 'Forever' can be delayed!", *Tech. Digest ISSCC* (2003), pp. 20–3; available at *http://www.intel.com/technology/mooreslaw/index.htm*
7. International Technology Roadmap for Semiconductors, 2005.
8. D. Varghese, D. Saha, S. Mahapatra, K. Ahmed, F. Nouri, and M. Alam, "On the dispersive versus Arrhenius temperature activation of NBTI time evolution in plasma nitrided oxides: Measurements, theory, and implications," *Tech. Digest IEDM* (2005), pp. 684–7.
9. L. Washington, F. Nouri, S. Thirupapuliyur, *et al.*, "pMOSFET with 200% mobility enhancement induced by multiple stressors," *IEEE Electron Dev. Lett.* **27**, 511 (2006).
10. A. S. Brown, "Winner: Flat, cheap, and under control," *IEEE Spectrum* **42**, 34 (2005).
11. D. Hsieh, "DisplaySearch 2006 TFT LCD and component market trend," (May, 2006).
12. K. E. Drexler, *Engines of Creation*, New York: Bantam Doubleday Dell, 1986.

32 nm: Lithography at a Crossroad

J. P. H. Benschop
ASML Inc., Veldhoven, The Netherlands

1. Introduction

Microlithography is used to define patterns of integrated circuits (ICs). The workhorse of lithography is optical projection lithography, whereby a pattern on a mask is imaged on a wafer with a 4:1 reduction ratio.

Using water immersion objectives, a production-worthy lithography system using the 193 nm wavelength and a numerical aperture $NA = 1.2$ has been shipped in early 2006.[1] This system is capable of printing lines and spaces below 45 nm. The next step will be a water-based immersion system with $NA = 1.35$, capable of printing lines and spaces below 40 nm. Several options are currently being pursued to extend the lithography roadmap down to 32 nm dense lines and spaces. Leading candidates are extreme ultraviolet (EUV) lithography and double patterning using water-based and non-water-based immersion. Opportunities and challenges of these technologies will be discussed in this chapter.

2. Lithography roadmap

The resolution R of optical lithography, defined as the half-pitch of a dense lines and spaces pattern, is determined by:

$$R = k_1\lambda/NA \tag{1}$$

where λ is the wavelength of light in vacuum, NA is the numerical aperture of the lens given by the refractive index (of the surrounding medium) multiplied by the sine of the angular semi-aperture of the lens, and k_1 is an imaging enhancement-dependent parameter. Lower k_1 generally leads to lower modulation of the light at the wafer and thus to lower process margins. Image enhancing technologies can be applied to partly mitigate this effect, but the physical limit of single exposure lithography is $k_1 = 0.25$.

Table 1 below shows various combinations of NA, wavelength and k_1 that can potentially be used to meet the required resolution over time. Current state-of-the-art optical lithography is water-based immersion lithography using an ArF laser with 193 nm wavelength and an objective lens with a numerical aperture of 1.2. Using aggressive dipole illumination, 42 nm lines and spaces can be imaged over 950 nm depth-of-focus, illustrated in Fig. 1. For this system, $k_1 = 0.26$, which is very close to the physical limit of 0.25.

Future Trends in Microelectronics. Edited by Serge Luryi, Jimmy Xu, and Alex Zaslavsky

ISBN 0-471-48

	half pitch [nm]	65	45	32	22	16	11
	year	2005	2007	2009	2011	2013	2015
[nm]	NA						
193	0.93	0.31					
	1.20	0.40	0.28				
	1.35		0.31	0.22	0.15		
	1.55			0.26	0.18		
13.5	0.25			0.59	0.41		
	0.35				0.57	0.41	
	0.45					0.53	0.37

Table 1. The calculated k_1 values for each half pitch (top line, correlated with year of production) are shown as a function of λ and *NA* combinations (left columns).

Figure 1. 42 nm lines and spaces imaged through focus using $NA = 1.2$ and $\lambda =$ 193 nm. Effective $k_1 = 0.26$.

The next sections will explain the challenges and opportunities for the three remaining options for the 32 nm node as depicted in Table 1:

- double patterning using 193 nm wavelength and water-based immersion;
- single patterning using 193 nm wavelength and high-index liquids and lens materials;
- EUV lithography using 13.5 nm wavelength.

3. Double patterning[2,3]

The maximum practical *NA* of water-based immersion is around 1.35. This corresponds to a maximum angle of 70 degrees of the rays in the liquid with respect to surface normal. A larger angle would lead to exploding lens size and cost.

As can be seen in Table 1, the k_1 value for $\lambda = 193$ nm, $NA = 1.35$ and 32 nm half pitch is below 0.25. Note however that this physical limit is related to the half pitch, whereas the actual size of printed structures can be arbitrarily smaller. Figure 2 shows a double patterning scheme, whereby a first exposure prints features on the order of a quarter pitch that are subsequently transferred onto a second layer (the so-called hard mask), that can later be used as an etch mask. Next a second exposure is shifted compared to the first and also transferred into the hard mask layer. This image transfer is essential. Since each exposure produces a sinusoidal modulation in the resist, the net effect of two such exposures without image transfer would be adding two sinusoidal functions shifted by a quarter period, resulting in no net modulation. Figure 3 shows the experimental results of such a two-step double patterning procedure.

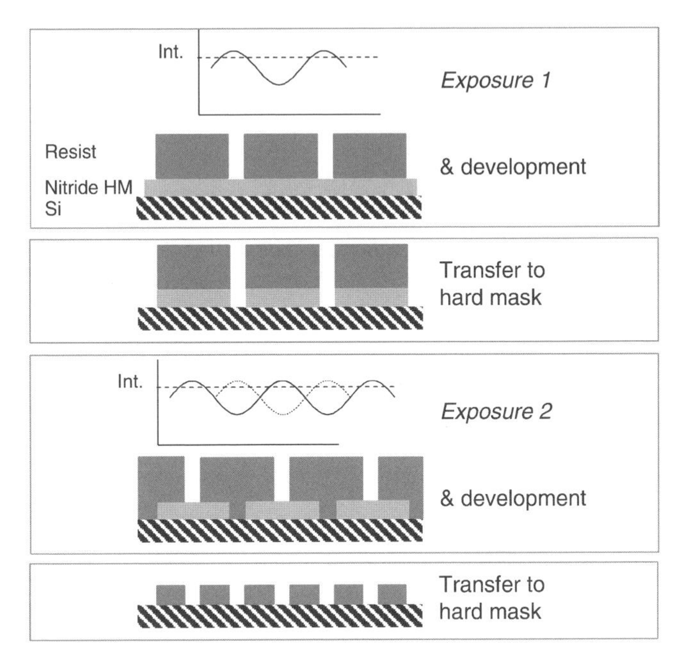

Figure 2. Schematic illustration of the double patterning procedure.

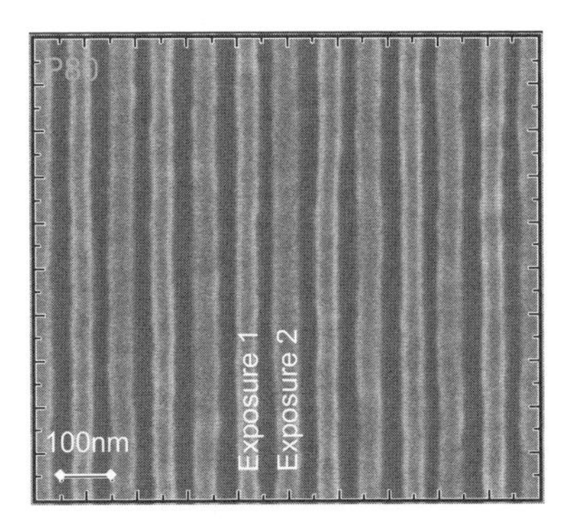

Figure 3. These 40 nm half-pitch lines and spaces have been obtained with 193 nm wavelength and *NA* = 0.93. The resulting k_1 equals 0.20.

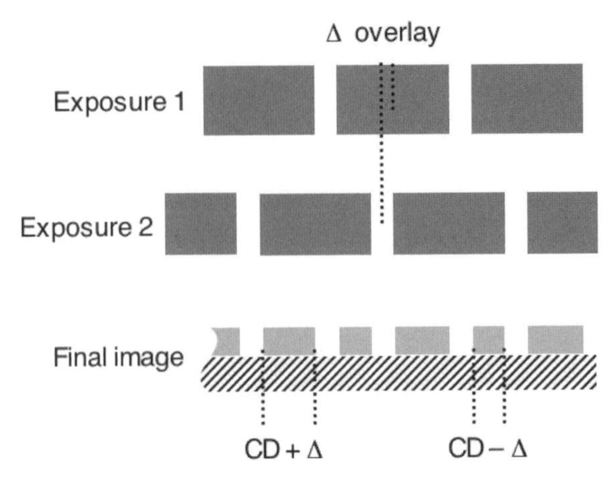

Figure 4. In the double exposure scheme, overlay error is translated into a critical dimension (CD) change.

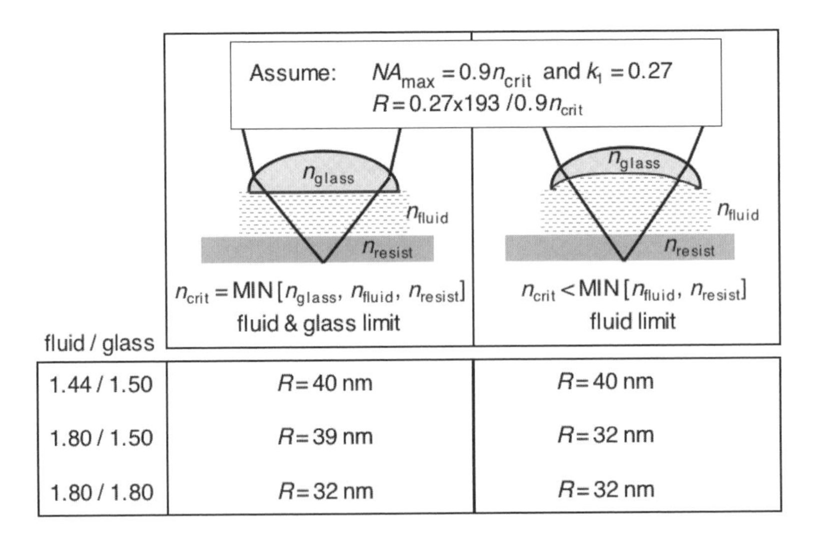

fluid / glass	fluid & glass limit	fluid limit
1.44 / 1.50	$R = 40$ nm	$R = 40$ nm
1.80 / 1.50	$R = 39$ nm	$R = 32$ nm
1.80 / 1.80	$R = 32$ nm	$R = 32$ nm

Figure 5. Resolution limits for two types of lenses: flat (left) and concave (right) last lens elements for various combinations of the refractive indices of the fluid and the last lens element.

However, since each pattern is built from two exposures, effective productivity is cut in half and the number of masks is doubled. Cost is therefore a critical issue for this option. Furthermore the overlay of the two exposures becomes crucial because in the double patterning procedure overlay error affects the critical dimension (CD) – see Fig. 4.

4. Immersion beyond water[4,5]

Liquids with refractive index higher than water exist and have the potential to increase the *NA* beyond 1.35.

Furthermore, two classes of lenses exist: those with a flat last lens surface and those with a concave last lens surface. As explained in Fig. 5, the maximum *NA* is only marginally increased if water (n = 1.44) is replaced by a hypothetical high-index liquid (n = 1.80), as long as the last lens element is flat and made of quartz (n = 1.57). This is due to the fact that the refractive index of the lens material is close to the refractive index of water. Only a concave last lens element combined with a high-index lens material produces a substantially higher *NA*.

Based on Fig. 5 it is tempting to conclude that concave last lens elements are attractive since they hold promise of substantially higher *NA* without changing the lens material. Unfortunately, concave last lens elements also suffer from a significant drawback. Absorption of the light passing through the liquid causes significant apodization. Worse still, this apodization is field-dependent. Furthermore, the absorption of light will heat the liquid, leading to an optical path change and thus image aberration. These effects are illustrated and quantified in Fig. 6. The temperature would need to be controlled to within 0.1 mK in order to have acceptable optical path changes. Control with 0.1 mK precision seems an insurmountable challenge and thus limits practical designs to flat last lens elements.

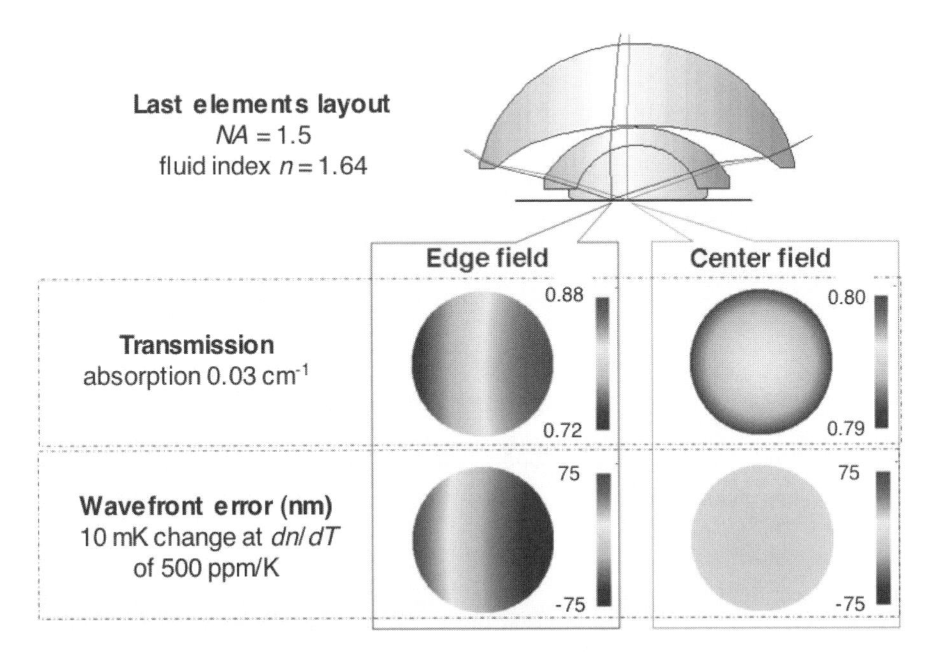

Figure 6. The transmission through the liquid is dependent upon angle and position in the field. In addition, the absorbed light leads to heating, resulting in a refractive index change that leads to a wavefront error.

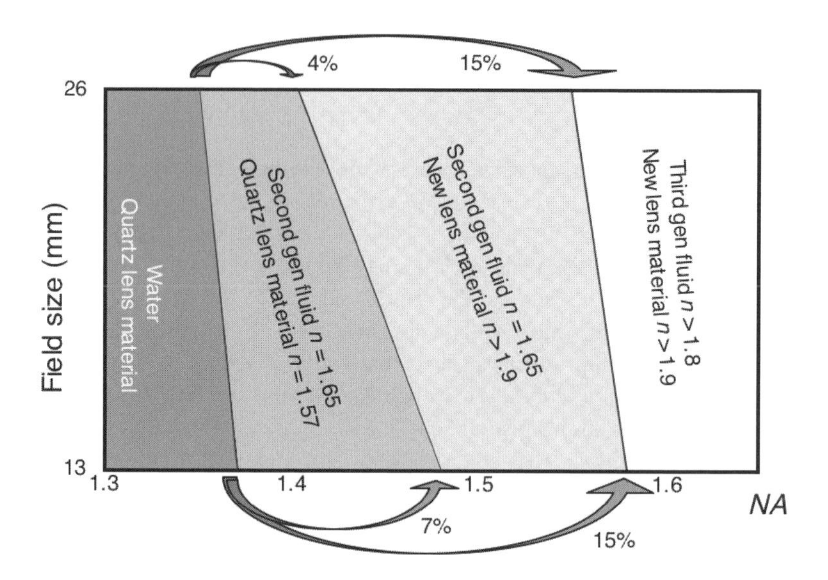

Figure 7. Achievable maximum *NA* for various combination of high-index liquid and lens material as function of field size.

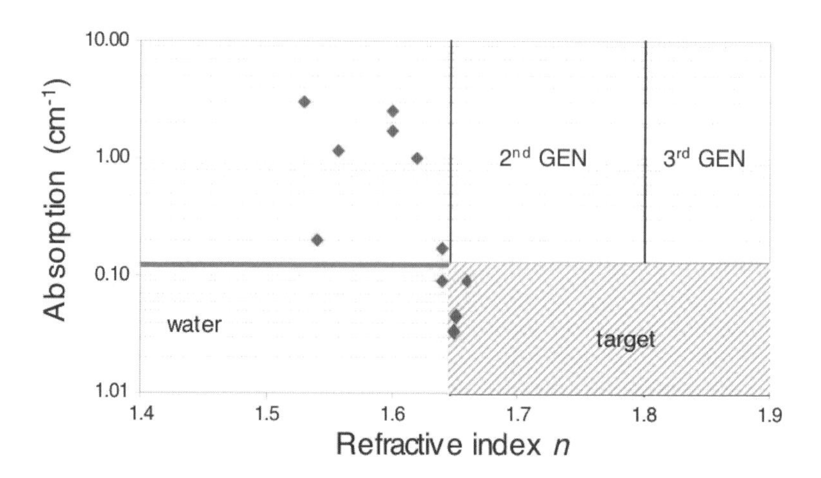

Figure 8. Overview of refractive index and absorption of various liquids. Target specifications include a refractive index $n > 1.65$ and absorption $< 0.15\ \text{cm}^{-1}$.

Today's field size is typically 26 mm wide. Using a smaller field size would result in lower throughput and thus in higher cost. Figure 7 shows that changing from water to an existing high-index liquid and using a quartz last lens element provides only a 4% *NA* increase for a 26 mm field system. A larger 15% increase in *NA* requires changing both the liquid and the lens material beyond quartz.

Figure 8 indicates status of various high-index liquids. Significant progress has been made on the absorption; today various candidates meet the target specification. However the maximum refractive index at present is limited to n < 1.67. Using a two-beam interferometer, images in resist have been shot using several leading candidates for the second generation high-index liquid. The results are shown in Fig. 9.

Leading candidates of high-index material for last lens-element are ceramic materials like spinel and LuAG. The currently available high-index lens materials are compared to target specifications in Table 2.[6] Today, the most critical parameters are strain birefringence and absorption.

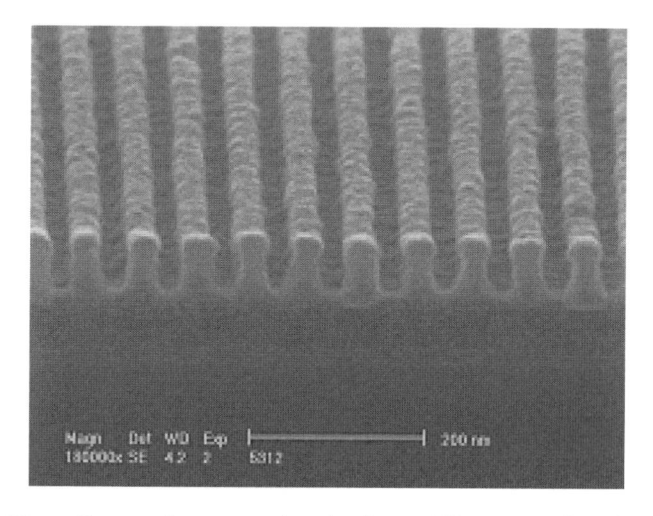

Figure 9. 32 nm lines and spaces printed using a 193 nm wavelength and a liquid with refractive index $n = 1.64$. Effective $NA = 1.50$.

	index	Intrinsic birefringence [nm/cm]	Strain birefringence [nm/cm]	scattering	A_{10} [1/cm]
Target specifications	>1.7	<10	< 1nm/cm	uncritical	<0.01
Ceramics (e.g. Spinel)	>1.85	None	~5	critical	2.7
LuAG	2.1	30	<5	uncritical	0.6

Table 2. Current status of high-index lens materials.

5. EUV lithography[7]

EUV lithography uses plasma sources generating the 13.5 nm wavelength and all reflective optics in vacuum. Since late 1980's, EUV has been pursued by a number of dedicated research centers, consortia and companies worldwide.

The multilayer coated optics need to be atomically flat over many decades of spatial frequency as depicted in Fig. 10 below. In addition to the formidable challenge of making these atomically flat aspheric optics, they need to be kept highly reflective: any oxidation should be avoided and carbon build-up should be slow and removable as needed.

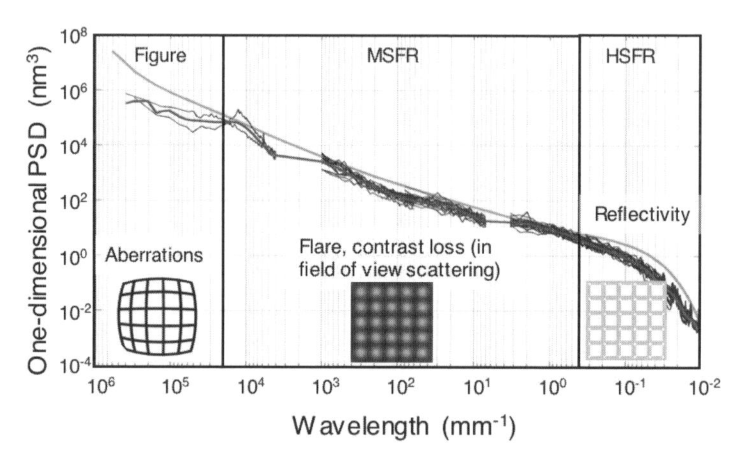

Figure 10. Power spectral density of EUV mirror measured over 12 decades. The spectrum is split into three ranges: the low frequency "figure" range that impacts aberrations; the mid spatial frequency range (MFSR) that impacts contrast and hence process latitude; and the high spatial frequency range (HSFR) that impacts reflectivity and hence productivity.

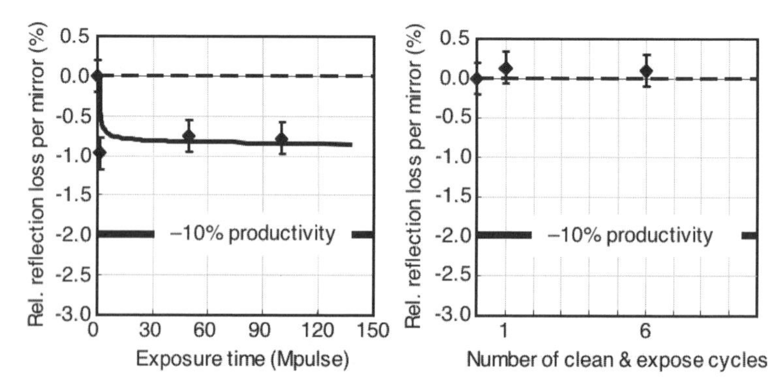

Figure 11. Relative reflectivity drop in multilayer-coated EUV mirrors as a function of number of pulses (left); removal of contamination by cleaning, indicating it is due to carbon growth rather than oxidation (right).

Figure 11 shows that the relative mirror reflectivity drops by about 1% when exposed to an EUV dose in vacuum corresponding to few million shots in an EUV tool, but then remains stable for 100 million shots (which corresponds to five hours in case of a 5 kHz source). Furthermore it can be seen that any reflection loss can be recovered by cleaning the mirror, indicating carbon growth is mechanism for reflection loss. Multiple cleaning cycles are possible without any loss of reflection.

To make EUV attractive, the productivity should be at least 100 wafers per hour. Assuming a 10 mJ/cm^2 resist sensitivity and realistic projections of tool parameters like mirror reflectivity leads to conclusion that 180 W of EUV power is required.

Figure 12 shows that three orders of magnitude in EUV power have been gained over the last five years by switching from Xe to Sn as the radiating material.[8,9] The 100 W output is measured in burst mode, meaning the source has demonstrated the capability to generate 100 W but not the thermal engineering to sustain this for a longer period.

Power is not the only hurdle EUV needs to overcome in order to enter mass production. An EUV resist meeting simultaneously the resolution, line-edge roughness (LER), and sensitivity specifications needs further development. Figure 13 illustrates the tradeoff between sensitivity and LER for a number of EUV resists – clearly, no existing EUV resist meets the specs for 32 nm node lithography.[10]

The first full-field EUV scanners, using $\lambda = 13.5$ nm and $NA = 0.25$, have become operational in early 2006. A photograph of one of these early machines is shown in Fig. 14, while Fig. 15 shows 40 nm dense lines and spaces and 55 nm contact holes printed with this tool.

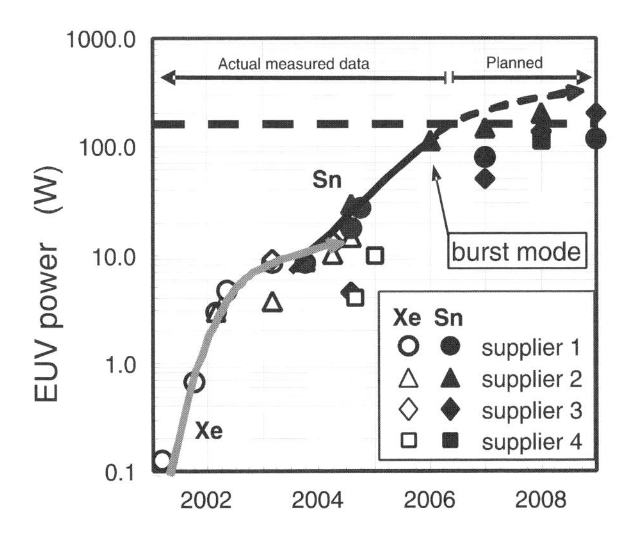

Figure 13. Progress in EUV source power in recent years, combined with projections for the near future.

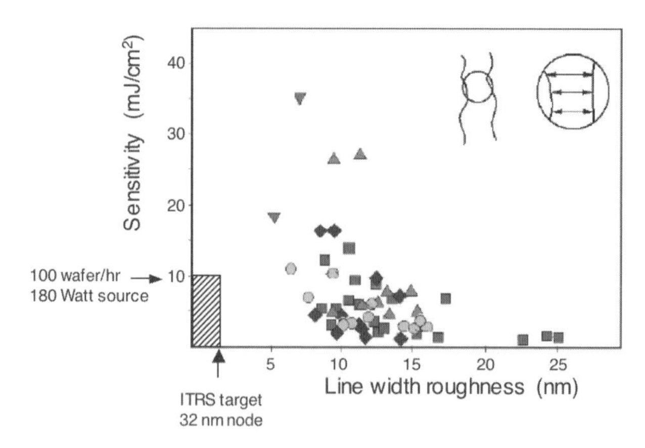

Figure 14. Line edge roughness *vs*. sensitivity of various EUV resists.

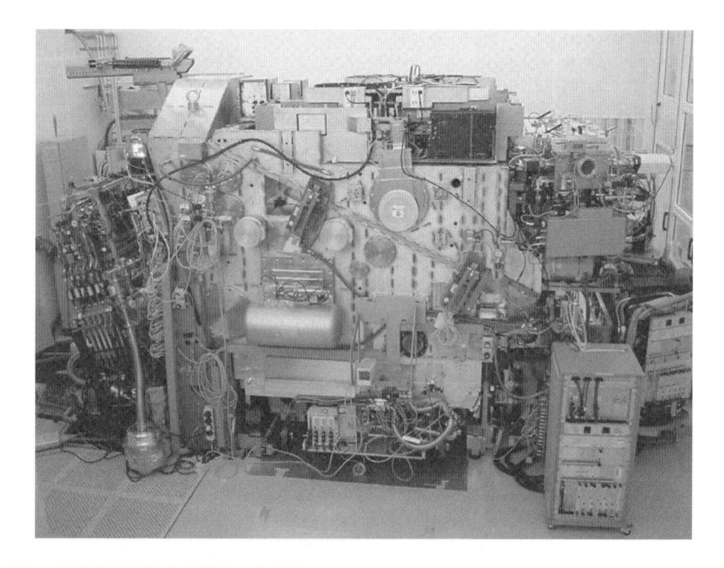

Figure 15. EUV full-field *NA* = 0.25 scanner.

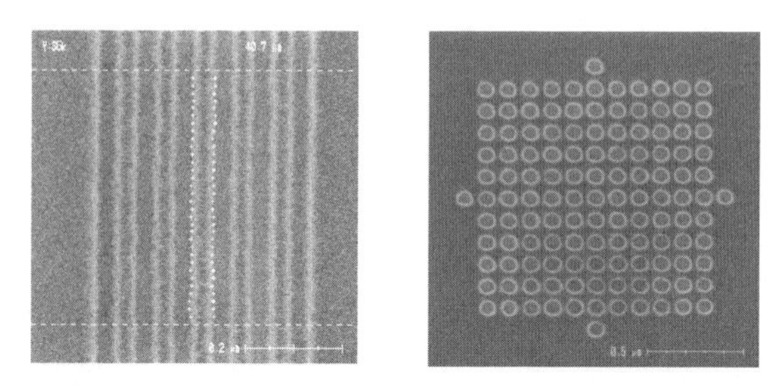

Figure 16. 40 nm lines and spaces and 55 nm contact holes printed with 13.5 nm wavelength (EUV) and *NA* = 0.25 optics.

6. Summary and conclusion

Three leading candidates for 32 nm half pitch lithography have been discussed in this chapter.

Double patterning using water-based immersion is the most straightforward extension, but it could almost double the cost of lithography and has a severe impact on tool requirements like overlay. Alternatively, non-water based immersion liquids offer the prospect of increasing the *NA* beyond 1.5, but will require currently unavailable lens materials.

Significant progress has been made in EUV technology over the last few years: first full-field tools became available early 2006. Clearly, EUV has the advantage of being the most extendable technology. However, additional progress will be needed on EUV sources, resists and masks before this technology is suitable for mainstream production.

Acknowledgments

The author would like to thank the many teams within ASML and Carl Zeiss whose work has been used throughout this publication.

References

1. H. Jasper, T. Modderman, M. van de Kerkhof, *et al.*, "Immersion lithography with and ultrahigh-*NA* in-line catadioptric lens and a high-transmission flexible polarization illumination system," *Proc. SPIE* **6154**, 61541W (2006).
2. C.-M. Lim, S.-M. Kim, Y.-S. Hwang, *et al.*, "Positive and negative tone double patterning lithography for 50 nm flash memory," *Proc. SPIE* **6154**, 615410 (2006)
3. M. Maenhoudt, J. Versluijs, H. Struyf, J. Van Olmen, and M. Van Hove, "Double patterning scheme for sub-0.25 $k1$ single damascene structures at *NA* = 0.75, λ = 193 nm," *Proc. SPIE* **5754**, 1508 (2005).
4. H. Sewell, J. Mulkens, D. McCafferty, L. Markoya, B. Streefkerk, and P. Graeupner, "The next phase for immersion lithography," *Proc. SPIE* **6154**, 615407 (2006).
5. R. H. French, W. Qiu, M. K. Yang, *et al.*, "Second generation fluids for 193 nm immersion lithography," *Proc. SPIE* **6154**, 615415 (2006).
6. J. H. Burnett, S. G. Kaplan, E. L. Shirley, *et al.*, "High-index optical materials for 193 nm immersion lithography," *Proc. SPIE* **6154**, 615418 (2006).
7. H. Meiling, H. Meijer, V. Banine, *et al.*, "First performance results of the ASML alpha demo tool," *Proc. SPIE* **6151**, 615108 (2006).
8. J. Jonkers *et al.*, "Integration of the tin source," *Proc. Fourth Intern. Symp. EUV Lithography*, San Diego (2005).

9. U. Stamm *et al.*, "Development status of EUV sources for use in alpha-, beta- and high volume chip manufacturing tools," *Proc. Fourth Intern. Symp. EUV Lithography*, San Diego (2005).
10. H. B. Cao, J. M. Roberts, J. Dalin, *et al.*, "Intel's development of EUV resists," *Proc. SPIE* **5039**, 484 (2003).

Physical Limits of Silicon CMOS: Real Showstopper or Wrong Problem?

M. Brillouët
CEA-LETI, 17 rue des Martyrs, 38054 Grenoble, France

1. Introduction

Digital information processing has become a key driver for the growth of the world economy. It has been fueled by the progress of microelectronics that is usually described by an exponential growth of some performance metric, often named Moore's Law. Actually, the early papers of Gordon Moore[1-3] only stressed the continuing search for a higher integration density of circuits, mostly through feature size reduction, while their electrical behavior was not even mentioned. It was not until the mid-1970's that Robert Dennard formalized the benefits of downscaling device dimensions:[4] the present paradigm that miniaturization makes integrated circuits denser, faster, less power-hungry, cheaper and more reliable was born. The question then is how long will this trend last?

For more than two decades, technical papers announced the imminent dismissal of Si CMOS technology, stressing first the 1 μm barrier, then the 100 nm brick walls, and recently the 10 nm limit: to date they all proved to be wrong. After looking at some similar claims, we will focus on the so-called physical limits of the processing unit, trying to outline the underlying assumptions of such assertions and their possible shortcomings.

Assuming that classical Si CMOS will encounter some practical limits in the future, the latest version of the ITRS roadmap gives a thorough analysis of possible candidates for the "beyond CMOS" era. We will review the necessary criteria for a successful replacement of the Si CMOS gate for information processing and make a critical assessment of some of the proposed approaches.

However the real question is: are we looking at the right problem? The focus on logic gates may just be the tree hiding the forest of issues to be addressed in information processing. We should also consider major pending problems, like interconnecting those gates or manufacturing complex circuits useful for specific applications.

As a word of caution, this paper will restrict its analysis to digital processing units, leaving other semiconductor technologies, such as memories and analog devices, outside of the scope of this discussion.

Future Trends in Microelectronics. Edited by Serge Luryi, Jimmy Xu, and Alex Zaslavsky
ISBN 0-471-48 © 2007 John Wiley & Sons, Inc.

2. A brief history of Si MOS limits

From the early days of microelectronics, an abundant literature explored the potential limits of the silicon technology: this chapter will mention a few milestones in this quest.

In the 1960's and 1970's, the focus was on the integration density – as exemplified by Moore's paper[1] – and the associated metric was the minimum achievable feature size. Among others, J. T. Wallmark published a detailed analysis of potential limits in shrinking dimensions[5-7] and concluded that "the end of the road to smaller size has already been reached"[8] – the smaller size being typically 5–10 microns in his early work. Pinpointing the flaws in his reasoning is not straightforward, but one can make a few remarks that may also apply to more recent papers on Si limits:

- the model of the elementary device is very crude (cube resistor, no signal restoration, *etc.*);
- the statistics used to assess circuit yield is questionable (*e.g.* cumulative independent events);
- some physical concepts (*e.g.* Heisenberg uncertainty principle) are misused.

In the 1980's and 1990's, the focus shifted towards speed and power consumption. Following these metrics, J. D. Meindl established a hierarchy of constraints in designing an integrated circuit:[9]

- at the most fundamental level, one faces physical limits dictated by quantum mechanics, thermodynamics and electromagnetism (*i.e.* relativity);
- more constraints are added when considering the materials used for fabricating the transistor, mostly associated with electrostatics (dielectric constant), carrier dynamics, thermal behavior and parameters fluctuations;
- the device architecture, the design style of the logic building blocks, and the system architecture and packaging bring additional constraints that limit considerably the design space in terms of power and delay.

However, considering the later levels of this hierarchy of constraints, as stressed by R. W. Keyes, "the potential tradeoffs [and the underlying assumptions] are too numerous and complex [...] and they obscure the quantitative significance of [deriving] performance limits from purely physical reasoning".[10]

The most recent works concentrate on the so-called "fundamental physical limits". The general analysis is along the following lines[11] (see Fig.1):

- thermodynamics (more specifically, the so-called "Landauer principle"[12]) suggests a minimum dissipated energy per operation, $E_{\mathrm{min}} \sim k_{\mathrm{B}}T\ln 2 \sim 17$ meV at room temperature;

- from this minimum energy and from the first Heisenberg uncertainty relationship one derives a minimum device size, $x_{min} \sim 1$ nm;
- from the same minimum energy and from some time–energy relationship in a quantum system (using mostly the second Heisenberg uncertainty relationship) one deduces a minimum switching time, $\tau \sim 40$ fs;
- these combined "limits" gives an upper value of the power density dissipated in a circuit whose transistors are packed at a maximum integration density and switch at the same time at the maximum speed. This estimate then gives a rather unrealistic ~4 MW/cm^2 value of power consumption.

However, each step of this logical chain can and perhaps should be challenged. First of all, implicit assumptions related to the thermodynamic limit are numerous and potentially questionable, casting some doubt about the unavoidability of this barrier. Some of these questionable assumptions are made explicit below:

- *the computing system interacts with its surroundings* (*e.g.* thermal bath): computation is typically characterized by a relaxation time in most systems, but experimental work on quantum computing – among others – tends to indicate that some computation may be performed in a quasi-isolated system;

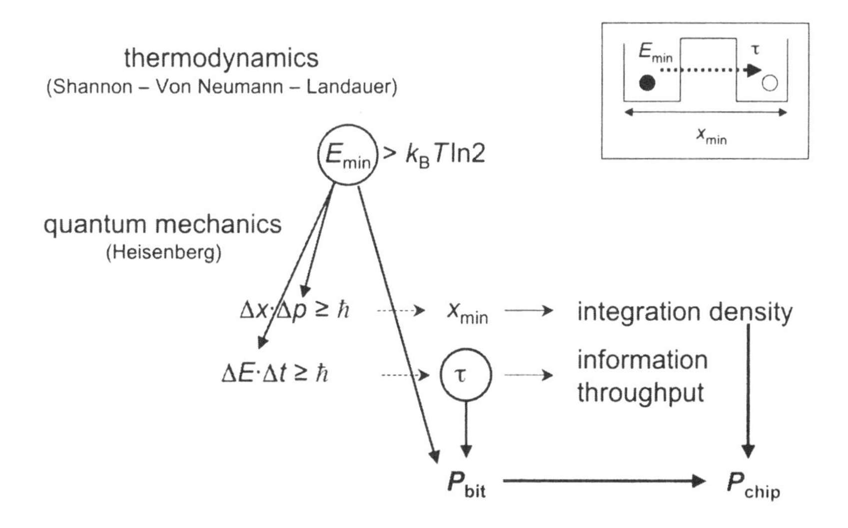

Figure 1. Combination of possible "fundamental physical limits" for a digital information processing mechanism. From the minimum switching energy E_{min} and from the Heisenberg uncertainty relationships, one derives the minimum size of the device x_{min}, and the minimum switching time τ, which translates into a maximum dissipated power per bit P_{bit}, and per area P_{chip}.

- *information needs to be destroyed during computation* (and its energy will be transferred to the thermal bath, *i.e.* dissipated): while this does happen in most of the present microelectronic circuits, reversible "adiabatic" computing[13] aims to develop an alternative computing scheme where this statement does not apply. The applicability of such adiabatic computing to complex systems is still debated, however, as it is unclear if the associated energy benefit will outweigh the complexity overhead of a reversible computer;

- more fundamentally, the $k_B T \ln 2$ energy associated with a bit of information applies to a statistical ensemble of states in quasi-equilibrium: it has to be revisited for systems with few particles and states far from equilibrium (see, for example, Ref. 14).

Further, it should be noted that the first Heisenberg uncertainty relationship,

$$(\Delta x)(\Delta p) \geq \hbar/2, \tag{1}$$

is used with the underlying assumption that the information carrier is localized. This imposes a much stronger constraint on the physical implementation of the computing mechanism than just assuming that the system should evolve among distinguishable (*i.e.* mutually orthogonal) quantum states. Furthermore many device models (see, for example, Ref. 11) assume a free quasi-particle as an information carrier in a quasi-static system, using a semi-classical approximation and without quantum confinement of the carrier: how far the resulting conclusions apply to a more realistic physical device needs to be explored. Finally, in these models Δx is loosely defined as the physical width of the energy barrier in the device and at the same time as the linear size of the transistor. A more rigorous discussion of the applicability of this limit in the most general instance of a physical computing device is still lacking.

The relationship between energy and time in a quantum system, using for example the so-called second Heisenberg uncertainty relationship,

$$(\Delta E)(\Delta t) \geq \hbar/2, \tag{2}$$

is still more doubtful. Actually, two models may apply to the discussion:

- the classical interpretation[15] of Eq. (2) states that in observing a quantum system during a time window Δt, the spread (*i.e.* standard deviation) in the measured energy of the system ΔE will be at least of the order of $\hbar/2\Delta t$. There is no clear reason to interpret ΔE as the energy dissipated during the transition between two distinguishable states, nor Δt as the transition time.

- N. Margolus and L. Levitin derived another inequality[16] giving an upper bound of the speed of evolution τ^{-1} of an isolated quantum system with respect to the average energy $\langle E \rangle$ of the orthogonal states of the system

$$\tau \geq h/2\langle E \rangle\,. \tag{3}$$

- *information needs to be destroyed during computation* (and its energy will be transferred to the thermal bath, *i.e.* dissipated): while this does happen in most of the present microelectronic circuits, reversible "adiabatic" computing[13] aims to develop an alternative computing scheme where this statement does not apply. The applicability of such adiabatic computing to complex systems is still debated, however, as it is unclear if the associated energy benefit will outweigh the complexity overhead of a reversible computer;

- more fundamentally, the $k_B T$ ln2 energy associated with a bit of information applies to a statistical ensemble of states in quasi-equilibrium: it has to be revisited for systems with few particles and states far from equilibrium (see, for example, Ref. 14).

Further, it should be noted that the first Heisenberg uncertainty relationship,

$$(\Delta x)(\Delta p) \geq \hbar/2, \tag{1}$$

is used with the underlying assumption that the information carrier is localized. This imposes a much stronger constraint on the physical implementation of the computing mechanism than just assuming that the system should evolve among distinguishable (*i.e.* mutually orthogonal) quantum states. Furthermore many device models (see, for example, Ref. 11) assume a free quasi-particle as an information carrier in a quasi-static system, using a semi-classical approximation and without quantum confinement of the carrier: how far the resulting conclusions apply to a more realistic physical device needs to be explored. Finally, in these models Δx is loosely defined as the physical width of the energy barrier in the device and at the same time as the linear size of the transistor. A more rigorous discussion of the applicability of this limit in the most general instance of a physical computing device is still lacking.

The relationship between energy and time in a quantum system, using for example the so-called second Heisenberg uncertainty relationship,

$$(\Delta E)(\Delta t) \geq \hbar/2, \tag{2}$$

is still more doubtful. Actually, two models may apply to the discussion:

- the classical interpretation[15] of Eq. (2) states that in observing a quantum system during a time window Δt, the spread (*i.e.* standard deviation) in the measured energy of the system ΔE will be at least of the order of $\hbar/2\Delta t$. There is no clear reason to interpret ΔE as the energy dissipated during the transition between two distinguishable states, nor Δt as the transition time.

- N. Margolus and L. Levitin derived another inequality[16] giving an upper bound of the speed of evolution τ^{-1} of an isolated quantum system with respect to the average energy $\langle E \rangle$ of the orthogonal states of the system

$$\tau \geq h/2\langle E \rangle\ . \tag{3}$$

Beside these explicit criteria the device is expected to have a number of implicit qualities. For example, the device should:

- operate at room temperature;
- perform order(s) of magnitude better than the Si CMOS "at the end of the roadmap": no emerging device may expect to challenge the legacy of 40+ years of research, development and manufacturing of Si information processing systems if it does not offer significant advantages in some respect;
- be scalable, *i.e.* be applicable for more than just one or two technology generations, owing to the huge effort needed to displace Si CMOS;
- give a clear path for integrating other functions like memories, mixed-signal devices, and interface capability on the same chip;
- offer some perspective of mass-producing circuits at an affordable yield and cost (though it may be difficult to assess manufacturing issues at the very beginning of the research phase).

Unfortunately, according to the ITRS, no emerging logic devices[21] passed the exam so far. In fact, many new devices have no chance of matching, even in the distant future, the integration density and/or speed of the CMOS technologies presently in production! On the other hand, one cannot exclude the discovery of a new structure or a revolutionary concept in the future.

One common pitfall of many disruptive device proposals is that, starting from an interesting physical effect for a unit information processing mechanism, it fails to show the capability to integrate complex functions in a way competitive with the present or future CMOS systems. Quantum-dot cellular automata (QCA, for short) provide an instructive example.

The concept of QCA was initially proposed by Tougaw, Lent and Porod[22] in 1993. The basic idea relies on a bistable cell (*e.g.* quantum dots positioned at the 4 corners of a square and close enough to allow excess electrons to tunnel between neighboring dots) that appears to promise an extremely fast and low-power device. The equilibrium state (or polarization) of this cell depends on the states of the nearest-neighbor cells that are arranged in a way to map logic functions (*e.g.* majority gate) and allow signal propagation along a line. The computation is performed by forcing the states of the input cells at the periphery of the device and by letting the system relax to the ground state. The computation result is read out looking at the final state of the output cells at the periphery of the device. However, it became rapidly clear that this relaxation mechanism led to unpredictable computation time and possibly wrong results if the system got stuck into metastable states. This led to the need to implement a complex multiphase "adiabatic" (*i.e.* slow) clocking scheme.[23] A comparison of QCA with CMOS for logic blocks of small complexity concluded[24] in 2001 that an ultimately scaled molecular QCA would barely compete with advanced CMOS in terms of operational frequency. Analyzing the achievable integration density would lead to the same conclusion. The application of QCA to complex circuits would also need more in-depth analysis of other potential issues:

- what is the actual power consumption of a clocked QCA, taking into account the power dissipated by the clock tree, and how it compares to standard CMOS for the same computational throughput?
- how scalable is the QCA approach?
- what is the noise immunity of the QCA scheme? Are there crosstalk issues and layout dependency?
- how manufacturing fluctuations will affect very complex systems?
- what is the set-up procedure of a complex QCA, *i.e.* how to control the exact number of excess electrons (or equivalent) in each of the hundreds of millions of QCA cells?

In short, after more than 12 years of research, QCA failed to provide a significant competitive advantage with respect to CMOS in terms of integration density, speed, power consumption, and implementation of complex systems. This interim conclusion has led to QCA using excess electrons in quantum dots and electrostatic interaction (so-called e-QCA) to be dropped from the list of potential emerging devices in the latest version of the ITRS. Surprisingly enough, magnetic and molecular variants of QCA did not share that fate despite facing the same basic issues.

Another, more subtle pitfall of many emerging devices and architectures lies in the difficulty of integrating the constraints of the whole chain from system architecture to complex circuits design, to manufacturing, test and assembly. In the recent years, there were many articles exploring the FPGA concept applied to nanoscale devices characterized by a very high fault rate. The basic idea[25] resides in fabricating regular arrays of extremely miniaturized active devices, then in mapping the defects of the resulting structure, and finally in compiling and implementing the intended algorithm into the working elements. This potentially very attractive approach raises many questions, however.

- Rather than testing once a fixed design for its functionality and discarding the faulty devices, one needs to sequentially extract some defect map(s) of the chip, then compile the function into the working parts of the circuit and finally retest the implemented algorithm. In short, one trades off the yield issue of a potentially costly collective manufacturing with an expensive die per die multi-step functional test and programming. How far the efficiency of the test and programming algorithm may overcome the resulting increase in the test cost needs to be proven in realistically complex cases.
- Each of these "test & program" steps may lead to the impossibility to proceed further, *e.g.* too many defects may prevent further testing of the circuit if the test probe does not find any percolation path. A realistic

yield model of this approach which would be compared with the yield of a more conventional random logic circuit having the same functionality is still lacking.

- It should also be noted that the proposed mesh-type structure of these nano-arrays may not be optimal in allowing the test of a significant part of the circuit: random and scale-free networks, such as the Internet, are known to be more resistant to random defects than a grid structure. This architectural aspect seems to be rather weakly addressed.
- Most of the nanoscale arrays rely on a diode-resistor type of logic. This design style was abandoned many years ago as it did not provide an easy method of signal restoration that was needed in the circuit (it does mean that a fine grained co-integration with restoring devices like CMOS is mandatory) and as functional gates were not cascadable. Added to that, one may expect a rather high static power consumption for complex circuits. How the new nanoscale crosspoint devices may address this architectural weakness has not been clearly discussed.

In conclusion, up to now all the emerging logic devices failed to show the capacity to be integrated into complex processing units. Part of the present situation may be attributed to the fact that the researchers who propose disruptive approaches are often marginally aware of the complexity of designing and manufacturing complex functions into an integrated system. Conversely, CMOS circuit developers are usually unable to explicitly enumerate all the criteria for a "good" device implemented into a "good" circuit architecture: convergence of both worlds is a challenge in itself.

4. Are we looking at the right problem?

The main focus of the major microelectronics conferences is the elementary processing unit (*i.e.* transistor or switch), the memories and their association into complex circuits.

More specifically, in the quest for higher performances, many publications look at different ways to increase the on-current I_{ON} of the transistor (or reduce the associated metrics CV/I_{ON}). Using the approximate formula,

$$I_{ON} \approx \mu C_G / L_G, \tag{4}$$

where μ is the carrier mobility of the carrier, C_G the gate capacitance and L_G the physical gate length, one may infer three main directions in the progress of MOS transistors, enumerated below.

- *Scaling of the critical L_G dimension of the transistor*

This is the conventional way initiated by Dennard and others. However, in moving to smaller L_G and in packing the different elements more closely together, unwanted interactions and parasitic effects become increasingly detrimental to the behavior of the classical planar MOS transistor. Alternative structures, like SOI, 3D fin-FETs, multiple-gate or multiple-channel devices, have been proposed for many years and in some limited cases are already implemented in production – *e.g.* SOI in some microprocessors. The rather slow technological insertion of these alternative architectures is not only related to the added processing complexity and its potential impact on yield. One should not forget that most of these "unconventional" MOS devices require a rethinking of the whole design flow of complex circuits, which is a major effort most companies cannot afford (even assuming the new technology significantly outperforms the traditional approach).

- *Increasing the capacitive coupling C_G of the gate and the conduction channel*

The better electrostatic control of the conducting channel by the gate electrode was conventionally achieved by scaling the gate oxide thickness. Unfortunately, we reached dimensions where direct tunneling through the gate dielectrics leads to increasingly higher gate currents. To move away from this dead end, dielectrics with a high permittivity (the so-called high-κ materials) were proposed. However one cannot expect that the quasi-perfect Si-SiO_2 system will be easily replaced: mobility degradation, Fermi level pinning and instabilities, among other drawbacks, are the present keywords of those high-κ dielectrics. As a result, these new materials are not expected to be introduced in the next CMOS generation.

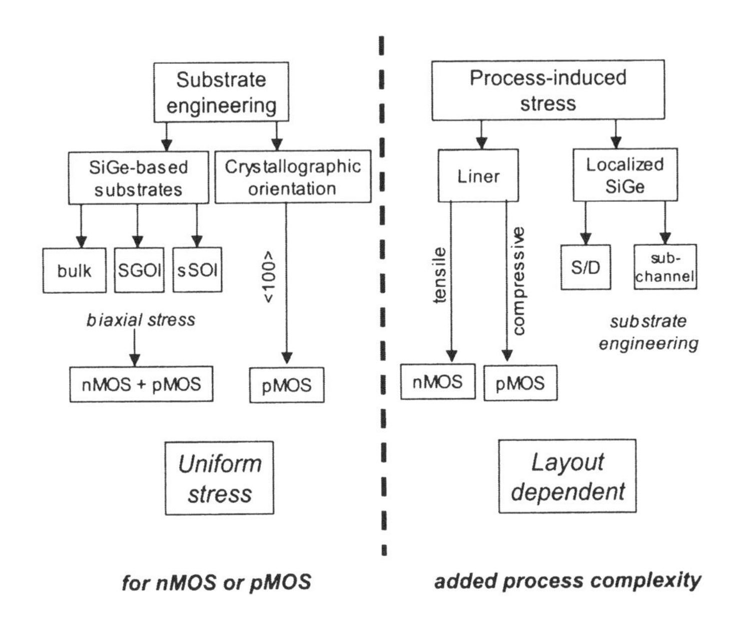

Figure 2. By applying different techniques for mechanically stressing the conductive channel and using different crystal orientations for the nMOS and pMOS, the carrier transport in the MOS transistors can be significantly enhanced.

- *Enhancing carrier transport*

As increasing transistor speed via traditional scaling becomes more and more difficult, enhanced transport of the carriers has been explored.

In the short term, the use of strain applied to the piezoresistive silicon and/or of other crystal orientations is a hot topic at device conferences. The combination of different techniques (see Fig. 2) allows balancing the benefits and drawbacks of each specific approach without adding too much complexity to the manufacturing process.

In the longer term, some researchers have proposed new channel materials to replace Si, such as Ge, III-V compound semiconductors, or carbon nanotubes. As in the case of high-κ dielectrics, one may expect significant processing difficulties and a diminishing return in the introduction of these materials into very complex integrated circuits.

The excitement about these "enhanced" MOS devices diverts attention from other equally important issues. In fact, one observes a widening gap between the research community and the real concerns of the microelectronics industry. The development engineer does not ask for breakthroughs in transistors and has more mundane questions, like:

- how can I use the same transistor architecture for high performance and low power applications and at different supply voltages?
- are adaptive power supplies, threshold voltages and clock frequencies applicable to the proposed technology?
- how does it allow input/output implementations?
- what will be the SRAM stability and its noise margin?
- is it possible to co-integrate digital logic, mixed signal circuitry and memories with the minimum increase in process complexity?

In fact, the present bulk planar MOS transistor provides legacy solutions even for the 45 nm technologies. As long as new transistor architectures do not answer all these questions related to their integration into complex circuits, they will remain "exotic" solutions for the future.

Interconnecting the processing units and memories in an efficient way is probably a more critical issue than building faster transistors: at this point, the wiring technology lags transistor performance. Optical interconnections, carbon nanotubes or even ultra-low-κ dielectrics are not expected to bring orders of magnitude improvement in performance requested by the transistor roadmap. Instead, further progress will likely require a wise combination of design techniques and advanced manufacturing.

The microelectronics is clearly at a turning point in its history.

- In the last decades, digital technology – and especially dimensional scaling – drove the progress of this industry, looking at a lower cost and a better performance per gate. As no disruptive breakthrough is presently in sight, the likely path is to stay with a Si CMOS-based logic, adding new materials and device architectures for enhancing the overall circuit performance. One may imagine a situation where downscaling will no longer serve as the right metric of progress in microelectronics technology (just as the speed of commercial aircraft is no longer the measure the progress in the aeronautics industry[26]).

- The value of the functions co-integrated on the same circuit will become an ever more powerful industry driver. This will call for a diversification of devices integrated on the same chip – a trend we might label "More than Moore", – and of the associated technologies that do not necessarily scale with the dimensions. The perceived limits might well change qualitatively as the design space is redefined.
- At the end, the real value will be what is perceived by the end user – *i.e.* enhanced services, specific features, *etc.* – with the actual technology behind the application being of lesser importance. Added value through software and system design may play a preeminent role alongside pure technology, adding more degrees of freedom for further progress.

5. Conclusions

While the future of microelectronics is fuzzier than ever, it is too early to conclude that we are (almost) at the end of the road. Too many bright scientists predicted insurmountable limits and declared some physical implementations impossible, from airplanes to nuclear power, for us to be sure of the progress in Si logic is over. I would suggest applying a creative skepticism to any claim related to approaching limits, by clarifying all the underlying assumptions and trying to surmount or sidestep the obstacles.

Looking back to the past decades there is no reason to be pessimistic and it is safe to say that one cannot imagine what technology will be possible a few decades from now. On the other hand, it is also wise not to be too optimistic: the recent hype about new switches or new computational approaches may prove to be just hype. One should never underestimate the real complexity of building a working information processing technology and of replacing the accumulated knowledge and know-how of the present microelectronic industry by something totally new.

Acknowledgments

I would like to specifically thank Daniel Bois, Sorin Cristoloveanu, Hervé Fanet, Paolo Gargini, James Hutchby, Hiroshi Iwai, Thomas Skotnicki, Paul Solomon and Claude Weisbuch, who raised my interest in exploring the limits of Si CMOS and helped me (some of them unconsciously) to put this work together.

References

1. G. E. Moore, "Cramming more components onto integrated circuits," *Electronics* **38**, 114 (1965); reproduced in *Proc. IEEE* **86**, 82 (1998).
2. G. E. Moore, "Progress in digital integrated electronics," *Tech. Digest IEDM* (1975), pp. 11–13.
3. G. E. Moore, "Lithography and the future of Moore's law," *Proc. SPIE* **2437**, 2 (1995).
4. R. H. Dennard, F. H. Gaensslen, H.-N. Yu, V. L. Rideout, E. Bassous, and A. R. LeBlanc, "Design of ion-implanted MOSFETs with very small physical dimensions," *IEEE J. Solid-State Circ.* **9**, 256 (1974).
5. J. T. Wallmark, "Basic considerations in microelectronics," chapter 2 in: E. Keonjian, ed., *Microelectronics,* New York: McGraw-Hill, 1963, pp. 10–96.
6. J. T. Wallmark, "Fundamental physical limitations in integrated electronic circuits," in: *Solid State Devices 1974,* Conf. Series No. 25, London: Institute of Physics, 1975, pp. 133–167.
7. J. T. Wallmark, "A statistical model for determining the minimum size in integrated circuits," *IEEE Trans. Electron Dev.* **26**, 135 (1979).
8. J. T. Wallmark and S. M. Marcus, "Maximum packing density and minimum size of semiconductor devices," *IRE Trans. Electron Dev.* **9**, 111 (1962).
9. J. D. Meindl, Q. Chen, and J. A. Davis, "Limits on silicon nanoelectronics for terascale integration," *Science* **293**, 2044 (2001).
10. R. W. Keyes, "Physical problems and limits in computer logic," *IEEE Spectrum* **6**, 36 (1969)
11. V. V. Zhirnov, R. K. Cavin, J. A. Hutchby, and G. I. Bourianoff, "Limits to binary logic switch scaling – a gedanken model," *Proc. IEEE* **91**, 1934 (2003).
12. R. Landauer, "Irreversibility and heat generation in the computing process," *IBM J. Res. Develop.* **5**, 183 (1961).
13. M. P. Frank and T. F. Knight, Jr., "Ultimate theoretical models of nano-computers," *Nanotechnology* **9**, 162 (1998).
14. J. Casas-Vásquez and D. Jou, "Temperature in non-equilibrium states: A review of open problems and current proposals," *Rep. Prog. Phys.* **66**, 1937 (2003).
15. L. I. Mandelstam and I. E. Tamm, "The uncertainty relation between energy and time in non-relativistic quantum mechanics," *J. Phys. (USSR)* **9**, 249 (1945).
16. N. Margolus and L. B. Levitin, "The maximum speed of dynamical evolution," *Physica D* **120**, 188 (1998).
17. I implicitly exclude the fundamental work of A. Turing and others, particularly Turing's halting problem, that brought Gödel's theorem into the computer field: the practical limitation it imposes on actual computing schemes is, however, not obvious.
18. R. W. Keyes, "Physics of digital devices," *Rev. Mod. Phys.* **61**, 279 (1989).
19. R. W. Keyes, "The cloudy crystal ball: Electronic devices for logic," *Phil. Mag. B* **81**, 1315 (2001).

20. *The International Technology Roadmap for Semiconductors: 2005 Edition – Emerging Research Devices*. This document can be downloaded from *http://www.itrs.net/Links/2005ITRS/ERD2005.pdf*
21. I exclude 1D FETs that do not offer any conceptual breakthrough and are merely aggressively scaled MOS transistors potentially using an alternative channel material (Ge, III-V or carbon nanotube).
22. P. D. Tougaw, C. S. Lent, and W. Porod, "Bistable saturation in coupled quantum-dot cells," *J. Appl. Phys.* **74**, 3558 (1993).
23. C. S. Lent and P. D. Tougaw, "A device architecture for computing with quantum dots," *Proc. IEEE* **85**, 541 (1997).
24. K Nikolić, D Berzon, and M. Forshaw, "Relative performance of three nanoscale devices – CMOS, RTDs and QCAs – against a standard computing task," *Nanotechnology* **12**, 38 (2001).
25. A. DeHon and K. K. Likharev, "Hybrid CMOS/nanoelectronic digital circuits: Devices, architectures, and design automation," *Proc. ICCAD 2005*, pp. 375–382.
26. H. Iwai, "Recent advances and future trends of ULSI technologies," *Tech. Digest ESSDERC* (1996), pp. 46–52.

Will the Insulated Gate Transistor Concept Survive Next Decade?

O. Engström
Dept. of Microtechnology and Nanoscience
Chalmers University of Technology, SE-412 96 Göteborg, Sweden

1. Introduction

The increasing influence of electronics on human life in the past couple of decades has promoted the MOS transistor to a device of similar significance for cultural change as, for example, the rotating electric motor and the combustion engine. A key property of the MOSFET, which has made this ranking possible, is its potential for fast and steady improvement. In the early 1970's, when the current-voltage characteristics of the device reasonably well could be described by the one-dimensional Ihantola-Moll model,[1] the gate channel was long enough that the electrostatic influence from source and drain contacts could be neglected or considered as a slight perturbation at higher drain voltages. The downscaling process, which has been going on since then, is presently at a stage where the electric field distribution between source and drain has become too complicated for simple analytical expressions in characterizing the device. Bulk CMOS technology has reached a point where further development includes a considerable "squeezing" of parameter values by small geometrical changes and by the introduction of novel materials, combined with the acceptance of numerous trade-offs in the design process.

Many of these problems in the scaling of bulk MOSFETs have a straight-forward solution in silicon-on-insulator (SOI) technology. Especially when using fully depleted materials, double-gate or fin-FET devices are very attractive for solving the electric field problems mentioned above due to their excellent control of the field distribution in the channel.

For bulk CMOS devices, solutions seem to exist to satisfy the requirements of the 45 nm node, with physical gate lengths between 18 and 25 nm depending on the application.[2] Beyond this milestone, however, common wisdom among specialists points in the direction of SOI technology for use in the 32 and 22 nm nodes, which are planned to be developed by 2013 and 2016, respectively.[3,4] The title of this chapter, therefore, could be reformulated more specifically: "Will it be possible within a decade to reach the 22 nm node with currently used MOSFET concepts?" A couple of the most important problems in this context will be discussed below.

 Future Trends in Microelectronics. Edited by Serge Luryi, Jimmy Xu, and Alex Zaslavsky

2. The downscaling problem of bulk CMOS

For the classical MOS transistor, the channel region meets the source and drain via the depletion regions occurring between the two different doping types constituting the three building blocks of source, drain and bulk, as shown in Fig 1. As long as the gate length is much larger than the depleted distances, the influence of the field distribution from source and drain on the threshold voltage is negligible. This makes the electric vector field from the gate point perpendicular to the channel. However, when decreasing the gate length of bulk CMOS transistors, a pronounced mixing with the fields from source and drain complicates the field distribution under the gate and lowers the threshold voltage. Also, current leakage may occur between source and drain due to the shrinking distance between the depletion edges surrounding the two contacts. This has been a recurring problem, labelled the "short channel effect", during the entire history of MOSFET downscaling.[5] For bulk CMOS technology, it has forced designers to invent new geometries for the doping profiles of the source and drain. Also, in order to decrease the source and drain depletion regions, the doping concentration under the gate has been continuously increased in the process of shortening the gate. This move, however, lowers the capacitive coupling between gate and channel, which is detrimental for the switching behaviour of the transistor.

When the transistor is switched on and off by changing the gate voltage, it is desirable that as much as possible of the voltage change be transferred to the semiconductor. This means that the capacitance of the gate insulator must be large compared to the series-connected capacitance of the semiconductor region under the gate. The gate capacitance can be increased by decreasing the thickness of the gate insulator. With present SiO_2 and SiON dielectrics such adjustments

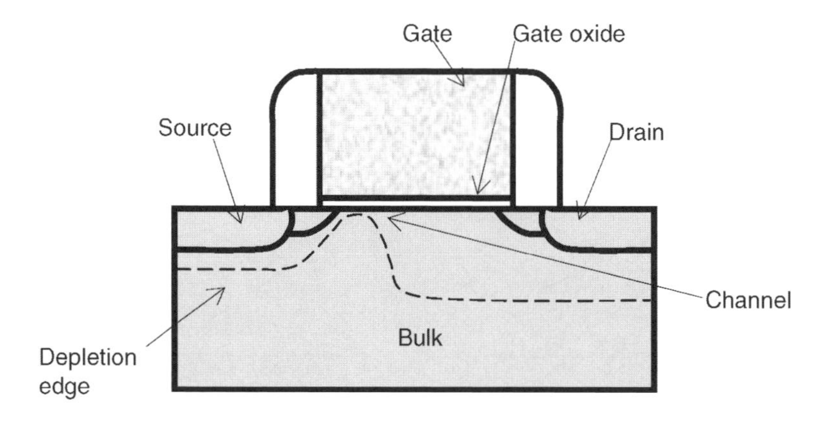

Figure 1. Schematic cross-section of a bulk MOSFET. The dashed curve schematically indicates the depletion edge for a high drain voltage.

are approaching the limits set by unacceptable gate leakage due to tunneling. This has motivated a chase for novel dielectric materials with high dielectric constants ("high-κ"), able to maintain the capacitive coupling for thicker insulator layers and thus decrease tunneling currents.[6] The numerical calculations for MOS capacitors in Fig. 2 correlate the gate voltage and oxide voltage for different doping concentrations and κ-values, respectively. For a channel doping concentration between 10^{18} and 10^{19} cm^{-3}, which will be required for a possible bulk CMOS technology at the 22 nm node, it is seen that a κ-value higher than 15 is needed for an insulator thickness of 20 Å. The effective bending of the energy bands in the semiconductor for a given change in gate voltage is necessary in order to obtain a reasonable value of the "subthreshold swing". This is the change in gate voltage needed to reduce the drain current by one order of magnitude, which is an important parameter determining the on/off behaviour of the transistor.

From this reasoning, the scaling problem for bulk CMOS can be described as a balance between high doping levels and gate capacitive coupling. In order to improve the field distribution under the gate, higher doping is necessary, which leads to a need for increased gate coupling. This in turn requires a thinner gate insulator, which eliminates SiO_2, used so far in MOSFET history, as a dielectric material because of unacceptable gate leakage current.[7]

Another possibility for improving the electric field distribution would be to create a doping distribution under the gate, a "super-halo", that would compensate for the influence from the source and drain depletion regions.[8] Such distributions have been suggested by simulation studies and seem possible to obtain by using advanced ion implantation techniques. However, the complexity of such a solution must be compared to a possible changeover to SOI technology, as described below. In both cases, a high-κ gate dielectric will probably be necessary in order to reach the 22 nm node.

In the historical process of improving the field distribution under the gate to avoid short channel effects, the depths of source and drain diffusions have been decreased. This gives rise to an increasing series resistance and encroaches on the achievable values of the on-current and switching capability. In recent years,

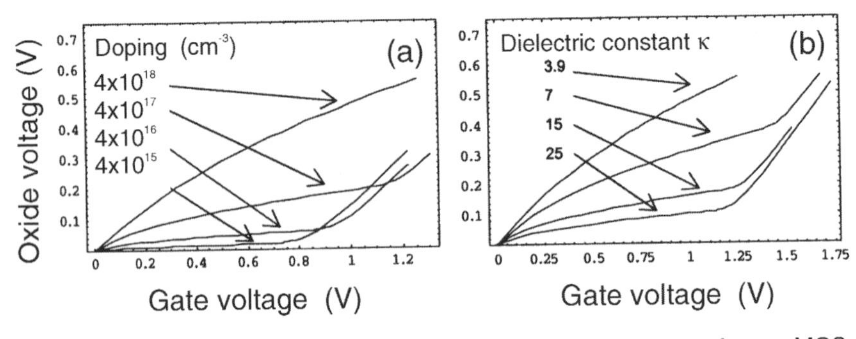

Figure 2. Voltage across the oxide as a function of gate voltage for an MOS structure *vs.* varying semiconductor doping concentrations (a) and *vs.* dielectric constant κ of the gate insulator (b).

novel source/drain solutions have been investigated by using Schottky contacts instead of the traditional *pn* junctions.[9] Increased series resistance is not only a problem for bulk CMOS; it appears similarly in SOI technology.

3. High-κ dielectrics and bulk CMOS technology

From the processing point of view, one of the most important advantages of silicon technology has been the existence of silicon dioxide, used as a masking and insulating material in the production of integrated circuits. This is especially true for the use of this material as a gate insulator of MOSFETs. The requirements on a replacement material, therefore, are very stringent. Not only the physical properties – interface and bulk charges, scattering of charge carriers in the channel, and energy band offsets compared to silicon – but also the chemistry that provides thermal stability during the processing sequence must be controlled with high precision. The fundamental current leakage is determined by tunneling that, in turn, depends on the thickness and the band offset barriers for electrons and holes between the conduction and valence bands of the silicon and the dielectric. Consequently, if the dielectric constant of the material is not high enough to allow for a small enough thickness *in combination* with high enough offset values, a chosen material cannot be used even if all other properties are promising.

In order to compare the thickness requirement of different gate dielectrics, the concept of "equivalent oxide thickness" (EOT) is used. This quantity expresses the thickness of silicon dioxide, with dielectric constant κ_{SiO2}, that is needed to give the same capacitance as an alternative high-κ insulator with a dielectric constant κ_H and thickness t_H: EOT = $t_H(\kappa_{SiO2}/\kappa_H)$. For the 22 nm node and low standby power (LSTP) technology, the ITRS roadmap forecasts a physical gate length L_G = 13 nm and an EOT of 5–6 Å for the gate insulator of bulk CMOS devices.[10] The problem of finding a suitable high-κ material has inspired much research and development activity at laboratories around the world.[6,7]

The gate leakage current requirement given by the ITRS roadmap[10] is expressed in terms of current per gate length. For a transistor, gate leakage includes leakage not only to the channel, but also to the source and drain due to gate overlap. Corresponding values expressed as a current density have been estimated and for LSTP in the 22 nm node, a leakage current of 10^{-2} A/cm^2 has been considered as reasonable.[11] Using this value and EOT = 0.5 nm as stated above, it is possible to calculate a limit line for the combination of dielectric constant κ_H and offset value ΔE_C that will meet the gate leakage requirement. In a recent paper this was done for different oxides, considered as candidates for future CMOS dielectrics, in the κ_H and ΔE_C parameter space.[12] The starting point of this investigation was the finding that La_2O_3 fulfils the requirements mentioned above.[13] Using this material as a reference, the limit line for the 22 nm node was estimated for direct and Fowler-Nordheim tunneling under the EOT = 0.5 nm condition and the result is depicted in Fig. 3. From this graph, only a few dielectrics are of interest for 22 nm LSTP devices. The data in Fig. 3 only take

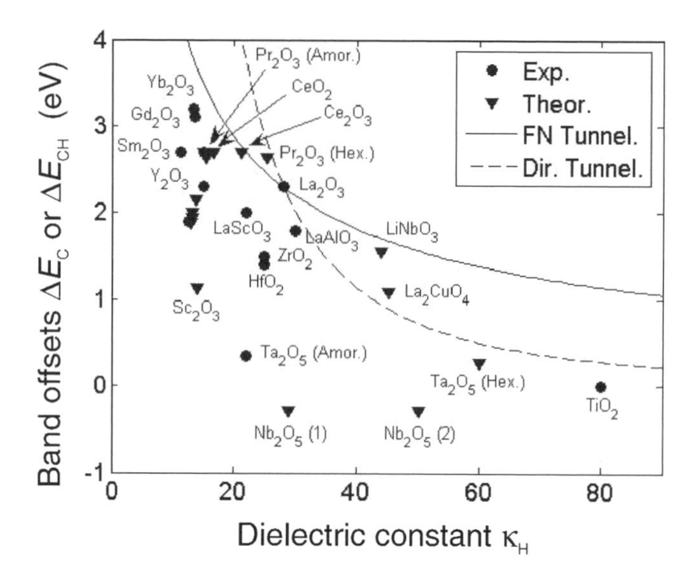

Figure 3. Offset barrier heights between conduction bands for different oxides and silicon plotted as a function of dielectric constant κ_H. The two curves represent estimated borders for the ITRS low stand-by power (LSTP) requirements at the 22 nm node: solid curve is for Fowler-Nordheim tunneling, dashed curve for direct tunneling. Dots represent experimental data (ΔE_C) taken from literature; triangles are data from empirical relations (ΔE_{CH}) for physical trends (from Ref. 12).

into account the conduction band energy offset values ΔE_C. From the bandgap values of the different oxides and of silicon, the corresponding offset values for the valence bands were also estimated. This left only La_2O_3 and $LaAlO_3$ as clearly interesting dielectrics from the fundamental requirement of tunneling current leakage. Taking into consideration that the estimates giving rise to the limit lines in Fig. 3 are relatively rough and that available experimental data may suffer from varying inaccuracies, some other materials like Pr_2O_3 and Gd_2O_3 close to the boundaries may serve as additional candidates. Even so, it should be emphasized that Fig. 3 only represents the basic condition of tunneling leakage. Other properties, like reduced mobility due to carrier scattering at the oxide-silicon interface, oxide bulk charges influencing threshold voltage variations, and thermal stability[6,7] are obstacles on the road to technically acceptable high-κ oxides.

4. Fully-depleted multiple-gate SOI technology

Silicon-on-insulator technology inherently improves the complicated electric field distribution under the gate.[3] Transistors fabricated by using SOI as a starting material are made in a thin silicon film on top of a silicon dioxide layer that, in turn, rests on a silicon wafer.[14] This allows for a better field control, especially for a "fully depleted" geometry in which the silicon layer thin enough that the depletion region from the gate reaches the buried oxide on the wafer underneath. Also, a number of ideas have been put forward for increasing the gate control by using multiple gate structures, including double-gate, triple-gate and "gate-all-around" geometries, where the names correspond to the number of directions from which a gate is applied to a thin slice of Si or a Si wire.[15] Controlling the channel electric field in a thin slice or wire from opposite planes held at the same potential, drives the potential along the channel extension towards a constant value. From a fabrication point of view, the double-gate structure seems to be the most likely to enter production over the next decade, especially when designed as a standing slice to build the so-called fin-FET illustrated in Fig. 4.[15] This is believed by many to be the least complicated method to make a fully-depleted double-gate MOS transistor on SOI (FD-DGSOI) and appears as the most promising candidate for future production if bulk CMOS reaches its limits.[4]

Even if the thin silicon layer between the two gates of a FD-DGSOI transistor provides better electric field control along to the channel, a competition with the lateral fields from the depletion regions of source and drain exists also in this structure.[16] This originates from the relations between, on the one hand, oxide properties like dielectric constant κ and thickness t_{ox}, and, on the other hand, the thickness of the silicon film, t_{Si}. For a shorter gate length L_G, a thinner silicon layer will be needed. A "natural length" λ sets the length scale from which the smallest possible dimensions can be estimated. Expressions for λ values valid for multiple-gate geometries have been derived by a number of authors.[15, 17] For the central part of a FD-DGSOI intrinsic Si transistor channel at zero gate voltage and zero flatband voltage, the electric potential $\Phi(x)$ along the channel coordinate x, is described by:[15, 17]

$$d^2\Phi(x)/dx^2 - \Phi(x)/\lambda^2 = 0\,, \tag{1}$$

where

$$\lambda = [(\kappa_{SiO2}/2\kappa_H)t_{Si}\,t_{ox}]^{1/2}\,. \tag{2}$$

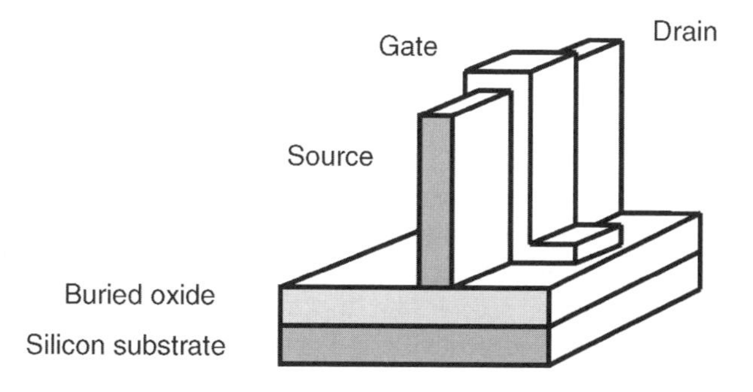

Figure 4. Double gate SOI transistor shaped as a fin-FET.

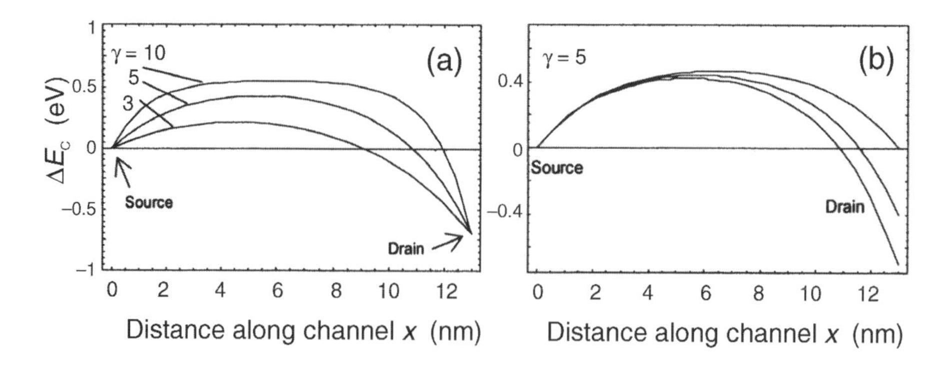

Figure 5. Conduction band variation along the intrinsic channel of a double-gate SOI transistor with L_G = 13 nm gate length for a drain voltage of 0.7 V with three different values of $\gamma \equiv L_G/\lambda$ (a); and for $\gamma = 5$ with three different drain voltages V_D = 0, 0.4 and 0.7 V (b).

Using the boundary conditions $\Phi(x=0) = 0$, $\Phi(x=L_G) = 0.7$ V, we find the standard solution of Eq. (1) as a combination of *sinh* terms. Further, expressing the silicon conduction band edge $E_C(x) = -q\Phi(x)$, we find the plot shown in Fig. 5(a). As long as the conduction band edge in this graph has a flat region, one may expect that the change in gate voltage needed to increase the drain current will be close to the ideal value $(k_BT/q)\ln 10 \approx 60$ mV/decade at room temperature. We notice in Fig. 5(a) that this criterion is fulfilled when the ratio $\gamma \equiv L_G/\lambda = 10$, but this is no longer the case when γ approaches 3.

A second important property for a short channel MOSFET is "drain-induced barrier lowering" (DIBL), which is the decrease in maximum potential along the channel due to an increased drain voltage as seen by a charge carrier entering the channel from source. Figure 5(b) shows the conduction band edge for $\gamma = 5$ and for V_D = 0, 0.4 and 0.7 V. The calculated DIBL is about 50 mV, which may be considered as acceptable. These ideas can now be used to estimate the necessary thickness of the SOI silicon film in order to fulfil the 22 nm LSTP requirements.

Looking into experimental literature data, one finds that the EOT values for La_2O_3, HfO_2 and SiO_2 needed to limit the leakage current to 10^{-2} A/cm^2 at 1 V across the insulator is 0.6, 1.2 and 2.1 nm respectively.[6, 12] Taking the dielectric constants for these three dielectrics optimistically as 28, 22 and 3.9, respectively, we get the corresponding physical oxide thicknesses of 4.3, 6.5, and 2.1 nm. Plugging these t_{ox} values into Eq. (2), we can calculate t_{Si} as a function of γ for fully-depleted double-gate SOI transistors at the 22 nm LSTP node with a physical gate length L_G = 13 nm. The result, shown in Fig. 6, indicates that extremely thin silicon films are needed. Taking, in accordance with the discussion above, $\gamma = 5$ as a lower limit, we find maximum silicon film thicknesses for La_2O_3, HfO_2 and SiO_2 to about 8, 4 and 2–3 nm, respectively. For a gate-all-around geometry,[15]

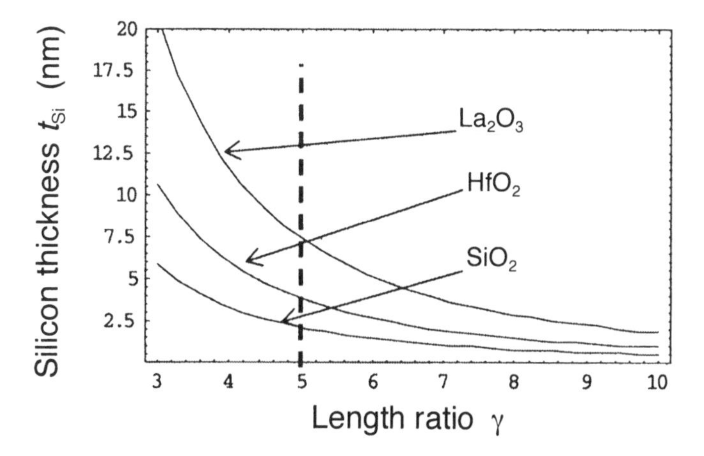

Figure 6. Silicon film thickness *vs.* $\gamma = L_G/\lambda$ for double-gate fully-depleted SOI transistors with L_G = 13 nm. The vertical dashed line at $\gamma = 5$ marks the maximum possible Si thickness t_{Si} to be used with three different gate insulator materials for 22 nm LSTP. For a gate-all-around geometry, the t_{Si} values are doubled.

including a silicon wire with square section, the factor of 2 in the denominator of Eq. (2) increases to 4. This doubles the corresponding maximum t_{Si} values in Fig. 6 to 16, 8 and ~5 nm. For this calculation, we should remember that the κ values for La_2O_3 and HfO_2 were optimistically set very high, at 28 and 22, respectively. Even if these values are valid for the pure oxides, the effective values for oxides deposited on silicon often are lower, due to the appearance of a thin SiO_x interlayer between the high-κ oxide and the silicon surface.[7] Silicon films with thicknesses in the range discussed here will certainly have increased charge carrier scattering, not only due to an increased influence by the two adjacent silicon/oxide interfaces but also due to defects[18] and, for very thin layers, due to scattering by confined acoustic phonons. In the latter case, theoretical treatment predicts a steep decrease in electron mobility due to combined surface and phonon scattering for thicknesses below about 6 nm.[19] This is an additional problem, specific to SOI technology.

From these considerations, it seems that SiO_2 as a gate insulator will not be suitable for double-gate devices, although it may still work for gate-all-around geometries. For double-gate structures, a dielectric with the properties of La_2O_3 seems necessary. An interesting observation is that HfO_2 might be possible to use for the gate-all-around case if silicon wire dimensions of about 8 nm can be prepared.

5. Conclusions

For a long time, the complicated electric fields occurring under the MOSFET gate have been a key obstacle to downscaling. In order to reach the 22 nm node, new materials and novel geometries are being studied, which pose a number of challenging development problems in the next decade. For bulk devices, it may not be enough to find a high-κ dielectric with sufficient capacitive coupling between gate and semiconductor. To control the potential distribution between source and drain, doping the semiconductor volume for example in a super-halo pattern as suggested in Ref. 9 also appears essential. This requires a precise ion implantation process step, which complicates the process and thus increases production cost.

Multiple-gate SOI structures have long been proposed as superior from this point of view. The estimates presented in this chapter, however, indicate that the same issue of finding a suitable high-κ dielectric exists as for bulk technology. Furthermore, specific new SOI problems crop up due to the need for extremely thin silicon films. This conclusion agrees with investigations carried out by others using similar methods.[15]

More advanced simulations, in Ref. 3 and 4, have pointed out the possibility of making FD-DGSOI transistors for the 22 nm LSTP node by using silicon dioxide as a gate insulator material. However, in both cases t_{ox} was chosen thinner (1.0 and 1.5 nm, respectively) than can satisfy the LSTP requirements set by this node. Furthermore, similar to the present results, the simulations in Ref. 3 lead to a silicon film thickness of 3 nm for a DG MOSFET with 13 nm gate length, again in the risk zone for low intrinsic mobility.[19] In Ref. 4, a subthreshold slope close to ideal was found by simulation for a 13 nm device with a silicon film thickness as high as 10 nm. Looking at the material parameters used in this simulation, one finds a natural length $\lambda = 4$ nm, resulting in $\gamma = L_G/\lambda \sim 3$, which would be expected to give a high DIBL. For the 22 nm node, therefore, SOI technology does not seem to offer a clear advantage compared with that of bulk CMOS.

The results above suggest that changing from bulk CMOS to SOI technology, the gate dielectric problem will mainly be the same while the problem of electric field distribution in the bulk is supplanted by the problem of increased charge carrier scattering in the channel. A convenient but superficial conclusion would be that traditional silicon technology will never reach the 22 nm node within the next ten years, but will instead be replaced by new methods based on for example carbon nanotubes or molecular structures.[20] Considering the fact that transistor functionality of 5 nm bulk devices has already been demonstrated,[21] even if most technical device requirements have not been fulfilled, it is obvious that silicon technology has a far better point of departure from the present stage than any other. In the history of products based on the MOS transistor, starting in the 1970's, a decade is a long period of time. The present analysis emphasizes that the drawn-out competition between bulk and SOI technologies will continue, which increases the probability that the MOS transistor concepts of today will survive the next decade.

References

1. H. K. J. Ihantola and J. L. Moll, "Design theory of a surface field-effect transistor," *Solid State Electronics* **7**, 423 (1964).
2. T. Sugii, "High performance bulk CMOS technology for the 65/45 nm node," *Solid State Electronics* **50**, 2 (2006).
3. J. G. Fossum, L.-Q. Wang, J.-W. Yang, S.-H. Kim and V. P. Trivedi, "Pragmatic design of nanoscale multi-gate CMOS," *Tech. Digest IEDM* (2003), p. 613.
4. L. Risch, "Pushing CMOS beyond the roadmap," *Solid State Electronics* **50**, 527 (2006)
5. H. Iwai and S. Ohmi, "Silicon integrated circuit technology from past to future," *Microelectronics Reliability* **42**, 465 (2002).
6. J. Robertson, "High dielectric constant oxides," *Eur. Phys. J. Appl. Phys.* **28**, 265 (2004).
7. H. Wong and H. Iwai, "On the scaling issues and high-κ replacement of ultrathin gate dielectrics for nanoscale MOS transistors," *Microelectronics Eng.* **83**, 1867 (2006).
8. Y. Taur, C. H. Wann and D. J. Frank, "25 nm CMOS design considerations," *Tech. Digest IEDM* (1998), p. 789.
9. M. Zhang, J. Knoch, Q. T. Zhao, St. Lenk, U. Breuer, and S. Mantl, "Impact of dopant segregation on fully depleted Schottky-barrier SOI-MOSFETs," *Solid State Electronics* **50**, 594 (2006).
10. ITRS Roadmap, 2004 edition, see *http://public.itrs.net*
11. T. Kauerauf, B. Govoreanu, R. Degraeve, G. Groeseneken, and H. Maes, "Scaling CMOS: Finding the gate stack with the lowest leakage current," *Solid State Electronics* **49**, 695 (2005).
12. O. Engström, B. Raeissi, S. Hall, *et al.*, "Navigation aids in the search for future high-κ dielectrics: Physical and electrical trends," *Proc. 7th Eur. Workshop Ultimate Integr. Silicon (ULIS)*, Grenoble (2006), submitted to *Solid State Electronics*.
13. H. Iwai, S. Ohmi, S. Akama, *et al.*, "Advanced gate dielectric materials for sub-100 nm CMOS," *Tech. Digest IEDM* (2002), p. 625.
14. G. K. Celler and S. Cristoloveanu, "Frontiers of silicon-on-insulator," *J. Appl. Phys.* **93**, 4956 (2003).
15. J.-P. Colinge, "Multiple gate SOI-MOSFETs," *Solid State Electronics* **48**, 897 (2004).
16. L. Chang and C. Hu, "MOSFET scaling into the 10 nm regime," *Superlatt. Microstruct.* **28**, 351 (2000).
17. R.-H. Yan, A. Ourmazd, and K. F. Lee, "Scaling Si MOSFET: From bulk to SOI to bulk," *IEEE Trans. Electron Dev.* **39**, 1704 (1992).
18. T. Ernst, D. Munteanu, S. Cristoloveanu, *et al.*, "Investigation of SOI MOSFETs with ultimate thickness," *Microelectronics Eng.* **48**, 339 (1999).
19. L. Donetti, F. Gamiz, J. B. Roldán and A. Godoy, "Acoustic phonon confinement in silicon nanolayers: Effect on electron mobility," *J. Appl. Phys.* **100**, 013701 (2006).

20. B. Yu and M. Meyyappan, "Nanotechnology: Role in emerging nano-electronics," *Solid State Electronics* **50**, 536 (2006).
21. H. Wakabayashi, S. Yamagami, N. Ikezawa, *et al.*, "Sub-10-nm planar-bulk-CMOS devices using lateral junction control," *Tech. Digest IEDM* (2003), pp. 989–991; H. Wakabayashi, T. Ezaki, M. Hane, *et al.*, "Transport properties of sub-10-nm planar-bulk-CMOS devices," *Tech. Digest IEDM* (2004), pp. 429–432.

Scaling Limits of Silicon CMOS and Non-Silicon Opportunities

Y. Nishi
Dept. of Electrical Engineering, Stanford University, Stanford, CA 94305, U.S.A.

1. Introduction

As we look at the future of silicon based CMOS, it is becoming more difficult to be as optimistic as we used to be. A variety of problematic issues and, more importantly, the magnitude of anticipated challenges, may not be as easy to circumvent as those we have successfully overcome over the past two decades. Not only difficulties in cost-effective processing equipment and process parameter controls, but also perceived limits in device operation improvement itself will pose more problems than in the past. Rapidly increasing MOSFET leakage currents at the gate, source, and drain terminals, and the diminishing improvement in the drive current with reduced channel length are now pushing us to consider alternative materials for the channel of MOSFETs. At the integration level, the interconnect delay that has been identified as the stumbling block for large high-performance chips will certainly become worse, coupled with power consumption and heat dissipation management challenges. As we will have more areas dedicated to high-speed embedded memory, mostly consisting of static random access memory (SRAM), memory standby power will become a major problem, especially with MOSFETs operated at a much lower I_{ON}/I_{OFF} ratio than they used to be. Furthermore, general trends for system-on-a-chip design driven by significant growth of wireless systems, mostly for consumer products, will require heterogeneous integration of digital, analog/RF, power management and more, which will make the magnitude of challenges greater than ever.

This chapter will discuss present and future perspectives of nanoelectronic materials and devices both in evolutionary and revolutionary "nano" electronics domains. It is likely that the evolutionary progress in all problem areas of silicon-based CMOS will keep its pace, despite a variety of technical challenges. Revolutionary nanoelectronics, such as nanowires and nanotubes, could provide unique advantages, but only when engineering breakthroughs for control of growth, placement, and integration are achieved.

2. Silicon-based CMOS scaling

"Scaling" has provided continuous improvement of digital integrated circuits (ICs) through cost reduction per bit/gate combined with improved current drive without an increase in the leakage current. This has permitted the semiconductor industry to stay with the same material, silicon, and ensured evolutionary progress of

Future Trends in Microelectronics. Edited by Serge Luryi, Jimmy Xu, and Alex Zaslavsky
ISBN 0-471-48

processing equipment with continuous productivity improvement. However, as illustrated in Fig. 1, we are coming to the juncture at which scaling may cease to provide the necessary performance improvement. It is evident from Fig. 1 that the scaling-driven current drive improvement is saturating, while subthreshold current leakage as well as band-to-band tunneling leakage at the drain junction edge are both increasing. Furthermore, the fundamental tool for geometry scaling, photolithography, is facing its greatest challenges ever. Not only a variety of resolution enhancement techniques, but a hardware revolution involving liquid immersion lithography and extreme UV sources will be needed soon, with many issues to be solved before practical utilization. Finally, one-dimensional silicon scaling for digital ICs is giving way to multi-functional integration with digital, ananlog/rf, power management, non-volatile memories, and even MEMS with 2D and 3D configurations.

Of course, much ongoing research continues to be invested in overcoming the obstacles to the continued growth of integration density. They can be classified into four areas, as shown in Table 1. Improvement in electrostatic control of field-effect devices has fostered interest in new structures that promise better control over the channel current flow, such as double gate, fin-FET, ultra-thin body (UTB) SOI, and metal gate devices. Combinations of metal gates and high-κ dielectrics, shown in Table 2, have been aggressively pursued, leading to a large number of publications, but it appears more effort will be needed before these technologies will hit the manufacturing floor.

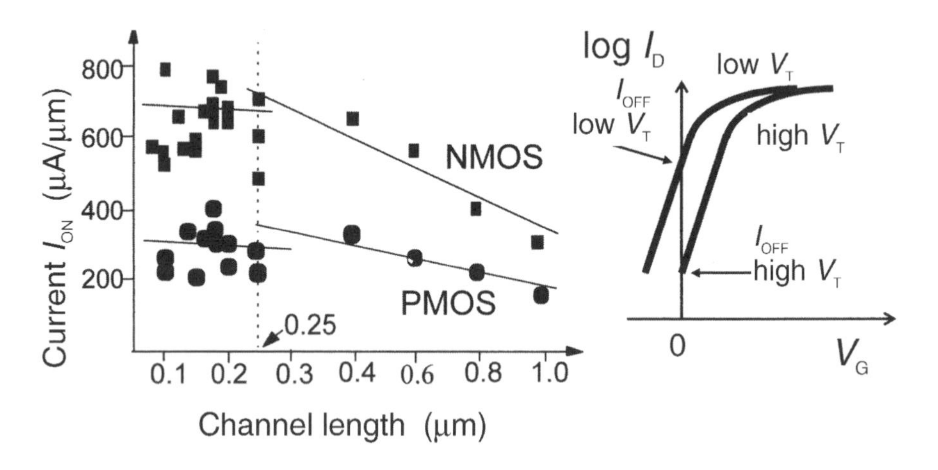

Figure 1. Saturating trends of MOSFET performance improvement with scaling: (a) current drive I_{ON} *vs.* gate length at constant I_{OFF} (data from IBM, TI, Intel, AMD, Motorola and Lucent);[1] (b) contrasting requirements for high $I_{ON} \sim (V_{DD} - V_T)^{\eta}$ provided by low V_T and low I_{OFF} provided by high V_T.

Electrostatics double gate, fin-FET, ultra-thin body (UTB) SOI, metal gate
Transport band splitting of Si: (1) strain, (2) UTB quantum confinement new channel materials: (1) Ge, (2) III-V
Leakage current management gate leakage: high-κ dielectrics with metal gates source-drain leakage: DIBL control drain-substrate leakage: band-to-band tunneling
Source/drain parasitic resistance metal source/drain for reduced series resistance

Table 1. Possible alternative materials and structures for future MOSFETs.

Gate structure	Who	Metals	Work function	Tunability
Dual metal	Many	Many	Individual Φ_m	No
Alloying	NCSU, AIST	Ru/Ta, Ni/Al	4.2-5.2 eV	Yes
Ion implanted metal	UCB	Mo (N)	4.5-4.9 eV	Limited
FUSI	IBM, TI, others	NiSi, CoSi, TiSi	4.2-5.1 eV	Limited
Substituted metal	Nat. U of Singapore	Al(Si), Pt (Si)	4.25, 4.9 eV	No
Laminated structure	UT Austin	HfN/Ti/TaN	4.8-5.2 eV	Limited
Bilayer structure	Stanford, others	Ti/Pt, Al/TaN, others	3.9-5.2 eV	Yes

Table 2. Possible solutions for metal gates with adequate work functions.

As for the transport in the channel, a couple of approaches for the energy band engineering have been studied, using both strain and quantum confinement. Also, new materials that have higher electron and/or hole mobilities have been explored, resulting in Ge and III-V compound semiconductor channel MOSFETs. Finally, source-drain leakage mechanisms, including drain-substrate leakage, have become hot topics. Recently, more attention has been paid to source resistance, as the source/drain junction depth is getting smaller in order to reduce drain-induced barrier lowering (DIBL) and short channel effects, There is now increasing research into metal source/drain structures.[2]

Taking a closer look at band engineering of silicon *n*-MOSFETs for better electron mobility, the leading approaches involve biaxial or uniaxial tensile strain on (100) surface or making channel thickness enough small as compared to the electron wave function spread, *e.g.* ultra-thin body UTB-SOI.[3] Both of these techniques split the six constant energy conduction band ellipsoids into 2-fold and 4-fold degenerate ellipsoids, and the preferentially populated 2-fold ellipsoids with lighter transport mass m_T* enhance the mobility. Compressive strain, on the other hand, makes the conductivity effective mass of holes smaller, thereby improving the *p*-MOSFET.

A number of technological procedures for such band engineering have been proposed and are now in production, especially the use of silicon nitride as a uniaxial strain-generating film for *n*-channel and SiGe heterostructure source and drain for *p*-channel. Research into using high-mobility III-V compound semiconductor channels is now on its way. As shown in Fig. 2, this approach is attractive because most of III-V materials have significantly higher electron mobility than silicon. However, when the high mobility is based upon low effective mass of electrons, the density of states is also low, and the drain current – the product of mobility and carrier concentration – may not be as high as one might expect from the mobility improvement alone.

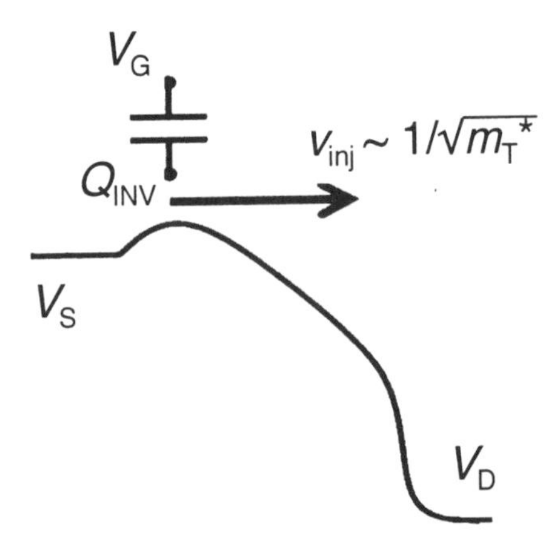

Material	Si	Ge	GaAs	InAs	InP	InSb
Electron μ	1,600	3,900	**9,200**	**40,000**	**5,400**	**77,000**
Mass m_T*	0.19	0.08	0.067	0.023	–	0.014
Hole μ	430	**1,900**	400	500	200	850
E_G (eV)	1.12	**0.66**	1.424	**0.36**	1.34	**0.17**
ε_r	11.8	**16**	12	**14.8**	–	**17**

Figure 2. Possible high-mobility channel materials. Note that low E_G increases leakage due to band-to-band tunneling, while high ε_r increases short-channel effects.

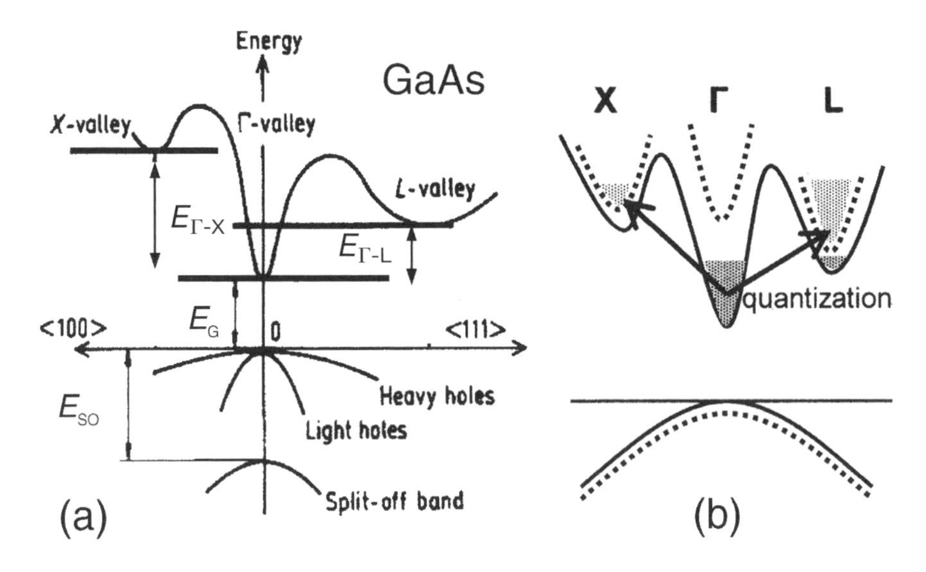

Figure 3. (a) GaAs bulk band diagram, showing the three conduction band valleys; (b) schematic illustration of electron spillover from high mobility Γ valley to lower mobility X and L valleys due to quantum confinement.[4]

Figure 3 shows the band diagram of GaAs and a schematic example of a quantum-confined GaAs band structure in which the Γ valley is pushed up to higher energy and electrons in that valley spill out to X and L valleys.[4] This increases the contribution of X and L valley electrons to the overall channel conduction, but the mobility of X and L valley electrons is much lower. The same type of intervalley transfer in quantum-confined channels can happen in other channel materials.

If alternative III-V channel materials can achieve higher electron mobility than strained silicon for *n*-MOSFETs, the next challenge would be to do the same for *p*-MOSFETs, because the hole mobilities in III-V materials are no higher than in silicon. This is perhaps where Ge channels could play a major role. The challenge for Ge channel devices is both control of the interface to high-κ dielectrics and also that narrow Ge bandgap that can cause undesirable *pn* junction leakage and band-to-band tunneling at the drain edge. In order to cope with these issues, hetero-structured channel devices have been proposed and verified by both simulation and experiments.[5] Generally speaking, we certainly need much more experimental data to further investigate these new material-based options. In the case of III-V semiconductors, in particular, technological promise requires the bold assumption that we can achieve highly controlled III-V semiconductor-insulator interfaces.

3. Nanoelectronic materials and devices

There are a number of opportunities on the horizon beyond silicon CMOS scaling. There are many attractive candidates such as carbon nanotube devices, nanowire devices, molecular/organic devices and spintronic devices. Both carbon nanotubes and semiconductor nanowires have been extensively studied already, and have exhibited promising electrical characteristics. Figure 4 shows an example of carbon nanotube MOSFET with Schottky source and drain electrodes.[6]

However, at this moment the biggest challenge for both nanotubes and nanowires lies not in the device characteristics, but rather in achieving highly controlled growth of tubes and wires at specified locations. The growth of semiconducting single-wall carbon nanotubes with anything near 100% yield has not been established. Also nanotube-specific characteristics, such as ambipolar conduction, present a number of design challenges if and when nanotubes reach the stage of practical deployment in ULSI circuits of the future.

Another important area outside of high-performance switching devices is nonvolatile memory. A number of nonvolatile memory technologies based on phase change, resistance change, dielectric polarization, magnetic polarization and nanoparticle floating gate devices have been attracting attention. Molecular/organic memories may eventually be added to this list.

Both ferroelectric and phase-change memories have a long research history since the pioneering original demonstrations, but are only now finally reaching technological insertion as high density embedded nonvolatile memory, standalone nonvolatile memory, and possibly even nonvolatile logic elements. The conductance bridge memory is another example of an emerging "nano" memory. Switching characteristics of the conductance bridge memory are shown in Fig. 5,[7] where on-resistance is set by nanofilamentary conduction and is area independent, whereas off-resistance is determined by electronic conduction across the memory material and depends on area – scaling of such memory actually improves the on/off ratio, which is a promising feature.

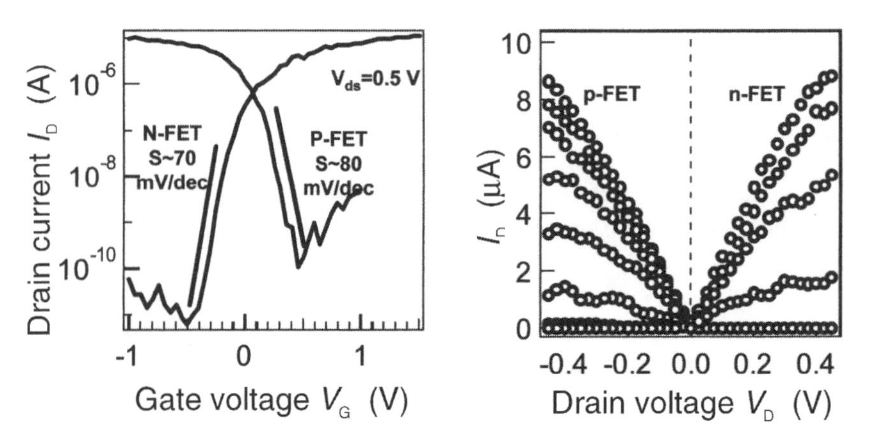

Figure 4. Complementary carbon nanotube MOSFETs.

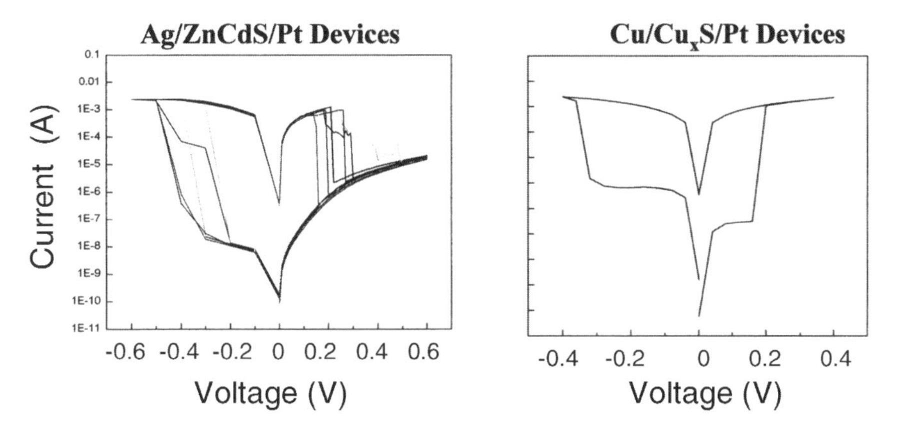

Figure 5. Conductance bridge memory characteristics.[7]

4. Three-dimensional (3D) integration

Despite a number of trials, the practical realization of 3D circuits has proven difficult because everything we have developed over the past four decades was focused on 2D integration, *i.e.* design tools, circuits testing, failure diagnostics, and, of course, the scaling paradigm itself. Today we are finally beginning to see a rudimentary shift towards 3D integration, but only at the packaging level. The fundamental advantages of true 3D integration are many, including the reduction of the chip footprint, the minimization of interconnect length and hence minimization of interconnect *R*, *L*, *C* which would reduce power consumption and delay, and possible integration of heterogeneous technologies, such as logic, memory, sensors, I/O's and even MEMS. Recent progress toward monolithic integration has been achieved by using either wafer stacking or monolithic stacking as shown in Fig. 6 for the case of Ge single-crystal recrystallization in vias on top of silicon.

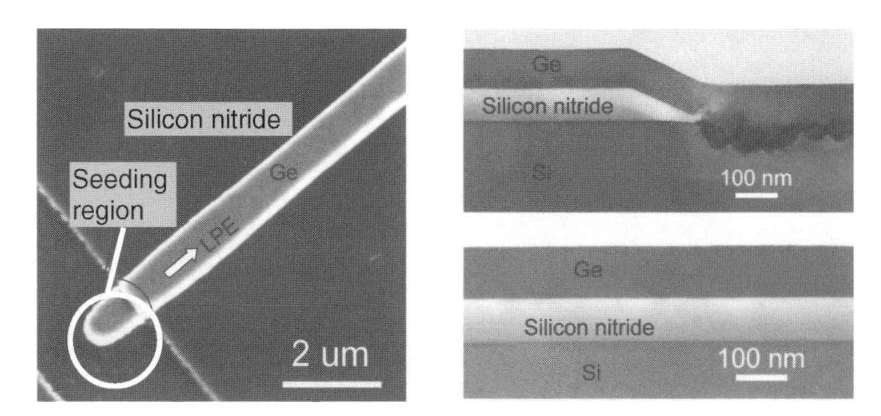

Figure 6. Germanium-on-insulator film by rapid melt growth.[8]

Unfortunately, true 3D monolithic integration appears unlikely in the foreseeable future. However, if such integration did arrive, one of the most promising areas of application would be in FPGAs. Furthermore, heterogeneous material combinations, such as a silicon-based CMOS platform coupled with Ge-channel MOSFETs or similar combinations with other materials, could become possible, as in Fig. 6. Such 3D integration could become a cornerstone technology for integrating multifunctional capabilities on a single chip, though testability, redundancy, reliability and manufacturing cost issues remain areas for future research and development.

5. Summary

As we look at available and/or soon-to-be available technologies from a practical viewpoint, there is no doubt that we still have a lot of room to explore before departing from silicon-based CMOS by expanding to new materials and device structures, and new ways to design and test chips. There is hope that the world of "nano" will keep providing revolutionary/evolutionary solutions that would significantly benefit our society directly or indirectly.

References

1. C. Choi, "Modeling of nanoscale MOSFETs," PhD thesis, Stanford University (2002).
2. For example, J. Larson and J.P. Snyder, "Overview and status of metal S/D Schottky-barrier MOSFET technology," *IEEE Trans. Electron Dev.* **53**, 1048 (2006).
3. K. Uchida, T. Krishnamohan, K. Saraswat, and Y. Nishi, "Physical mechanisms of electron mobility enhancement in uniaxial stressed MOSFETs and impact of uniaxial stress engineering in ballistic regime," *Tech. Digest IEDM* (2005), pp. 135–8.
4. A. Pethe, T. Krishnamohan, D. Kim, S. Oh, H. S. P. Wong, Y. Nishi, and K. Saraswat, "Investigation of the performance limits of III-V double-gate *n*-MOSFETs," *Tech. Digest IEDM* (2005), p. 619.
5. T. Krishnamohan, Z. Krivokapic, K. Uchida, Y. Nishi, and K. Saraswat, "High mobility ultrathin strained Ge MOSFETs on bulk and SOI with low band-to-band tunneling leakage: Experiments," *IEEE Trans Electron Dev.* **53**, 990 (2006).
6. A. Javey, J. Guo, Q. Wang, M. Lundstrom, and H. Dai, "Ballistic carbon nanotube field effect transistors," *Nature* **424**, 6949 (2003).
7. Z. Wang, P. B. Griffin, J. McVittie, S. Wong, P. C. McIntyre, and Y. Nishi, "Resistive switching mechanism in $Zn_xCd_{1-x}S$ nonvolatile memory devices," to appear in *IEEE Electron Dev. Lett.* (2007).
8. Y. Lu, M. D. Deal, and J. D. Plummer, "High quality single-crystal Ge-on-insulator by liquid-phase epitaxy on Si substrate," *Appl. Phys. Lett.* **84**, 2563 (2004).

Carbon-Nanotube Solutions for the Post-CMOS-Scaling World

P. M. Solomon
IBM Research, SRDC, T. J. Watson Research Center
Yorktown Heights, NY 10598, U.S.A.

1. Introduction

The CMOS research world has been up-ended in the past few years with the realization that the end of scaling is indeed approaching fast[1] and that other, more radical solutions need to be found. Much work has been focused on investigating a radical (for CMOS) set of new materials, strain engineering, and new geometries such as two-dimensional (2D) electrostatically confined structures and 3D heterogeneous integration. These approaches extend the technology and provide a more powerful end-point, but do not drastically alter the scaling scenarios. The capabilities projected for future CMOS are enormously larger than even today's gigascale integrated circuit (IC) chips, so the question arises as to the need for any CMOS follow-on at all. An interesting feature, pertaining especially to the silicon-on-insulator (SOI) approaches, is that the "silicon" technology is becoming divorced from the bulk silicon material. Indeed, for these technologies any convenient material could, in principle, be used for the substrate.

Of the various alternative "nano-offerings" investigated in the past few years,[2,3] none can be seen as a serious competitor to ultimate CMOS. Device characteristics are poor, unreliable and noisy, and manufacturing methodologies are uncertain at best. A possible exception, at least in terms of the intrinsic device, is the carbon nanotube. Carbon nanotubes (CNTs) can be thought of as the perfect electron waveguides, where electrons can be transported through unimaginably small channels without scattering off the boundaries. In addition, phonon interactions are weak. This results in transport properties far superior to silicon per unit channel area, with the additional advantage that these properties are symmetric for both *p* and *n* channel transport. However, challenges to implementing this technology and realizing these intrinsic advantages in terms of system performance are formidable. Firstly, how do we place, or grow, CNTs of desired properties on the circuit substrate. Then, given such placement, how do we fabricate FETs with specific and tightly controlled characteristics, and with low parasitic capacitances and resistances? Last, but not least, how do we best use CNTs in circuits and systems, and what performance advantages are expected.

Future Trends in Microelectronics. Edited by Serge Luryi, Jimmy Xu, and Alex Zaslavsky

2. Intrinsic properties

It is for their intrinsic properties as an FET that CNTs arouse the most excitement in those looking for future CMOS solutions. Scaling limitations of MOSFETs due to boundary scattering of electrons from imperfect interfaces[4,5] are solved naturally in CNTs, as they have a smooth, well-coordinated graphene structure[6] with no bonds to the outside. This enables CNTs to retain excellent transport properties down to much smaller lateral dimensions than silicon. Their small radius and the possibility of completely surrounding the CNT by a gate can provide excellent electrostatic confinement of the channel electrons, enabling the channel length to be scaled down to very small dimensions. Their small size would enable high packing densities. Bandstructure calculations[6,7] of CNTs show that conduction and valence bands are mirror images of each other, *i.e.* both electrons and holes should share equally good transport properties. This indicates the suitability of CNTs for a general-purpose high-performance complementary circuit technology.

As is now well known, CNTs can be either metallic or semiconducting, depending on their chirality,[6,7] and the semiconducting tubes have a bandgap E_G that is inversely proportional to their diameter d_{CNT}. Specifically, tight binding calculations lead to the following useful relation between d_{CNT} and bandgap E_G:

$$E_G = \gamma(2d_{CC}/\sqrt{3}d_{CNT}) \quad (1)$$

where γ is the hopping matrix element and d_{CC} is the carbon-carbon bond distance. Inclusion of electron-electron interactions increases E_G significantly,[6] and for a 1 nm nanotube the bandgap is roughly 1 eV. From an FET scaling perspective, a large bandgap is desirable to minimize band-to-band tunneling, as well as the tunneling breakdown of the drain contact that can result in undesirable ambipolar[8] characteristics. In practice, however, the best transport measurements and device characteristics have been obtained on rather larger diameter nanotubes, in the d_{CNT} = 1.7–3 nm range. The idealized electron/hole dispersion relation is hyperbolic in shape, with a quasi-parabolic "effective mass" regime at lower energies and a linear "constant velocity" regime at higher energies, where the limiting velocity v_{lim} reaches ~5–10×10^7 cm/s.[9]

Transport properties are further enhanced by the weak coupling of the charge carriers to acoustic phonons and the fact that the optical phonons have large energies of ~0.15eV.[10] All of these factors lead to extraordinarily large mobilities, reported at ~10^5 cm^2/V·s at room temperature![11] Granted, as we have learned over the years from development of circuit technologies using III-V materials, high mobilities do not lead to proportionate increases in device and circuit performance. Nevertheless, the high mobility of CNTs is an important factor in estimating their overall performance.

Unlike in silicon, reliable doped contacts for CNTs are not available, and contacts are typically made using Schottky-barrier metals. For instance Pd will make *p*-type and Al *n*-type contacts[12] to CNTs. To date, good contacts have only been made to CNTs of diameter ~1.7 nm and larger, which have bandgaps less than ~0.7 eV. These smaller band gaps pose a problem for FETs. Since they lower the Schottky barrier height for both electrons and holes, while they permit good contacts for the one carrier type at the source, they also permit tunneling of the opposite type of carriers at the drain. This is the "ambipolar" effect[8] and is deleterious because it opens a path for flow of the opposite carrier type under conditions where the FET should be switched off.

The CNT band-structure, while facilitating electron and hole transport, also facilitates band-to-band tunneling.[13] Band-to-band tunneling in FETs is only operative when conduction and valence bands are brought into alignment, *i.e.* when the bandgap is comparable to or less than the supply voltage (~1 V). This places a technological premium on being able to successfully contact small diameter CNTs.

3. Device and circuit benchmarks

The CNT technology is still at a very primitive state where, for the most part, placement, CNT type (metallic or semiconductor), and diameter and length of the nanotubes cannot be controlled.[6] Devices and circuits are made, for the most part, by building devices around existing tubes, or blindly fabricating large arrays of structures and picking the statistically good ones.

Typically, e-beam lithography is used in the former approach because of its flexibility in shaping patterns around existing tubes. Important benchmarks have been established. The first CNT FETs[14] were *p*-FETs using a silicon wafer as a back-gate. Complementary circuits have been made with selective doping.[15] In another approach, both *p* and *n*-type FETs[16] have been demonstrated with tens of microamps of current per tube, using Pd contacts for the *p*-type and Al contacts for the *n*-type FETs. A transconductance g_m of 26 μS has been achieved, which is to be compared with the maximum possible theoretical transconductance of 167μS,[17] for the ground-state subband of a single tube. Once the gate and drain voltages

		n-MOS[18]	*p*-CNT[19]
gate length	(nm)	40	50
oxide (HfO_2) thickness	(nm)	2	8
$V_{G,ON} - V_{G,OFF}$	(V)	0.8	1.2
transconductance g_m	(nS/nm)	2000	8000
I_{ON}	(nA/nm)	1400	4000
I_{OFF}	(nA/nm)	0.1	0.15–1.5
subthreshold slope S	(mV/dec)	90	70
capacitance	(aF/nm^2)	0.038	0.029

Table 1. Best of breed, *n*-MOS *vs.* *p*-CNT. The I_{OFF} in *p*-CNT is limited by ambipolar conduction at V_D = 0.3 V.

go above ~0.5 V, optical phonon emission can limit device current.[20] Self-aligned Pd source/drain contacts were demonstrated[19] as a means of minimizing parasitic capacitance. An integrated ring-oscillator has been made,[21] with all of the circuits laid out along a single CNT. The ring oscillator achieved an average delay of down to 2 ns per stage, but this delay reflected mainly the parasitic overlap and wiring capacitance contributions rather than the intrinsic speed of the CNT itself. High frequency measurements have also been made on CNTs,[22] but care should be taken in interpreting these results, because of parasitics and contact effects.

4. Manufacturing paradigm: Top down or bottom up?

For the CNT to succeed as a technology it has to overcome the placement, diameter and type selection problems. While work is proceeding in growing nanotubes *in-situ*, from placed catalysts,[6] it seems that there might be a more elegant way to solve this problem, leading to a very different paradigm for integrated circuit manufacture. Today's paradigm would prepare locations on a semiconductor chip, grow nanotubes in these locations, coat them with dielectrics, gate metals *etc.* and then lithographically define the nanotube FETs. The new approach would be to disperse the nanotubes in solution (liquid or gas), filter them according to type, diameter and length, coat them with dielectric and perhaps metal layers, and then deposit them in prepared locations on the IC chip. Subsequently the rest of the FET, including source and drain contacts, are made using conventional IC fabrication techniques. The advantages are not only that of a mass fabrication and selection process, but also that dielectric and metal layers are easier to coat conformally on a CNT while it is in solution, than when it is already lying on a substrate. Furthermore, the coating process provides a natural way of separating the nanotubes, so they might be stacked next to each other on the substrate. While this scheme in all its details has not been articulated previously, to my knowledge; parts of it are subjects of active research, such as filtration techniques for nanotubes in solution,[23] methods of conformal coating of CNTs in solution,[24] and methods of placing CNTs in pre-determined locations.[6] These might well lead to an exciting revolution in future IC technology.

5. Ballistic transport

In this section we examine fact and fiction, opportunities and pitfalls, concerning ballistic transport. The intrinsic properties of CNTs make them good candidates for ballistic transport, and several signatures for ballistic transport have been reported.[20,25,26] Ballistic transport is uncritically thought to lead to large improvements in device performance, but in reality it matters little, in terms of device current, whether carriers have or have not scattered a few times in transiting the channel. Figure 1 illustrates a situation where carriers are either partially or

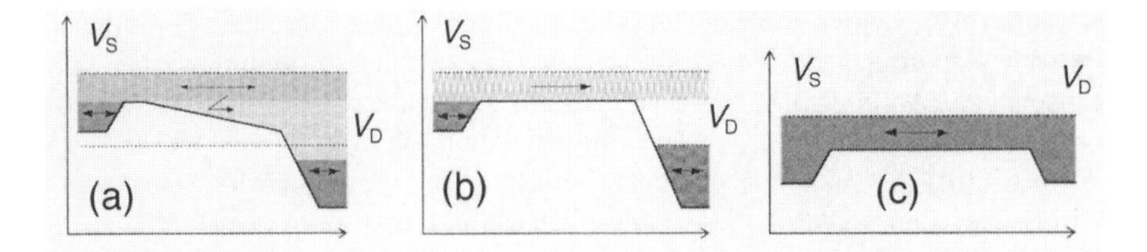

Figure 1. Ballistic transport: schematic band diagram showing half-degenerate state occupancy for large drain bias with partially (a) or fully (b) ballistic transport in the channel; (c) shows full occupancy at zero drain bias.

fully ballistic – Fig. 1(a) and (b), respectively. One sees that the current injected into the channel is about the same, while the charge in the ballistic channel has been reduced because of the absence of back-scattering and scattering to lower energies. Thus the performance is improved. This advantage is lost, however, in switching from the "on" state[27] to the "off" state, because when the transistor is on and $V_D = 0$, both directions are fully occupied – see Fig. 1(c).

Ballistic transport is seen as a panacea to avoid energy loss. In a transistor, even with a ballistic channel, dissipation in the drain contact provides a dissipation mechanism. However, it is instructive to consider a case where this mechanism is absent. If there were no contact with the thermal bath, could not computation be dissipationless? We will not debate the broader issue[28] here, but concentrate on a simple logic transition involving the discharge of a capacitor through a transistor, illustrated in Fig. 2. The whole system could, in principle, be ballistic since both capacitor and transistor could be part of the same CNT, eliminating the drain contact. The charged capacitor is represented by an empty potential well. Were the system lossy, the capacitor would discharge as the potential well filled up. For a purely ballistic system, however, the electrons from the source cannot thermalize into the well and are reflected back. The above example is a cautionary lesson that, contrary to expectations, ballistic transport might well slow down logic transitions.

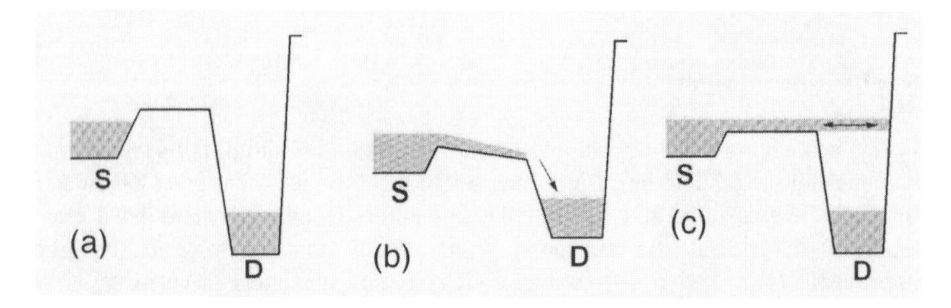

Figure 2. Discharge of a ballistic capacitor through a ballistic FET: transistor is off in (a) and on in (b) and (c); (b) is non-ballistic while (c) is ballistic.

	ballistic	electromagnetic
velocity	v_{lim}	c
characteristic impedance	r_0	Z_0
inductance	r_0/v_{lim}	Z_0/c
capacitance	$1/(r_0 v_{lim})$	$1/(Z_0 c)$

Table 2. Analogy between a ballistic conductor and a transmission line.

Table 2 illustrates an interesting analogy between a ballistic system described by a limiting velocity v_{lim} and an electromagnetic transmission line, for the case where the capacitance is dominated by the intrinsic quantum capacitance[29] of the CNT wire, rather than the gate capacitance.[30] In Table 2, $r_0 = h/4e^2$.[17] Both the quantum capacitance and kinetic inductance[31] are accounted for in this analogy. The resistance r_0 is usually described as a contact resistance, but a more correct description is wave impedance, since it is a non-local property of the traveling electron wave rather than a local property of the contact. Like its electromagnetic counterpart, the wave impedance characterizes the impedance under pulsed conditions and the dc input resistance will be influenced by reflections from the far contact. Although instructive, this analogy is not perfect. While electrons may be reflected from a barrier at the far end, this does not cause voltage doubling, as in an electromagnetic transmission line, but rather charge doubling, because now states below the Fermi level are filled by electrons traveling in both directions. Further, there is no easy equivalent to a shorted transmission line, where reflections are negative.

The kinetic inductance plays a major role if one tries to use nanotubes as interconnects.[32] It is larger than the electromagnetic transmission line inductance by the ratio of $Z_0 v_{lim}/r_0 c$. This is a very large ratio since $r_0 >> Z_0$ (by a factor of ~120) and $c >> v_{lim}$ (by a factor of ~300). Whether the line acts as a resistor or as an inductor depends on the time scale, set by L/R. For a ballistic line, this time scale is l/v_{lim}, where l is the line length, otherwise it is set by the momentum relaxation time τ_p. Since most switching times are expected to be longer than τ_p (~6 ps for a mobility of $10^5\,cm^2/V{\cdot}s$) the inductance, though large, would be unimportant compared to the wire's resistance. For long wires, many nanotubes would have to be bundled together (or thicker, multiwalled nanotubes[6,31] should be used), which would reduce both the resistance and the inductance.

The same arguments apply to the use of kinetic inductors as part of resonant systems, for instance in energy recovery circuits. For most applications where the operating frequency $f << 1/\tau_p$, the Q of the inductor will be low and not much will be gained.

The ballistic nature of transport in nanotubes does introduce the possibility of new types of solid-state devices. For example, the correlated nature of ballistic transport can be used to achieve amplification at frequencies well beyond transit-time limitations. These are the direct descendents of vacuum tubes, where ballistic transport is the norm, but on a vastly smaller scale. A schematic of such a device, along with a crude ballistic simulation, is shown in Fig. 3. The simulation shows that the kinetic energy in the channel may be increased beyond the applied gate voltages by suitably timing the gate signals with respect to the ballistic transit time between gates. When distances between gates are less than 1 μm, terahertz amplification might be possible.

6. Logic applications

Looking toward the future, both CMOS and CNT transistors will be scaled toward their limits,[1] although, for practical reasons, they will stop well short of their ultimate limits. A schematic drawing of the active areas of these hypothetical devices is shown in Fig. 4, with the corresponding electrical parameters given in Table 3. End capacitances have been estimated by assuming 5 nm silicon dioxide end-spacers for the CNTs. These parameters are realistic for CMOS but are "best case" for CNTs, since at present such high currents cannot be extracted from 1 nm sized CNTs.

We see that the current drive of ultimate CMOS and CNTs is comparable, but the CNT has a much lower capacitance and a somewhat lower operating voltage. The CNTs may be organized into logic arrays[33] stacked side by side to make use of their higher intrinsic packing density. Even a short length of wire will overwhelm the ~10 aF gate (including ends) capacitance of the CNT. Typical wire capacitance is ~0.2 aF/nm so that a wire length of greater than ~50 nm will already dominate the capacitance. So where is the advantage of CNTs over CMOS?

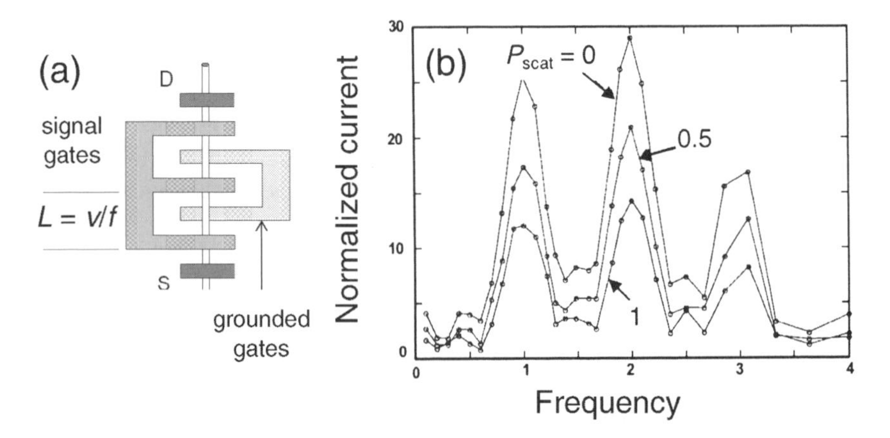

Figure 3. Traveling-wave type ballistic device: (a) schematic layout and (b) simulation showing the normalized output current *vs.* frequency in units of inverse transit time between the gates, parametrized by the scattering probability P_{scat} per transit between two signal gates.

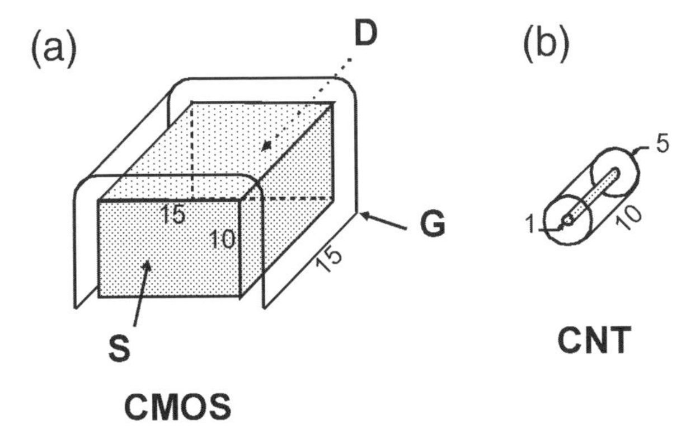

Figure 4. Active areas of CMOS and CNT transistors extrapolated to the future.

		CMOS	CNT
V	(V)	0.8	0.5
C_G	(aF)	20	3
C_{end}	(aF)	7	2.8
I	(μA)	40	40
$(C_G+2C_{end})V/I$	(fs)	680	107

Table 3. Intrinsic parameters of scaled FETs.

In today's power constrained environment, achieving competitively high performance at a low switching energy per logic transition is more important than achieving the highest switching frequency. In general the dynamic switching energy U is given by

$$U = ½(C_{dev} + C_{wire})V^2 \ , \tag{2}$$

where C_{dev} and C_{wire} are the device and wiring capacitance contributions, respectively. To minimize U, minimum-sized devices and corresponding interconnect lengths should be used, as well as the minimum supply voltages consistent with ensuring an acceptable on/off current ratio. Here the CNT has a considerable advantage, of up to a factor of ~6, because of its smaller intrinsic capacitance and size. The trade-off between switching energy and performance can be examined in more detail by considering the CMOS switching energy *vs.* transition rate plot as modeled by Frank and co-workers,[34] where we have superimposed some hypothetical CNT curves – see Fig. 5. In CMOS, the switching energy increases with switching rate for two main reasons. Firstly, to

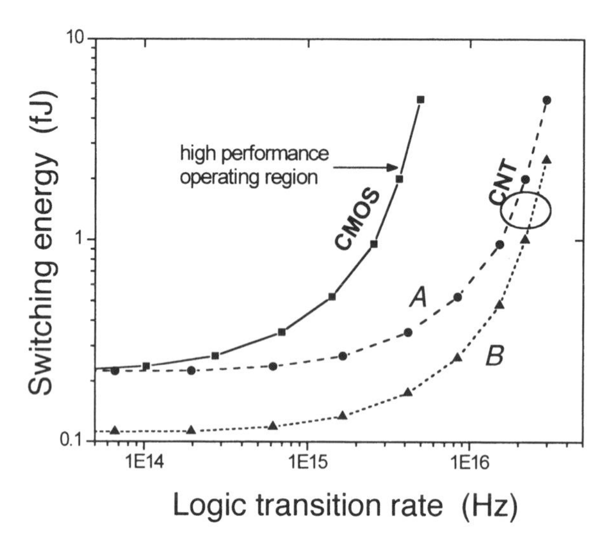

Figure 5. Loaded switching energy *vs.* logic transition rate for a CMOS processor core (adapted from Ref. 34).

drive the wires at higher speeds, wider FETs with larger capacitance have to be used. This not only increases C_{dev}, but also C_{wire}, because of lower device density. Secondly, because of the low CMOS mobility, voltages have to be increased to get the speed. For CNTs, the device size is small, so that wire lengths will be shorter; the intrinsic capacitance is so small that devices may be added in parallel to get more current without impacting the overall capacitance much; and because of the high CNT mobility, voltages do not have to be increased get higher speeds. These properties are illustrated by means of the superimposed curves in Fig. 5. The curve marked *A* assumes the same wire length and voltage (at low performance) as CMOS, but the smaller intrinsic capacitance of the CNT and constant supply voltage enables one to stay at the "low performance" switching energy up to much higher switching rates than CMOS. Furthermore, as shown by curve *B*, when these constraints are relaxed, the CNT benefits further, *i.e.* the starting switching energy is lower because of the shorter wire length and lower voltage. Thus, the most competitive arena for CNT logic is at roughly the same performance as high performance CMOS, but with much lower switching energies.

7. Conclusions

The CNT is a revolutionary device because, for the first time, the dimensions of the conducting channel are controlled by chemical bond lengths and not by some arbitrary manufacturing process. Electron transport down a nanotube is extraordinary. To make the CNT a contender in the post-CMOS arena, difficult challenges have to be overcome and hard questions need to be answered concerning the integration CNTs onto an IC chip, the contacting of small diameter CNTs, and the elimination of ambipolar conduction. The small size and low capacitance of CNTs lend them to the critical area of logic applications at low switching energy. The expected ballistic nature of transport at sub-micrometer dimensions opens up opportunities for unique high frequency applications in the THz regime.

Acknowledgments

I wish to acknowledge the help of the following in the form of discussions and material supplied: Tom Theis, Phaedon Avouris, Joerg Appenzeller, Zhihong Chen, Yu-Ming Lin and David Frank.

References

1. W. Haensch, E. J. Nowak, R. H. Dennard, *et al.*, "Silicon CMOS devices beyond scaling," *IBM J. Res. Dev.* **50**, 339 (2006).
2. S. Luryi, J. M. Xu, and A. Zaslavsky, eds., *Future Trends in Microelectronics: The Nano, the Giga, and the Ultra*, New York: Wiley, 2004.
3. ITRS "Emerging Research Devices" (2005), *www.itrs.net/Links/2005ITRS*
4. T. Ando, A.B. Fowler and F. Stern, "Electronic properties of two-dimensional systems," *Rev. Mod. Phys.* **54**, 437 (1982).
5. K. Uchida, H. Watanabe, A. Kinoshita, J. Koga, T. Numata, and S. Takagi, "Experimental study on carrier transport mechanism in ultrathin-body SOI *n*- and *p*-MOSFETs with SOI thickness less than 5 nm," *Tech. Digest IEDM* (2002), pp. 47–50.
6. An excellent review carbon nanotube theory, devices and technology can be found in: P. Avouris and J. Chen, "Nanotube electronics and optoelectronics," *Mater. Today* **9**, 46 (2006).
7. J. W. Mintmire, D. H. Robertson, and C. T. White, "Properties of fullerene nanotubules," *J. Phys. Chem. Solids* **54**, 1835 (1993).
8. V. Derycke, R. Martel, J. Appenzeller, and P. Avouris, "Controlling doping and carrier injection in carbon nanotube transistors," *Appl. Phys. Lett.* **80**, 2773 (2002).
9. G. Pennington and N. Goldsman, "Semiclassical transport and phonon scattering of electrons in semiconducting carbon nanotubes," *Phys. Rev. B* **68**, 045426 (2003); Yung-Fu Chen and M. S. Fuhrer, "Electric-field-dependent charge-carrier velocity in semiconducting carbon nanotubes," *Phys. Rev. Lett.* **95**, 236803 (2005).
10. V. Perebeinos, J. Tersoff, and P. Avouris, "Electron-phonon interaction and transport in semiconducting carbon nanotubes," *Phys. Rev. Lett.* **94**, 086802 (2005).

11. T. Dürkop, S. A. Getty, Enrique Cobas, and M. S. Fuhrer, "Extraordinary mobility in semiconducting carbon nanotubes," *Nano Lett.* **4**, 35 (2004).
12. Z. Chen, J. Appenzeller, J. Knoch, Y. Lin, and P. Avouris, "The role of metal-nanotube contact in the performance of carbon nanotube field-effect transistors," *Nano Lett.* **5**, 1497 (2005).
13. J. Appenzeller, Y.-M. Lin, J. Knoch, and P. Avouris, "Band-to-band tunneling in carbon nanotube field-effect transistors," *Phys. Rev. Lett.* **93**, 196805 (2004).
14. R. Martel, T. Schmidt, H. R. Shea, T. Hertel, and P. Avouris, "Single- and multi-wall carbon nanotube field-effect transistors," *Appl. Phys. Lett.* **73**, 2448 (1998).
15. V. Derycke, R. Martel, J. Appenzeller, and P. Avouris, "Carbon nanotube inter- and intramolecular logic gates," *Nano Lett.* **1**, 453 (2001).
16. A. Javey, Q. Wang, W. Kim, and H. Dai, "Advancements in complementary carbon nanotube field-effect transistors," *Tech. Digest IEDM* (2003), pp. 31.2.1–4.
17. This value comes from $4e^2/h$, with a factor of 2 arising from the two-fold degeneracy of the lowest bands and the other factor of 2 from the spin degeneracy.
18. J. Kavalieros, B. Doyle, S. Datta, *et al.*, "Tri-gate transistor architecture with high-κ gate dielectrics, metal gates and strain engineering," *Tech. Digest VLSI Symp.* (2006), pp. 50–51.
19. A. Javey, D. Farmer, R. Gordon and H. Dai, "Self-aligned 40 nm channel carbon nanotube field-effect transistors with subthreshold swings down to 70 mV/decade," *Proc. SPIE* **5732**, 14 (2006).
20. A. Javey, J. Guo, M. Paulsson, Q. Wang, D. Mann, M. Lundstrom, and H. Dai, "High-field quasiballistic transport in short carbon nanotubes," *Phys. Rev. Lett.* **92**, 106804 (2004).
21. Z. Chen, J. Appenzeller, Y-M. Lin, *et al.*, "An integrated logic circuit assembled on a single carbon nanotube," *Science* **311**, 1735 (2006).
22. D. V. Singh, K. A. Jenkins, J. Appenzeller, D. Neumayer, A. Grill, and H. S. P. Wong, "Frequency response of top-gated carbon nanotube field-effect transistors," *IEEE Trans. Nanotechnol.* **3**, 383 (2004); J. Appenzeller and D. J. Frank, "Frequency dependent characterization of transport properties in carbon nanotube transistors," *Appl. Phys. Lett.* **84**, 1771 (2004); A. A. Pesetski, J. E. Baumgardner, E. Folk, J. X. Przybysz, J. D. Adam, and H. Zhang, "Carbon nanotube field-effect transistor operation at microwave frequencies," *Appl. Phys. Lett.* **88**, 113103 (2006).
23. M. S. Arnold, A. A. Green, J. F. Hulvat, S. I. Stupp, and M. C. Hersam, "Sorting carbon nanotubes by electronic structure using density differentiation," *Nature Nanotechnol.* **1**, 60 (2006).
24. D. B. Farmer and R. G. Gordon, "ALD of high-κ dielectrics on suspended functionalized SWNTs," *Electrochem. Solid State Lett.* **8**, G89-G91 (2005).

25. A. Javey, J. Guo, Q. Wang, M. Lundstrom and H. Dai, "Ballistic carbon nanotube field-effect transistors," *Nature* **424**, 654, (2003); A. Javey, J. Guo, D. B. Farmer, *et al.*, "Self-aligned ballistic molecular transistors and electrically parallel nanotube arrays," *Nano Lett.* **4**, 1319 (2004).
26. J. Kong, E. Yenilmez, T. W. Tombler, W. Kim, and H. Dai, "Quantum interference and ballistic transmission in nanotube electron waveguides," *Phys. Rev. Lett.* **87**, 106801 (2001).
27. The "on" state refers to a CMOS logic inverter where, in steady state, the voltage drop across the "on" transistor is zero and the full voltage appears across the "off" transistor. The switching delay and switching energy for the "on/off" transition is determined by the amount of charge that has to be removed from the "on" transistor in order to turn it off.
28. These broader issues involve the use of reversible computation rather than conventional logic circuits.
29. For the lowest subband occupancy, linear dispersion and a degenerate electron distribution, the quantum (degeneracy) capacitance per unit length is given by $C_Q = n_s n_v n_b g_0 / v_{lim}$, where n_s, n_v and n_b are the spin, valley and ballistic degeneracy factors ($n_v = 1$ or 2), g_0 is the quantum of conductance and v_{lim} the limiting velocity of the linear dispersion branch.
30. For the opposite limit, where the gate is farther away, the velocity will be increased concomitantly with the reduction in capacitance.
31. P. J. Burke, "An rf circuit model for carbon nanotubes," *IEEE Trans. Nanotechnol.* **2**, 55 (2003).
32. A. Naeemi, R. Sarvari, and J. D. Meindl, "Performance comparison between carbon nanotube and copper interconnects for GSI," *Tech. Digest IEDM* (2004), p. 699.
33. P. M. Solomon and C. R. Kagan, "Understanding molecular transistors," in: S. Luryi, J. M. Xu, and A. Zaslavsky, eds., *Future Trends In Microelectronics: The Nano Millennium*, New York: Wiley/IEEE Press, 2002.
34. D. J. Frank, W. Haensch, G. Shahidi, and O. Dokumaci, "Designing CMOS for maximum chip performance," *IBM J. Res. Dev.* **50**, 419 (2006).

Alternatives to Silicon: Will Our Best Be Anywhere Good Enough in Time?

M. J. Kelly
Centre for Advanced Photonics and Electronics, Dept. of Engineering
University of Cambridge, 9 JJ Thomson Avenue, Cambridge CB3 0FA, U.K.

1. Introduction

The 2005 edition of the International Technology Roadmap for Semiconductors[1] draws attention to the need for post-CMOS devices within a decade. I want to argue from several premises that we are most unlikely to be ready in time. If we assume that the ITRS will be as generally accurate in the near future as it has proved in the recent past, then 45 nm and even 32 nm CMOS seem destined to arrive on time. (See Section 5 if this assumption is incorrect.) The individual transistors have been demonstrated for some years and large-scale integration has also been demonstrated. It is only the further grind of refining the technology that is left, and no one has shown any show-stoppers: the predicted ones seem to have been overcome, even if the heat dissipation means that not all devices are clocking as fast as they might.

The Roadmap is fairly sanguine about all the radical alternative technologies, but the longer CMOS continues, the harder it will be for any alternative device ideas to be good enough to take up from CMOS and continue the progress of Moore's law for computing power. Only two, engineered tunnel barriers and nano-floating gates, seem to offer even the prospects of complementing CMOS in the future. All the others will fall away because of cost, complexity, performance, or manufacturability, or indeed all of these at once.

The engineered tunnel barriers and the nano-floating gate technologies build on much research into tunneling and quantum dots over the last decade, and it is from these studies that I will extrapolate here. I am not convinced that we have enough time to beat either of these technologies into shape to make higher performance circuits and systems at ever lower cost.

2. Tunneling

In Si CMOS today, tunnel currents contribute to gate and other leakage, something to be suppressed. If tunnel currents are to be exploited, we will have to engineer the barriers that control them. Many attractive prototype devices have been demonstrated over the last three decades that use tunnel currents as integral to their operation, but none has entered low-cost volume production. The main problem is

 Future Trends in Microelectronics. Edited by Serge Luryi, Jimmy Xu, and Alex Zaslavsky

our inability to achieve satisfactory reproducibility of device performance within and between wafers. Several efforts have been made to produce small runs of tunnel devices, and while the in-wafer uniformity of the dc *I*–*V* characteristics can be controlled to ±5%, the wafer-to-wafer reproducibility of the same characteristics is still woefully inadequate.[2] I have studied a single barrier diode structure in the AlAs/GaAs materials system that has immediate applications in automotive radar systems: low sensitivity to ambient temperature of the range from -50 °C to +80 °C, low added noise and a detection efficiency and dynamic range that matches Schottky diodes – see Fig. 1. There are typically 30% variations in the current densities for fixed forward bias between wafers prepared a month apart, even under the regime of a rapid *ex-situ* calibration of the MBE machine to improve the precision of the actual growth of tunnel barriers.[2] This variability has to come down to below 10% before automated pick-and-place manufacture will be contemplated. Already this implies that tunnel barriers have to be grown with an accuracy, precision and reproducibility of order ±0.1 monolayer.

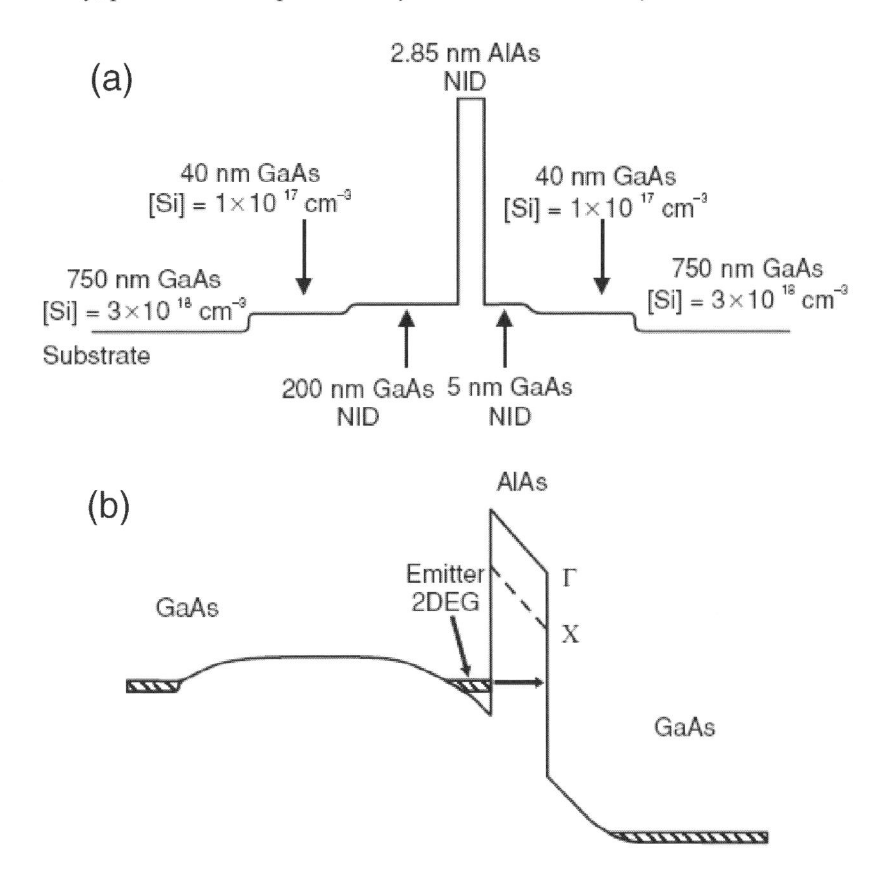

Figure 1. The conduction band profile of the ASPAT diode (a) in the absence of an applied bias, showing the doping densities and thicknesses of the epitaxial layers, and (b) under forward bias.

Most recently, I have examined the interface roughness of tunnel barriers. We consider an ideal AlAs tunnel barrier, *e.g.* exactly an integer number N of monolayers thick in GaAs. Next we consider one interface layer that has a composition $Al_xGa_{1-x}As$, where x is the Al content. We denote by T_N and T_{N+1} the transmission probability of a typical electron of a given energy through an ideal barrier of N and (N+1) monolayers, respectively. If the Al is uniformly distributed across the interface layer, the tunnelling probability will be $T_N \exp[-x\ln(T_{N+1}/T_N)]$. If the Al is segregated into areas of full coverage and areas of no coverage, the transmission probability will be $(1-x)T_N + xT_{N+1}$. The difference between these two values can be as much as 10–15% for values of $(T_{N+1}/T_N) \sim 3$ appropriate to AlAs barriers when $x \sim 0.5$, as illustrated in Fig. 2. The lateral scale of this segregation is also important. The typical bias across the tunnel barrier is of order 0.1 V during operation. From this, the de Broglie wavelength of the tunnelling electron is order 10 nm, and in turn the coherence area of the tunnelling electron is of order (10 nm)^2. This is the lateral scale on which the segregation of Al must occur, and this is not unreasonable for 500 °C growth, where lateral movement of order 0.1 μm is observed. If we consider both interfaces as having fractional

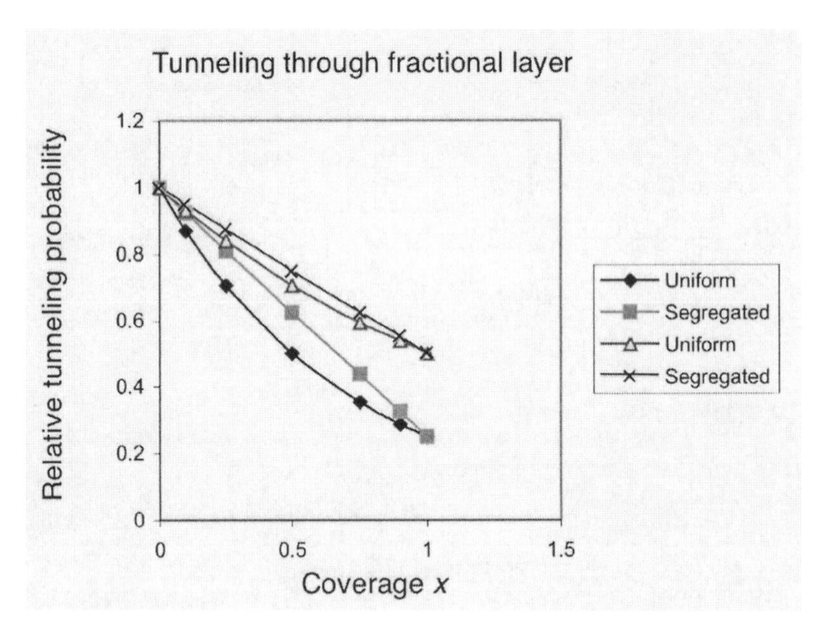

Figure 2. A schematic of the variation of relative tunneling probability through a single fractional layer of AlGaAs as one interface of an AlAs barrier in GaAs. It is assumed that one interface of the barrier is perfectly sharp, while the other barrier includes a monolayer of composition $Al_xGa_{1-x}As$. Curved lines correspond to statistically uniform Al coverage; straight lines to Al segregation into regions large compared to coherence coherence area of the tunneling electron. Calculations are for T_{N+1}/T_N=0.5 and 0.25, normalized to tunneling through an ideal AlAs barrier with two ideal interfaces ($x = 0$).

coverage, we are examining columns of AlAs that could be between N and (N+2) atoms deep, and the above analysis goes through using $(T_{N+1}/T_N)^2$. This exacerbates the degree of variability, and the only hope is to aim to grow integer layer thickness tunnel barriers, a study of which is now underway. Unless and until this project works, we will not be able to begin to engineer tunnel barriers as a low-cost process for mainstream silicon. Or course, if either interface is more than one monolayer wide, the variability of tunnel current gets even worse.

Finally, test diodes for these studies have been >10 μm in diameter, which means that the segregated model is well defined. If the tunnel barrier has a diameter of only 0.02 μm, and is in the Si/SiO_2 system, the segregated model is not well defined, and studies of 1D resonant tunnelling shows a much greater degree of irreproducibility than encountered in the studies described above.[3]

3. Quantum dots

Nanometer-scale floating gates are nothing but quantum dots buried in the gate dielectric. A survey of the literature on quantum dots indicates that for dots of order of 5 nm in diameter, there is a 15% standard deviation in the volume of quantum dots in most materials systems, and this value has remained remarkably constant for 15 years.[4] To get the best out of quantum dots in lasers, photon sources, nano-floating gates, *etc.*, this standard deviation has to come to nearer 5%. This is equivalent to about ±1 monolayer in just one of the spatial dimensions. We are going to have to do better than that. One way might be to pattern the substrate and rely on this to grow dots in hollows, so that the control over dot area is at least determined by lithography rather than surface accretion. This will leave the dot height to be the focus of precision growth. Studies of precision etching will be required to see if adequate area control can be delivered. Again, until and unless the standard deviation in quantum dot volume is brought closer to 5%, we will not have nano-floating gates of sufficient uniformity to feature in the beyond-CMOS era. Is there a Gödel-like theorem here, based on kinetics and thermodynamics, that precludes quantum dots of less than a lower limit of variability from being grown at 500 °C?

4. Split-gate devices

Future CMOS transistors at 0.02 μm gate length will have lithographic features on the same scale as the split-gate devices used to study one-dimensional transport in semiconductors. Recently, studies[5] of split-gate devices have show that the standard deviation of the threshold voltage in GaAs/AlGaAs HEMTs is of order 6% in the dark (rising to 20% when illuminated) and the standard deviation of the width of the first quantisation plateau in the conductance is of order 15% in light or dark. Although these results are compounded by the statistical position of dopants in the supply layer, the contribution from lithographic variations is at least 50%, and again this value is unacceptably large.

5. How many gates?

Finally, if we imagine a post-CMOS regime taking over at or after the 32 nm generation of devices, by this stage there will be of order 10^{10} gates per chip. Both the engineered tunnel barrier and the quantum dots will have to be engineered to 6σ precision to this level of complexity, starting form where we are now. We will need to be thinking of $>10^{12}$ gates if the alternative technology is going to last for more than one or two generations.

6. What happens if we are not ready in time?

There are at least two preceding technologies that have progressed under a Moore's law regime, the shaft horsepower of marine turbines during 1895–1942 and the "people momentum" of jet passenger aircraft during the 1950's and 1960's. In both cases, it was a combination of cost and technical difficulty that together brought a halt to progress, as illustrated in Fig. 3. The last two ships (USS Iowa and USS

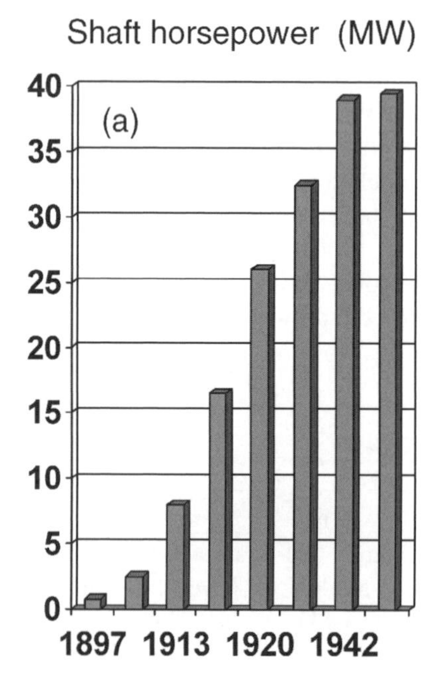

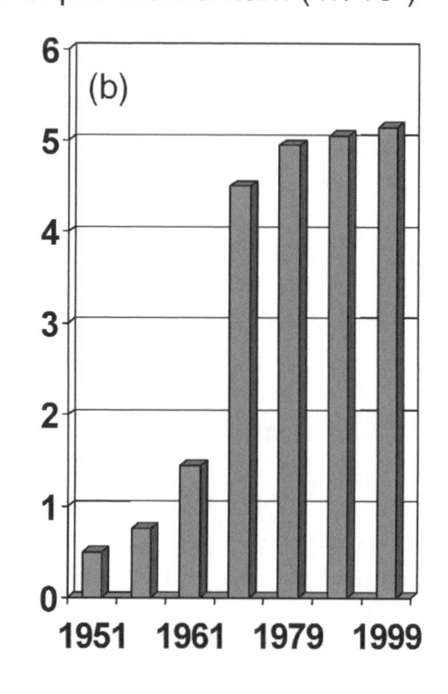

Figure 3. (a) The shaft horsepower of marine turbines, and (b) the people momentum of jet passenger aircraft (maximum passenger number *n* multiplied by velocity *v* in km/hour), showing a Moore's law phase for earlier technologies.

Midway) were built for the Pacific theatre in World War II, and bombed by enemy aircraft. The 40 MW level has never been exceeded. In the case of the aircraft, the speed of sound limited the growth in the 1960's to passenger numbers, and with the jumbo jet that has been flat since 1969. While supersonic aircraft fly faster, they have a smaller figure of merit.

In both cases, the marine and aircraft industries did not come to a grinding halt, but rather diversified, so that there is a proliferation of types of ship and plane, more carefully chosen for specific applications. For flying, the discriminators now are fuel efficiency, the in-flight experience and the eco-friendliness of the aircraft.

This diversification is already happening with the ITRS Roadmap with the "More than Moore" axis, where the raw computing power is secondary to the range of applications and services that can be provided on a single handset. The "More than Moore" axis is enabled almost entirely by semiconductors other than silicon, using superior materials properties to achieve higher power and frequency operation.

It is likely that the endpoint of computational power available to the consumer might come from improvements of software, architecture, and hardware-software co-design. One can also ask where hardware research is best placed in a post-CMOS regime. Will the discriminating feature between electronic systems migrate from the processor to the display?

The basics of quantum computing seem to be making strides, but the moves towards manufacturability have not begun in earnest, so that the effort may yet not bear fruit.

Acknowledgments

I thank my many co-workers on the topics raised here.

References

1. See *http://www.itrs.net/Common/2005ITRS/Home2005.htm*
2. V. A. Wilkinson, M. J. Kelly, and M. Carr, "Tunnel devices are not yet manufacturable," *Semicond. Sci. Technol.* **12**, 91 (1997); K. Billen, V. A. Wilkinson, and M. J. Kelly, "Manufacturability of heterojunction tunnel diodes: Further progress," *Semicond. Sci. Technol.* **12**, 894 (1997); R. K. Hayden, A. E. Gunnaes, M. Missous, R. Khan, M. J. Kelly and M. J. Goringe, "Ex-situ re-calibration method for low-cost precision epitaxial growth of heterostructure devices," *Semicond. Sci. Technol.* **17**, 135 (2002).
3. M. Tewordt, V. J. Law, M. J. Kelly, *et al.*, "Electron-state lifetimes in submicron diameter resonant tunneling diodes," *Appl. Phys. Lett.* **59**, 1966 (1991).
4. N. Saucedo-Zeni, A. Y. Gorbatchev, and V. H. Mendez-Gracia, "Improvement on the InAs quantum dot size distribution employing high-temperature GaAs(100) substrate treatment," *J. Vac. Sci. Technol. B* **22**, 1503 (2004); T. J. Krzyzewski and T. S. Jones, "Ripening and annealing effects in InAs/GaAs (001) quantum dot formation," *J. Appl. Phys.* **96**, 668 (2004); Y. Ebiko, S. Muto, D. Suzuki, *et al.*, "Scaling properties of InAs/GaAs self-assembled quantum dots," *Phys. Rev. B* **60**, 8234 (1999).
5. Z. Yang, G. A. C. Jones, M. J. Kelly, H. Beere, and I. Farrer, "Manufacturability of split-gate transistor devices – initial results," *Semicond. Sci. Technol.* **21**, 558 (2006).

MRAM Downscaling Challenges

F. Arnaud d'Avitaya and V. Safarov
CRMCN–CNRS, Campus de Luminy, 13288 Marseille cedex 9, France

A. Filipe
Spintron, Technopôle de Château-Gombert – BP100, 13382 Marseille, France

1. Introduction

Memories[i] play a major role in achieving superior computer performance. Indeed, the execution rate of a code is not only related to the processor clocking, but greatly depends on the writing and reading rate of the data to and from memory. Aside from the large hard-disk memories, which now reach capacities and performance unimaginable only ten years ago, the memory chips presently used in computer cores can be classified in different ways depending on the data retention mode and the technology used, but they essentially rest on two concepts: i) the core of the device is essentially based on MOS transistors, and ii) there is random access to the data (hence the acronym RAM, for random access memory).

Within the RAM family, we can distinguish two main classes: the dynamic RAM (DRAM) that has a simple single-transistor architecture but needs to be refreshed regularly (because reading erases the data stored in a capacitor), and static RAM (SRAM) that need not be refreshed but requires a more complex architecture detrimental to the integration density. Despite the fact that they are extremely fast and therefore well-adapted to the computer's requirements, both these technologies suffer from a major drawback: memory is erased when the power is switched off. This leads to the frustration of losing precious time whenever the machine is restarted, familiar to every computer user. Many thousands or even millions of hours are lost each day, throughout the world, for launching operating systems!

And yet, from the early 1950's until the mid-1970's, we have made use of magnetism properties to fabricate "fast" nonvolatile memories (magnetic core and magnetic bubble memories). But, at that time mega- or gigabit memories were simply science fiction! We had to wait until the early 1990's to see the first dense nonvolatile memories, the "flash memories" based on semiconductors. Today, because of their gigantic capacity housed in very small areas, they allow for the rapid development of mobile devices that need to handle millions or billions of bits (cameras, video cameras, USB keys, MP3 players, *etc.*).

Nevertheless, these flash memories cannot replace classical DRAMs or SRAMs due to two major drawbacks: their writing and reading time and their limited number of read/write cycles. Hence, for a decade now, we have been

Future Trends in Microelectronics. Edited by Serge Luryi, Jimmy Xu, and Alex Zaslavsky
ISBN 0-471-48

witnessing the revival of studies involving magnetic properties of materials that use a new degree of freedom – the electron spin. This science is usually known as spintronics.

The discovery of giant magnetoresistance[ii,iii,iv] (GMR) has boosted these studies and immediately led to important industrial applications for consumer electronics (read heads of high capacity hard disks). Also, GMR gave rise to renewed interest in devices involving spin. About ten years ago, based on these new results, researchers began investigating the fabrication of magnetic RAMs, or MRAMs. Year after year, this interest has led to an exponentially increasing number of papers and patents in this field, as well as great industrial research effort.

Why such an interest? We may find the answer in numerous papers and conference presentations where it was (and still is) declared that the MRAM is a universal memory that combines the advantages of all other memories, *i.e.* high integration density (like DRAM, flash, and phase-change PCRAM), high speed (like DRAM and SRAM) and nonvolatility (like flash and PCRAM). Since 2003, major industrial players fabricated MRAM prototypes from 64 kb to 16 Mb and they all had announced plans for MRAM mass production before the end of 2005. However, except for Honeywell[v] that posts a data sheet for a 1 Mb MRAM chip (HXNV0100) on its website and Freescale that, very recently, announced the production of a 4 Mb MRAM, other companies like Cypress and Micron have exited the MRAM arena.

This leads us to ask whether MRAM really is the universal memory we have been waiting for, because of a number of technical questions, such as:

- can MRAM integration density compete with DRAM and flash?
- can MRAM write and read rates compete with DRAM and SRAM?
- are MRAM write and read voltages and currents compatible with existing technologies?
- are the technological steps reliable enough for mass production?

This chapter aims to answer, as honestly as possible, all of these questions. To this end, we will first discuss the magnetic tunnel junction (MTJ),[vi] which is the basic element of an MRAM, and the physics that governs the electronic transport in such structures. We will then comment on the reversal of the magnetization by using the simple model known as "spin-flip by coherent rotation".[vii] We will extend this analysis to a network of memory cells.

In the light of this model, we will analyze various memory parameters, such as the retention time, the write and read currents, and the endurance, as a function of the cell size and the number of memory cells. We will then discuss the various alternatives (loosely termed second generation memories) that have been proposed to remedy the problems inherent in the first generation of MRAM. Finally, we will attempt to formulate the most promising approach towards building a "universal memory".

2. What is a tunnel junction?

As seen in Fig. 1, a magnetic tunnel junction is composed of two thin layers of ferromagnetic metals, FM1 and FM2, separated by a very thin insulating layer through which the electrons can tunnel. The two metals are chosen such that their coercive fields, *i.e.* the magnetic fields required to reverse the magnetization, are different. In this way, by applying an external field we can reverse the magnetization of one layer, for example FM1, without altering the other.

Let $\rho 1\uparrow$ and $\rho 1\downarrow$ represent the spin-up (majority carrier) and spin-down (minority carrier) densities of states in the FM1 layer, and $\rho 2\uparrow$ and $\rho 2\downarrow$ the corresponding quantities in the FM2 layer. If we assume that there are two independent conducting channels that correspond to spin-up and spin-down states and that the current is proportional to the product of densities of states in FM1 and FM2 (the Julliere model[8]), the conductance of the MTJ structure is given by:

$$G = k(\rho 1\uparrow \rho 2\uparrow + \rho 1\downarrow \rho 2\downarrow)\,. \tag{1}$$

As a result, when the magnetizations of the two layers are parallel, the resulting conductance G_P is given by

$$G_P = k(\rho 1\uparrow \rho 2\uparrow + \rho 1\downarrow \rho 2\downarrow)\,, \tag{2}$$

whereas for the anti-parallel case one obtains G_{AP} to be:

$$G_{AP} = k(\rho 1\uparrow \rho 2\downarrow + \rho 1\downarrow \rho 2\uparrow)\,. \tag{3}$$

As spin densities of states are different, there is a conductivity difference between the parallel and anti-parallel states. Then we can define a characteristic dimensionless magnetoresistance (MR) that is given by:

$$\mathrm{MR} = (R_{AP} - R_P)/R_P = (G_P - G_{AP})/G_{AP}\,. \tag{4}$$

We note that according to this definition, MR may take values higher than 100%. If we define the polarization P of a given material as

$$P = (\rho\uparrow - \rho\downarrow)/(\rho\uparrow + \rho\downarrow)\,, \tag{5}$$

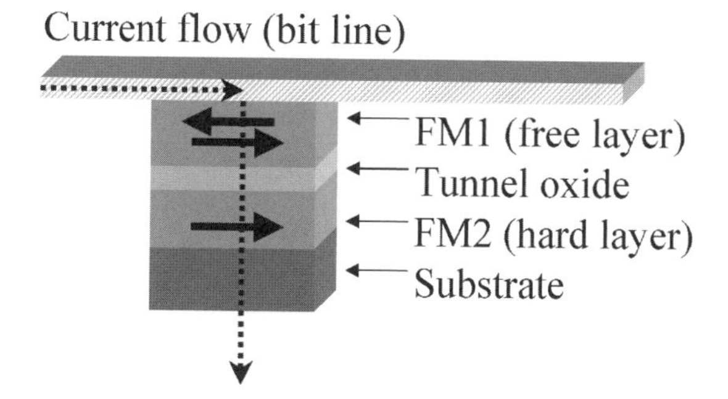

Figure 1. Schematic of a magnetic tunnel junction (MTJ).

we can express the magnetoresistance as

$$MR = 2P_1P_2/(1 - P_1P_2)\,. \tag{6}$$

However, although simple MTJs (FM1/insulator/FM2) may exhibit relatively high MR values (20–25%), they are not ideally suited for memory applications because of layer coupling and the difficult problem of accurately controlling the switching field over a large network of imperfect cells.

So, all the practical realizations involve, for the hard layer (FM2 for example), a bi-layer composed of an antiferromagnetic (AFM) and a ferromagnetic (FM) layer. The exchange coupling[9] between the two layers has the effect of pinning the ferromagnetic layer, which serves as a reference. The exchange field H_{EX}, which determines the immunity of the FM/AFM structure to the external field, is given by:

$$H_{EX} = E_{EX}(1 - T/T_B)/(M_S t) \tag{7}$$

where E_{EX} is the exchange energy, M_S the saturation magnetization of the AFM layer, t the thickness of the FM layer, and T_B the blocking temperature, *i.e.* the temperature above which the antiferromagnetic order vanishes.

The value of these parameters depends, of course, of the materials used. However, as memory chips can be expected to face an increase in temperature to above 150°C (evaluated for the case of 1 Gb memory), it is prudent to choose an antiferromagnetic alloy with a rather high blocking temperature (PtMn or IrMn for example).

3. Reading the memory cell: Control of the tunnel barrier

Reading a memory cell requires discriminating between the two resistivity states of the cell. The simplest technique detects this variation by using a simple MOS transistor or a combination of transistors. Of course, the greater the difference between the two states (*i.e.* the higher the MR value), the easier the cell is to read. When looking at the magnetic tunnel junction of Fig. 1, the current flowing through the structure is essentially determined by the resistance of the tunnel oxide barrier. This resistance must be much larger than the source-drain resistance of the transistor in saturation (~ kΩ), but sufficiently low to allow for currents larger than noise level. Values in the 5–50 kΩ range are the most suitable.

One of the difficulties in producing a reliable MTJ is the control of the tunnel oxide. For large (> 1 μm) lateral dimension of the memory cell, these resistance values are obtained with relatively thick oxide layers (3 to 4 nm of Al_2O_3, MgO, *etc.*) that are not very difficult to fabricate by traditional techniques (plasma oxidation). However, when the dimensions of the memory cell reach the deep sub-micron regime, a drastic decrease of the oxide thickness is necessary (2 to 3 atomic layers). As the tunnel current varies exponentially with thickness (0.3 nm leads to one decade current variation), it becomes very difficult to control thickness uniformity over large 300 mm wafers.

Accordingly, it becomes necessary to make the cell control circuitry more complex by introducing reference memory cells. This implies a lowering of the integration density. For technology nodes below 90 nm, we will probably need more sophisticated deposition or oxidation techniques (*e.g.*, physical atomic layer deposition). Moreover, defects in the oxide layer or at the interface may considerably affect transport mechanism leading to a huge reduction of MR. Often, these defects are cured by annealing, but this can lead to interdiffusion of metals that becomes more crucial as the thickness of the oxide is decreased.

4. Writing to the memory cell: Ferromagnetic layer magnetization reversal

Now we turn to another major problem related to the writing of the memory cell. Writing requires the creation of a local magnetic field sufficient for the reversal of the free magnetic layer. The simplest method of generating this magnetic field relies on passing current through a wire located above and/or below the memory dot. The wire(s) will be oriented in such a way that the resulting field makes a given angle with the easy axis of magnetization of the dot. Notice that the ferromagnetic layers are often deposited in the presence of a magnetic field to establish an easy magnetization axis. To simplify the architecture, the reading connection (bit line) is often also used as the write wire, as illustrated in Fig. 1.

Let us analyze the magnetization reversal mechanism. We want to estimate the order of magnitude of the writing parameters and the way they vary as the cell size is reduced. For this purpose we will use the model of coherent rotation,[7] which considers single-domain ferromagnetic dots (namely macrospins). The validity of this model improves as the size of the dot is reduced. Of course, more precise analysis could be realized using micromagnetic modeling, but we believe that there is no great discrepancy between the models, especially as regards the scaling to smaller sizes.

Consider an isolated magnetic dot. We suppose that the easy axis of magnetization is oriented along the bit line. Let M_S be the magnetization at saturation and H the applied magnetic field, with α and φ denoting the angles of M_S and H with the easy axis, respectively.

The total energy E_{TOT} of the system is then given by:

$$E_{TOT} = KV\sin^2\alpha - \mu_0 H M_S V\cos(\varphi-\alpha)\ , \tag{8}$$

where V is the volume of the dot and μ_0 the vacuum permeability. In the case where $\varphi = \pi$, Eq. (8) becomes:

$$E_{TOT} = KV\sin^2\alpha + \mu_0 H M_S V\cos(\alpha)\ . \tag{9}$$

Equation (9) has an extremum when $dE_{TOT}/d\alpha = 0$, which happens when $H = 2K\cos(\alpha)/\mu_0 M_S$. In the particular case of the magnetization oriented along the easy axis ($\alpha = 0$), we can define the reversal field as H_a, where

$$H_a = 2K/\mu_0 M_S\ . \tag{10}$$

The corresponding energy barrier ΔE_{TOT} is then:

$$\Delta E_{TOT} = KV(1 - H/H_a)\,, \tag{11}$$

which goes to zero for $H = H_a$. Analogous reasoning shows that the energy barrier is equal to KV in the case where H is applied perpendicularly to the easy axis. Figure 2 shows the barrier evolution for various values of the applied field H.

We can also use this theory to evaluate the optimal angle between H and M_S to reverse the magnetization. The value of the reversal field H_0 as a function of the arbitrary angle φ is given by:

$$H_0 = H_a\,[\cos^{2/3}\varphi + \sin^{2/3}\varphi]^{-3/2}\,. \tag{12}$$

This leads to the Stoner-Wohlfarth[7] astroid representation, which shows that reversal is easier when the angle between H and M_S is 45°. The value of H_0 is then given by $H_0 = H_a/2$.

However this analysis does not consider the finite temperature of the system. In fact, thermal agitation will favor a random reversal. The mean reversal time τ is given by

$$\tau = \tau_0 \exp[KV/k_B T]\,, \tag{13}$$

where τ_0 is the characteristic time of gyromagnetic precession ($\tau_0 \sim 10^{-9}$ s), KV the energy barrier, k_B the Boltzmann constant, and T the temperature in Kelvin.

In the light of this model, we will now analyze the performances of an idealized magnetic memory chip as a function of its capacity (number of bits) and the size of the memory cell.

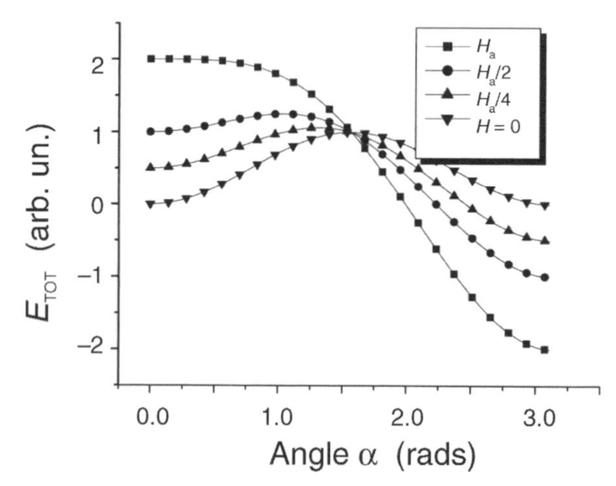

Figure 2. Reversal energy barrier as a function of 〈 for various applied magnetic fields, where 〈 is the angle between magnetization and the easy axis.

5. Actual performance and comparison with calculations

We will take as our reference the only available data sheet as of the time of writing for a 1 Mb HXNV0100 memory chip by Honeywell, summarized in Table 1.

Although we do not know all the technical details, we know this memory is based on first-generation technology, *i.e.* it is composed of a pinned hard layer, a tunnel oxide and a free layer. Chip area is 81 mm^2, memory cell size is 25 μm^2, transistor technology node $F = 150$ nm, and16 bits are read or written in parallel.

As illustrated in Fig. 3, cell writing is realized by means of two perpendicular wires (bit and word lines) that make a 45° angle with the easy axis (to be in the most favorable energetic configuration, *i.e.* that of Stoner-Wohlfarth model). Indeed, the top write line is in contact with the cell and it also acts as the read line. With this configuration, only the cells that are at the intersection of bit and word lines are written. All the other cells located along either the bit or word line see the field created by a single line, which is equal to $H_a/2\sqrt{2}$.

How great a current I is needed in the write line to generate the reversal field? In the case of the word line, which is isolated from the cell, we may simply use the Biot-Savart law, which implies that the field is proportional to the current and inversely proportional to the distance between the wire and the free layer. In the case of the bit line, where the wire is touching the top metal contact of the free layer, if we suppose that the thickness of the wire is 1.7 times its width (according to 2003 International Technology Roadmap for Semiconductors), then the field is given by $H = I/4.8F$.

Read cycle time	< 60 ns
Write cycle time	< 100 ns
Average writing current	260 mA (@3.3V) + 15mA (@1.8V)
Average reading current	3 mA (@3.3V) + 15mA (@1.8V)
Retention	> 10 years
Endurance	> 10^{15} Cycles
Magnetic field limit	50 Oe

Table 1. Specifications of Honeywell's HXNV0100 memory chip (2005).

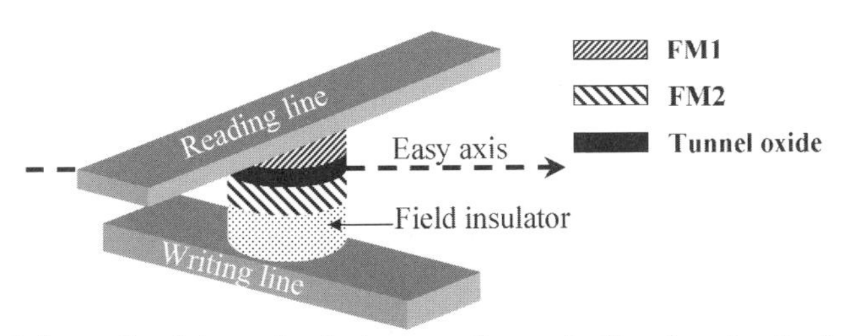

Figure 3. Schematic picture of a first generation cell. Read and write lines are also known as bit and word lines.

Does the performance promised by Honeywell represent a mature device that could potentially replace all the other memories in the foreseeable future? To answer this question, below we analyze six performance metrics and compare them to existing dynamic, static and flash memories. These metrics are: read current, read cycle time, retention, write current, write cycle time, and endurance.

- *Read cycle*

The mean read time given by Honeywell is ≤ 60 ns. This time is longer, by a factor of 10 to 20, than that of a DRAM or SRAM. However, it does not seem that this is due to some intrinsic limit of the MTJ. Indeed, looking in detail at the data sheet, the duration of the current pulse that creates the writing field is only 10 ns. Furthermore, an error correction code is applied. This increases the cycle time. Recently, Freescale announced a shorter value for its 4 Mb memory chip, launched in 2006 (read/write time of 35 ns). However, the Freescale chip is a second-generation memory with a more complex MTJ cell and a writing scheme very different from Honeywell's. Still, the read time is markedly higher than that of a DRAM or SRAM, but much lower that that of a flash memory.

The power dissipated when reading a 16 bit word, announced to be 40 mW, is relatively close to that of other memories.

- *Retention*

Honeywell guarantees a retention time greater than 10 years. This is equivalent to EEPROM and flash memories. Figure 4 shows retention time, as a function of the KV/k_BT ratio, evaluated from Eq. (13) for a single isolated bit, and 1 Mb and 1 Gb memories. Here we have taken the temperature of the chip to be 50 °C.

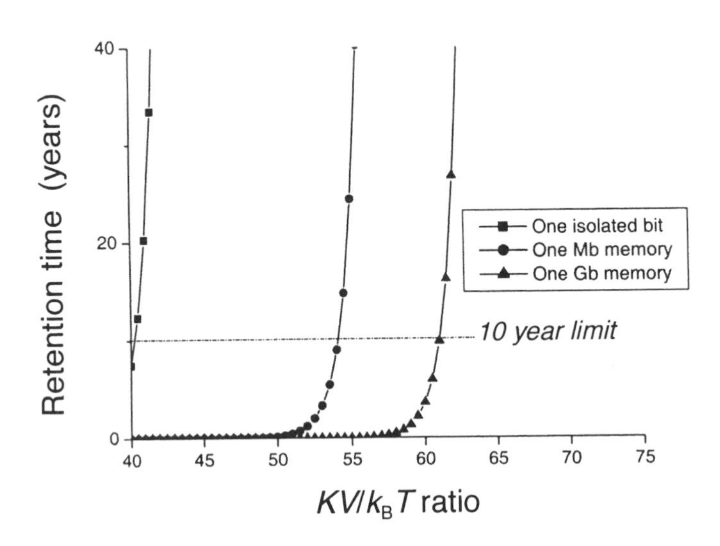

Figure 4. Retention time as a function of the KV/k_BT ratio.

This is the result of a simple statistical analysis postulating that the unintentional reversal of "at least 1 bit among N" is inversely proportional to the number of bits N. In the case of 1 Mb and 1 Gb chips, the ten-year limit is obtained for KV/k_BT ratios greater than 55 and 65 respectively. Are such ratios achievable? From Eq. (10) we can solve for KV as a function of H_a:

$$KV = H_a V \mu_0 M_S / 2 \tag{14}$$

Now if we consider an elliptic magnetic dot with a long axis L that is twice the short axis and a thickness t we can estimate H_a as

$$H_a = 4\pi M_S (t/L)(n_y - n_x) \tag{15}$$

where $(n_y - n_x)$ is the demagnetization factor equal to 0.6 for this geometry. Figure 5 gives the values of the KV/k_BT ratio and H_a as a function of the dot size (where the short axis of the dot is set equal to the technological node F), for two values of the free layer thickness t.

We see that high KV/k_BT ratios can be obtained by increasing the magnetic dot volume, but this implies an increase of the switching magnetic field. For F = 150 nm, the KV/k_BT = 65 corresponds to a field of around 80 Oe, so that both write lines have to provide a field of about 30 Oe each.

- *Write cycle*

Concerning the write time, Table 1 gives a < 100 ns figure (the Freescale 4 Mb chip promises a considerably lower value). This write time is considerably higher than that of SRAM and DRAM. However, writing is always preceded by a read cycle to establish whether the state of the memory cell needs to be changed. This process tends to increase the write cycle time.

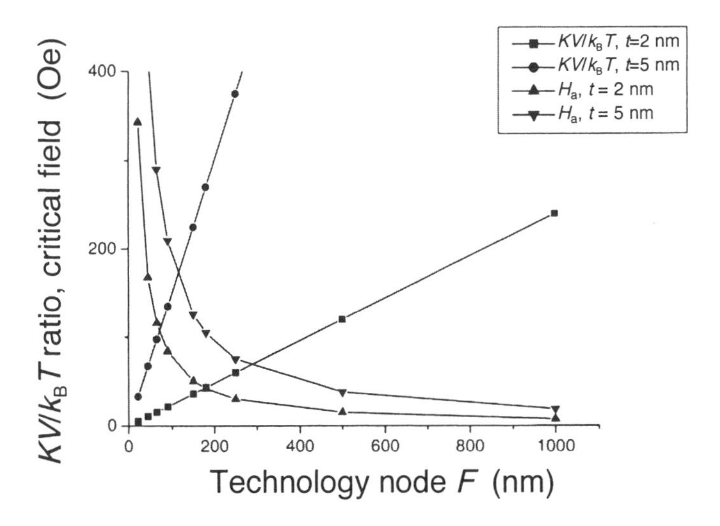

Figure 5. The KV/k_BT ratio and reversal field as a function of lateral cell size for two thicknesses of the free layer. The cell is assumed to have an elliptic shape with a long axis/short axis ratio of 2.

	ISOLATED WIRE						WIRE IN CONTACT					
	15 mA		10 mA		5 mA		15 mA		10 mA		5 mA	
F (µm)	H (Oe)	Current density (10^6A/cm2)	H (Oe)	Current density (10^6A/cm2)	H (Oe)	Current density (10^6A/cm2)	H (Oe)	Current density (10^6A/cm2)	H (Oe)	Current density (10^6A/cm2)	H (Oe)	Current density (10^6A/cm2)
0,045	667	436	444	290	222	145	873	436	582	290	291	145
0,09	333	109	222	73	111	36	436	109	291	73	145	36
0,18	167	27	111	18	56	9,1	218	27	145	18	73	9,1
0,25	120	14	80	9	40	4,7	157	14	105	9,4	52	4,7
0,6	50	2,5	33	1,6	17	0,8	65	2,5	44	1,6	22	0,8
1	30	0,9	20	0,6	10	0,3	39	0,9	26	0,6	13	0,3

Table 2. Generated field and current density in wires as a function of size and current flow. The cells in grey correspond to unreachable values.

The given value of the write current of a 16 bit word (260 mA) is extremely high, about 100 times that of a classic memory. This gives a value of ~15 mA per write line. Is this value a fundamental limitation?

Let us consider that the wire has a section of $1.7F^2$, as suggested by the ITRS. Table 2 gives the values, as a function of the smallest dimension of the magnetic dot (*i.e.* the technology node F), of the generated magnetic field and current densities in both write lines using the Biot-Savart law for the word line (supposed to be at a distance of 1 µm from the free layer) and the $H = I/4.8F$ relation for the bit line (supposed to be in contact with the free layer).

These values are calculated for three different currents flowing through each writing wire. Of course, for a given current, the generated magnetic field increases drastically if the cross-sectional size of the wire is decreased, but this leads to an explosion in current density. Indeed, if we consider that a state-of-the-art interconnect cannot withstand current densities higher than $2x10^6$ A/cm^2 (ITRS 2005), the field necessary to reverse dot magnetization is reached for a wire that is 600 nm wide.

Smaller sizes exceed the limiting current density values. Although we do not know the exact architecture of the memory fabricated by Honeywell, we may imagine that the writing wires are much larger than the dot size, in agreement with the quoted bit area of 25 µm^2. This analysis shows unambiguously that, in terms of bit size and integration density, the Honeywell MRAM architecture represents a kind of limit.

Indeed, if we decrease the lateral size of the memory cell we have to face several problems. First, the energetic barrier is lowered, which is detrimental to the retention time. Moreover, the magnetic field we have to apply increases dramatically. This implies an increase in the write current. In order to improve the integration density, it is necessary that all cell components scale down with the technology node, including the write wires. Table 2 shows that this is almost impossible, in the absence of some unforeseen technological breakthrough.

Moreover, the Honeywell chip cannot be exposed to external magnetic fields higher than 50 Oe. This is a major limitation.

- *Endurance*

The endurance given in Table 1 is greater than 10^{15} cycles. This is similar to performance obtained with SRAM and DRAM.

Is it possible to avoid the use of error correction codes? As before, let us consider the writing of a 16 bit word in the case of a 1 Mb memory chip, *i.e.* a memory organized as a matrix of 1000 lines and columns. This writing supposes that we supply at the same time 16 lines and 1 column (or, conversely, 16 columns and 1 line). This makes a total of 16,984 cells that see a local magnetic field. Only the cells located at the intersection of a line and a column will be affected by the field sufficiently strong to reverse magnetization. All the others will see a field whose value is $H_a/2\sqrt{2}$, assuming the two lines produce the same magnetic field. This implies that during writing all these half-selected bits will see a lowering of the energetic barrier by a factor of 4. Thermal agitation will be able to reverse magnetization of these half-selected bits in a random way. Table 3 gives the number of reading cycles (16 bits at a time) of a 1 Mb memory we can hope to achieve before a single writing error occurs. Note that now we need KV/k_BT ratios higher than 190. This value is potentially reachable by increasing the volume of the free layer, but again this would be detrimental to the integration density.

5. Are there solutions?

We will only discuss three of the solutions under development in industry. The first is based on a scheme proposed by L. Savchenko[10] that lies at the origin of the recent 4 Mb MRAM by Freescale. Here the free magnetic layer is replaced by an artificial antiferromagnetic layer composed of a nonmagnetic coupling layer sandwiched between two ferromagnetic layers. By varying the thickness of the

KV/k_BT	Isolated bit	1 Mb memory
40.00	2.4×10^2	1.3×10^{-1}
45.00	4.5×10^4	4.5×10^{-1}
50.00	5.2×10^6	1.6
65.00	1.7×10^{13}	6.7×10^1
70.00	**2.5×10^{15}**	2.3×10^2
80.00	5.5×10^{19}	2.9×10^3
100.00	2.7×10^{28}	4.2×10^5
120.00	1.3×10^{37}	6.3×10^7
150.00	1.4×10^{50}	1.1×10^{11}
170.00	6.8×10^{58}	1.7×10^{13}
190.00	3.3×10^{67}	**2.5×10^{15}**

Table 3. Number of cycles before random bit reversal for a single isolated bit and a 1 Mb memory taking into account the barrier lowering for half-selected bits.

coupling layer we observe an oscillation between ferro- and antiferromagnetic states. The period of this oscillation (in terms of thickness) depends on the material used as coupling layer. Ruthenium is commonly used because it exhibits a relatively large period and allows for a relaxed constraint on thickness uniformity. Of course, this sandwich is grown on the tunnel oxide barrier that is itself deposited on the ferromagnetic hard layer already pinned by exchange coupling with an antiferromagnetic layer. The writing process also uses two write lines oriented at 45° with respect of the axis of the artificial antiferromagnet. To reverse the magnetization of this system we use the so-called "toggle" sequence illustrated on Fig. 6.

This process is attractive because it overcomes the problem of half-selected bits. Indeed, it is almost impossible for the field produced by a single write line to reverse the magnetization of the two layers that form the artificial antiferromagnet. However, the required magnetic fields remain very high (> 50–60 Oe per write line), leading to the rather high writing currents close to the current density limits already discussed in the context of first-generation memories. An innovation used by Freescale is the cladding of write wires by a ferromagnetic material. However, the gain is only a factor of two, which according to our estimates limits the memory size to the 90 nm technology node.

The second alternative is know as spin torque transfer (STT) and relies on a spin transport model introduced by Slonczewski[11] in 1996. The structure consists of a three-layer stack: ferromagnetic metal FM1/nonmagnetic metal (for example Cu)/ferromagnetic metal FM2. Here FM1 is the hard layer and FM2 is the free layer. As we inject electrons from the top of the structure (*i.e.* from the hard layer), FM1 also acts as a polarizer that redirects the spins of the incoming electrons. In this model, we suppose that the magnetizations of the two ferromagnetic layers (M_1 for FM1 and M_2 for FM2) differ by an angle θ. Thus, the polarization of the electrons that have crossed the FM1 hard layer has an orientation intermediate between M_1 and M_2. The transverse component (with respect to M_2) of these electrons is then transferred to FM2 when the electrons align on it. This phenomenon is equivalent to a torque that acts on M_2. If the injected current is high enough, this torque can rotate the magnetization M_2 of the free FM2 layer.

Companies like Grandis and Sony, have used this effect to fabricate MRAM prototypes, replacing the nonferromagnetic layer by a tunnel oxide in order to get higher series resistances for improved detection. What are the benefits and

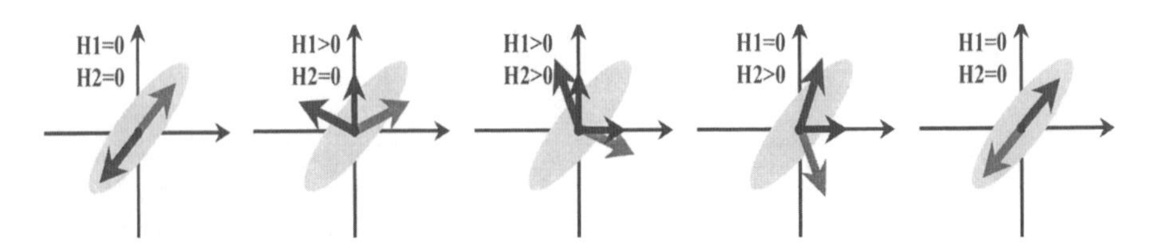

Figure 6. Schematic of magnetization reversal using the toggle sequence.

drawbacks of STT? In STT, no external magnetic field is required, since the rotation is due to the current flowing through the structure. Thus, the problem related to half-selected bits no longer exists. Furthermore, the memory cell structure is greatly simplified, permitting higher integration density compared to classic MRAMs. However, a major drawback lies on the current density needed for reversing the magnetization of the free layer. This value is around 10^7 A/cm^2, a factor of 10 higher than industry can achieve. However, at constant current density, the writing current decreases in proportion to the cell area, *i.e.* as $2F^2$ if we take elliptical cells with a long to short axis ratio of two.

Since the main problem of STT cells is the high current density, some authors[12] have proposed structures in which the free layer is replaced by a hybrid layer composed of two ferromagnetic layers sandwiching an insulating layer pierced with nanochannels filled with the same ferromagnetic metal. It is then possible to increase considerably the local current density, so that the magnetization of the free layer can be totally reversed through the displacement of domain walls. However, the understanding of the mechanism involved in STT is only partial, so it is unclear that magnetization reversal by much lower currents will actually work.

Finally, a third alternative, called thermally assisted switching[13] (TAS) is currently being developed by Renesal Technologies, NVE and a new French company CROCUS. The structure of the device is similar to a first-generation MRAM. However, the free layer is pinned by an antiferromagnetic layer with a relatively low blocking temperature $T_B \sim 250–300$ °C. The switching is assisted by heating the cell with a current pulse. If we go back to Eq. (9) for the total energy of the free layer, the first term is now zero and the anisotropy field is then replaced by the exchange field. As we approach T_B, the energy barrier is drastically reduced, making it possible to write the cell with a weak magnetic field (10–100 times lower than for a standard first generation cell). Furthermore, after the cell is written and cooled down, the energy barrier to magnetization reversal recovers, reaching $KV/k_BT \geq 200$ at the 100 nm technology node. Then, only very intense fields may reverse the magnetization (~1000 Oe). Consequently, there is no risk of half-selected bit reversal, leading to excellent retention.

All of the above-discussed structures rely on memory cell fabrication as part of the back-end process. There have been some attempts[14] to integrate the memory cell at the front-end-of-line level. This supposes we are able to inject spin-polarized electrons into the semiconductor and that they do not lose their polarization during transport in the semiconductor medium. Injection and transport have already been demonstrated in the case of III-V compounds. What happens in silicon? Since the spin-orbit interaction is weaker in silicon than in III-V compounds, we can hope to obtain larger mean free paths. A proposed cell would consist of two identical structures: the injector is a ferromagnetic layer deposited on a tunnel oxide on top of *p*-type silicon; whereas the collector has the same structure but is deposited onto *n*-type silicon at a distance of a few tenths of a micron from the injector. Once the electrons are injected into the semiconductor substrate (in this case Si), they are accelerated towards the collector by the *pn* junction electric field. A change in the current will then be observed, depending on the relative magnetizations of injector and collector. The main advantage of this structure is that it is simple to fabricate, at least in theory. The same ferromagnetic material can be used for the injector and collector. The magnetization reversal is realized by a single connection located in the vicinity of either the injector or collector. Furthermore, one can imagine the same collection cell for many injection cells. However, cell writing requires high currents, just as in the case of first-generation memories. It may be possible to implement solutions developed for classic MRAM cells (*e.g.* TAS).

6. Conclusions

In conclusion, in this chapter we have shown that magnetic memory is a very attractive device because of its properties that make use of the new degree of freedom inherent in the electron spin. Will this device become the universal memory long dreamt of by personal computer users? The model we have used to deduce some of the physical quantities is, of course, very simple, perhaps even too simple. But we believe that it gives the correct order of magnitude of the observed phenomena and future trends. It explains why it takes time to jump over the barrier between the laboratory and production.

Of course, MRAM memories are already commercially available, but they have not yet demonstrated adequate performance in terms of integration density, speed, and power consumption compared to their competitors. We are at the beginning of the story. A lot of work is still necessary to understand all the mechanisms involved in such devices and to quantify them. Already, this memory can be appropriate for some niche markets (*e.g.* automotive applications or applications requiring radiation hardness). We can still hope that MRAM may break out of these niches and surpass its rivals in the future.

References

1. E. W. Pugh, R. A. Henle, D. L. Critchlow, and L. A. Russel, "Solid state memory development in IBM," *IBM J. Res. Develop.* **25**, 585 (1981).
2. L. L. Hinchey and D. L. Mills, "Magnetic properties of superlattices formed from ferromagnetic and antiferromagnetic materials," *Phys. Rev. B* **33**, 3329 (1986).
3. M. N. Baibich, J. M. Broto, A. Fert, *et al.*, "Giant magnetoresistance of (001)Fe/(001)Cr magnetic superlattices," *Phys. Rev. Lett.* **61**, 2472 (1988).
4. S. S. P. Parkin, N. More, and K. P. Roche, "Oscillations in exchange coupling and magnetoresistance in metallic superlattice structures: Co/Ru, Co/Cr, and Fe/Cr," *Phys. Rev. Lett.* **64**, 2304 (1990).
5. See *http://www.ssec.honeywell.com/aerospace/datasheets/hxnv0100_mram.pdf*
6. W. J. Gallagher and S. S. P. Parkin, "Development of the magnetic tunneling junction MRAM at IBM: From first junctions to a 16-Mb demonstrator chip," *IBM J. Res. Develop.* **50**, 5 (2006); *erratum ibid.* **50**, 333 (2006).
7. E. C. Stoner and E. P. Wohlfarth, "A mechanism of magnetic hysteresis in heterogeneous alloys," *Phil. Trans. Roy. Soc. London A* **240**, 599(1948).
8. M. Jullière, "Tunneling between ferromagnetic films," *Phys. Lett. A* **54**, 225 (1975).
9. A. Chaiken, G. Prinz, and J. Krebs, "Spin-valve. Magnetoresistance of uncoupled Fe-Cu-Co sandwiches," *Proc. MMM-Intermag* (1991), pp. 5864–66.
10. M. Durlam, D. Addie, J. Akerman, *et al.*, "A 0.18 μm 4 Mb toggling MRAM," *Tech. Digest IEDM* (2003), pp. 995–7.
11. J. C. Slonczewski, "Current-drive excitation of magnetic multilayers," *J. Magn. Magn. Mater.* **159**, L1 (1996).
12. H. Meng and J.-P. Wang, "Composite free layer for high density magnetic random access memory with lower spin transfer current," *Appl. Phys. Lett.* **89**, 152509 (2006)
13. R. Beech, J. Anderson, A. Pohm, and J. Daughton, "Curie point written magnetoresistive memory," *J. Appl. Phys.* **87**, 6403 (2000).
14. See *http://www.emacproject.com/*

Atomically Controlled Processing for Future Si-Based Devices

Junichi Murota and Masao Sakuraba
Laboratory for Nanoelectronics and Spintronics
RIEC, Tohoku University, 2-1-1 Katahira, Aoba-ku, Sendai 980-8577, Japan

Bernd Tillack
IHP, Im Technologiepark 25, 15236 Frankfurt (Oder), Germany

1. Introduction

Atomically controlled processing for group IV semiconductors has become indispensable for the fabrication of ultrasmall MOS devices for ULSI, because high performance devices require atomically abrupt heterointerfaces and doping profiles, as well as strain engineering due to introduction of Ge into Si. In the fabrication of high performance devices, there are many surface reaction processes, typically by chemical vapor deposition (CVD). Further progress in CVD process technology requires atomic-order surface reaction control, as well as low-temperature processing, in order to suppress thermal degradation such as unexpected reactions and impurity diffusion. Improvements in the quality of gases and equipment have enabled ultraclean low-temperature processing for atomic-order control.[1-3]

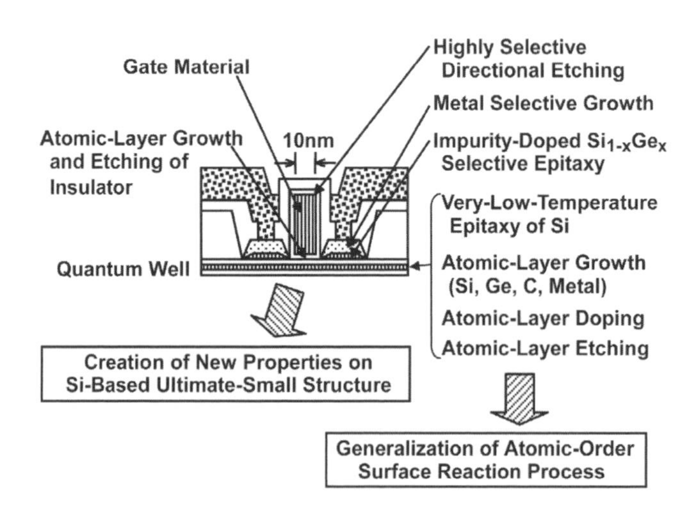

Figure 1. Atomically controlled processing for group IV semiconductors for ultrasmall and nanodevices.

 Future Trends in Microelectronics. Edited by Serge Luryi, Jimmy Xu, and Alex Zaslavsky

Our concept of atomically controlled processing for group IV semiconductors is based on atomic-order surface reactions. The final goal is the generalization of the atomic-order surface reaction processes and the creation of new properties in Si-based ultimate small structures, leading to nanometer scale Si devices as well as Si-based quantum devices – the process flow is illustrated in Fig. 1. Based on the investigation of surface reaction processes, the concept of atomic layer process control[3-7] has been demonstrated for high performance $Si_{0.65}Ge_{0.35}$ channel *p*-type MOSFETs with a 0.12 μm gate length by utilizing *in-situ* impurity-doped $Si_{1-x}Ge_x$ selective epitaxy on the source/drain regions at 550 °C,[8] for ultrathin P barriers in infrared SiGe/Si heterojunction internal photoemission detectors,[9] and for B and P base doping in *npn* and *pnp* HBTs.[7,10] Additionally, for *in-situ* doped $Si_{1-x}Ge_x$ epitaxial growth on the (100) surface in a SiH_4–GeH_4–dopant (PH_3 or B_2H_6 or SiH_3CH_3)–H_2 gas mixture, the deposition rate, the Ge fraction and the dopant concentration have been expressed quantitatively by modified Langmuir-type rate equations.[4,5]

In this chapter, surface reactions of hydride gases on Si(100) and Ge(100) for atomic-order growth is reviewed based on the Langmuir-type adsorption and reaction scheme. Further, we will discuss typical atomic-layer doping followed by epitaxial growth of Si or SiGe on the (100) surface using ultraclean low-pressure CVD, as well as atomically controlled CVD processing. We will conclude by discussing atomically controlled processing using electron-cyclotron resonance (ECR) plasma.

2. Atomically controlled CVD processing

- *Atomic-order reaction of hydride gases on Si(100) and Ge(100) surfaces*

Hydrogen termination of Si(100) and Ge(100) surfaces can be desorbed by heating the substrate.[11] Hydrogen desorption initiates adsorption and reaction of reactant gases, with self-limiting characteristics due to Langmuir-type kinetics observed in many cases. Self-limiting conditions for some hydride gases on the Si(100) and Ge(100) surfaces are summarized in Table 1.

In the case of SiH_4 and GeH_4 adsorption on monohydride Si(100) and Ge(100) surfaces respectively,[3] the adsorption proceeds according to Langmuir-type kinetics illustrated in Fig. 2(a), but the adsorbed SiH_4 and GeH_4 do not react and desorb when SiH_4 and GeH_4 gases are removed from the reactor. The reaction must be induced using a method such as flash heating during an interval sufficiently short to prevent the next adsorption cycle.

In the other cases shown in Table 1, hydride molecules are adsorbed and react simultaneously on the surface as shown in Fig. 2(b) according to Langmuir-type kinetics. In the case of low-temperature atomic-order surface nitridation of Si using NH_3 at 400 °C shown in Fig. 2(c),[4] the experimental data are well described by the equations in Fig. 2(b). It has been suggested that NH_3 molecules are adsorbed on the hydrogen-terminated Si surface after wet cleaning, following the

Gas/substrate	**Conditions** (AL = atomic layer)
SiH_4 / Si	385 °C, 100-500 Pa, flash, adsorption/desorption equilibrium
SiH_4 / Ge	260 °C, 10-500 Pa, thermal, 1 AL
GeH_4 / Ge	268 °C, 2.9-13 Pa, flash, 1 AL
NH_3 / Si	400 °C, 124-1400 Pa, thermal, 2 AL, Flash, 4 AL
PH_3 / Si	450 °C, 0.26 Pa, thermal, 3 AL
PH_3 / Ge	300 °C, 0.26 Pa, thermal, 1 AL
CH_4 / Si	600 °C, 50-1600 Pa, thermal, 2 AL
SiH_3CH_3 / Si	450 °C, 18 Pa, thermal, 1 AL
SiH_3CH_3 / Ge	450 °C, 18 Pa, thermal, 1 AL

Table 1. Typical self-limiting conditions for hydride gases on Si(100) and Ge(100).[4]

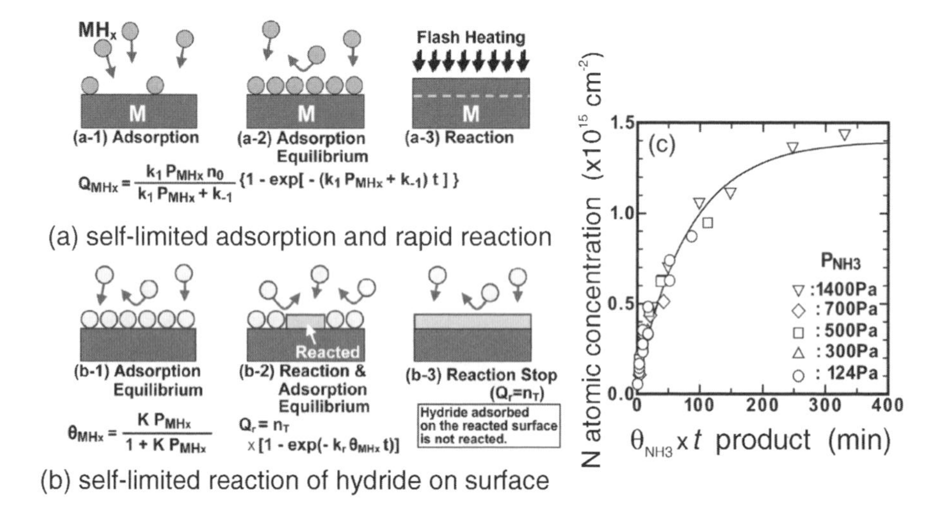

Figure 2. Schematic images of (a) self-limited adsorption and (b) self-limited reaction of hydride for atomic-order growth based on Langmuir-type model. (c) Product of NH_3 coverage θ_{NH3} and time *t vs.* N atomic concentration on the wet-cleaned Si(100) at various NH_3 pressures. Substrate temperature is 400 °C. Solid curve is calculated from the modified Langmuir-type equations in Fig. 2(b).

Langmuir's adsorption isotherm, but do not react without hydrogen desorption. On a hydrogen-free Si surface formed by preheating in Ar at 650 °C, the N atomic concentration increases spontaneously up to about 2×10^{14} cm^{-2} and further nitridation proceeds with hydrogen desorption. This is similar to NH_3 reaction on a hydrogen-terminated surface. The total reaction site density appears to depend on the temperature, because the self-limited N atomic concentration is 2.7×10^{15} cm^{-2}

at 500–650 °C and 5.4×10^{15} cm^{-2} at 750–800 °C (4 and 8 atomic layers, respectively).

Self-limited reactions of SiH_4 on Ge(100),[12] CH_4 on Si(100)[13] and SiH_3CH_3 on Si(100) and Ge(100)[14] are also described by the equations in Fig. 2(b). In particular, in the case of the SiH_4 reaction, it was found that single atomic layer growth of Si occurs for the hydrogen free surface formed by preheating at 350 °C in Ar, and for the hydrogen-terminated surface with the dimer structure formed by preheating at 350 °C in H_2. The density of the SiH_4 reaction sites on the hydrogen-terminated Ge surface with the dimer structure is lower than that on the hydrogen-free surface. In the case of SiH_3CH_3 reaction, it appears that SiH_3CH_3 is adsorbed without breaking the Si-C bond at 400–500 °C.

In the case of PH_3 reaction on Si(100) and Ge(100),[15] we find that the P concentration formed on the surface depends on the PH_3 exposure temperature as shown in Fig. 3. The PH_3 reaction is suppressed on the hydrogen-terminated Si and Ge surfaces, but PH_3 is adsorbed dissociatively on the hydrogen-free Si and Ge surfaces at 300 and 200 °C, respectively. As a result, the P concentration on the surface tends to saturate below one AL. On the Ge surface at 300–450 °C, the P concentration tends to saturate at about one AL. Furthermore, P desorption from Ge surface occurs at 450 °C, but not at 300 °C.[15] On the Si surface at 450–750 °C, the P concentration tends to saturate at about 2 or 3 ALs. At 450 °C, the P concentration is independent of PH_3 partial pressure (0.087–0.78 Pa). Looking at Fig. 3 in more detail, the P concentration is about one AL in early stage. Additionally, at 650 °C, thermal desorption of P and reduction of hydrogen decreases the P concentration to about one AL.[15] Therefore, we believe that PH_3 is self-limited to one-atomic-layer adsorbed on the Ge surface sites.

These results indicate that the reaction site density depends on not only substrate surface structure, but also hydride gas species and substrate temperature.

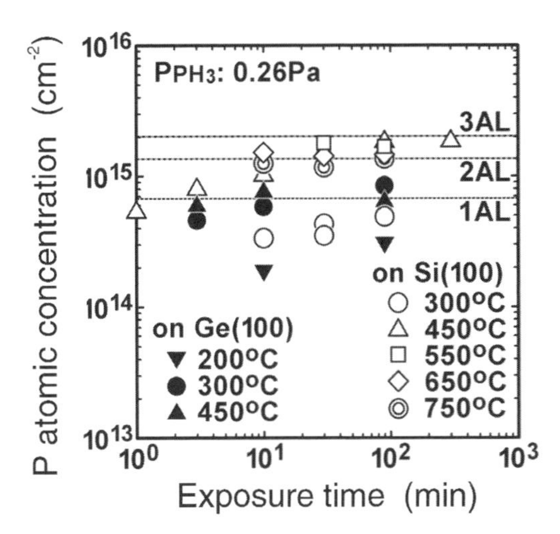

Figure 3. PH_3 exposure time dependence of the P atomic concentration on Si(100) and on Ge(100).

- *Atomic layer doping in Si and SiGe epitaxial growth*

Atomic layer doping is performed by epitaxial growth over the material formed on Si(100) or SiGe(100) surface.[4-7,16-19] The epitaxial growth of Si/0.5 AL of N/Si(100),[5,17] Si/0.5 AL of P/Si(100),[5,18] SiGe/1 AL of B/SiGe(100),[7,16] SiGe/0.8 AL of C/SiGe(100),[6] and Si/0.03 AL of W/Si(100)[19] have been reported. In this section, we review the atomic layer doping of N, P and B.

In the case of Si/0.5 AL of N/Si(100) growth, the Si film is epitaxially grown by SiH_4 exposure at 500 °C on the nitrided Si(100) surface formed by NH_3 reaction at 400 °C. Most of the N atoms are buried in the initially nitrided region within the thickness of about 1 nm within the measurement accuracy, as shown in Fig. 4(a). It should be noted that N atoms tend to segregate at the grown surface with increasing Si growth temperature. If the N amount is 6×10^{14} cm^{-2}, amorphous Si is grown, probably caused by the generation of Si_3N_4. We have achieved high-quality epitaxial growth of multi-layer N-doped Si films composed of 3×10^{14} cm^{-2} N layers and 3.0 nm thick Si spacers. In such N AL-doped Si films, the N atoms act as donors. The typical donor activation ratio is about 0.4% at the N amount in a 5×10^{13} cm^{-2}/layer. Ionization energy of the donor level is about 150–180 meV. The Hall mobility is much larger than the ~100 cm^2/V·s value expected at 300 K from uniformly P-doped Si with P concentration of 10^{19} cm^{-3}, reaching instead values typical of 10^{16}–10^{17} cm^{-3} bulk doping, see Fig. 4(b). Since the local concentration of the ionized donors is about 10^{18}–10^{19} cm^{-3} in 1 nm-thick N doped region, we believe that scattering in electron transport is reduced by AL doping. Moreover, there is a possibility that the carrier mobility is enhanced by highly

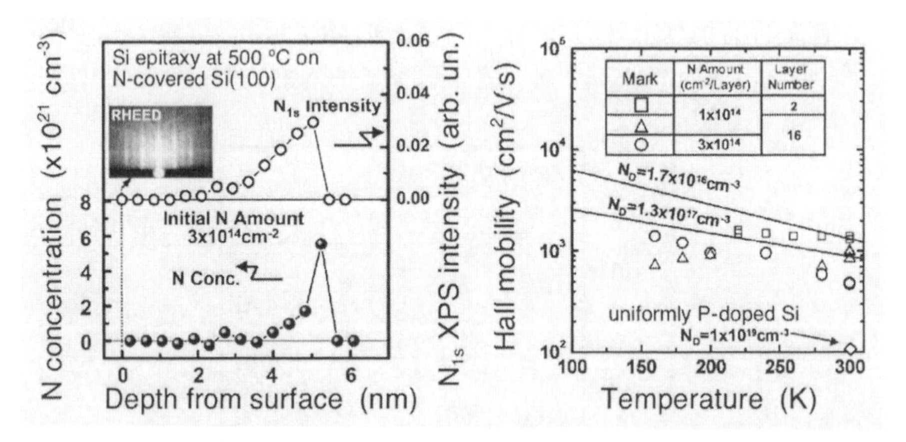

Figure 4. (a) Depth dependence of the N_{1s} intensity measured by XPS and depth profile of N concentration for the AL-doped Si film with the 3×10^{14} cm^{-2} N concentration. The total N amount for the doped region is roughly 3×10^{14} cm^{-2}. (b) Measured temperature dependence of electron Hall mobility in the N AL-doped Si films. Solid lines indicate Hall mobility of the bulk doped *n*-Si.[20] Mobility of uniformly P-doped Si epitaxial film grown by LPCVD on Si(100) with 10^{19} cm^{-3} concentration is also shown for comparison (open diamond).

concentrated strain near the N AL-doped region. Since N atoms in delta-doped regions tend to diffuse and a part of them segregates at the surface at 750 °C, very low temperature subsequent processing is required for the device fabrication.

In the case of Si growth on the (100) surface with P amount of 2×10^{15} cm^{-2} at a SiH_4 partial pressure of 6 Pa at 500 °C,[18] the surface P concentration decreases with increasing SiH_4 exposure time without Si deposition. After the P concentration falls below one AL due to the SiH_4 reaction, Si growth begins and P atoms segregate onto the surface. As a result, AL P doping below 10^{14} cm^{-2} can be achieved in $Si_{1-x}Ge_x$ epitaxial growth at above 500 °C by using reduced pressure CVD in a single wafer reactor.[16] At a rather low temperature of 450 °C and a rather high SiH_4 partial pressure of 220 Pa, P incorporation into the Si film and the growth of heavily P delta-doped Si film have been achieved without an incubation period of Si growth,[18] although the tailing towards surface was also observed. The SiH_4 reaction is suppressed in the P-doped Si deposition by using SiH_4-PH_3, but not by Si_2H_6-PH_3.[21] The Si_2H_6 is produced due to polymerization of SiH_4 and increases in proportion to the square of the SiH_4 partial pressure.[22] In the case of 220 Pa partial pressure of SiH_4, Si deposition may be enhanced by Si_2H_6 produced from SiH_4, because almost no Si deposition is observed at a SiH_4 partial pressure of 220 Pa. By this technique, heavily P-doped epitaxial Si films on Si(100) with average P concentration of 6×10^{20} cm^{-3} can be formed with 7 nm Si spacers, as shown in Fig. 5(a). The average carrier concentration reaches a value as high as 3.6×10^{20} cm^{-3} and the resistivity as low as 2.7×10^{-4} Ω·cm, as shown in Fig. 5(b).

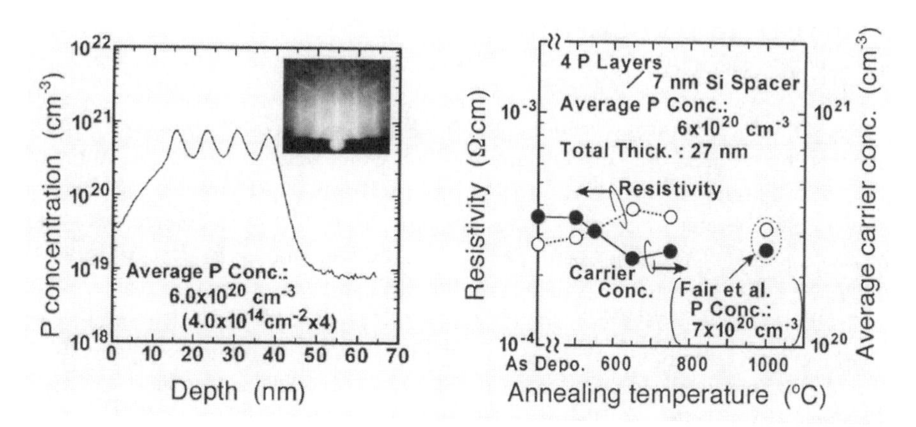

Figure 5. (a) Depth profile of the P concentration in the four layers of P-doped epitaxial Si film measured by SIMS. Atomic amount of adsorbed P is 1.4×10^{15} cm^{-2} for each layer. For subsequent Si growth, the SiH_4 partial pressure was 220 Pa and the temperature was 450 °C. (b) Annealing temperature dependence of carrier concentration and resistivity of the P-doped epitaxial Si films from the same batch as in (a); annealing time was 60 minutes.

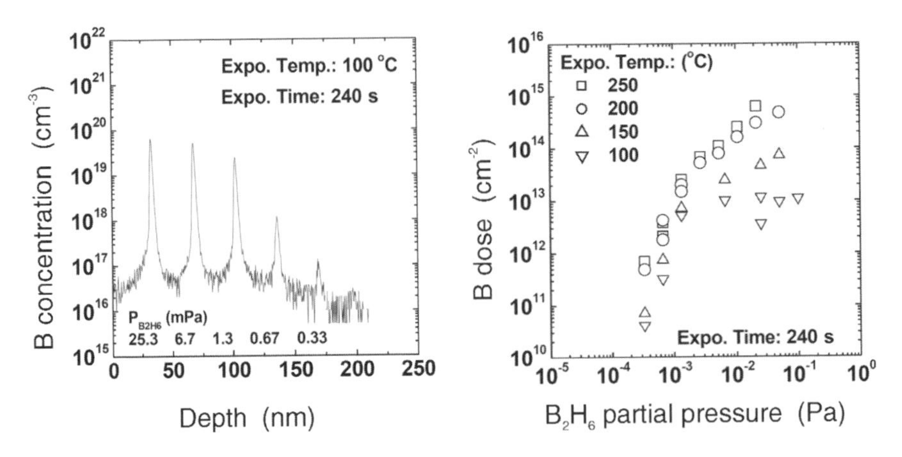

Figure 6. (a) SIMS profile of B peaks in $Si_{0.8}Ge_{0.2}$ grown at different B_2H_6 partial pressures during 240 s exposure at 100 °C in N_2 (total pressure was 10640 Pa). (b) B dose in $Si_{0.8}Ge_{0.2}$ *vs.* B_2H_6 partial pressures for 240 s exposure between 100 and 250 °C (total pressure was 10640 Pa).

However, annealing the material above 550 °C causes a decrease in carrier concentration and an increase in resistivity. They become close to those of the P-doped Si film formed by P diffusion at 1000 °C.[23] We believe that low-resistivity P-doped Si films result from a higher Si deposition rate that suppresses electrically inactive P formation mechanisms, such as P clustering. Using AL doping, a very low contact resistivity of about 5×10^{-8} $\Omega\cdot cm^2$ between Ti and the Si film has been obtained.[25]

Atomic level control of B doping of $Si_{1-x}Ge_x$ and $Si_{1-x}Ge_x$:C films can also be obtained by reduced-pressure CVD in a single wafer reactor. At temperatures of 350 °C and higher,[6,25] the B AL process is not self-limited. Boron atomic concentration incorporated into the SiGe epitaxial films (B dose) increases with increasing B_2H_6 partial pressure without saturation, and a B dose of several monolayers can be obtained.[10] This also implies that B adsorption occurs at Si, Ge and B surface sites, consistent with the fact that B tends to form clusters in $Si_{1-x}Ge_x$ at high B concentration.[25] If the B dose is higher than one monolayer, B tends to be incorporated as electrically inactive. Figure 6(a) shows the secondary ion mass spectroscopy (SIMS) profile of B peaks fabricated at different B_2H_6 partial pressures at exposure temperature of 100 °C. To increase the B dose at this low temperature, the hydrogen termination of the Si surface was prevented by cooling down under N_2 to 100 °C after the $Si_{1-x}Ge_x$ epitaxy and performing the exposure at low temperature in N_2.[7] Very steep B profiles of 1 nm/decade have been obtained. Figure 6(b) shows the B dose as a function of the B_2H_6 partial pressure for different exposure temperatures. In contrast to the exposure at higher temperature, where no self-limitation is observed, there are indications of saturation at low temperature. By lowering the temperature of the adsorption step from 250 °C to 100 °C, saturation behaviour of the B dose at high B_2H_6 partial pressures is visible.

The process tends to become self-limited at the lowest temperature used. Compared with the adsorption at higher temperature, the adsorption of B_2H_6 at Si and Ge sites seems to be favourable. As a result of this change in the adsorption mechanism, one would expect a higher proportion of electrically active B at very low temperatures.

3. Atomically controlled plasma processing

- *Si and Ge epitaxial growth on Si(100) by ECR Ar plasma enhanced CVD*

Si and Ge epitaxial growth was performed using ultraclean ECR Ar plasma enhanced reaction without substrate heating. The system is schematically shown in Fig. 7(a).[26-28] Argon plasma was generated by supplying microwave power (2.45 GHz) at Ar pressure of 2.1 Pa. For 200 W microwave power, flux density and peak energy of the incident Ar ions were about 3×10^{15} cm^{-2} s^{-1} and 2 eV, respectively, and substrate temperature was below 50 °C during the 10 minute plasma exposure. For 800 W microwave power, the corresponding values were 8×10^{16} cm^{-2} s^{-1} and 3 eV, respectively, and substrate temperature increased to above 300 °C at 10 minutes.[26]

Typical RHEED patterns of the deposited Si and Ge films are shown in Fig. 7(b). Clear streaks are observed. It should be noted that crystallinity degradation tends to proceed with incorporation of Ar atoms (above 2×10^{21} cm^{-3}) in the deposited Si film.[26] In the case of GeH_4 reaction on Si(100) at lower microwave power, the deposited Ge film has an atomically flat surface with a roughness of about 0.1–0.2 nm. For the higher microwave power, the surface roughness is relatively larger. We believe the rough surface is due to the higher substrate temperature during the deposition.

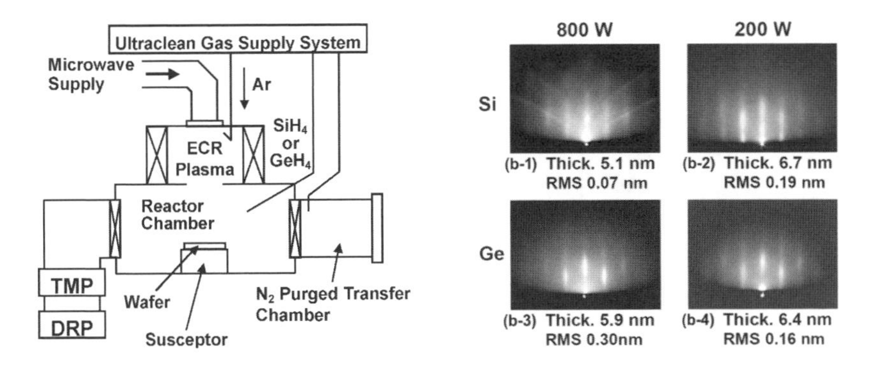

Figure 7. (a) Schematic diagram of an ultraclean ECR plasma enhanced CVD system (<10^{-5} Pa ultimate vacuum, TMP and DRP are turbomolecular and dry pumps). (b) RHEED patterns from [011] azimuth of Si and Ge films grown on Si(100). Thickness and root mean square (RMS) value of surface roughness are also shown.

	Pressure	5.12 Pa	2.56 Pa	1.28 Pa
Indirect	Energy distribution (eV)	0–5	0–12	0–18
	Peak energy (eV)	1.5	5	3, 10
	J_i ($cm^{-2}\,s^{-1}$)	3.6×10^{9}	9.2×10^{10}	5.1×10^{11}
	I_N (arb. unit)	1	4	16
Direct	Energy distribution (eV)	0–16	0–28	0–37
	Peak energy (eV)	3, 10	10, 21	9, 28
	J_i ($cm^{-2}\,s^{-1}$)	1.8×10^{12}	4.4×10^{13}	4.6×10^{14}
	I_N (arb. unit)	90	400	1300

Table 2. Energy distribution, peak energy of the incident ions, the flux density (J_i), and the relative optical emission intensity (I_N) *vs.* ECR nitrogen plasma conditions. When directional incidence of ions is suppressed ("indirect exposure"), a shutter in front of the substrate was closed compared to "direct exposure".

- *Atomic-order nitridation of Si(100) by ECR nitrogen plasma*

Atomic-order plasma nitridation of Si(100) was performed by ECR nitrogen plasma.[29,30] Energy distribution, peak energy of the incident ions, flux density and relative optical emission intensity I_N from the typical N_2 radical (337.1 nm: $C^3\Pi_u$–$B^3\Pi_g$ second positive system) are summarized in Table 2. The dependence of the N atomic concentration on the product of I_N and plasma exposure time t is shown in Fig. 8. Below 1–2 AL the N concentration correlates with the I_N x t product. Therefore, in the initial stages of nitridation, the N radicals are the major species to contribute to the plasma nitridation of Si. Assuming that the radicals

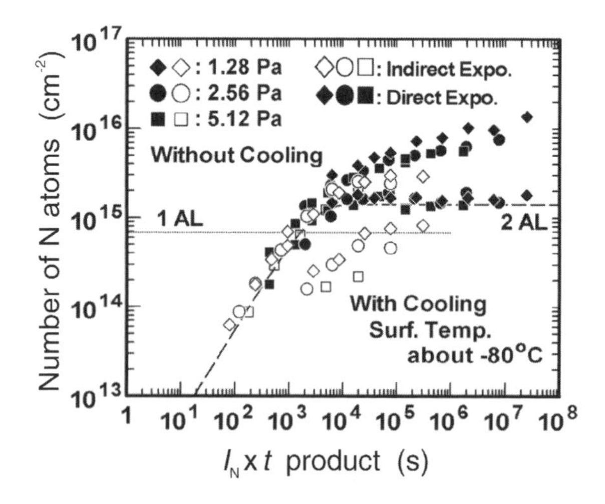

Figure 8. Dependence of the N atomic amount on the product of the relative optical emission intensity I_N and the plasma exposure time t. Dashed line corresponds to Eq. (1), 1 AL (6.78×10^{14} cm^{-2}) and 2 AL (1.36×10^{15} cm^{-2}) of Si(100) are also indicated.

react only with surface atoms and the reaction proceeds according to the Langmuir-type kinetics neglecting desorption, the N atomic concentration (n_N) is given by:

$$n_N = n_S \cdot \{1 - \exp(-k_N \cdot \alpha \cdot I_N \cdot t)\} , \quad (1)$$

where n_S is the saturated N atomic amount, k_N is reaction rate constant of the radicals, and α is the constant converting I_N to number of the incident radicals. The broken line in Fig. 8 is obtained by fitting the experimental data to Eq. (1), where n_S is set to 1.36×10^{15} cm^{-2} (2 AL). In all the cases, good agreement is observed in the initial nitridation region provided the substrate is cooled. Without cooling, the saturated N amount is over 2 AL, and deviates from Eq. (1). The deviation can be caused by contribution of the incident ions, because the incident ions are expected to react with deeper atoms below the surface.

- *Epitaxial growth of N delta-doped Si films by ECR plasma enhanced CVD*

Atomic-order nitridation of Si(100) and subsequent Si epitaxial growth on the nitrided Si(100) were carried out without substrate heating. Argon and SiH_4 were supplied into the plasma-generation and reactor chambers, respectively. Depth profiling of N and Ar concentrations was obtained from repeated XPS measurements combined with sub-nanometer wet etching. The results for Si epitaxial films

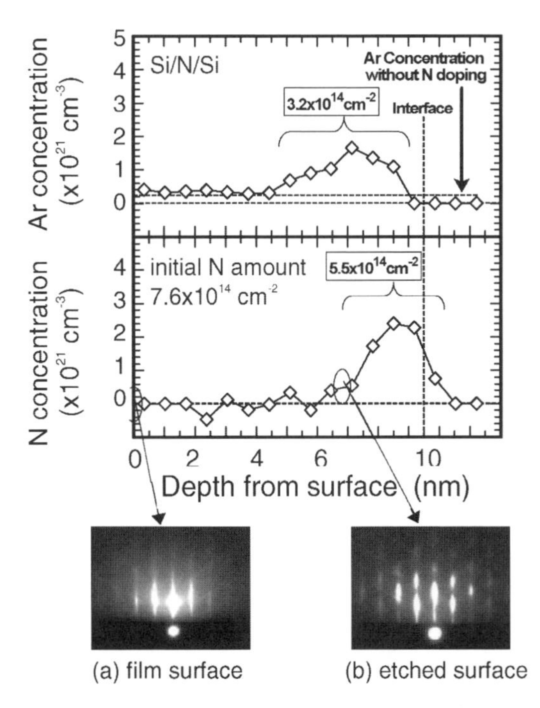

Figure 9. Depth profile of N and Ar concentration for the Si epitaxial film deposited on the 7.6×10^{14} cm^{-2} (1.1 AL) nitrided Si(100). Bottom pictures are RHEED patterns taken from (a) the deposited Si film surface and (b) the partially etched surface at a depth of 6.8 nm.

deposited on the nitrided Si(100) are shown in Fig. 9.[31] Most of the N atoms are confined to a roughly 3 nm thick region. Since peak position for the N concentration is shifted towards the Si capping layer, we believe N atoms tend to segregate during the Si deposition. The incorporated N atoms are confined to a 3 nm thick region and the total N dose reaches as high as 5.5×10^{14} cm^{-2} (0.8 AL). This amount is much larger than the value obtained by the thermal CVD described above. The Ar concentration peak shifts further than the N concentration peak, and the incorporated Ar dose is much lager than that for a Si film without N doping.

4. Conclusions

Self-limiting formation of 1–3 atomic layers of group IV or related atoms by thermal adsorption and reaction of hydride gases on Si(100) and Ge(100) is generalized based on a Langmuir-type model. Atomic layer doping of N, P and B can be achieved by the epitaxial growth of Si and SiGe over the material formed on (100) surfaces. Atomic layer doping results indicate that new group IV semiconductors of very high carrier concentration and higher mobility can be prepared compared to doping under equilibrium conditions. Our results indicate that atomic layer-by-layer epitaxy of group IV materials as well as atomic layer doping are possible with well-controlled initiation of the reaction governed by Langmuir-type self-limited kinetics in many cases, and open the way to atomically controlled CVD technology for ULSI.

By using ECR Ar plasma-enhanced CVD, Si and Ge epitaxial growth from SiH_4 and GeH_4, respectively, are achieved on Si(100) without substrate heating. Results of nitrogen plasma irradiation of Si(100) suggest that the nitridation of the Si atoms below the surface is enhanced with increasing ion energy as well as the Si surface temperature. Silicon epitaxy on atomic-order nitrided Si(100) can also achieved without substrate heating, with the result that about 0.8 atomic layers of N can be confined within an approximately 3 nm thick region. These results open the way to room-temperature atomically controlled processing.

References

1. B. S. Meyerson, "Low-temperature silicon epitaxy by ultrahigh vacuum/ chemical vapor deposition," *Appl. Phys. Lett.* **48**, 797 (1986).
2. J. Murota, N. Nakamura, M. Kato, N. Mikoshiba, and T. Ohmi, "Low-temperature silicon selective deposition and epitaxy on silicon using the thermal decomposition of silane under ultraclean environment," *Appl. Phys. Lett.* **54**, 1007 (1989).
3. J. Murota and S. Ono, "Low-temperature epitaxial growth of $Si/Si_{1-x}Ge_x/Si$ heterostructure by chemical vapor deposition," *Jpn. J. Appl. Phys.* **33**, 2290 (1994).

4. J. Murota, T. Matsuura, and M. Sakuraba, "Atomically controlled processing for group IV semiconductors," *Surf. Interface Anal.* **34**, 423 (2002).
5. J. Murota, M. Sakuraba, and B. Tillack, "Atomically controlled processing for group IV semiconductors by chemical vapor deposition," *Jpn. J. Appl. Phys.* **45**, 6767 (2006).
6. B. Tillack, B. Heinemann, and D. Knoll, "Atomic layer doping of SiGe – Fundamentals and device applications," *Thin Solid Films* **369**, 189 (2000).
7. B. Tillack, Y. Yamamoto, D. Bolze, *et al.*, "Atomic layer processing for doping of SiGe," *Thin Solid Films* **508**, 279 (2006).
8. D. Lee, S. Takehiro, M. Sakuraba, J. Murota, and T. Tsuchiya, "Fabrication of 0.12- m pMOSFETs on high Ge fraction $Si/Si_{1-x}Ge_x/Si(100)$ heterostructure with ultrashallow source/drain formed using B-doped SiGe CVD," *Appl. Surf. Sci.* **224**, 254 (2004).
9. R. Banisch, B. Tillack, M. Pierschel, K. Pressel, R. Barth, D. Krüger, and G. Ritter, "A novel infrared SiGe/Si heterojunction detector with an ultrathin phosphorus barrier grown by atomic layer deposition," *Proc. MRS Symp.* **450**, 213 (1997).
10. B. Tillack, Y. Yamamoto, D. Knoll, B. Heinemann, P. Schley, B. Senapati, and D. Krüger, "High performance SiGe:C HBTs using atomic layer base doping," *Appl. Surf. Sci.* **224**, 55 (2004).
11. M. Sakuraba, T. Matsuura, and J. Murota, "H-termination on Ge(100) and Si(100) by diluted HF dipping and by annealing in H_2," in: J. Ruzyllo and R. E. Novak, eds., *Proc. 5th Intern. Symp. Cleaning Technol. Semicond. Dev. Manufact.*, Pennington, NJ: The Electrochemical Society, 1997, p. 213.
12. T. Watanabe, M. Sakuraba, T. Matsuura, and J. Murota, "Atomic-layer surface reaction of SiH_4 on Ge(100)," *Jpn. J. Appl. Phys.* **36**, 4042 (1997).
13. A. Izena, M. Sakuraba, T. Matsuura, and J. Murota, "Low-temperature reaction of CH_4 on Si(100)," *J. Crystal Growth* **188**, 131 (1998).
14. T. Takatsuka, M. Fujiu, M. Sakuraba, T. Matsuura, and J. Murota, "Surface reaction of CH_3SiH_3 on Ge(100) and Si(100)," *Appl. Surf. Sci.* **162-163**, 156 (2000).
15. Y. Shimamune, M. Sakuraba, T. Matsuura, and J. Murota, "Atomic-layer adsorption of P on Si(100) and Ge(100) by PH_3 using an ultraclean low-pressure chemical vapor deposition," *Appl. Surf. Sci.* **162-163**, 388 (2000).
16. B. Tillack, "Atomic control of doping during SiGe epitaxy," *Thin Solid Films* **318**, 1 (1998).
17. Y. Jeong, M. Sakuraba, and J. Murota, "Electrical properties of N atomic layer doped Si epitaxial films grown by ultraclean low-pressure chemical vapor deposition," *Mater. Sci. Semicond. Process.* **8**, 121 (2005).
18. Y. Shimamune, M. Sakuraba, J. Murota, and B. Tillack, "Formation of heavily P-doped Si epitaxial film on Si(100) by multiple atomic-layer doping technique," *Appl. Surf. Sci.* **224**, 202 (2004).
19. T. Kurosawa, T. Komatsu, M. Sakuraba, and J. Murota, "Electrical properties of W delta-doped Si epitaxial films grown on Si(100) by ultraclean low-

pressure chemical vapor deposition," *Mater. Sci. Semicond. Process.* **8**, 125 (2004).
20. S. S. Li and W. R. Thurber, "The dopant density and temperature dependence of electron mobility and resistivity in *n*-type silicon," *Solid State Electronics* **20**, 609 (1977).
21. S. Nakayama, H. Yonezawa, and J. Murota, "Deposition of phosphorus doped silicon films by thermal decomposition of disilane," *Jpn. J. Appl. Phys.* **23**, L493 (1984).
22. T. Morie and J. Murota, "Trench coverage characteristics of polysilicon deposited by thermal decomposition of silane," *Jpn. J. Appl. Phys.* **23**, L482 (1984).
23. R. B. Fair and J. C. C. Tsai, "A quantative model for the diffusion of phosphorus in silicon and the emitter dip effect," *J. Electrochem. Soc.* **124**, 1107 (1977).
24. J. Noh, M. Sakuraba, J. Murota, S. Zaima, and Y. Yasuda, "Contact resistivity between tungsten and impurity (P and B)-doped $Si_{1-x-y}Ge_xC_y$ epitaxial layer," *Appl. Surf. Sci.* **212-213**, 679 (2003).
25. B. Tillack, P. Zaumseil, G. Morgenstern, D. Krüger, and G. Ritter, "Strain compensation in $Si_{1-x}Ge_x$ by heavy boron doping," *Appl. Phys. Lett.* **67**, 1143 (1995).
26. D. Muto, M. Sakuraba, T. Seino, and J. Murota, "Argon plasma irradiation effects in atomically controlled Si epitaxial growth," *Appl. Surf. Sci.* **224**, 210 (2004).
27. K. Sugawara, M. Sakuraba, and J. Murota, "Atomically controlled Ge epitaxial growth on Si(100) in Ar-plasma-enhanced GeH_4 reaction," *Mater. Sci. Semicond. Process.* **8**, 69 (2005).
28. K. Sugawara, M. Sakuraba, and J. Murota, "Thermal effect on strain relaxation in Ge films epitaxially grown on Si(100) using ECR plasma CVD," *Thin Solid Films* **508**, 143 (2006).
29. T. Seino, T. Matsuura, and J. Murota, "Contribution of radicals and ions in atomic-order plasma nitridation of Si," *Appl. Phys. Lett.* **76**, 342 (2000).
30. T. Seino, D. Muto, T. Matsuura, and J. Murota, "Thermal effects in atomic-order nitridation of Si by a nitrogen plasma," *J. Vac. Sci. Technol. B* **20**, 1431 (2002).
31. M. Mori, T. Seino, D. Muto, M. Sakuraba, and J. Murota, "Si epitaxial growth on atomic-order nitrided Si(100) using electron cyclotron resonance plasma," *Mater. Sci. Semicond. Process.* **8**, 65 (2005).

Ultimate VLSI Clocking Using Passive Serial Distribution

M. Banu and V. Prodanov
MHI Consulting LLC, Murray Hill, NJ 07974, U.S.A.

1. Introduction

The quest for absolutes is perhaps as old as scientific thought. As a popular tradition goes, thousands of years ago Archimedes recognized the great value of a reference by his celebrated observation that he could move the Earth if given a fixed point. In the modern age, an arguably even more powerful feat, the development of advanced digital machines, was made possible by designing massive synchronous logic switching in unison to the beat of a global timing reference: the system clock. The machine miniaturization into VLSI chips followed the same approach.

In the beginning of the VLSI era, when the chips were still small by today's standards and the system clock was running at sub- and low-MHz speeds, the practical realization of the timing backbone was relatively easy. The initial VLSI design methodology treated timing almost as an afterthought. As the chips got much larger and the clock speed increased to the GHz range, the correct logic timing became increasingly difficult, to the point of turning into a major design bottleneck. Currently, highly-specialized engineering teams helped by proprietary sophisticated computer programs labor long hours to configure clock distribution solutions known as "timing trees", which are customized and fine-tuned for every particular VLSI chip. The typical end result is a highly complex clocking network claiming substantial resources in chip area and power dissipation, and providing little room for modeling inaccuracies and manufacturing tolerances. Today, the traditional VLSI clocking technology operates close to its performance limits and clearly does not support the modern trend for rapid and inexpensive VLSI product development based on reusable intellectual property (IP).[1]

A new concept called bidirectional signaling or BDS was proposed recently,[2,3] enabling a revolutionary development in VLSI clock distribution. Theoretically, this development not only promises substantially better jitter and power dissipation performance, especially at sub-100 nm IC fabrication technology nodes, but also initiates a paradigm shift in VLSI design methodology. The successful practical demonstration of this new clock distribution technology will transform the VLSI clocking from a "black art" into a routine task, which could be automatically executed by a computer program.

The remaining of this chapter is organized as follows. We start with basic concepts. Then, we briefly review the VLSI clock distribution problem and

Future Trends in Microelectronics. Edited by Serge Luryi, Jimmy Xu, and Alex Zaslavsky
ISBN 0-471-48 © 2007 John Wiley & Sons, Inc.

identify some fundamental shortcomings of traditional designs. Next, we discuss recently proposed alternative schemes, which correct the traditional limitations but introduce new constraints. Finally, we describe the BDS technology and show its advantageous application to this area.

2. Clock synchronization, clock distribution, and timing

As a prerequisite, it seems necessary to clarify the concepts of clock synchronization, clock signal distribution (or simply "clock distribution"), and timing, which are frequently used loosely and interchangeably. These are related but separate concepts. Clock synchronization assumes the presence of independent time-measuring devices called clocks placed at different points in space. Classical and relativistic physics postulate that clocks can be synchronized and tests are prescribed to define/verify when this is the case. In effect, this assumes the existence of an absolute time, which can be measured consistently at any point in space. Clock distribution refers to a single time measuring device whose output is distributed to other points in space through communication channels, such that if the distributed local times had come from local clocks, those clocks would be synchronized. Timing in a clocked system, such as a digital VLSI chip, is switching every node of the system according to a set of absolute time beats.

In general, system timing can be accomplished either by using synchronized local clocks or by clock distribution. In the case of VLSI chips, it would be totally impractical to place millions of synchronized oscillators everywhere on the chip. Instead, clock distribution from a master oscillator is the natural way of ensuring proper timing.

3. Traditional VLSI clock distribution

Figure 1 illustrates a common principle used in VLSI clock distribution[4-6] for the simple case of only two client gates placed far away from the master clock generator and from each other. The object of this network is to switch the client gates synchronously, *i.e.* according to an absolute time reference. The distances between the generator and the gates are so long that simple wires cannot be used to interconnect the components due to excessive parasitic capacitances limiting the clock frequency severely. Hence, buffer/amplifiers are placed appropriately to regenerate the signals and transmit as high a clocking frequency as possible.

The key condition this network must meet is having equal propagation times through the two paths. Otherwise, the client gates would be clocked at different instances, the resulting timing error being known as the clock skew. The obvious way to obtaining equal propagation times or zero skews is to use two physically identical paths. The inherent capability of IC technology for excellent device matching is well documented in the case of side-by-side component placement. In

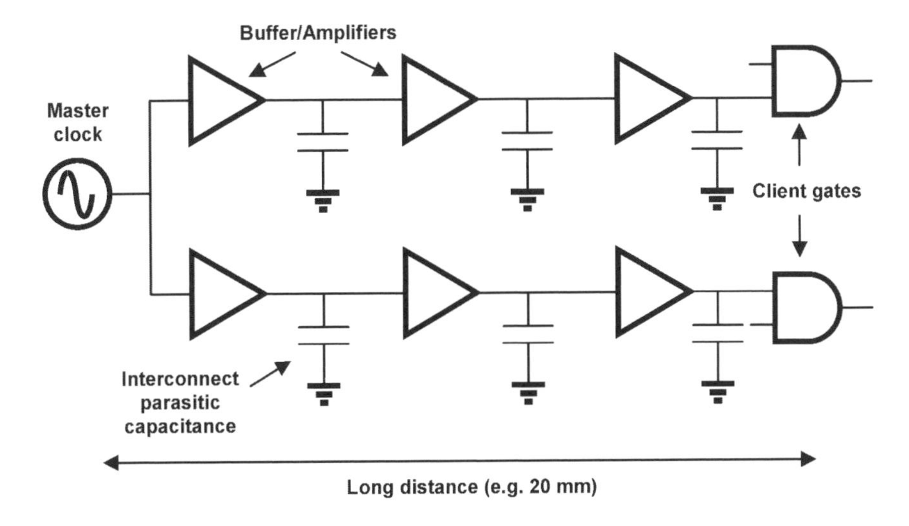

Figure 1. Traditional VLSI clock distribution approach.

this case, however, the two paths diverge further and further away form each other, accruing mismatches due to fabrication parameter, power supply voltage, and temperature gradients on the chip.[7,8] For this reason, intermediary timing equalization between the two paths, known as de-skewing, is usually attempted through sophisticated circuit techniques.[9,10] This is not an easy task considering the physical separation of the paths. The ultimate tool for path equalization is post-design calibration using special adjustable delay circuits.[10] Even with all these measures, final skews remain present due to limited precision in overall delay tuning and unpredictable local temperature and power supply voltage variations.

The matter is further complicated by the ubiquitous presence of noise, either generated inside the transmission paths by natural physical processes or coupled in from the surrounding switching logic through power supply lines, substrate, and other channels. Most of this noise appears at the buffer/repeaters, which are sensitive active circuits providing signal gain. It is known that the repeater jitter is approximately proportional to the delay it introduces in signal propagation[8] and the overall repeater delay is about half of the total path delay.[9] This explains why large distribution networks inherently have high jitter. As the clocking signals travel further and further away from the master generator, the clock transitions become more and more jittery, adding increasing uncertainty to the timing information they carry.

The timing error at the client gates is the sum of the skew and the peak-to-peak jitter. Naturally, the maximum digital processing speed of the system is directly limited by the final timing error, hence the emphasis on precise clock distribution.

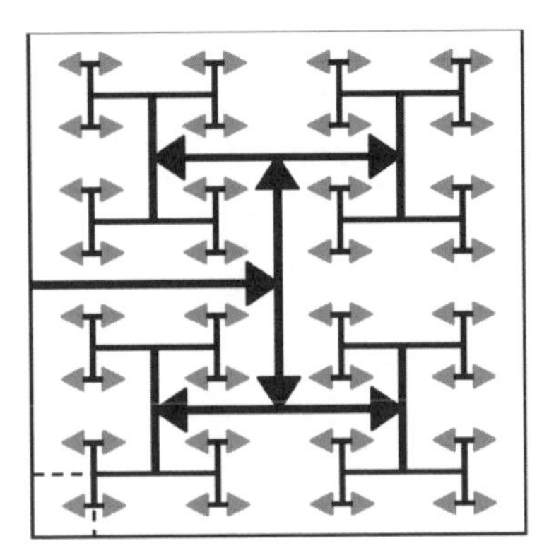

Figure 2. H-tree network with two-fold reflection symmetry (input branch excluded).

Author	Source	Before de-skewing	After de-skewing
Geannopoulos	*ISSCC-98*	**60 ps**	15 ps
Rusu	*ISSCC-00*	**110 ps**	28 ps
Kurd	*ISSCC-01*	**64 ps**	16 ps
Stinson	*ISSCC-03*	**60 ps**	7 ps

Table 1. Typical active tree performance (from Ref. 6).

While the two gate example was used to illustrate the VLSI clocking problem, in reality the VLSI clocking networks support millions of gates. The typical hierarchical structure of such a network follows a complex active-tree topology, which is even more prone to errors than previously suggested. Two-dimensional layout symmetry may be used as in Fig. 2 to produce theoretically matched paths at least up to a reasonably small local clocking area. As explained earlier, this approach is often complemented by intermediate de-skewing and use of calibrating circuits, jitter filters, *etc.* Without such techniques, native active tree timing networks would not support VLSI clocking above 1 GHz. This point was clearly demonstrated by Rusu,[6] who compiled the information in Table 1. The native active tree skews shown in the second column are 60 ps or larger, independently of technology scaling (quoted papers are from different years and technologies). They are produced mostly by device mismatches and supply voltage variations.[7,8]

The resulting customized clocking solutions are difficult to migrate to other technology nodes or reuse in other IC products. Furthermore, in some important cases, such as FPGAs, de-skewing may be largely ineffective since the final loading of the timing network is decided by the end users and is therefore *apriori* unknown to the IC designers.

4. Wave propagation to the rescue

The telegraph and telephone engineers of one hundred years ago would be quite appalled by our current approach to VLSI clock distribution, and rightly so! They had a similar problem in the early phases of their technology deployment, when their connecting wires were loaded by excessive parasitic capacitances. This problem was solved elegantly by resonating out the line capacitors with appropriate inductors. Furthermore, this led to the mass production and use of wide-band transmission lines (TL) with uniformly distributed capacitors and inductors, in which electromagnetic waves propagate long distances without repeaters.

Figure 3 illustrates the principle with lumped elements for eliminating most buffers in Fig. 1 and propagating the clock information all the way to the local areas passively, as electromagnetic waves. The electrical charge placed on the parasitic capacitors is extracted and preserved by the inductor currents. This is fundamentally better than the lossy transmission in Fig. 1, where charge is placed on line capacitors and then dumped to ground on every cycle. The losses incurred in wave propagation due to the wire resistance are orders of magnitude smaller than in the charge/discharge process of digital-logic-type circuits. Ironically, what makes the digital approach so valuable for information processing, *i.e.* on/off switching, is quite detrimental to signal propagation. The actual propagation speed of the clock transitions in a conventional active tree is a "far cry" 10–20% of the speed of light in Si.[11] The conventional clock distribution method is not unlike tying the legs of a thoroughbred horse before the race!

As we plan to free the VLSI clock signal to reach the speed of light through electromagnetic (EM) effects, we may be tempted to imagine that the skew ceases to be a problem due to very small on-chip distances. Unfortunately, a simple calculation clearly shows this not to be the case for multi-GHz clocks. Figure 4 illustrates the fact that a 2.5 GHz sinusoidal clock transmitted at the speed of light over a 20 mm TL shifts by approximately 25% of its period. Normally, only skews substantially less than 10% of the clock period are acceptable in synchronous logic. Therefore, even in the case of on-chip EM-transmission, it is still necessary to ensure time-of-flight equalization in the clocking network.

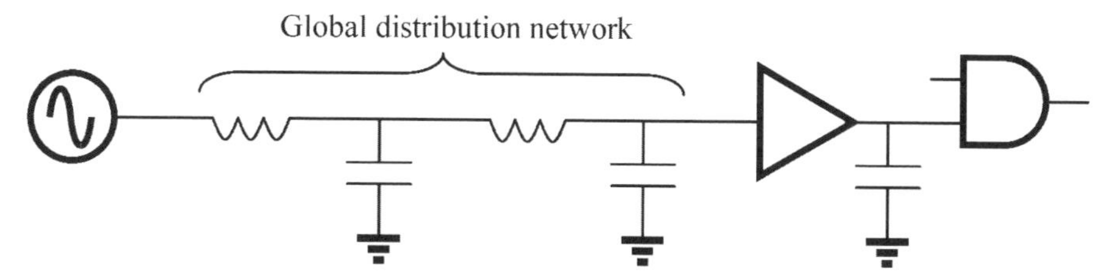

Figure 3. Resonating the parasitic capacitors with inductors.

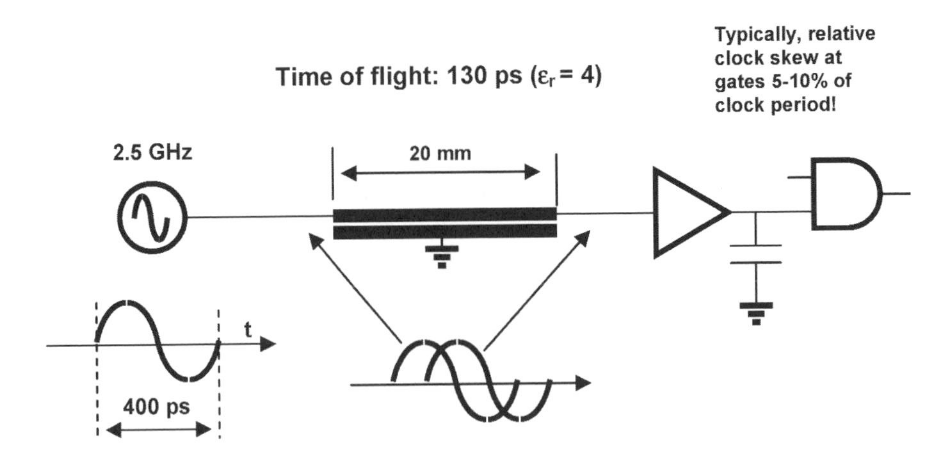

Figure 4. Transmission line clock distribution does not eliminate the skew problem.

We conclude that TL-based VLSI clock distribution is highly desirable due to passive ideally lossless transmission, provided we develop easy skew removal techniques. Before we address this topic, a few comments are necessary regarding TL implementations in CMOS.

5. CMOS transmission lines

The mainstream VLSI design methodology has long dismissed magnetic devices such as inductors and TLs as hard-to-integrate, not useful, and archaic circuit elements. When magnetic effects started showing up in the VLSI chips clocked at very high frequencies, they were seen as a nuisance and a parasitic evil.

The integration of inductors and TLs was indeed a difficult proposition for most generations of IC technology until recently, when multi-level metallization and thick metal technologies have become prevalent. Figure 5 shows the metal stacks of typical modern CMOS technologies, providing ample opportunities for integrating excellent inductors and TLs with reasonable physical dimensions.[11-14] For example a 50–100 Ω characteristic-impedance micro-strip line can be realized with 10–20 μm wide metal traces, given several μm of dielectric thickness, which is available in the higher levels of the metal stack. In addition, the thick Cu of these levels is good for low-loss transmission. Radio-frequency IC engineers and others have already welcomed back the magnetic effects into their designs with impressive results. It would be hard to imagine having high performance wireless RFICs inside so many modern portable devices without on-chip inductors!

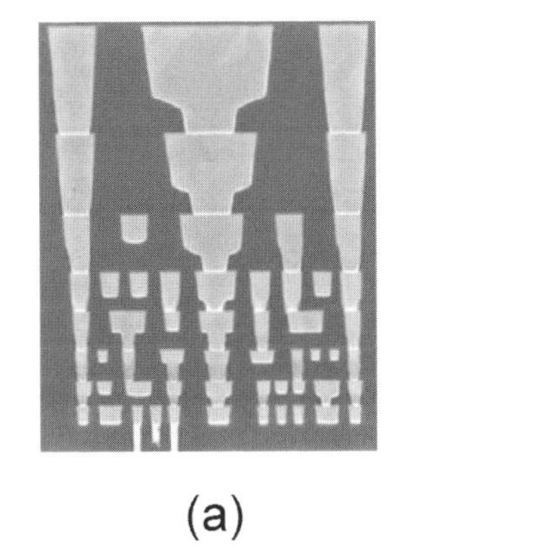

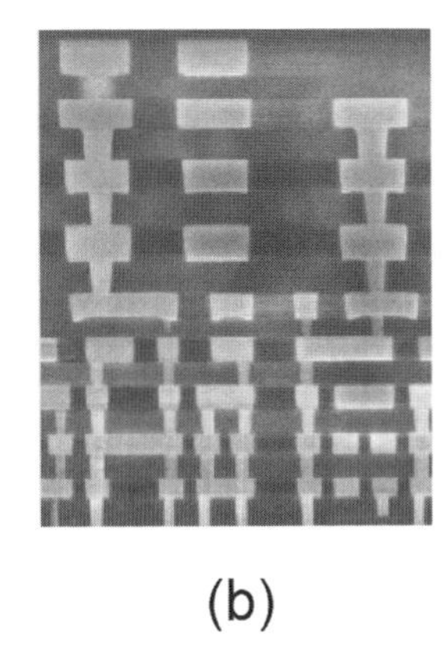

(a) (b)

Figure 5. Modern CMOS metal stacks: (a) Intel 65 nm technology; (b) AMD 130 nm technology.

In the opinion of these authors, an even greater impact will be produced by the introduction of TLs to VLSI clocking. Already, several research results have demonstrated a great potential in this direction,[15-17] to be discussed next.

6. First proposals for the use of TLs in clock distribution

Discrete passive-tree TL configurations called "corporate feeds" are commonly used by microwave engineers to distribute phase-synchronized RF signals in point-to-multipoint applications. An integrated version of this approach targeting VLSI clocking was demonstrated by Mizuno *et al.*[15] The tapered two-level passive H-tree operated at 5 GHz with respectable 20 ps uncompensated skew. However, this experiment included only 16 clocking zones over a 10 mm x 10 mm chip area, with a TL characteristic impedance of only about 4 Ω at the root of the tree.

The previous example illustrates the fundamental limitations of corporate feed networks applied to VLSI clocking. At every network node in the passive tree where the signals split, accurate matching must be ensured in order to avoid signal reflections. This is accomplished either by scaling up the characteristic impedance of the higher level branches or by introducing error-prone multiple-port matching networks. The clocking network in Ref. 15 used the first option and the resulting starting impedance at the root of the tree was very small with only two levels in the hierarchy. Generalizing this approach to higher-level hierarchies is obviously impractical. It should be noted that the highest final impedance at the top of the tree cannot be very large; certainly less than 300 Ω, the impedance of free space. Using multiple-port matching networks at the nodes is not a practical solution either, because these elements are hard to design with low loss and small phase errors.

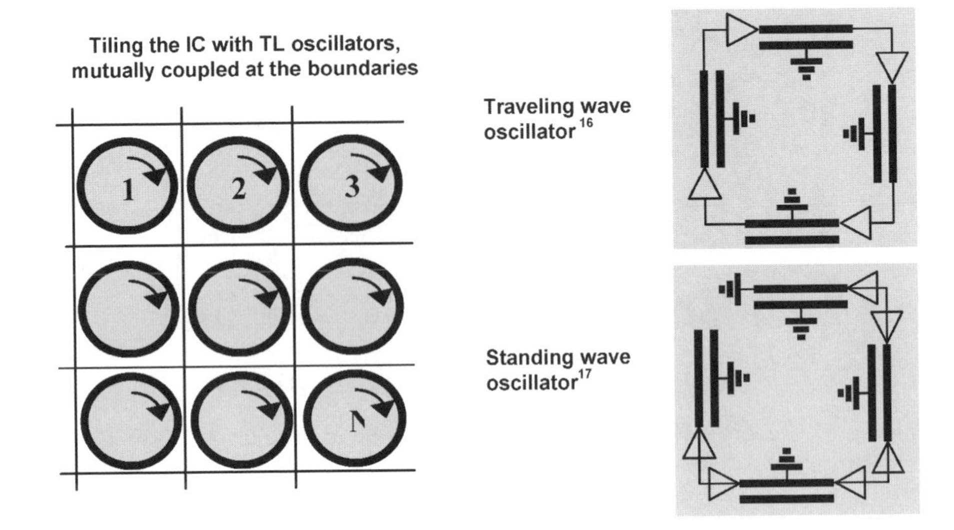

Figure 6. Traveling wave and standing wave VLSI clocking.

We conclude that on-chip TL-based corporate feeds mitigate the signal transmission problems of classical active trees but introduce new and difficult challenges. At this time, they do not offer a clear advantage over conventional clocking technology.

A superior use of TLs for VLSI clocking was reported in Refs. 16 and 17, where the circuits achieve impressive 1 ps skews at 10 GHz, with very low power dissipation. In addition, the proposed methods were capable of covering practical-sized clocking regions. Both schemes use many coupled oscillators interconnected in two-dimensional grids sustaining either traveling or standing waves, respectively. Figure 6 illustrates these concepts.

Despite excellent skew and power dissipation performance, the multiple TL oscillator approach has important fundamental limitations. First, the clock frequency is "hard-wired", determined by the TL characteristics and the capacitive loading. This is not desirable in many VLSI applications, where the capability to change the clock frequency is important. Moreover, clock dithering for EMI reduction is also difficult to implement for the same reasons. Second, the behavior of such a massive and complex autonomous network in a typically very noisy chip environment is largely unknown and will remain unpredictable for the foreseeable future. The current analog/RF simulation and modeling technologies are still very far from covering an entire VLSI chip. As a result, the risk of major architectural flaws cannot be mitigated during the design. This includes the possibility of exciting multiple oscillatory modes during field operation, creating a potential reliability problem.

The third fundamental limitation of the traveling/standing wave concept comes from the inability to yield constant phase and magnitude simultaneously. Under traveling-wave conditions, the signals derived at different locations along the TL have a constant magnitude but different phases, which vary linearly with position. De-skewing the clock drop-points is mandatory but fundamentally difficult without an absolute phase reference. Under standing-wave conditions, signals extracted at different locations along the TL have the same phase but different magnitudes. These vary as a sinusoidal function with position, creating sizable regions where the signal is too small for practical clock extraction.

For completeness, we also mention the technique proposed Chan *et al.*,[18] which uses inductor-based oscillators. This method suffers from the same drawback of a hard-wired frequency as Refs. 16 and 17.

7. Passive serial distribution

As we discussed the benefits and deficiencies of the current proposals for TL-based VLSI clock distribution,[15-17] four principles have clearly emerged, which will allow us to find yet a better solution: 1) passive transmission over TLs is very attractive for jitter, skew, and power-dissipation; 2) TL branching is no good due to introduction of error-prone impedance matching; 3) the clocking network must be driven, rather than autonomous (no oscillators), to have no frequency restrictions and to minimize the risk for unpredictable effects; and 4) it is important to realize clock distribution with constant phase and magnitude simultaneously. The first three criteria are met by the only possible architecture shown in Figure 7.

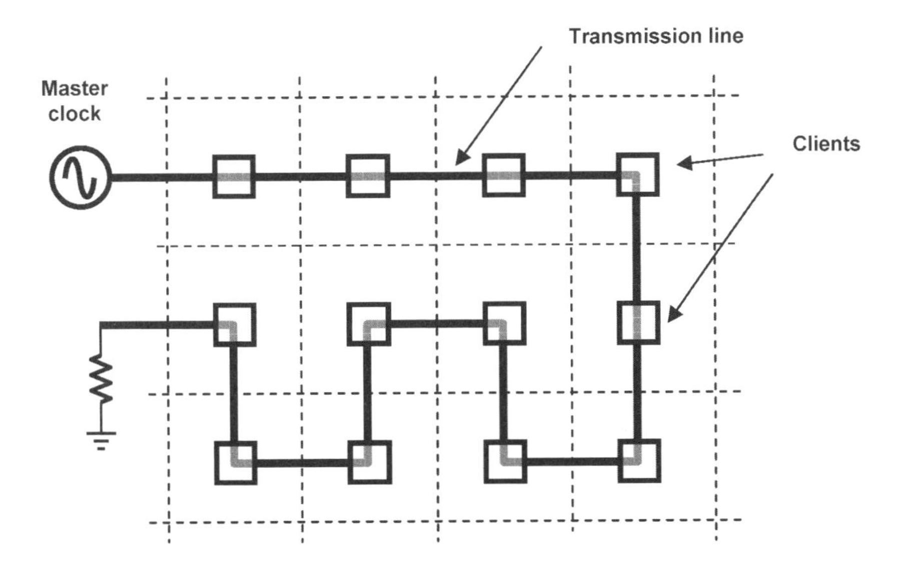

Figure 7. Passive serial connectivity using a single transmission line.

In the serial scheme proposed, an on-chip transmission line properly terminated at both ends meanders over the area to be clocked with client buffers tapping the line and deriving local clock signals. Ignoring for the moment the major issue of variable skew along the line as in Ref. 16, and violating for the time being the fourth criterion of constant phase and magnitude, we first review the fundamental reasons why this architecture is otherwise ideal for clock distribution.

As discussed before, the passive, low-loss nature of TLs eliminates the need for global repeaters and yields minimum power dissipation and jitter. The absence of branches maintains a constant characteristic impedance along the TL, assuming the clocking clients present no significant load to the TL. The clients just tap the line and sense the local voltages without the need for signal splitters. This assumption requires further clarifications.

The input impedance of the CMOS buffers tapping the line can be shown to be mostly capacitive in practice. The finite resistance "seen" into the transistor gates at the clock frequency f_{clock} is negligible, because the clock frequency is always low compared to the MOSFET transient frequency f_{T}. Otherwise, no useful digital circuits could be built. For example, typical "10 FO4 clocking" (a clock period ten times the delay through an inverter driving four similar inverters) gives an $f_{\mathrm{T}}/f_{\mathrm{clock}}$ ratio of about 40. Therefore, the buffer loading will not introduce any significant signal loss. Nevertheless, it is still important to avoid reflections, which could introduce magnitude variations.

If the buffers have small input transistors, the capacitive loading effects are negligible, as the typical input impedance of small transistors is an order of magnitude bigger than the TL characteristic impedance even at 10 GHz or beyond. For large buffer input transistors, the effective characteristic impedance of the line changes locally, but if all buffers are equal in size and the line is loaded evenly, there is only a global effective characteristic impedance change. This can be compensated by adjusting the line terminations during design. A more sophisticated solution would set the termination impedance adaptively by monitoring the signal levels at the TL ends, intentionally placed in close proximity to one another.

The remaining major issue is the severe problem of variable skews on serial distribution. These skews are predictable due to constant speed of EM waves in the TL. However, this does not help much with de-skewing locally extracted clocks. In the absence of a locally available absolute reference, which is the very purpose of clock distribution in the first place, the only theoretical option is to synchronize adjacent areas in a hierarchical manner with a large number of PLLs or DLLs. The resulting complexity and performance limitations would likely be worse than for active trees.

Our BDS concept[2,3] provides a simple and effective solution to the skew accumulation problem of a serial distribution network. It has equal capabilities for jitter and power dissipation performance as Refs. 16 and 17 because it uses TL concepts in a similar fashion. However, unlike these competing techniques, it is an open-loop method containing no oscillators and yields constant-phase and constant-magnitude clock signals simultaneously.

8. The bidirectional signaling (BDS) concept

Let us consider two identical TLs side-by-side, as shown in Fig. 8, and two respective pulses originating from opposite ends at the same time. For constant velocity propagation, *i.e.* assuming homogeneous TLs, the world lines of the propagating pulses are straight and mirror each other at the TL middle coordinate. If we sense the bidirectional pulses at any coordinate x, we detect two respective skews T_1 and T_2. The simple operations of adding or averaging these skew numbers yield absolute results independent of x. For example, the $\frac{1}{2}(T_1 + T_2)$ average equals the time of flight on any of the two pulses to the center of the line, which is a constant as illustrated in Fig. 8. Likewise, the $(T_1 + T_2)$ sum represents the total propagation time over each TL length. If multiple points on the bidirectional signaling (BDS) line extract and process the local skews as mentioned, all derived local signals will be in phase. As there is no physical superposition of incident and reflected waves like in Refs. 17 or 18, the magnitude of the extracted local reference is totally decoupled from the phase and in principle can be designed to be a constant throughout the system.

Figure 9 shows a straightforward application of the BDS concept to the serial clock distribution system of Fig. 7. The master clock sends identical pulses in different directions over the two side-by-side TLs making up the BDS line. This configuration will be called "line return to origin". All attractive properties of the network in Fig. 7 discussed previously remain valid. In fact, the new scheme is a superposition of two conventional serial TL networks, as in Fig. 7, with the added advantage that skew accumulation can be cancelled out. Another way of describing this capability is the creation of an absolute time reference anywhere in the BDS line by virtue of having symmetric world-lines.

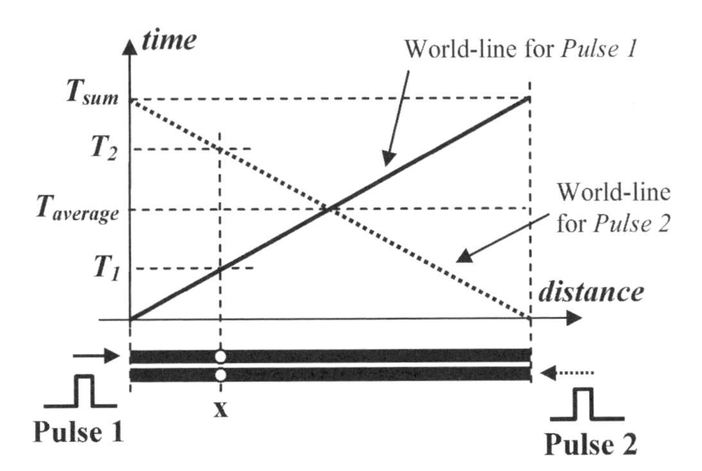

Figure 8. World-line diagram of a BDS system.

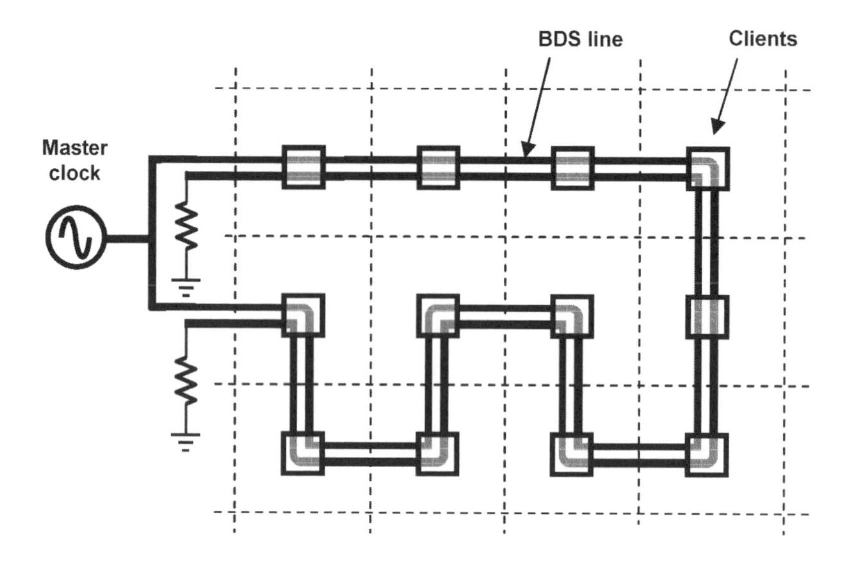

Figure 9. A BDS "line return to origin" system.

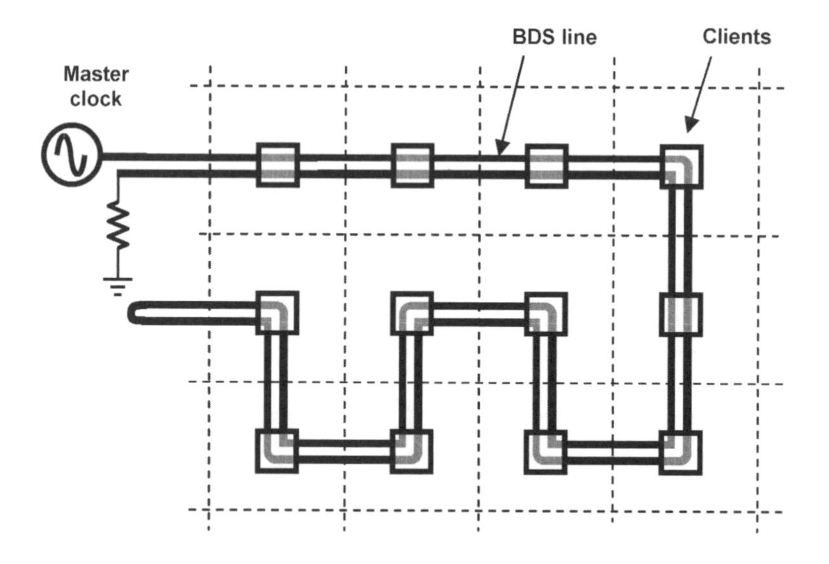

Figure 10. A BDS "signal loop back" system.

The BDS principle holds even if one adds an arbitrary fixed delay between the starting times of the two signals, as can be easily verified. Since there is no need to keep the two pulses in a predetermined phase relationship, the "line return to origin" is not the only possible BDS architecture. Another important possibility is to transmit on one TL and loop back the signal at the "far end" from the generator. The simplest "signal loop back" architecture is shown in Fig. 10 and consists of turning the TL around to create the BDS line from a single uninterrupted TL. Naturally, this doubles the losses but halves the power dissipation as compared to "line return to origin". Active "signal loop back" is also possible by regenerating the signal at the far end using a PLL or DLL. In any case, the "signal loop back" architecture separates the two ends of the BDS line, allowing more freedom in routing.

Furthermore, unlike tree networks, the BDS concept assumes nothing about geometrical conditions for line or client placement. The skew cancellation capability comes purely from electrical properties of wave transmission over two identical TLs. As a result, the BDS line can be bent in any ways resulting in irregular shapes without loosing its valuable properties. Figure 11 illustrates this important observation.

9. Local clock generation by average time extraction

A circuit extracting the absolute reference from the BDS line by time averaging is called an average time extractor or ATE. The realization of the ATE function would run into fundamental difficulties (non-causal system) if we had only a single

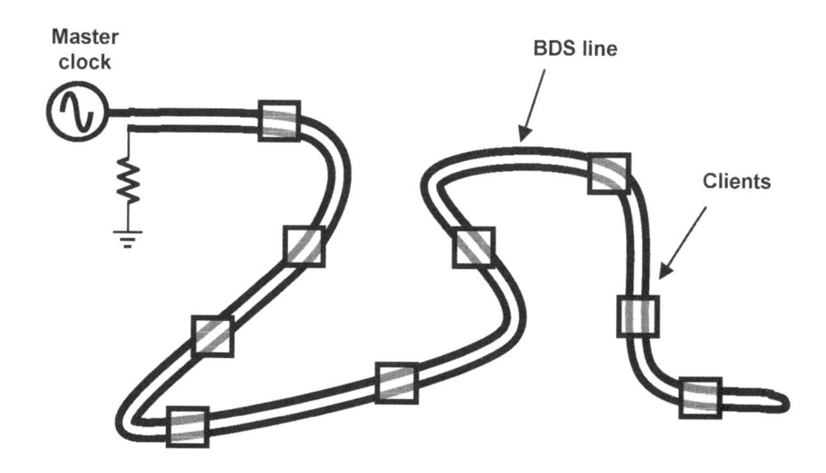

Figure 11. BDS system with arbitrary line and client placement.

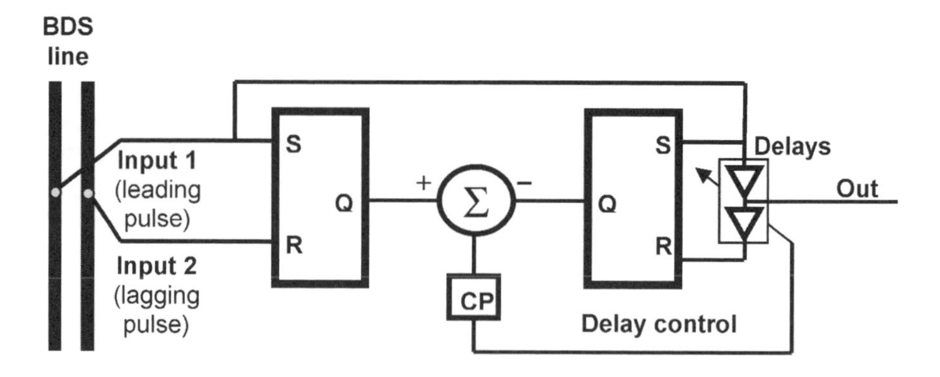

Figure 12. A DLL-based average time extractor (ATE).

pair of pulses. Fortunately, the distributed clock is a periodic signal and therefore the ATE can average "future" values from the previous clock period. Many ATE configurations are possible. The block diagram in Fig. 12 shows a topology based on a classical DLL. The feedback controls the delay elements forcing the total delay to be equal to the time interval between the arrival times of the two input pulses. In the lock state, the center tap of the delay line extracts the absolute time reference T_{average} (see Fig. 8). The delay control could be either analog or digital. In the latter case, the DLL would be disabled after lock.

The ATE function as defined for periodic signals cannot distinguish between delays separated by an integer number of the clock period. This gives rise to a phase reversal ambiguity problem. In other words, while all ATEs connected to the BDS line are synchronized in terms of clock transitions, some ATEs will have inverted signals with respect to others, depending on the ATE position along the TL. This consistent phase inversion error, occurring over large TL sections, can be easily detected during design and corrected with the addition of inverters. If it is desirable to allow blind ATE placement with no attention to potential phase reversals, it is possible to design simple circuits, which would correct this error automatically. For example, at the boundary of each clocking region, a phase detector would check if the two regional clocks are in phase or out of phase and communicate this to an appropriate local clock driver through a single control bit. This one time operation at turn on is considerably easier than complicated boundary phase comparison and correction for clock-transition edge alignment already used in VLSI with high precision DLLs.

A more problematic practical issue is the phase error introduced by pulse shape distortion due to dispersion or any other mechanism. This is an important potential problem for any ATE processing high-frequency pulse signals. As discussed previously, the clocking speed is low compared to the transistor switching speed, but still the demands of modern clocking systems are very stringent. Next, we show a superior solution to the clock extraction problem from a BDS line, which is considerably simpler than the ATEs discussed so far, has no phase ambiguity, and does not rely on edge detection. This is accomplished by using sinusoidal BDS and analog multipliers for clock extraction.

10. Local clock generation by multiplication

For sinusoidal signals, delay and phase summation are equivalent operations. Indeed, multiplying two sinusoidal signals of frequency f and phases φ_1 and φ_2 yields a dc term and a sinusoidal term of frequency $2f$ and phase $(\varphi_1 + \varphi_2)$. But φ_1 and φ_2 are proportional to corresponding time delays. As a result, ATE functionality is automatically included in the operation of analog signal multiplication.

We configure a very efficient BDS clock distribution system as in Fig. 13. A single sinusoidal signal of frequency f enters the TL at node A, passes sequentially through B, C, and D and exits into a termination resistor at node E. Analog multipliers are connected as shown. After multiplication, phase-synchronous local clocks are derived at points AA, BB, CC, or any additional similar points.

Extracting a clock signal at twice the transmitted frequency is beneficial since the line loss for half-rate distribution is lower (skin-effect limited operation). Half-rate distribution has been used successfully in commercial microprocessors,[10] albeit with a conventional active tree.

Unlike using DLLs or PLLs, analog multiplication is a simple memory-less (non-dynamic) operation. This classical, well-understood circuit function can be implemented in any semiconductor technology.[19] The dc term resulting from multiplication can be easily removed through ac coupling or standard dc removal feedback techniques.

We performed extensive transient and harmonic-balance simulations of the proposed sinusoidal BDS using the Mentor Graphics Eldo-RF simulator. We modeled the BDS line with level-4 TL elements, assuming an on-chip Cu microstrip design with 10 μm width, 10 μm dielectric (SiO_2) thickness, and 5 μm metal thickness. The characteristic impedance was approximately 80 Ω. The multipliers were modeled using 120 nm TSMC CMOS technology files. The

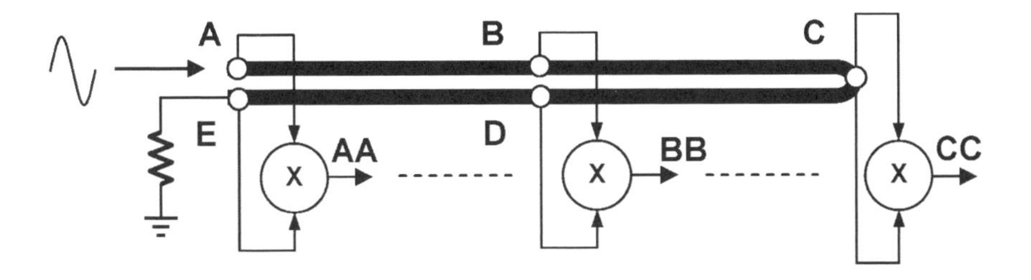

Figure 13. BDS system using multipliers for local clock extraction.

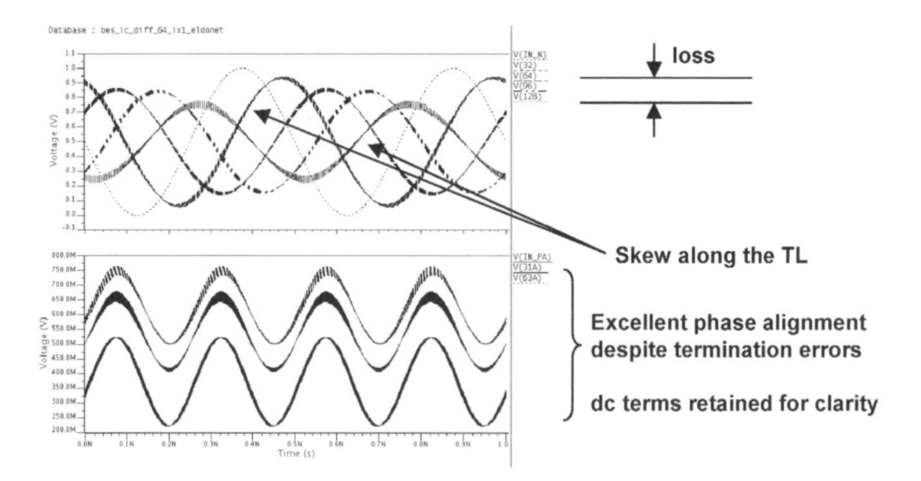

Figure 14. Typical simulation results.

total BDS line length in our simulations was 32 mm (64 mm total TL length) with 32 local clocking points, one per every millimeter. Such a BDS line can provide synchronous clocks to a 32 mm^2 area *of any shape* with 1mm × 1mm granularity. Four such lines, fed centrally, would cover a 128 mm^2 area.

Figure 14 shows a typical simulation result for 2 GHz distribution and 4 GHz local clocks. For clarity, only 3 out of 32 clock signals are shown. We notice the large skews at points B, C, D, and E compared to A, consistent with finite signal propagation speed. After multiplication, all 4 GHz local clocks are in phase synchronism. The different dc levels, a byproduct of the multiplication, are intentionally kept for demonstration purposes. They would be removed in practice, as discussed previously.

Several important nonideal conditions are included in these simulations. First, we have realistic TL loss: approximately 6 dB from point A to point E. This loss has little effect on the extracted clocks, confirming a theoretical observation that the extracted clock is independent of TL loss and position. We expect to have no phase errors due to signal loss but, at first glance, we would expect to see magnitude errors. However, even the magnitude is insensitive to loss. This important property is a consequence of the fact that the TL loss in dB is linear with the distance. As multiplication is equivalent to addition in dB, a loss-error cancellation occurs at every clock drop point. The inconsequential magnitude error visible in Fig. 14 is due to large loading mismatches in the line.

Second, notice that all signal traces are somewhat thick. This is because each one is composed of twenty superimposed traces resulting from a 25% termination resistance sweep. The synchronization errors are obviously small, proving our technique is resilient to termination errors.

Finally, we mention that at every TL-tapping point (there are 64 such points), the multipliers load the line with the equivalent of four minimum CMOS inverters.

11. Summary and conclusions

The traditional method for achieving correct timing in VLSI digital systems has been based on clock distribution via active electrical trees. This technique served older IC technologies well over many years, but at clock speeds above 1 GHz it encounters severe limitations in skew, jitter, power dissipation, and sensitivity to design and operational conditions. These limitations are fundamental and are related to two factors: i) the use of electrical rather than electromagnetic propagation; and ii) the use of a tree rather than serial network architecture.

The advent of modern IC technologies with many metal levels, including several thick Cu layers, makes it possible to replace the high-loss electrical clock distribution with low-loss passive electromagnetic clock distribution on monolithic transmission lines. This further opens the possibility of replacing the tree architecture with a serial architecture, albeit in the presence of skew accumulation. The latter problem is mitigated efficiently by our new BDS technique through double transmission and analog time averaging.

The resulting clock distribution system is a revolutionary new VLSI platform technology, qualifying as reusable IP for the VLSI "timing physical backbone". Due to wave transmission and RF-type signal processing, the speed limitations of clock distribution are pushed beyond the limitations of switching digital logic of practical depth at any future technology node. As a result, the introduction of the BDS technology will remove clocking as the limiting factor in VLSI design. In addition, the reusable IP aspect of this technology will create a new paradigm in design quality and development cycle time.

References

1. S. Sarkar, C. G. Subash, and S. Shinde, "Effective IP reuse for high quality SOC design," *Proc. IEEE Intern. SOC Conf.* (2005), pp. 217–224.
2. V. Prodanov and M. Banu, "GHz serial passive clock distribution in VLSI using bidirectional signaling," *Custom Integr. Circ. Conf.* (2006).
3. M. Banu and V. Prodanov, "Recent advances in VLSI routing – Serial passive clock distribution," *CMOS Emerging Technol. Workshop*, Banff, Canada (2006).
4. E. Friedman, "Clock distribution networks in synchronous digital integrated circuits," *Proc. IEEE* **89**, 665 (2001).
5. A. Mule, E. Glytsis, T. Gaylord, and J. Meindl, "Electrical and optical clock distribution networks for gigascale microprocessors," *IEEE Trans. VLSI Syst.* **10**, 582 (2002).
6. S. Rusu, "Clock generation and distribution for high-performance processors," *Proc. Intern. Symp. SOC* (2004), p. 207.
7. G. Geannopoulos and X. Dai, "An adaptive digital deskewing circuit for Clock distribution networks," *Tech. Digest ISSCC* (1998), pp. 400–401.

8. D. Harris and S. Naffziger, "Statistical clock skew modeling with data delay variation," *IEEE Trans. VLSI Syst.* **9**, 888 (2001).
9. J. D. Warnock, J. M. Keaty, J. Petrovick, *et al.*, "The circuit and physical design of the POWER4 microprocessor," *IBM J. Res. Develop.* **46**, 27 (2002).
10. P. Mahoney, E. Fetzer, B. Doyle, and S. Naffziger, "Clock distribution on a dual-core multi-threaded Itanium®-family processor," *Tech. Digest ISSCC* (2005), pp. 292–293.
11. R, T. Chang, N. Talwalkar, C. P. Yue, and S. S. Wong, "Near speed-of-light signaling over on-chip electrical interconnects," *IEEE J. Solid-State Circ.* **38**, 834 (2003).
12. V. Milanovic, M. Ozgur, D. C. DeGroot, J. A. Jargon, M. Gaitan, and M. E. Zaghloul, "Characterization of broad-band transmission for coplanar waveguides on CMOS silicon substrates," *IEEE Trans. Microwave Theory Techniques* **46**, 632 (1998).
13. M. Morton, J. Andrews, J. Lee, *et al.*, "On the design and implementation of transmission lines in commercial SiGe HBT BiCMOS processes," *Digest Topical Meeting Si Monolithic Integr. Circ. RF Syst.* (2004), pp. 53–56.
14. C. H. Doan, S. Emami, A. M. Niknejad, and R. W. Brodersen, "Millimeter wave CMOS design," *IEEE J. Solid-State Circ.* **40**, 144 (2005).
15. M. Mizuno, K. Anjo, Y. Sumi, *et al.*, "On-chip multi-GHz clocking with transmission lines," *Tech. Digest ISSCC* (2000), pp. 366–367.
16. J. Wood, T. Edwards, and S. Lipa, "Rotary traveling-wave oscillator arrays: A new clock technology," *IEEE J. Solid-State Circ.* **16**, 1654 (2001).
17. F. O. O'Mahony, C. P. Yue, M. A. Horowitz, and S. S. Wong, "A 10-GHz global clock distribution using coupled standing-wave oscillators," *IEEE J. Solid-State Circ.* **38**, 1813 (2003).
18. S. Chan, K. Shepard, and P. Restle, "Uniform-phase uniform-amplitude resonant load global clock distributions," *IEEE J. Solid-State Circ.* **40**, 102 (2005).
19. G. Han and E. Sanchez-Sinencio, "CMOS transconductance multipliers: A tutorial," *IEEE Trans. Circ. Syst. – II* **45**, 1550 (1998).

Origin of 1/*f* Noise in MOS Devices: Concluding a Noisy Debate

K. Akarvardar, S. Cristoloveanu, and P. Gentil
Institut de Microélectronique, Electromagnétisme et Photonique (IMEP)
Minatec, 3 Parvis Louis Néel, BP257, 38016 Grenoble Cedex 1, France

1. Introduction

The importance of 1/*f* noise in semiconductors is well documented, enthusiastically explored, and definitely feared. There are two distinct camps in the 1/*f* world. For McWhorter's partisans, there is no doubt that the noise originates from fluctuations in the carrier number.[1] In MOS devices, for example, carrier trapping is related to the presence of slow traps in the gate oxide. Data on MOSFETs collected for years are rather convincing in this respect, except for Hooge and his disciples. For them, there is even less doubt that the 1/*f* noise in semiconductor devices proceeds from carrier mobility fluctuations $\Delta\mu$.[2]

The aim of our paper is to show that the SOI four-gate transistor (G^4-FET) is a competent referee for this rivalry. The G^4-FET has four independent gates, which offer tremendous versatility of operation. In the G^4-FET, the cross-sectional size and position of the conducting channel can be adjusted via the appropriate biasing of the four gates. This feature enables an *in-situ* comparison of the interface and the volume conduction properties.[3] By comparing the bulk and the surface noise characteristics to the McWhorter and Hooge noise models, the domains of validity of these models can be determined. Also, the comparison of the noise amplitudes for the different conduction modes provides the optimal conditions for low-noise operation.

We describe the G^4-FET structure, operation and different conduction modes (surface and volume) in Section 2. In Section 3, the typical noise spectra for surface and volume currents will be separately presented and explained. In Section 4, the intrinsic difference between the surface and the bulk noise will be confirmed by using the G^4-FET as a depletion-mode MOSFET and by driving the front interface from depletion to accumulation. Section 5 consists of the quantitative and the qualitative comparison of the noise characteristics in surface and volume conduction modes.

2. Device structure and operation

The *n*-channel G^4-FET (Fig. 1) has the same structure as that of a *p*-channel, inversion-mode, partially-depleted SOI MOSFET with two independent body

Future Trends in Microelectronics. Edited by Serge Luryi, Jimmy Xu, and Alex Zaslavsky
ISBN 0-471-48 © 2007 John Wiley & Sons, Inc.

contacts, one on each side of the channel. The current direction and the roles of the terminals are modified in the G^4-FET. Drain current is comprised of majority carriers flowing in the perpendicular direction of the inversion-mode MOSFET usual current flow: from one body contact (G^4-FET source) to the other (G^4-FET drain). The source and drain of the regular MOSFET are used in the G^4-FET as two extra gates (junction-gates, JG1 and JG2), which squeeze laterally the channel via the reverse-biased junctions as in a JFET. The role of the polysilicon top gate (G1) depends on the operation mode: it may induce *accumulation* at the front interface leading to *surface* conduction due to the extra majority carriers attracted to the front interface by V_{G1}. Alternatively, V_{G1} may drive the front interface to *depletion* or *inversion* to enable *bulk* current. In this case, the conduction is ensured by the majority carriers whose maximum concentration is defined by the doping. The substrate emulates a fourth gate (G2): it is used to modulate the back surface potential and the location of the conducting channel.[3]

The G^4-FET's low-frequency noise is analyzed by distinguishing two different operation modes.[4] In the first one, which we will call the "surface mode", the G^4-FET is driven by the front-gate voltage V_{G1}, while the junction-gate voltage $V_{JG} = V_{JG1} = V_{JG2}$ (the junction gates are shorted to simplify the analysis), and the back-gate voltage V_{G2} are constant and fully deplete the body. The G^4-FET is normally off ($I_D = 0$ for $V_{G1} = 0$ V) and the drain current is dominated by the majority carriers accumulated at the front interface, as shown in Fig. 2(a). In the second mode of operation, the transistor is driven by the junction gate voltage V_{JG}, while the front and back gates induce either depletion or inversion at the interfaces. In this mode of operation, which we will call the "volume mode", the current flows through the bulk, far from the silicon/oxide interfaces – see Fig. 2(b). The static parameters (threshold voltage, subthreshold slope and transconductance) are defined with respect to the front-gate in surface mode and with respect to the junction-gates in volume mode.

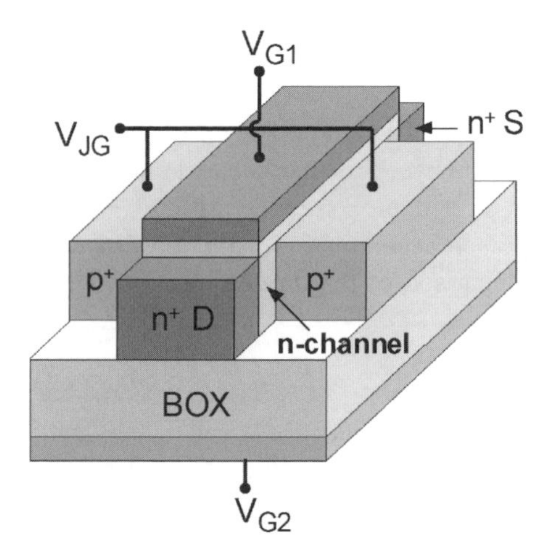

Figure 1. The *n*-channel G^4-FET structure.

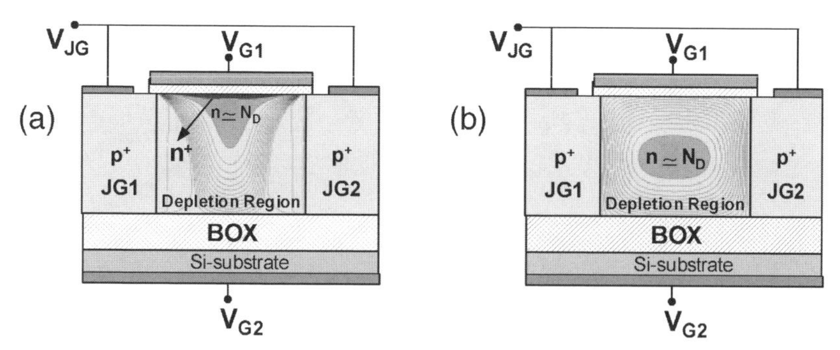

Figure 2. Cross-sections describing the surface (a) and volume (b) conduction modes for an *n*-channel G^4-FET. The drain current flow is perpendicular to the figures; *n* and N_D designate respectively the electron and the doping concentrations.

3. Noise power spectral density (PSD) characteristics in surface and volume conduction modes

We use our *n*-channel G^4-FETs, fabricated in a conventional 0.35 μm partially-depleted SOI process, for low-frequency noise measurements. The transistors were 0.35 μm wide and 3.4 μm long, with 20 channels in parallel to achieve a relatively high drive current. Channel doping was in the 10^{17} cm^{-3} range. The silicon film, front-gate oxide and buried-oxide thicknesses were 150 nm, 8 nm and 400 nm, respectively. Drain-to-source voltage was 50 mV for all measurements. Static and noise characteristics of the devices were measured using the "Programmable Point Probe Noise Measuring System" (3PNMS).[5] The noise floor of the system is 2.5×10^{-27} A^2/Hz.

- *Surface conduction mode*

The measured static characteristics in surface mode operation are given in Fig. 3(a): the variation of the drain current and transconductance (defined as $g_{m,G1} = \partial I_D/\partial V_{G1}$) as a function of the front-gate voltage is similar to that observed in inversion-mode MOS transistors. For $V_{G1} = 0$, the channel is fully depleted by the junction gates and the back gate ($V_{JG} = -2$ V, $V_{G2} = -9.5$ V). For $0 < V_{G1} < V_{T,G1} = 0.7$ V, the G^4-FET operates in the subthreshold region. Here, $V_{T,G1}$ is the threshold voltage in surface mode and is defined as the front-gate voltage providing flatband condition at the front interface. In strong accumulation ($V_{G1} >> V_{T,G1}$), drain current is dominated by the electrons accumulated at the front interface.

The drain current noise power spectral density (PSD), S_I, is presented for various front-gate voltages in Fig. 3(b). Above the threshold voltage, 1/*f* noise is systematically observed (uppermost spectrum). The similarity of spectra in strong accumulation with those of inversion-mode MOSFETs suggests that the origin of the noise is the same in both cases, *i.e.* dynamic electron trapping and

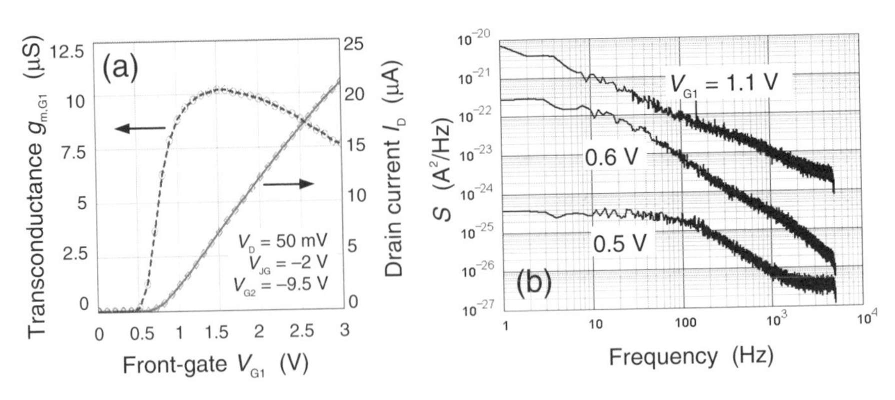

Figure 3. Surface mode operation of the G[4]-FET (V_{JG} = −2 V, V_{G2} = −9.5 V, V_D = 50 mV): (a) measured drain current and transconductance as a function of the front-gate voltage; (b) measured power spectral density of the drain current noise for various front-gate voltages.[4]

release of electrons from oxide traps close to the interface.[6,7] In the subthreshold operation, we observe either $1/f$ noise or Lorentzian spectra like those shown in Fig. 3(b) for V_{G1} = 0.5 and 0.6 V.

- *Volume conduction mode*

Drain current and transconductance (here defined as $g_{m,JG} = \partial I_D/\partial V_{JG}$) of the G[4]-FET operating in volume mode are shown in Fig. 4(a). The front-gate and back-gate voltages are such that both interfaces are depleted ($V_{G1} = 0$, $V_{G2} = -9.5$ V), rendering characteristics similar to those of JFETs. For $V_{JG} = -1.4$ V, the channel is pinched-off. As V_{JG} approaches zero, the depletion regions shrink in the lateral direction and the electron concentration increases in the middle of the channel, giving rise to an increase in both the current and the transconductance

The PSD of the drain current noise is presented in Fig. 4(b) for various junction-gate voltages. We again note the presence of Lorentzian noise for the subthreshold region and of $1/f$ noise above threshold ($V_{JG} > V_{T,JG} = -0.7$ V). Nevertheless, in the volume mode of operation above threshold, the frequency exponent α (in $1/f^{\alpha}$) varies between 0.6 and 0.8, systematically lower than its value in the surface mode ($\alpha = 1$). This indicates that the predominant fluctuation mechanism and/or noise source is not the same for these two modes. In the volume mode, we believe that the noise sources intrinsic to JFETs always exist and determine the minimum noise level as long as they are not masked by the surface noise mechanisms, *i.e.* carrier capture into and emission from the oxide traps. In JFETs, the low-frequency noise is attributed to (i) carrier trapping and detrapping by the centers within the depletion regions (giving rise to channel-section fluctuations) and/or within the channel (carrier number fluctuations) and (ii) to correlated mobility fluctuations.[8]

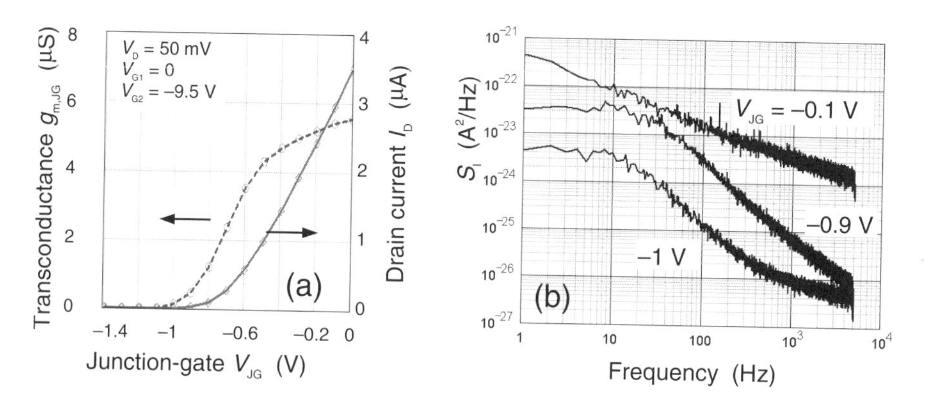

Figure 4. Volume mode operation of the G^4-FET (V_{G1} = 0, V_{G2} = −9.5 V, V_D = 50 mV): (a) measured drain current and transconductance as a function of the junction-gate voltage; (b) measured power spectral density of the drain current noise for various junction-gate voltages.[4]

4. Transition from volume noise to surface noise

The use of the G^4-FET as a depletion-mode MOSFET enables the observation of the direct transition from volume to surface noise.[9] We biased the G^4-FET with a constant V_{JG} and V_{G2}, while sweeping V_{G1} to drive the front surface from depletion to strong accumulation. In Fig. 5 the resulting drain current noise power S_I at 10 Hz is shown as a function of the drain current. Related current-voltage characteristics are plotted in the inset of Fig. 5 on a semilogarithmic scale.

For V_{G1} = −2 V, transistor is in the off state because the depletion regions induced by the front, back and the junction-gates fully deplete the body. Until the flatband condition is reached, *i.e.*, for $-2\ V < V_{G1} < 0.56\ V$ or $I_D < 0.5\ \mu A$, the front-depletion region shrinks gradually and enables *volume* conduction, far from the front interface. In this region, the noise power follows the increase in current.

The striking aspect occurs around the onset of accumulation at the front surface (flatband condition), for $I_D = 0.5\ \mu A$: a very abrupt kink is observed in the noise amplitude. This kink is as large as *one order of magnitude*, due to the excess 1/f noise generated at the surface. The excess noise is explained by the dynamic capture/release of the carriers by the oxide traps located in the vicinity of the interface.[3] The transition between the two modes of conduction (from volume to surface channel) which occurs at flatband voltage is clearly confirmed by the visible change in subthreshold slope (inset of Fig. 5) for the same current value.[9]

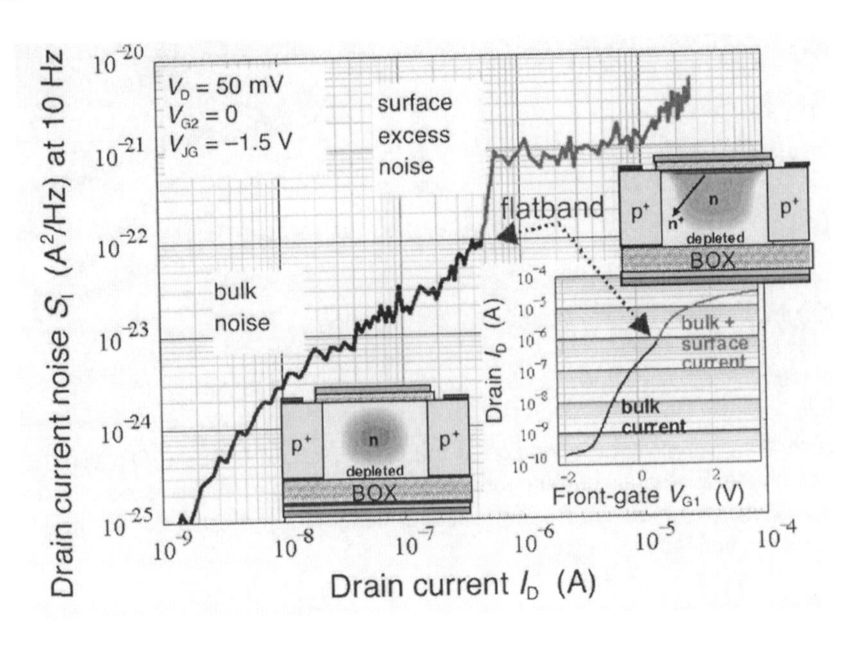

Figure 5. Drain current noise of the *n*-channel G⁴-FET. The kink corresponds to the onset of the front accumulation current that adds to the bulk component. Inset: related current-voltage characteristics.[9]

5. Bias dependence of the normalized current noise in surface and volume conduction modes

We studied the bias dependence of the G⁴-FET noise by monitoring the "*total* noise current" in a given frequency range $\langle i_T^2 \rangle$, defined as:[10,11]

$$\langle i_T^2 \rangle = \int_{f_{min}}^{f_{max}} S_I(f)\, df \cong \sum_{j=0}^{n} S_I(f_j)\, \Delta f, \tag{1}$$

where $\Delta f = f_{j+1} - f_j$ is the bandwidth, $f_{min} = f_0$ and $f_{max} = f_n$ are the minimum and maximum frequencies, respectively. As can be seen from Eq. (1), $\langle i_T^2 \rangle$ depends on the multiple noise PSD values measured between f_{min} and f_{max}, resulting in smoother variations of the noise power and more reproducible trends than an analysis based on a single frequency. Throughout the rest of this chapter we use Δf = 1 Hz, f_{min} = 1 Hz and f_{max} = 500 Hz for the total noise calculations.

The variations of the "normalized total noise current", $\langle i_T^2 \rangle / I_D^2$, as a function of the drain current are often used for the noise spectroscopy. These characteristics make it possible to identify the noise sources starting from the experimental trends.[13,14] In Fig. 6, we compare the normalized total noise current for the surface mode (top curve) and the volume mode with the front surface in depletion (middle curve) or in strong inversion (bottom curve).[12] For all three cases in Fig. 6, the G^4-FET operates in the subthreshold region for I_D lower than 0.2–0.3 μA, and above threshold for a larger I_D. This comparison unveils quantitative and qualitative features.

- *Noise reduction*

For a given drain current, $<i_T^2>/I_D^2$ is the highest in the surface mode, where the conducting channel is formed by electrons accumulated at the front interface. Here the current fluctuations are dominated by the intense exchange of majority carriers between the near-interface oxide traps and the channel. When the conducting channel is moved away from the surface towards the volume (volume mode, bottom two curves of Fig. 6), a substantial decrease in $<i_T^2>/I_D^2$ is observed due to the reduced amount of electron exchange between the conducting channel and the oxide traps. This is especially true when the front surface is in strong inversion: the exchange is almost completely suppressed because the top interface is "shielded" by the minority carriers (holes) and its effect on the bulk channel is minimized.[15] This leads to a reduction of noise amplitude that may be *larger than one decade* compared to surface mode.

The low-noise feature of the volume conduction mode becomes more apparent when the noise is referred to the input.[9,12] The total input-referred noise voltage, given by $<v_T^2> = <i_T^2>/g_m^2$, is a more significant parameter than $<i_T^2>$ for the

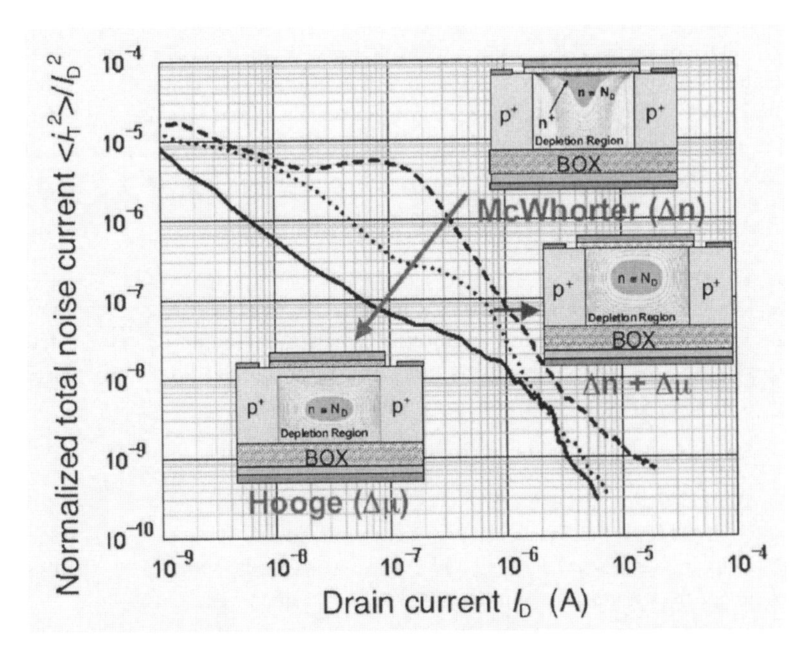

Figure 6. Normalized total noise current as a function of drain current in surface mode (dashed curve, V_{JG} = –4 V), in volume mode with front surface in depletion (dotted curve, V_{G1} = 0) and in volume mode with front surface in strong inversion (solid curve, V_{G1} = –3 V). For all curves V_{G2} = 0, V_D = 50 mV.[12]

analog circuit design, since it takes into account the noise level *and* the gain of the transistor at the same time. In the volume conduction mode, the transconductance $g_{m,JG}$ is not degraded by the gate electric field, as is $g_{m,G1}$ in the MOSFET mode above threshold – compare Fig. 4(a) and Fig. 3(a). This indicates a substantially lower $<v_T^2>$ in volume conduction mode, especially when the front interface is in inversion.[9,12]

- *Noise transformation*

In the surface mode, the $<i_T^2>/I_D^2$ *vs.* I_D characteristic in Fig. 6 presents a plateau in the subthreshold region followed by a $g_m^2/I_D^2 \cong 1/I_D^2$ roll-off above threshold. Such a behavior is similar to that observed in inversion-mode MOSFETs and is consistent with the McWhorter's carrier number fluctuation model.[13,14]

In MOSFETs, the noise power spectral density normalized to drain current is given by :

$$S_I/I_D^2 = (g_m/I_D)^2 S_{VG} = \frac{q^4 \lambda_{ox}}{\eta^2 kTWL} \times \frac{N_t}{(C_{ox}+C_D+qD_{it}+C_{inv})^2} \times f^{-1} , \qquad (2)$$

where S_{VG} is the gate voltage spectral density, λ_{ox} is an average tunneling length, W and L are the device width and length, N_t is the density of the slow oxide traps, and C_{ox}, C_D, qD_{it} and C_{inv} are oxide, depletion, interface trap and inversion layer capacitances respectively. The factor $\eta = 1$ in weak inversion and 2 in strong inversion.[16]

By contrast, when the current flows sufficiently far from the interfaces and the interaction between the oxide traps and the conducting channel is suppressed (volume mode with front interface in strong inversion, bottom curve of Fig. 6), the plateau observed in surface conduction disappears and $<i_T^2>/I_D^2$ decreases as $1/I_D$ over about three decades in current. This dependence is predicted by Hooge's empirical relation:[17]

$$S_I/I_D^2 = \alpha_H/fN , \qquad (3)$$

where α_H is the Hooge constant and N is the total number of carriers proportional to I_D (assuming a bias-independent mobility when the conduction is far from the surfaces). Hooge's equation can be adapted to normalized total noise current via an integration of the both sides of Eq. (3) from f_{min} to f_{max}:[12]

$$<i_T^2>/I_D^2 = (\alpha_H/N) \ln(f_{max}/f_{min}) . \qquad (4)$$

In the volume mode, we estimate $N = 4.5\times10^5$ for $I_D = 6\ \mu A$, leading to $\alpha_H \cong 2\times10^{-5}$ when the front interface is in strong inversion and $\alpha_H \cong 4\times10^{-5}$ when the front interface is in depletion. These values are consistent with those predicted by Hooge for perfect materials (with negligible defect densities).[17]

When the conduction is in the volume, but not completely isolated from the front-interface (JFET mode with the front interface depleted, middle curve of Fig. 6), the behavior of $<i_T^2>/I_D^2$ is a combination of trends implied by the McWhorter and Hooge models. This suggests a transition from the carrier number fluctuations to mobility fluctuations as the predominant noise mechanism when the conduction region is moved from the front surface to the volume of the G^4-FET.[12]

The difference between the top and bottom curves in Fig. 6 is the first experimental evidence of the paradigm shift between the two classical $1/f$ noise models. These models are not mutually exclusive because they can coexist in the same device. More studies are, however, necessary to fully confirm this conclusion because:

i) the type of G^4-FET noise at low currents is not necessarily $1/f$ and it may present Lorentzian spectra as well (whereas the models developed by McWhorter and Hooge apply only to $1/f$ noise);

ii) the $1/I_D$ noise variation seen for the volume conduction (volume mode with the front interface in strong inversion) that suggests agreement with Eq. (3) is no longer observed at relatively high currents (the slope deviates from $1/I_D$ for $I_D > 2$ μA and becomes the same as that of the surface current, as seen in Fig. 6, bottom curve). One possible explanation of this issue is the dominance of the noise generated by the back surface traps over the volume noise at high currents (note that in Fig. 6, $V_{G2} = 0$, making the back interface only slightly depleted).

Although further investigation is needed to appropriately model the characteristics presented above, there is no doubt concerning the intrinsic difference between the current dependence of the surface noise and the volume noise.[12]

6. Conclusion

The low-frequency noise characteristics of the G^4-FET were investigated by distinguishing the surface and volume modes of operation. We observe $1/f$ noise in both surface and bulk conduction modes when the transistor is operated above threshold. The frequency exponent is higher at the surface than in the volume ($1/f$ *versus* $1/f^{0.7}$). For a fixed drain current, the noise related to the volume of the transistor could be more than one decade lower than that generated at the surface. The G^4-FET is a device where the intrinsic nature of the noise changes: the experimental results suggest a transition from McWhorter's carrier number fluctuation model to Hooge's mobility fluctuation model when the conducting channel is moved away from the front-interface towards the device center.

What about the conclusion of the noisy competition between the two camps? One set each.

References

1. A. L. McWhorter, "1/f noise and germanium surface properties," in: R. H. Kingston, ed., *Semiconductor Surface Physics*, Philadelphia: University of Pennsylvania Press, 1957, pp. 207–228.
2. F. N. Hooge, "l/f noise is no surface effect," *Phys. Lett. A* **29**, 139 (1969).
3. B. Dufrene, K. Akarvardar, S. Cristoloveanu, B. J. Blalock, P. Gentil, E. Kolawa, and M. M. Mojarradi, "Investigation of the four-gate action in G^4-FETs," *IEEE Trans. Electron Dev.* **51**, 1931 (2004).
4. K. Akarvardar, B. Dufrene, S. Cristoloveanu, J. A. Chroboczek, P. Gentil, B. J. Blalock, and M. M. Mojarradi, "Surface *vs.* bulk noise in SOI four-gate transistors," *Proc. IEEE Intern. SOI Conf.* (2005), pp. 161–162.
5. J. A. Chroboczek, "Automatic, wafer-level low frequency noise measurements for the interface slow trap density evaluation," *Proc. IEEE Intern. Conf. Microelectronics Test Structures (ICMTS)* (2003), pp. 95–98.
6. R. A. Wilcox, J. Chang, and C. R. Wiswanathan, "Low-temperature characterization of buried-channel NMOST," *IEEE Trans. Electron Dev.* **36**, 1440 (1989).
7. J. Chang, C. R. Wiswanathan, and C. Anagnostopoulos, "Flicker noise measurements in enhancement mode and depletion mode NMOS transistors," *Proc. Intern. Symp. VLSI Technol. Syst. Applications* (1989), pp. 217–221.
8. S.-H. Ng and C. Surya, "A model for low-frequency excess noise in Si-JFETs at low bias," *Solid State Electronics* **35**, 1803 (1992).
9. K. Akarvardar, S. Cristoloveanu, B. Dufrene, *et al.*, "Evidence for reduction of noise and radiation effects in G^4-FET depletion-all-around operation," *Proc. ESSDERC* (2005), pp. 89–92.
10. P. R. Gray and R. G. Meyer, *Analysis and Design of Analog Integrated Circuits*, 3rd ed., chapter 11, New York: Wiley, 1993.
11. B. Razavi, *RF Microelectronics*, chapter 7, New York: Prentice-Hall, 1998.
12. K. Akarvardar, B. Dufrene, S. Cristoloveanu, P. Gentil, B. J. Blalock, and M. M. Mojarradi, "Low-frequency noise in SOI four-gate transistors," *IEEE Trans. Electron Dev.* **53**, 829 (2006).
13. J. Brini, "Low-frequency noise spectroscopy in MOS and bipolar devices," *Microelectronics Eng.* **40**, 167 (1998).
14. G. Ghibaudo, O. Roux, Ch. Nguyen-Duc, F. Balestra, and J. Brini, "Improved analysis of the low frequency noise in field-effect MOS transistors," *Phys. Stat. Sol.* **124**, 571 (1991).
15. T. Elewa, B. Boukriss, H. S. Haddara, A. Chovet, and S. Cristoloveanu, "Low-frequency noise in depletion mode SIMOX MOS transistors," *IEEE Trans. Electron Dev.* **38**, 323 (1991).
16. S. Cristoloveanu and S. S. Li, *Electrical Characterization of Silicon-On-Insulator Materials and Devices*, chapter 9, Dordrecht: Kluwer Academic Publishers, 1995.
17. F. N. Hooge, "1/f noise sources," *IEEE Trans. Electron Dev.* **41**, 1926 (1994).

Quasiballistic Transport in Nano-MOSFETs

E. Sangiorgi, S. Eminente, C. Fiegna
IU.NET and *ARCES-DEIS, University of Bologna, Bologna, Italy*

P. Palestri, D. Esseni, and L. Selmi
IU.NET and *DIEGM, University of Udine, Udine, Italy*

1. Introduction

The International Roadmap for Semiconductors has predicted that starting from the 45 nm technology node, an increase in the basic transport properties (called a "ballistic technology booster") must be achieved to reach the target specifications of high-performance devices.[1] The on-current I_{ON} of the MOSFETs is limited to a maximum value I_{BL} that is reached in the ballistic transport regime.[2,3] Hence, improvements in I_{ON} demand an increase of either I_{BL} or of the ballisticity ratio BR defined as I_{ON}/I_{BL}, which is a metric of how close I_{ON} comes to the ballistic limit. In recent technologies, the observed BR values have not significantly improved[4] and are thought to have only a modest dependence on the technology node (TN) and on the gate length L_G for conventional silicon MOSFETs. Further improvements in I_{BL} requires ultra-thin silicon-on-insulator (SOI) with silicon thickness below 10 nm or non-conventional channel composition.[5]

In this chapter we extend previous analyses on the role of scattering in the channel and drain of decananometer MOSFETs and our results demonstrate that, for the explored gate lengths values L_G, scattering still holds I_{ON} far below the ballistic limit.[6,7]

Accurate modeling of MOSFETs with L_G comparable to or shorter than 25 nm (corresponding to the 45 nm TN),[1] must take into account both quantization and non-equilibrium transport effects. Ballistic transport models[2,3] can account for quantum mechanical effects but, since scattering is neglected, they provide only an upper estimate of the drain current. The nonequilibrium Green's function (NEGF) method[8] is too computationally demanding and still impractical when all the major scattering mechanisms that are expected to play an important role in nanoscale MOSFETs – phonons, surface roughness, Coulomb scattering, *etc.* – are included. The fully quantum approach is still computationally prohibitive for a realistic two-dimensional (2D) MOSFET.

In our study, a Monte Carlo simulator including quantum corrections to the potential is used to study electronic transport in double gate (DG) SOI bulk MOSFETs with L_G down to 14 nm. Our results demonstrate that, for the explored L_G values, scattering still controls the on-current I_{ON}, keeping it much lower than

Future Trends in Microelectronics. Edited by Serge Luryi, Jimmy Xu, and Alex Zaslavsky
ISBN 0-471-48

the ballistic limit I_{BL}. By monitoring the back-scattered electrons at the source, we discuss the role of scattering in different parts of the device.

We have also investigated the impact of technology scaling on quasi-ballistic transport by computing the *BR* for MOSFETs designed according to the 2003 Roadmap, down to the 45 nm node. Our results show that in the case of devices for which phonon scattering is dominant, *e.g.* double-gate (DG) MOSFETs with lightly-doped channels, *BR* exhibits an universal dependence on L_G for the TNs we have considered. The impact of quasi-ballistic transport increases for L_G below ~50 nm and contributes most of the I_{ON} improvements related to scaling. Thanks to a lower transverse electric field, the DG SOI MOSFETs with low channel doping get closer to the ballistic limit than their bulk counterparts.

2. Device design and simulation

We simulated both bulk and DG SOI devices representative of the high-performance transistors at the 130, 90, 65, and 45 nm TNs; the devices have been designed to comply with the specifications of the ITRS 2003 (or ITRS 2001, for bulk devices of the 130 nm TN). The main device characteristics are reported in Table 1 and correspond to the devices presented in Refs. 6, 7. The doping profiles of the bulk MOSFETs have been tailored in order to keep short channel effects (SCE) and gate leakage below the limits set by the ITRS for each TN. For the SOI transistors, we assumed lightly-doped silicon films and used a DG device architecture, where the specification on the drain-induced barrier lowering (DIBL) was met by scaling the thickness of the silicon layer t_{Si}. Since $t_{Si} > 10$ nm for all the simulated devices, the dependence of the scattering rates and of the mobility on the thickness of the silicon layer is essentially negligible.[9] The donor concentration profiles of the source and drain regions for the 90 and 65 nm TNs are the same adopted for the corresponding bulk MOSFETS, but no halo implants are included. In the case of the 45 nm TN, the profiles were obtained by scaling the doping profile parameters in order to keep SCE under control, without scaling t_{Si} below 10 nm. Ideal *n*-polysilicon gate electrodes have been assumed in the case of bulk transistors, while the work-function of the gate electrodes of DG SOI MOSFETs was tailored in order to get a correct value for the threshold voltage. All devices feature a moderately high relative dielectric constant ($\varepsilon_{ox} = 7$). The length of the S/D access regions is set to 50 nm for all the simulated devices.

The simulation model consists of a full-band self-consistent ensemble Monte Carlo (MC) simulator for the free-electron gas that has been adapted to the study of nanoscale MOSFETs. The effect of carrier quantization on the inversion charge is taken into account by using the effective potential approach.[10] The scattering model includes phonons, ionized impurities, carrier-plasmon and surface roughness (SR) interactions. The SR scattering been implemented as an additional scattering mechanism, whose rate is related to the vertical effective field following an original approach that adapts the scattering rate of the 2D electron gas to the full-band simulator for free carriers.[11]

Technology node		130 nm	90 nm	65 nm	45 nm
Supply voltage V_{DD} (V)		1.2	1.2	1.1	1.0
Nominal gate length (nm)		65	37	25	18
Relative dielectric constant		7	7	7	7
t_{ox}/EOT (nm)		2.87/1.6	2.15/1.2	1.6/0.9	1.26/0.7
DGSOI	t_{Si} (nm)		17	12	10
	DIBL (mV/V)		91	97	112
	I_{off} (nA/μm)		4	7	10
	Gate WF (eV)		4.6	4.6	4.6
bulk	N_{sub} ($\times 10^{18}$ cm^{-3})	2.5	2 + halo	3 + halo	
	DIBL (mV/V)	105	108	120	
	I_{off} (nA/μm)	2.5	30	60	
	Gate WF (eV)	4.05	4.05	4.05	

Table 1. Main technological parameters of the simulated devices. For each TN the nominal gate length for high-performance transistors is reported.

3. Scattering in the channel and in the drain

In a previous publication,[6] we compared I_{ON} and its ballistic limit I_{BL} (computed by neglecting scattering in the channel) of a 25 nm DGSOI MOSFET designed according to the 45 TN for $V_{GS} = V_{DS} = 1$ V. The results demonstrated the important role of scattering even in this ultra-short device, since I_{BL} is approximately 1.5 times larger then I_{ON}. In order to understand the large discrepancy between I_{ON} and I_{BL}, Fig. 1 plots the total number of scattering events per unit length and time suffered by electrons moving from source to drain a with positive component of the group velocity (dashed line in Fig. 1). This is equivalent to an inverse mean-free-path for the forward-moving electrons, and the value in the channel is $1/\lambda^+ \sim 0.1$ nm^{-1}, meaning that for this biasing and device geometry, forward-moving carriers suffer approximately one scattering event every 10 nm. The solid line in Fig. 1 indicates the contribution to the back-scattered current at the *virtual source* x_{inj} (*i.e.* the point where the maximum of the source-channel barrier is located) due to scattering events at different positions within the channel. The spatial source of the back-scattered current is sharply peaked at x_{inj} and decays exponentially for $x > x_{inj}$, with a characteristic decay length hereafter denoted as L_{scat}. This confirms that the dominant contribution to the back-scattered current is given by scattering events close to the potential barrier, in agreement with Lundstrom and Ren.[12]

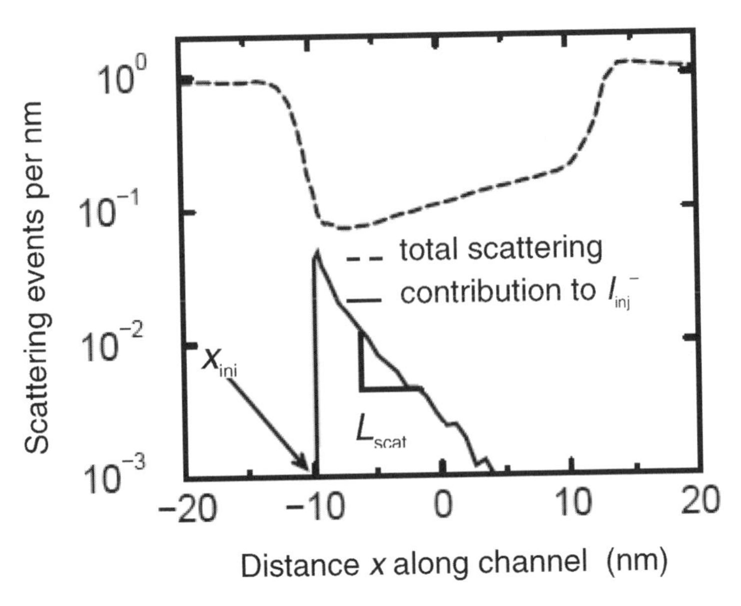

Figure 1. Dashed line: number of scattering events per unit length and time, suffered by electrons moving from source to drain, *i.e.* with positive component of the group velocity. Solid line: number of scattering events contributing to the back-scattered current at the virtual source I_{inj}^-, per unit length and time. Both curves are normalized to the current injected at the virtual source (calculations for DGSOI, L_G = 25 nm, $V_{GS} = V_{DS}$ = 1 V).

In Fig. 2 the ballistic ratio *BR* is related to the reflection coefficient *r*, defined as $r = I_{inj}^-/I_{inj}^+$, where I_{inj}^- and I_{inj}^+ are the carrier fluxes at the virtual source in the simulation with scattering. The *BR* is lower than $(1-r) = (I_{ON}/I_{inj}^+)$, because I_{inj}^+ increases when the scattering is turned off. This can be understood by invoking the assumption[12] that the inversion charge at the virtual source is controlled only by the gate capacitance and voltage, and is therefore the same with and without scattering, so that without scattering we have more carriers moving with a positive group velocity. With the additional assumption that the velocities of carriers contributing to I_{inj}^+ and I_{inj}^- fluxes are equal, it follows that $BR \approx (1-r)/(1+r)$.[12]

It has been suggested that in ultra-short devices, backscattering at the drain could have a noticeable impact on I_{ON}.[13] However, our results in Fig. 1 indicate that the contribution to I_{inj}^- due to scattering at the drain is negligible, and Fig. 3 (filled squares) confirms that this is the case also in a much shorter device with L_G = 14 nm. Figure 3 also shows that when the scattering inside the channel is turned off, as done in Ref. 13, the contribution to I_{inj}^- due to back-scattering at the drain becomes much higher (filled circles), and even larger if plasmon scattering in the drain is switched off (open circles). This is because particles injected in a ballistic

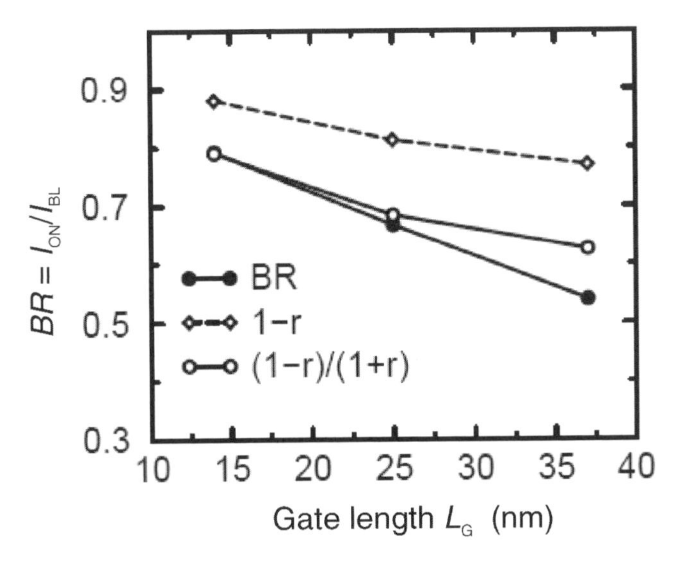

Figure 2. Filled symbols: ratio *BR* between the drain current I_{ON} and its ballistic limit I_{BL}. Open symbols: (1−*r*) (diamonds) and (1−*r*)/(1+*r*) (circles), where $r = I_{inj}^{-}/I_{inj}^{+}$ (calculations for 45 TN SOI devices, $V_{GS} = V_{DS} = 1$ V).

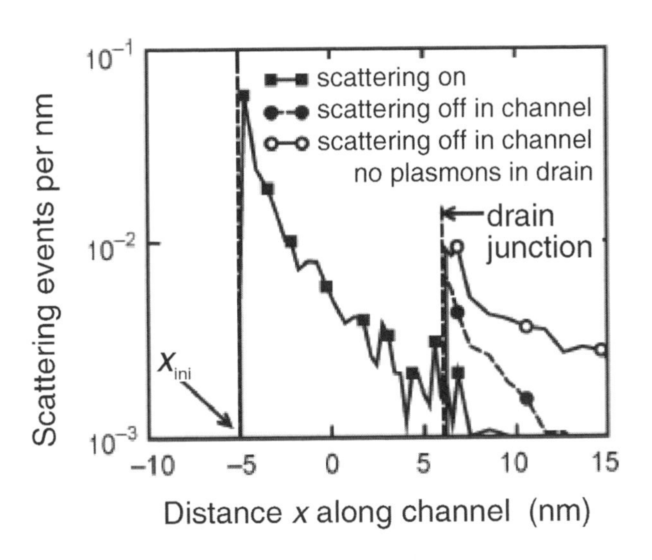

Figure 3. Contribution to I_{inj}^{-} given by the scattering events at a given position along the channel. Filled squares: scattering is activated in the whole device; filled circles: scattering is turned off inside the channel; open circles: scattering is turned off inside the channel, and plasmon scattering is turned off inside the drain. The device is similar to the one in Fig. 1, except that L_G = 14 nm ($V_{GS} = V_{DS} = 1$ V).

channel enter the drain with an energy equal to or higher than the source barrier. Therefore, when backscattered by elastic collisions, they have a high probability of traveling back to the source and contributing to I_{inj}^{-}. Plasmons reduce the back-scattered flux, because they dissipate a significant fraction of energy of the hot electrons entering the drain. On the other hand, if inelastic scattering in the channel (*e.g.* optical phonon scattering) reduces the carrier energy, then carriers entering the drain do not have enough energy to go back to the source even if they are back-scattered toward the source. This result points out that it is methodologically incorrect to estimate the impact of scattering at the drain by switching off scattering in the channel.

4. Ballistic ratio and scaling

Figure 4 shows the simulated I_{ON} and I_{BL} for nominal gate length transistors of the different TNs. Unlike I_{BL}, the simulated I_{ON} values increase significantly as the gate length is scaled down, which necessarily implies an improvement of the ballistic ratio *BR*, as illustrated explicitly in Fig. 5.

It is interesting to observe that in the case of DG SOI MOSFETs with an almost undoped silicon film, the *BR* values fall essentially on the same curve for all TNs, so that *BR* is a unique function of L_G. We have previously shown[7] that the reflection coefficient at the virtual source (and thus *BR*) depends on the ratio between L_{scat} and a suitably defined carrier mean free path λ_{mfp}: smaller λ_{mfp}/L_{scat} leads to a larger reflection coefficient at the virtual source and hence lower *BR*. For the ITRS scaling strategy, L_{scat} is uniquely related to the channel length,

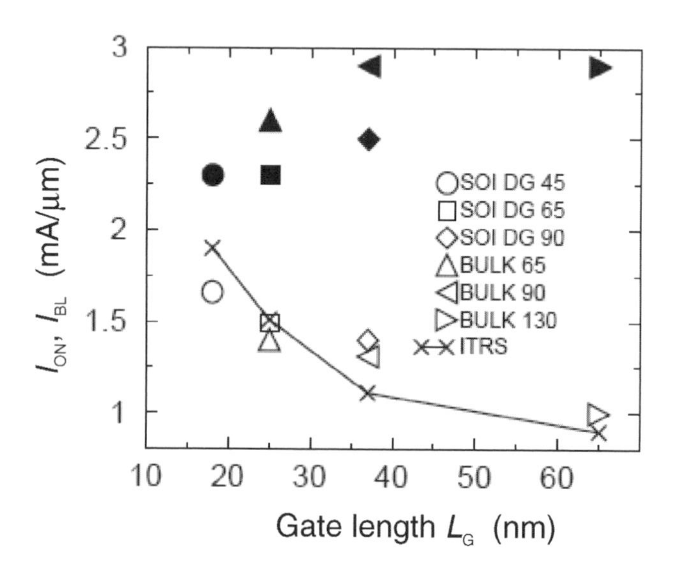

Figure 4. Current *vs.* nominal gate length at $V_{GS} = V_{DS} = V_{DD}$ for each TN. Filled symbols: ballistic current. Open symbols: current with scattering.

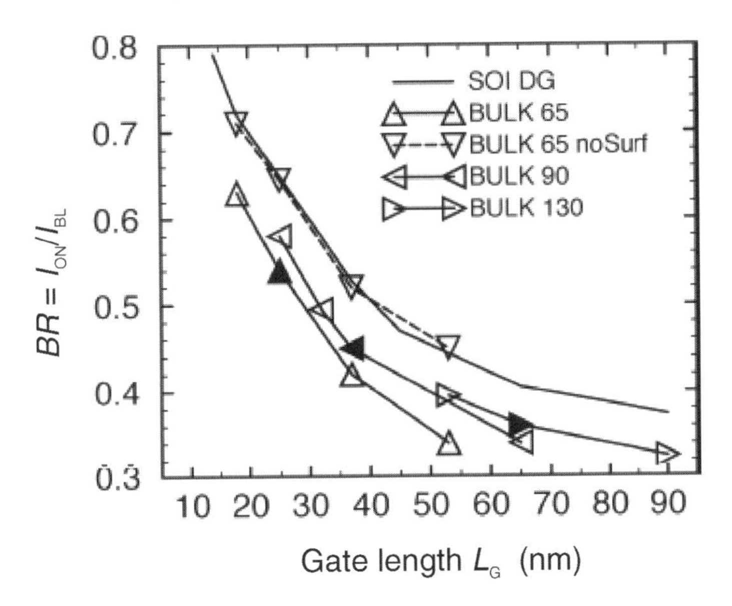

Figure 5. Ballisticity ratio *BR vs.* L_G at $V_{GS} = V_{DS} = V_{DD}$ for devices with different L_G values and different TNs. The solid line represents the results for DG MOSFETs, which fall on the same "universal" curve. Symbols report results for bulk MOSFETs. Filled symbols denote devices with the nominal L_G for the high-performance MOSFET of each TN. Downward-pointing triangles joined by a dashed line simulate bulk 65 nm TN devices with surface scattering turned off.

whereas for a given L_G the devices belonging to different TNs have very similar vertical effective fields and scattering rates at the virtual source, and hence similar λ_{mfp}.[7] Consequently, the ratio λ_{mfp}/L_{scat} becomes a unique function of L_G, justifying qualitatively the *BR* to L_G correlation of Fig. 5.

It should be emphasized that such a correlation results from the combination of EOT and V_{DD} values given by the ITRS for each L_G and should not be construed as a universal curve in a physical sense.

The correlation between *BR* and L_G breaks down in the case of bulk MOSFETs, because the change in channel doping concentration affects λ_{mfp} through ionized impurity scattering and surface-roughness scattering, the latter enhanced by the larger vertical field induced by increased doping concentration.

This view is supported by Fig. 5, showing that as we move from one bulk TN to the following one, the increase of doping necessary to counteract SCE also affects the *BR*. This keeps the *BR* scaling curve for bulk devices below that of the SOI devices, due to increased scattering. This reduction in performance occurs mainly through the increase of surface roughness scattering, as proven by the fact that when this scattering mechanism is switched off in bulk devices (downward-pointing triangles in Fig. 5), the scaling trend of DG SOI devices is recovered.

These results indicate that scattering will prevent fully ballistic transport down to at least L_G = 14 nm; nonetheless, nanometric MOSFETs are expected to get progressively closer to the ballistic limit at the TNs of the near future. In this sense, quasi-ballistic effects will be a booster for the overall device performance of future TNs, provided that other effects (*e.g.* increased source-drain series resistances) do not reduce the ballistic upper limit. Double-gate SOI MOSFETs, thanks to low doping concentration in the channel, allow for larger *BR* values, whereas in the case of scaled bulk technologies, ballistic transport appears to be limited by increased surface roughness scattering.

5. Conclusions

By using Monte Carlo simulations we have shown that scattering still controls the MOSFET on-current for L_G down to at least 14 nm. The main role is played by scattering events near the source barrier. The effect of scattering on I_{ON} is not simply proportional to the number of backscattering events and this is in agreement with the picture presented in Ref. 12.

A significant increase in the ballistic effects is expected for the future technology nodes, which follows a very tight correlation to L_G, more evident for lightly-doped DG SOI MOSFETs than for bulk MOSFETs.

Acknowledgments

This work was partially funded by the Italian MIUR through the PRIN 2004 and FIRB (RBNE012N3X) projects and by the U.E. SINANO Network of Excellence (FP6, IST-1-506844-NE) and EUROSOI (IST-1-506653-CA).

References

1. International Technology Roadmap for Semiconductors (ITRS), 2004 Update Emerging Research Devices (ERD).
2. K. Natori, "Ballistic metal-oxide-semiconductor field effect transistor," *J. Appl. Phys.* **76**, 4879 (1994).
3. M. Lundstrom, "Elementary scattering theory of the Si MOSFET," *IEEE Electron Dev. Lett.* **18**, 361 (1997).
4. M. Lundstrom, "Device physics at the scaling limit: What matters," *Tech. Digest IEDM* (2003), pp. 789–792.
5. S. Takagi, "Device reexamination of subband structure engineering in ultra-short channel MOSFETs under ballistic carrier transport," *Tech. Dig. VLSI Symp.* (2003), pp. 115–116.

6. P. Palestri, D. Esseni, S. Eminente, C. Fiegna, E. Sangiorgi, and L. Selmi, "Understanding quasiballistic transport in nano-MOSFETs: Part I—Scattering in the channel and in the drain," *IEEE Trans. Electron Dev.* **52**, 2727 (2005).
7. S. Eminente, D. Esseni, P. Palestri, C. Fiegna, L. Selmi, and E. Sangiorgi, "Understanding quasiballistic transport in nano-MOSFETs: Part II—Technology scaling along the ITRS," *IEEE Trans. Electron Dev.* **52**, 2736 (2005).
8. S. Datta, "The non-equilibrium Green's function (NEGF) formalism: An elementary introduction," *Tech. Digest IEDM* (2002), pp. 703–706.
9. D. Esseni, M. Mastrapasqua, G. K. Celler, C. Fiegna, L. Selmi, and E. Sangiorgi, "Low field electron and hole mobility of SOI transistors fabricated on ultra-thin silicon films for deep sub-micrometer technology application," *IEEE Trans. Electron Dev.* **48**, 2842 (2001).
10. D. K. Ferry, R. Akis, and D. Vasileska, "Quantum effects in MOSFETs: Use of an effective potential in 3D Monte Carlo simulation of ultra-short channel devices," *Tech. Digest IEDM* (2000), pp. 287–290.
11. P. Palestri, S. Eminente, D. Esseni, C. Fiegna, E. Sangiorgi, and L. Selmi, "An improved semiclassical Monte-Carlo approach for nano-scale MOSFET simulation," *Solid State Electronics* **49**, 727 (2005).
12. M. Lundstrom and Z. Ren, "Essential physics of carrier transport in nanoscale MOSFETs," *IEEE Trans. Electron Dev.* **49**, 133 (2002).
13. A. Svizhenko and M. Anantram, "Role of scattering in nanotransistors," *IEEE Trans. Electron Dev.* **50**, 1459 (2003).

Absolute Negative Resistance in Ballistic Variable Threshold Field Effect Transistor

Michel I. Dyakonov
Laboratoire de Physique Théorique et Astroparticules
Université Montpellier II, France

Michael S. Shur
ECSE Department and Broadband Center
Rensselaer Polytechnic Institute, Troy, NY 12180, U.S.A.

We consider a ballistic variable threshold field effect transistor (BVTFET),[1-4] where the electron transport is ballistic (no collisions with impurities or lattice vibrations), see Fig. 1(a). We will use the hydrodynamic approach,[5] which is justified if the electron-electron collisions are much faster than collisions with impurities and/or phonons. Qualitatively, our results hold even if this condition is not met, so long as the mean free path is greater than the characteristic length scale for the variation of the electron concentration.

Within this approach, the situation we consider is similar to a water flow through a pipe with varying cross-section, see Fig. 1(b) and (c). If there is no flow, obviously the pressure is the same on both sides. In the presence of a stationary water flow, the Bernoulli law (which follows from energy conservation) tells us that the pressure will be lower in the narrow part of the pipe, where the velocity is larger. Note, that this fact does *not* depend on the direction of the flow. This means that when water flows from the narrow part towards the wide part, the direction of the flow is locally *opposite* to the pressure gradient. One can say that the interface region has an absolute negative resistance. We will show below, that a similar effect should exist for the BVTFET.

To be specific, we assume, as shown in Figure 1, that the device threshold voltages for the section with coordinates $0 < x < L_1$ and $L_1 < x < L$ are V_{T1} and V_{T2}, respectively (with $V_{T2} < V_{T1}$) and that the source contact at $x = 0$ is grounded. The gate and drain voltages are V_G and V_D, respectively. The voltage drop in the ballistic sections of the device is zero, so the entire voltage drop occurs at the interface between the two sections at $x = L_1$, see Fig. 1(a). We will treat this interface as abrupt, assuming that the Debye length and the separation between the gate and the channel are small compared to L_1.

The basic equations describing this device are the equation of motion and the continuity equation,

$$\partial v/\partial t + v\partial v/\partial x = (e/m)\partial V/\partial x \text{ , and} \tag{1a}$$

$$\partial n/\partial t + \partial(nv)/\partial x = 0 \text{ ,} \tag{1b}$$

Future Trends in Microelectronics. Edited by Serge Luryi, Jimmy Xu, and Alex Zaslavsky

as well as the equations relating the concentrations n to the local gate-to-channel voltage swing in the gradual channel approximation:

$$en_1 = C(V_G - V_{T1}), \tag{2}$$

$$en_2 = C(V_G - V_{T2} - V_D) = en_1 + C(V_{T1} - V_{T2}) - CV_D. \tag{3}$$

Here C is the gate-to-channel capacitance, e is the electron charge, V_G is the gate voltage, V_{T1} and V_{T2} are the threshold voltages for the left and right sides of the transistor respectively, $V_D = V_1 - V_2$ is the source-drain potential difference, V_1 and V_2 are the channel potentials on the left and on the right side of the device.

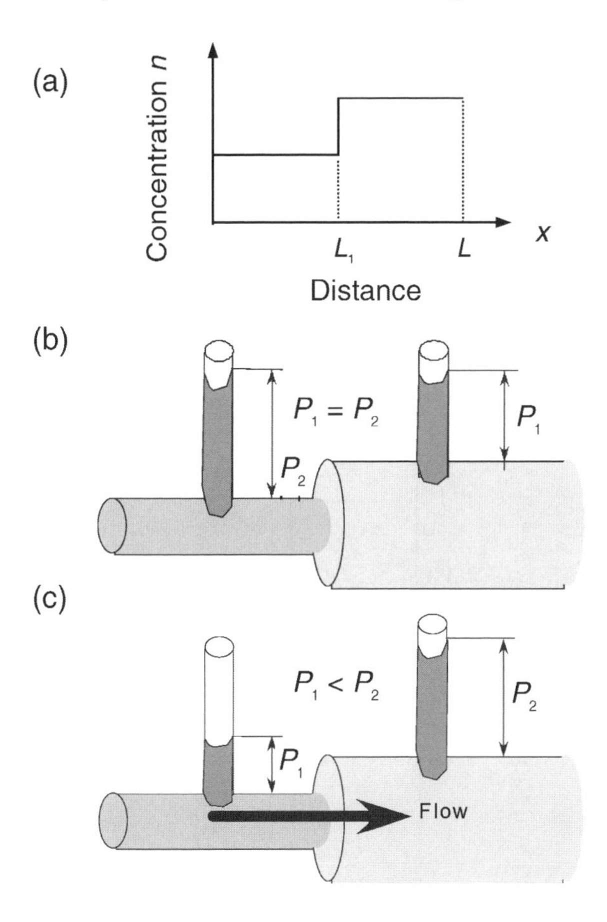

Figure 1. (a) Electron concentration distribution in a variable threshold field effect transistor; (b) a water pipe with varying cross section. In the absence of flow the pressures on both sides are obviously equal; (c) the same water pipe in the presence of flow – the pressure in the narrow part of the pipe is lower (Bernoulli law), regardless of the flow direction.

In the stationary situation that we are considering, Eqs. (1a) and (1b) yield:

$$mv_1^2/2 = mv_2^2/2 - eV_D \,, \; n_1v_1 = n_2v_2 \,, \tag{4}$$

where v_1 and v_2 are the left and right electron drift velocities, The first of Eqs. (4) is analogous to the Bernoulli law in hydrodynamics.

We denote by n_{10} and n_{20} the equilibrium electron concentration on the two sides of the interface when the current is absent. The concentration, velocity, and gate voltage swing profiles for the case $n_{10} < n_{20}$ are shown in Fig. 2. We will be interested also in the opposite case, when $n_{10} > n_{20}$. We will consider the situation when the left side of the device with the electron concentration n_{10} is grounded ($V_1 = 0$). Then the concentration on this side is always fixed: $n_1 = n_{10}$, while the concentration on the right side, n_2, will vary with changing current.

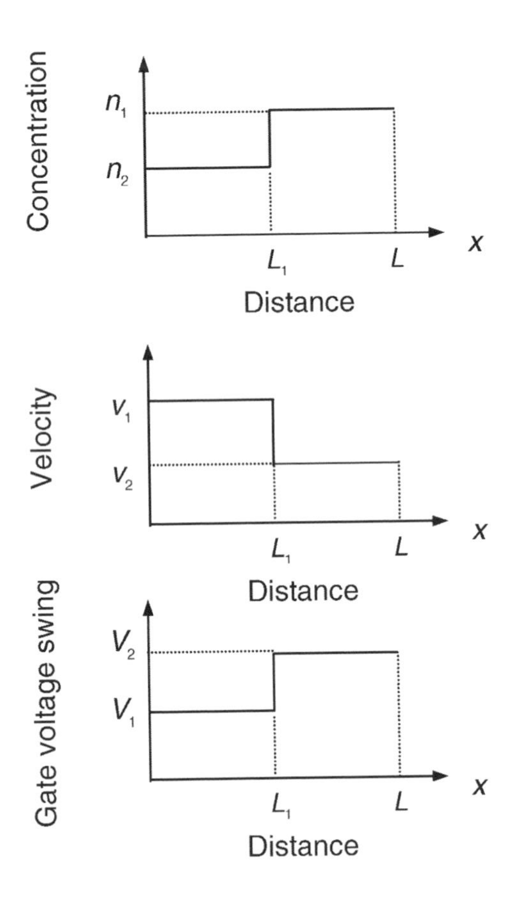

Figure 2. Concentration, velocity, and gate voltage swing profiles in BVTFET for the case $n_{10} < n_{20}$. Obvious modifications should be made for the opposite case.

If $n_{10} < n_{20}$ and $v_1 > v_2$ (see Figure 2), V_D is negative (meaning that the high concentration side has a more negative potential than the low concentration side with the electron concentration n_1). In the opposite case, when $n_{10} > n_{20}$ and $v_1 < v_2$, V_D is positive.

We now introduce dimensionless variables U for the source-drain voltage and J for the current, together with a dimensionless parameter N defined as the ratio of the carrier concentrations in the right section of the device to that in the left section in equilibrium (*i.e.* at zero current):

$$U \equiv C|V_D|/en_{10}\,,\quad J \equiv v_1/v_0\,,\; N \equiv n_{20}/n_{10} = 1 + C(V_{T1} - V_{T2})/en_{10}\,,$$

where the characteristic electron velocity v_0 is given by:

$$v_0 = \left(\frac{2e^2 n_{10}}{mC}\right)^{1/2} .$$

We find from Eqs. (2)–(4) for negative V_D:

$$n_2 = n_1 + C(V_{T1} - V_{T2})/e + C|V_D|/e \qquad (N > 1) \qquad (5a)$$

$$mv_1^2/2 = mv_2^2/2 + e|V_D| \qquad (N > 1)\,, \qquad (5b)$$

and for positive V_D, respectively:

$$n_2 = n_1 + C(V_{T1} - V_{T2})/e - C|V_D|/e \qquad (N > 1) \qquad (6a)$$

$$mv_1^2/2 = mv_2^2/2 - e|V_D| \qquad (N > 1)\,. \qquad (6b)$$

For typical values of an AlGaAs/GaAs ballistic HEMT, we can assume $V_G = 0$, $V_{T2} = -1$ V, $V_{T1} = -0.5$ V, $m = 0.067m_0$ (where m_0 is the free electron mass), and $C = 5\text{x}10^{-7}$ F/cm^2 (corresponding to approximately 20 nm gate-to-channel separation). These values yield a characteristic current density of approximately 4 μA/μm.

In the dimensionless units introduced above, Eqs. (5) and (6) can be rewritten as:

$$J^2 = \frac{U(N+U)^2}{(N+U)^2 - 1} \qquad (N < 1) \qquad (7a)$$

$$J^2 = \frac{U(N-U)^2}{1-(N-U)^2} \qquad (N > 1)\,. \qquad (7b)$$

For $N > 1$ (*i.e.* $n_{10} > n_{20}$), V_D is negative and increases the carrier concentration in the high-concentration section of the device. As seen from Eq. (7a), for $N > 1$ at large U, J^2 is proportional to U. At small $U << N$, the current-voltage characteristic for both cases becomes:

$$J = \pm N\left(\frac{U}{N^2 - 1}\right)^{1/2} \qquad (8)$$

Since typical current densities achieved in AlGaAs/GaAs HEMTs correspond to the range of J between 0.01 and 0.2, in many cases, Eq. (8) should apply.

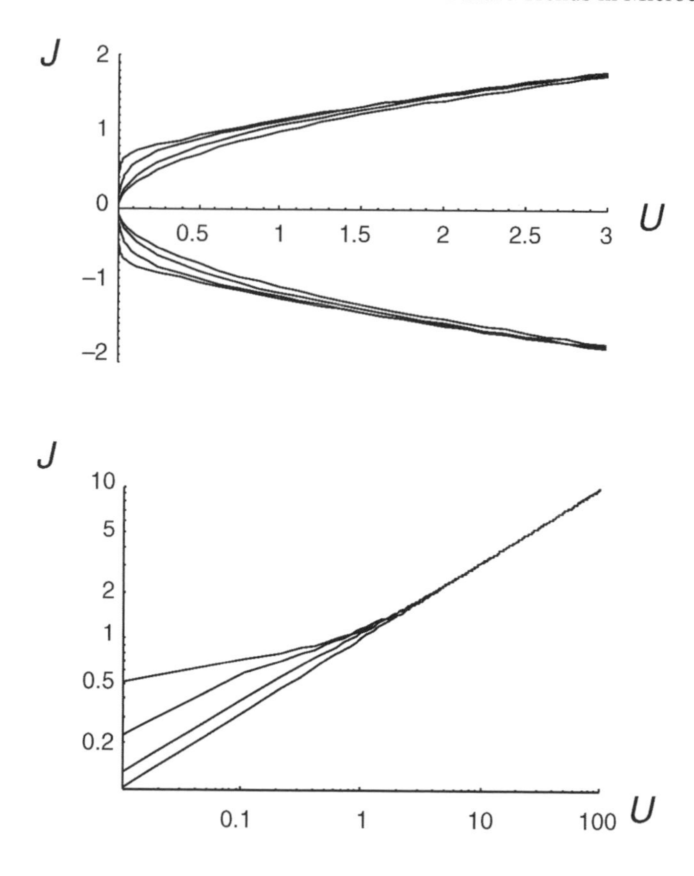

Figure 3. Dimensionless current-voltage characteristics of BVTFET for N = 1.01, 1.1, 1.6, and 10 plotted on linear (top) and double log (bottom) scales.

The resulting current-voltage characteristics for are shown in Fig. 3. For $n_{10} < n_{20}$, the carrier concentration in the right section of the device increases proportionally to the absolute value of V_D, as seen from Eq. 6(a). The dimensionless electron velocity in this section is given by

$$\frac{v_2}{v_0} = \frac{U^{1/2}}{(N+U)^2 - 1} \qquad (N > 1) \qquad (9)$$

For $n_{10} < n_{20}$, velocity v_2 reaches a maximum value at:

$$\left(\frac{v_2}{v_0}\right)_{max} = \left(\frac{(N^2-1)^{1/2}}{[(N^2-1)^{1/2} + N]^2 - 1}\right)^{1/2} \qquad (N > 1)\,. \qquad (10)$$

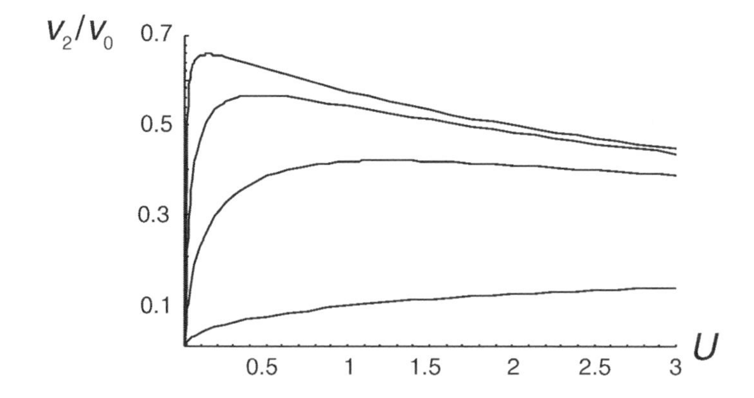

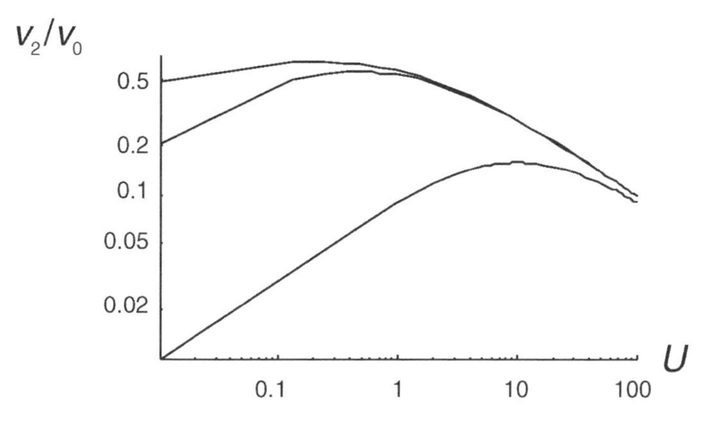

Figure 4. Dependence of v_2/v_0 on U for N = 1.01, 1.1, 1.6 and 10 plotted on linear (top) and double log (bottom) scales.

Figures 4(a) and 4b show the dependence of v_2/v_0 on U for $n_{10} < n_{20}$. In the opposite bias polarity, when $n_{10} > n_{20}$, we have $N < 1$ and V_D is positive. The resulting current-voltage characteristics for $n_{10} > n_{20}$ obtained from Eq. (7b) are shown in Fig. 5, where we also superimpose the dependence of the choking current J_{CH} on the choking voltage U_{CH} (found from the requirement that the electron velocity in region 2 for $n_{10} > n_{20}$, $N < 1$ is equal to the plasma wave velocity in this region):[6]

$$J_{CH} = (N - U_{CH})^{3/2}/\sqrt{2} \qquad (N < 1)\,. \tag{11}$$

As can be seen, the maxima of the current-voltage characteristics at $N > 0.75$ correspond to the choking current.

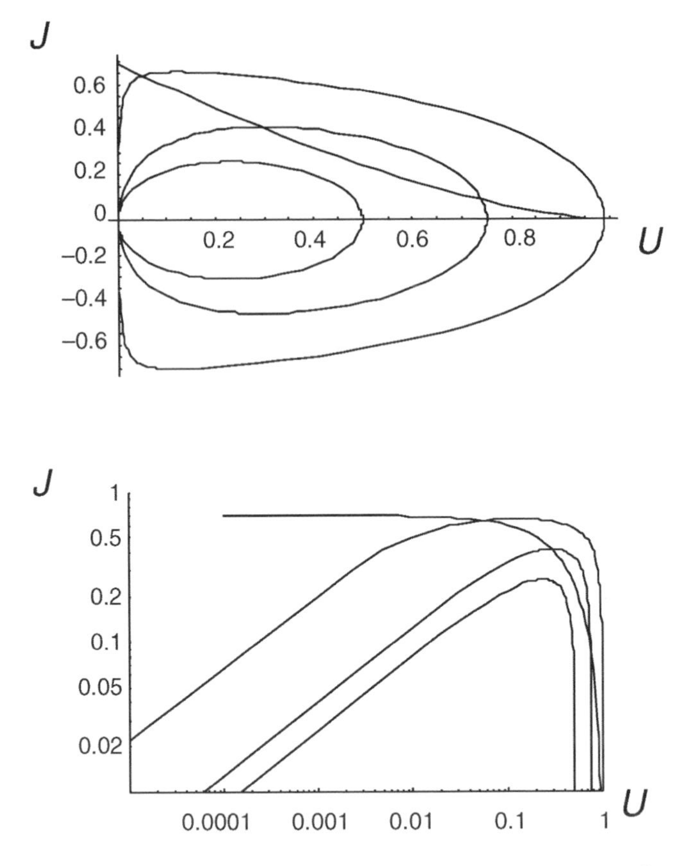

Figure 5. Dimensionless current-voltage characteristics of BVTFET for $n_{10} > n_{20}$ and $N = 0.5$, 0.75 and 0.99 plotted on linear (a) and double log (b) scales. The choking current dependence on the choking voltage is also shown.

Figure 6 shows the dimensionless electron velocity in region 2 as a function of U for $N < 1$, where we also superimpose the dependence of the choking velocity on the choking voltage (found from the requirement that the electron velocity in region 2 for $n_{10} > n_{20}$, $N < 1$ is equal to the plasma wave velocity in this region):

$$v_{2CH} = (N - U_{CH})^{1/2}/\sqrt{2} \qquad (N < 1) \qquad (12)$$

The physical meaning of the obtained results is as follows. The electrons crossing the boundary at $x = L_1$ must keep their energy, which can be only achieved by creating a dipole layer at the boundary leading to the voltage drop that changes the electron concentration in one section of the device and adjusts the electron velocities in both device sections accordingly. In the device section with the fixed potential, the carrier concentration is independent of the current. The other section accommodates any change in current by adjusting the carrier concentration. If the concentration in this section is higher, it increases further with an increase in the

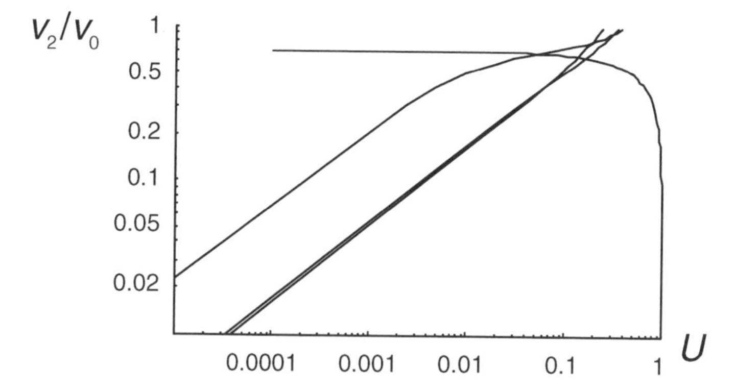

Figure 6. Dimensionless velocity v_2/v_0 in section 2 for $n_{10} > n_{20}$ for $N = 0.5$, 0.75 and 0.99 *vs*. U. The choking velocity dependence on the choking voltage is also shown.

device current. If the concentration in this section is smaller, it gets depleted with an increase in the device current. The striking result is that this voltage drop is independent of the direction of the current, so that, depending on the direction of the current, the device should exhibit equal positive or absolute negative resistance. The consequences of this absolute negative resistance will be discussed elsewhere.

Acknowledgments

This work was supported in part by ONR under EMMA MURI contract (project manager Dr. Colin Wood).

References

1. M. S. Shur, "Split-gate field effect transistor," U.S. patents 5,012,315 (April 30, 1991) and 5,079,620 (January 7, 1992).
2. G. Kelner and M. S. Shur, "Junction field effect transistor with lateral gate voltage swing (GVS-JFET)," U.S. patent 5,309,007 (May, 1994).
3. M. S. Shur, "Split-gate field effect transistor," *Appl. Phys. Lett.* **54**, 162 (1989).
4. G. U. Jensen and M. S. Shur, "Variable threshold heterostructure FET studied by Monte Carlo simulation," in: K. Hess, J. P. Leburton, and U. Ravaioli, eds., *Computational Electronics. Semiconductor Transport and Device Simulation*, Norwell, MA: Kluwer Academic Publishers, 1991, pp. 123–6.
5. M. Dyakonov and M. S. Shur, "Shallow water analogy for a ballistic field effect transistor. New mechanism of plasma wave generation by dc current," *Phys. Rev. Lett.* **71**, 2465 (1993).
6. M. Dyakonov and M. S. Shur, "Choking of electron flow – A mechanism of current saturation in field effect transistors," *Phys. Rev. B* **51**, 14341 (1995).

Formation of Three-Dimensional SiGe Quantum Dot Crystals

C. Dais, P. Käser, H. Solak, Y. Ekinci, E. Deckhardt, E. Müller, D. Grützmacher
Laboratory for Micro- and Nanotechnology
Paul Scherrer Institute, Villigen, Switzerland

J. Stangl, T. Suzuki, T. Fromherz, and G. Bauer
Inst. für Halbleiter und Festkörperphysik
Johannes Kepler University Linz, A-4040 Linz, Austria

1. Introduction

Silicon based quantum dot materials are gaining attention due their potential for ultimate CMOS devices[1] or for beyond-CMOS device architectures.[2-4] Germanium dots incorporated in Si are particularly promising because their fabrication can take advantage of the well-developed Si technology. The long spin coherence times in Si[5-7] represent another advantage of this material system, opening possible routes towards spintronics and quantum computation. The embedding of individually addressable Ge quantum dots into Si device technology might be crucial to open new paths for the fabrication of high speed electronics. Self-assembled Ge dots, which nucleate randomly on the surface and have typically a rather broad distribution in their sizes, have been studied intensively. Across a large window in deposition temperature a bimodal distribution of Ge islands is found, with small Ge huts coexisting with large Ge dome clusters.[8] However, many proposed devices, like the dot-FET[9] or dot arrays for qubit information technology,[2-4] require a uniform size distribution. Moreover, the addressing of individual dots will require the positioning of dots on predefined spots. In most approaches, a combination of pre-patterned substrates and self-assembled deposition was employed to achieve lateral ordering.[10,11] In the Si/Ge material system, patterns written by e-beam lithography and two-dimensional (2D) patterns fabricated by optical interference lithography have been used to achieve ordering of dome clusters.[12,13] Using CVD processes, Ge dots have also been selectively grown in oxide patterns.[14,15] However, these Ge islands were rather large and intermixed strongly with Si due to growth temperatures exceeding 600 °C.

In this chapter we present the lateral and 3D ordering of small Ge hut clusters on pre-patterned Si (100) surfaces. The pre-pattern was fabricated by X-ray interference lithography (XIL) using a dedicated beam line at the Swiss Light Source (SLS). XIL offers precise control over the periodicity of the pattern as well as patterns with periodicities in the sub-50 nm regime. The capability to expose an

Future Trends in Microelectronics. Edited by Serge Luryi, Jimmy Xu, and Alex Zaslavsky

area of 2x2 mm with a single shot, makes this technology a powerful alternative to e-beam lithography to fabricate structures in nanotechnology.

2. Fabrication and discussion

The process to fabricate ordered 2D and 3D arrays of Ge quantum dots starts with the formation of the pre-pattern on the (100) Si substrate. Silicon substrates were coated with PMMA resist and transferred into the vacuum chamber of the XIL beam line at the SLS.[16] Typically, areas of 0.7x0.7 mm were patterned in a single exposure. Multiple-beam diffraction was employed to achieve the exposure of 2D patterns. The gratings used for multiple-beam diffraction were fabricated using e-beam lithography from Si substrates coated with a SiN_x and a Cr film. After transferring the gratings into the Cr film, the Si underneath was removed, leaving the Cr gratings on free-standing SiN_x membranes. In our experiments, XIL was used to write patterns of either 90x100 nm or 250x250 nm into the PMMA resist. The pattern was transferred into the Si substrate by reactive ion etching (RIE). The depth of the pattern was restricted to 8–20 nm. After removal of the PMMA resist and cleaning of the patterned Si substrates, they were transferred into the solid source Si/Ge molecular beam epitaxy (MBE) system. The substrates were heated to 550 °C to remove the hydrogen from the surface and subsequently a 50 nm Si buffer layer was deposited at low temperatures. Note, that no high temperature anneal for *in-situ* cleaning of the Si substrate before the MBE process was employed. This requires special care during the wet chemical cleaning process to avoid build-up of contamination on the substrate. For 2D arrays of quantum dots, a single Ge island layer was deposited next; whereas for 3D quantum dot crystals, a multiple layer sequence of Ge dot layers and 10 nm Si spacer layers was grown.

The structures were analyzed using atomic force microscopy (AFM), transmission electron microscopy (TEM) and x-ray diffractometry. Figure 1 shows AFM images at different stages of the deposition process. In Fig. 1(a), a 2D hole array has been generated with a periodicity of 90x100 nm in the Si substrate using XIL exposure and subsequent RIE to transfer the pattern into the substrate. An exposure dose of 55 mJ measured at the location of the diffractive mask was used to expose the PMMA resist during XIL. After developing the resist, the checkerboard-like pattern consists of square holes with a depth of 8 nm and a periodicity of 90x100 nm in the *x-y* plane. The depressions appear dark in the AFM images of Fig. 1(a), (b), (d) and (e). The size of the holes is such, that they are still separated from each other, *i.e.* barely touching each other at their corners. Figure 1(b) shows this grating after the deposition of a 50 nm buffer layer. The holes shrunk in size and depth during the deposition of the buffer layer. The diameter is around 20 nm and the depth is about 4 nm. The Si surface around the holes is smooth. Figure 1(c), shows the final topography of the sample measured by AFM after the deposition of 7 monolayers of Ge. This thickness is above the critical thickness for Ge film deposition on Si, thus Ge islands have formed. The dots appear bright in the image of Fig. 1(c). A very regular array of Ge dots is formed with a density of $1.1x10^{10}$ cm^{-2}. The dots nucleate at the hole sites prepared by XIL and subsequent RIE. No islands are found on Si surfaces surrounding the holes. An island has nucleated in every hole, with no missing islands found within the inspected area. The islands have an almost square base and they are (105) faceted from bottom to the top, thus they belong into the category of Ge hut clusters.[8] Note that no elongated hut clusters with strongly rectangular bases have formed.

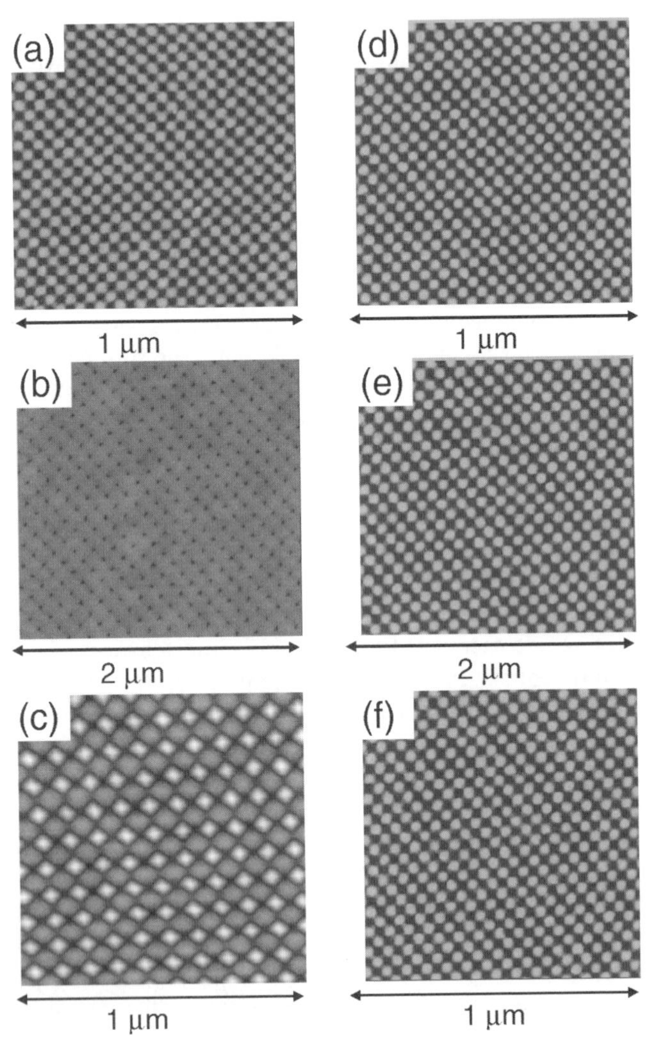

Figure 1. Formation of ordered arrays of Ge quantum dots described by AFM surface scans: (a) pre-patterned Si; (b) after the deposition of a 20 nm Si buffer, (c) ordered array of Ge quantum dots; (e)-(f) same experiment but using a higher exposure dose leading to twice the dot density.

Figures 1(d–f) show the same sequence for a sample that had been exposed with a higher dose during XIL, namely 75 mJ (measured at the mask). The higher dose leads to larger holes in the resist after the PMMA development. Again RIE was used to transfer the pattern into the Si substrate. Figure 1(d) shows that in this case the holes are touching each other. As a result, the substrate actually contains of an array of small Si pillars with a square base, which are bright in the AFM image of Fig. 1(d). The overgrowth of this pattern with a 50 nm thick Si buffer layer leads to the formation of a cross grating on the Si surface. In other words, the substrates has a periodic array of holes, however, these holes are linked to each other by grooves. Due to these grooves, the slight asymmetry of the 90x100 nm pattern in the *x-y* plane can be clearly seen in the AFM image of Fig. 1(e). The groves are aligned along the <110> directions of the Si (100) substrate. This pattern has also been overgrown by 7 monolayers of Ge using MBE. Again, a regular array of dots is formed, as shown in the AFM image of Fig. 1(f). Surprisingly, the islands do not nucleate in the holes, which are the deepest suppressions in the substrate, but within the grooves connecting the holes. Consequently, the dots have twice the density compared to those nucleating on the pattern shown in Fig. 1(a).

The dots have an elongated shape induced by the grooves. Notably, this is not the shape of elongated hut clusters,[17] which have a rectangular base along <001> directions. Here the grooves are along the <110> directions and accordingly the dots are stretched into this direction, leading to a rhombus-shaped base of the islands. Some islands appear to be formed from two adjacent nuclei. Most likely the sidewalls of the groves aligned along the <110> directions are (111) facets, since those are the slowest growing planes in Si. The fact that the islands nucleate in the grooves may indicate that corners of two adjacent (111) facets are preferred nucleation sites for Ge islands.

In order to fabricate 3D quantum dot crystals, the pre-patterned substrates have been overgrown by sequences of Ge and Si layers. Here, we used the process to fabricate the template Si substrate and the first island layer as described in Fig. 1(a–c). Figure 2 shows a cross-sectional TEM image (Z-contrast) of such a sample with 10 periods. The first period consisted of the 50 nm Si buffer layer and the first Ge dot layer (7 monolayers of Ge), whereas the subsequently grown nine periods contain 10 nm Si layers and 5 monolayer-thick Ge layers. The TEM micrograph clearly shows that the first Ge layer leads to the nucleation of Ge islands within the holes of the substrate. Thus this first Ge islands have an increased aspect ratio compared to conventional (105) faceted Ge huts. After the deposition of the first 10 nm Si spacer, the surface of the sample appears flat: the topography due to the pre-patterning and the island nucleation has vanished. However, the Ge islands are relaxed towards their apex, inducing tensile strain into the part of the Si layer covering the dot. Accordingly, the 10 nm Si spacer contains an undulating strain field reflecting the periodicity of the underlying Ge island layer. The next layer of Ge is affected by this strain field and Ge islands nucleate in those areas of the Si spacer layers that are under tensile strain, since the in-plane lattice constant is larger in these areas and hence better matched to the Ge lattice constant. This effect of vertical stacking of self-assembled Ge islands is well understood.[18] Here we make use of this effect to build up the 3D quantum dot crystal, since the lateral ordering of the Ge dots in the first dot layer will be transferred in the subsequently grown Ge layers by this self assembly.

The TEM micrograph in Fig. 2 nicely shows the stacking of the Ge islands. Neither Z-contrast nor conventional TEM micrographs reveal interface contamination or defects of the crystalline structure. In particular, the interface between the pre-patterned substrate and the first buffer layer is not visible.

The structural properties were investigated by x-ray diffractometry to gauge the crystalline perfection of the quantum dot crystal. Reciprocal space maps were recorded with a Seifert XRD 3003. The primary beam was prepared using a Göbel mirror, a single Ge (220) channel cut crystal, and a 1 mm slit. On the secondary side, a position sensitive detector (PSD) was used. Two maps in the [110] azimuth were recorded, the symmetrical (004) and the asymmetrical (224), as shown in Fig. 3.

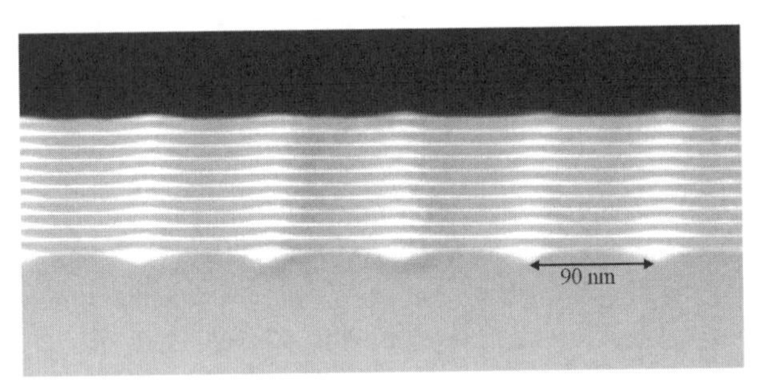

Figure 2. Z-contrast TEM image a ten-period 3D quantum dot crystal with a lateral periodicity of 90x100 nm and a vertical periodicity of 11 nm.

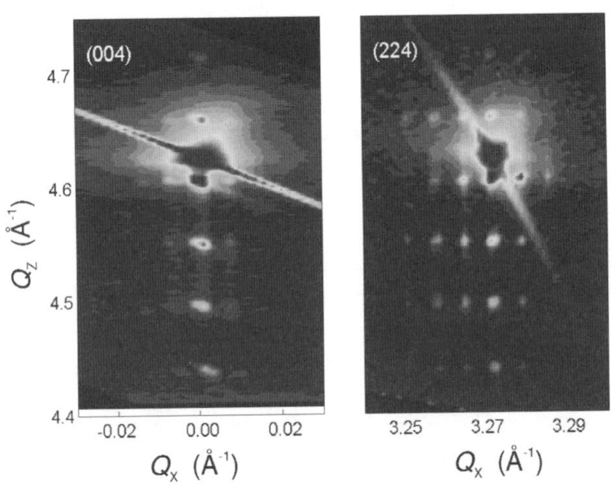

Figure 3. Reciprocal space maps of the Si/Ge quantum dot crystal taken at the symmetrical (004 – left) and the asymmetrical (224 – right) reflex.

From these maps, the lateral and vertical periods are obtained as 90.7 and 11.4 nm, respectively. By rotating the sample by 90° to change the azimuth, the second lateral periodicity of 100.2 nm is found. This is in good agreement with the nominal values. From the (224) map, one can see that the envelope of the satellites is shifted to small Q_X values, which is an indication of elastic relaxation in the Ge islands. Detailed analysis of this data, to gather insight into the Ge concentration of the islands, will be a subject of future study.

From the AFM and TEM data it can be found that the dots in the 3D dot crystal have a diameter of 20±5 nm, thus Ge dots exhibit a remarkably narrow size distribution and close to perfect ordering.

Figure 4 shows photoluminescence (PL) spectra obtained from structures containing a single Ge layer, two Ge dot layers and finally ten dot layers forming a 3D dot crystal. The single Ge layers of 3, 5.4 and 7 monolayer thicknesses were capped with 100 nm of Si, the double layer had a 10 nm cap layer and the crystal was uncapped, thus the topmost dot layer is exposed to air and most likely does not contribute to the PL spectrum. In order to interpret the PL spectra correctly, one has to consider that the radiative transitions in this Si/Ge system are indirect in both real and reciprocal space. Thus the complexity of the confinement situation for electrons and holes in these structures has to be considered. Isosurfaces of the wavefunctions around a SiGe quantum dot have been calculated by the *nextnano3* simulation package.[19] This simulation package provides quantitative information on quantized bound state energies for arbitrarily shaped 3D nanostructures, using a single-band $k{\cdot}p$ Hamiltonian. Elasticity theory is used to globally minimize the strain and local strain fields are incorporated into the band structure using deformation potential theory. Using the calculated energy levels and their dependence on the Ge concentration in the QD, the Ge content of the QDs can be estimated. Assuming a $Si_{0.5}Ge_{0.5}$ ($Si_{0.3}Ge_{0.7}$) dot and a QD shape as observed for the first QD layer (see Fig. 2), a calculated Δ_z^1–HH transition energy of 770 (580) meV and a Δ_{xy}–HH_1 transition energy of 810 (650) meV are obtained. Depending on the strain fields and on the composition gradients within the islands and considering the fact that at the large excitation densities used in these PL experiments several electron states around the QDs are likely to be occupied, a rather broad PL signal can be expected. This is true despite the narrow size distribution of the dots because many electron-hole transitions are optically allowed and the electrons are confined at different potential minima surrounding the dot having slightly different energies.

Furthermore, the stacking of Ge dots in the double layer and in the ten-period crystal will enhance the strain field between the dots and consequently increase the confinement energy for electrons.[20] In a stack of dots this strain field is strongest in the centre of the stack and gets weaker towards the bottom and top Ge dot layers. This change in the strain field in stacked dots leads to different confining potentials for the electrons[20,21] and, in turn, to a broadening of the PL line of stacked Ge dots. Thus, even if the Ge dots were all perfectly identical, a broadened PL line would be expected. These considerations should be kept in mind when interpreting the PL data shown in Fig. 4. All spectra are obtained from the pre-patterned areas of the sample. The dominant sharp line in the spectra at 1090 meV is the phonon-assisted replica of the Si substrate (Si-TO). The spectrum at the bottom of Fig. 4 is obtained for a three monolayer Ge layer embedded in Si, below the critical thickness for island formation, which is confirmed by the PL data showing a double peak below the Si-TO line typical for Ge wetting layer luminescence (marked WL in Fig. 4).[21] The deep broad peak (labelled D) around 730 meV is assigned to defects in the Si substrate, since this peak is observed for the as-received Si substrates. It is present in all single layer dot samples as well as in the double layer structure. It is not present in the quantum dot crystal, which has been deposited on a different Si substrate of higher quality. Increasing the Ge layer thickness to 5.4 monolayers leads to island formation. In the PL spectrum, the

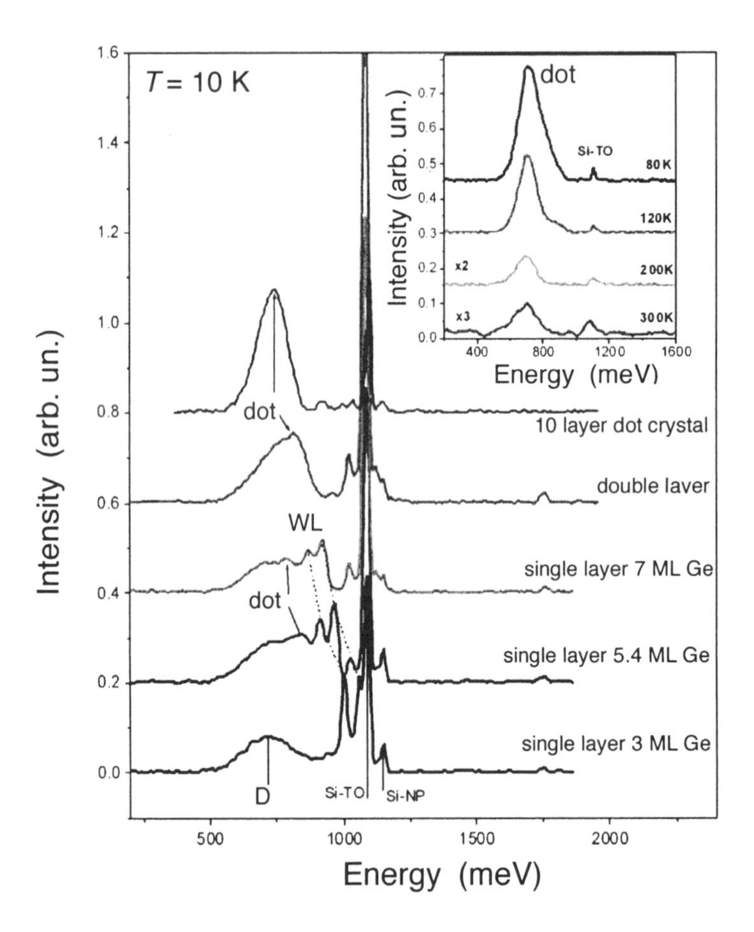

Figure 4. Comparison of low temperature (T = 10 K) photoluminescence of Si/Ge quantum dot crystals containing 1, 2 and 10 periods of dots. The inset shows high temperature photoluminescence (T = 80–300K) of a three-dimensional Ge quantum dot crystal.

wetting layer PL is shifted to lower energies due to the increased thickness and a new peak appears between the wetting layer PL and the defect band. As indicated in Fig. 4, this peak is assigned to the Ge dots. The wetting layer PL is shifted to a lower energy compared to literature data,[21] which might be attributed to Ge agglomeration in the areas between the dots. A further increase of the Ge layer thickness to 7 monolayers shifts the wetting layer PL only slightly and the dot PL to lower energies (784 meV), overlapping with the substrate defect band at 730 meV. Thus, experimentally, for a seven-monolayer single dot layer, the QD PL peak is observed around 784 meV. This value is very close to the calculated Δ_{xy}–HH_1 transition energy for a $Si_{0.5}Ge_{0.5}$ QD. Thus, from this comparison, a Ge concentration close to 50% is estimated. In a next step a second island layer (5 monolayers of Ge) has been added, separated by an 8 nm wide spacer layer from the first island layer (7 monolayers of Ge). The PL spectrum of the sample does not show resolved wetting layer luminescence, consistent with literature data,[21] where this effect has been explained by tunnel processes between closely spaced island layers. The PL line at 815 meV is attributed to the Ge hut clusters in the upper Ge layer. The peak is asymmetric with a tail towards lower energies, due to the overlap with the bigger Ge dots in the lower layer and the substrate defect line. Finally, the top spectrum in Fig. 4 shows the PL data of the ten-period Ge quantum dot crystal. The spectrum reveals a strong line centered at 746 meV. The shift in energy is explained by the increased confinement energy for electrons in stacked dots,[20] as well as by the vertical coupling of the dots shifting the quantized hole energy level. The inset in Fig. 4 shows the high-temperature PL of a ten-period Ge dot crystal. The luminescence of the Ge dots in persists up to room temperature, indicating a low defect density in the samples.

3. Conclusions

We have successfully fabricated 2D and 3D Si/Ge quantum dot crystals consisting of closely packed Ge islands. X-ray interference lithography (XIL) was employed to write arrays of 2D periods into PMMA resist and transfer the patterns into Si (100) substrates, which were used as templates for the deposition of ordered arrays of islands. Changing the x-ray exposure conditions and thereby the shape of the template makes it possible to control the shape and density of the Ge islands. Using the method of templated self-organization in combination with vertical stacking of Ge islands, 3D dot crystals were formed. Atomic force microscopy, x-ray diffraction and TEM analysis reveal the high perfection of those dot crystals. Photoluminescence measurements indicate a low defect density in the quantum dot structures.

Our results on the fabrication and properties of 2D and 3D Ge quantum dot crystals may open new routes towards the realization of nanoelectronic and spintronic devices as well as for quantum computing.

References

1. O. G. Schmidt, A. Rastelli, G. S. Kar, *et al.*, "Novel nanostructure architectures," *Physica E* **25**, 280 (2004).
2. D. Loss and D. P. DiVincenzo, "Quantum computation with quantum dots," *Phys. Rev. A* **57**, 120 (1998).
3. B. E. Kane, "A silicon-based nuclear spin computer," *Nature* **393**, 133 (1998).
4. M. Friesen, P. Rugheimer, D. E. Savage, *et al.*, "Practical design and simulation of silicon-based quantum-dot qubits," *Phys. Rev. B* **67**, 121301 (2003).
5. G. Feher and E. A. Gere, "Electron spin resonance experiments on donors in silicon. II. Electron spin relaxation effects," *Phys. Rev.* **114**, 1245 (1959).
6. C. Tahan, M. Friesen, and R. Joynt, "Decoherence of electron spin qubits in Si-based quantum computers," *Phys. Rev. B* **66**, 035314 (2002).
7. J. P. Gordon and K. D. Bowers, "Microwave spin echoes from donor electrons in silicon," *Phys. Rev. Lett.* **1**, 368 (1958).
8. A. Rastelli and H. von Känel, "Surface evolution of faceted islands," *Surf. Sci.* **515**, L493 (2002).
9. O. G. Schmidt and K. Eberl, "Self-assembled Ge/Si dots for faster field-effect transistors," *IEEE Trans. Electron Dev.* **48**, 1175 (2001).
10. E. S. Kim, N. Usami, and Y. Shiraki, "Control of Ge dots in dimension and position by selective epitaxial growth and their optical properties," *Appl. Phys. Lett.* **72**, 1617 (1998).
11. D. S. L. Mui, D. Leonard, L. A. Coldren, and P. M. Petroff, "Surface migration induced self-aligned InAs islands grown by molecular beam epitaxy," *Appl. Phys. Lett.* **66**, 1620 (1995).
12. O. G. Schmidt, N. Y. Jin-Phillipp, C. Lange, *et al.*, "Long-range ordered lines of self-assembled Ge islands on a flat Si (001) surface," *Appl. Phys. Lett.* **77**, 4139 (2000).
13. Z. Zhong, A. Halilovic, T. Fromherz, F. Schäffler, and G. Bauer, "Two-dimensional periodic positioning of self-assembled Ge islands on prepatterned Si(001) substrates," *Appl. Phys. Lett.* **82**, 4779 (2003).
14. T. I. Kamins and R. S. Williams, "Lithographic positioning of self-assembled Ge islands on Si(001)," *Appl. Phys. Lett.* **71**, 1201 (1997).
15. L. Vescan and T. Stoica, "Luminescence of laterally ordered Ge islands along <100> directions," *J. Appl. Phys.* **91**, 10119 (2002).
16. H. H. Solak, C. David, J. Gobrecht, *et al.*, "Sub-50 nm period patterns with EUV interference lithography," *Microelectronic Eng.* **67**, 56 (2003).
17. Y.-W. Mo, D. E. Savage, B. S. Swartzentruber, and M. G. Lagally, "Kinetic pathway in Stranski-Krastanov growth of Ge on Si(001)," *Phys. Rev. Lett.* **65** 1020 (1990).
18. F. Liu, A. H. Li, and M. G. Lagally, "Self-assembly of two-dimensional islands via strain-mediated coarsening," *Phys. Rev. Lett.* **87**, 126103 (2001).
19. J. A. Majewski, S. Birner, A. Trellakis, M. Sabathil, and P. Vogl, "Advances in the theory of electronic structure of semiconductors," *Phys. stat. sol. C* **1**, 2003 (2004).

20. O. G. Schmidt, K. Eberl, and Y. Rau, "Strain and band-edge alignment in single and multiple layers of self-assembled Ge/Si and GeSi/Si islands," *Phys. Rev. B* **62**, 16715 (2000).
21. O. G. Schmidt and K. Eberl, "Multiple layers of self-assembled Ge/Si islands: Photoluminescence, strain fields, material interdiffusion and island formation," *Phys. Rev. B* **61**, 13721 (2000).

Robust Metallic Interconnects for Flexible Electronics and Bioelectronics

D. P. Wang, F. Y. Biga, A. Zaslavsky, and G. P. Crawford
Division of Engineering, Brown University, Providence, RI 02912, U.S.A.

1. Introduction

Integrated interconnects for electronic applications have traditionally been restricted to planar rigid substrates that place negligible mechanical constraints on their performance. However, the continued interest in flexible electronic applications and convergence of biomedical implants with conventional electronics, necessitates the development of robust flexible interconnect schemes. Flexible electronics will encompass many applications, such as wearable electronics, large-area sensor arrays, bendable and conformal solar cell arrays, flexible displays, as well as flexible (and possibly stretchable) implantable biosensors.

In the case of next generation flexible displays, flexible interconnects are seen as one of the key enabling technologies that will bring the concept of rollable and bendable screens to fruition. Flexible display applications, which broadly include bendable, conformable and rollable displays,[1] will require flexible substrates, electronic components, optical layers as well as robust circuitry for functionality and reliable performance. Figure 1 shows a concept flexible display that allows for full rollability, similar to a conventional newspaper.

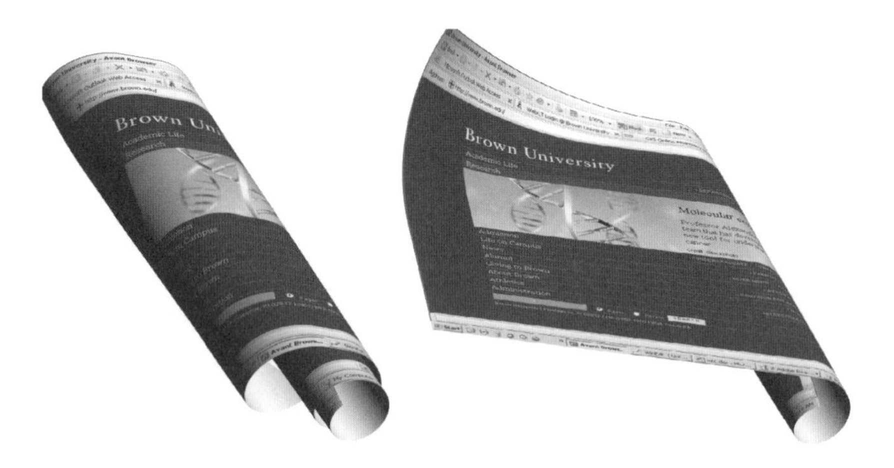

Figure 1. A concept flexible display device that can be rolled out to show an image and rolled back into a tube to be stowed away.

Flexible display devices will make it possible to roll out a screen for viewing images and roll it back into small form factors, delivering images larger than the area of the display itself. Flexible display applications have been demonstrated with liquid crystals,[2] electrophoretic inks,[3] organic light emitting diodes,[4] micro-electro-mechanical systems, and hollow fiber elements.[5] Regardless of the display technology incorporated into the device, an underlying commonality for this wide array of technologies is an interconnect scheme that maintains the electrical integrity of the interconnect material when the devices are bent, rolled, or flexed.[6]

Some possible candidates for flexible interconnect materials include conducting polymers, printed carbon pastes, carbon nanotube dispersions, and thin metal films. Although all the aforementioned candidates have their merits and demerits, thin metal film interconnects appear most suitable for flexible display circuitry because of their inherently high conductivity. The conductivity of conducting organic materials is still orders of magnitudes less than metals, whereas other technologies, like carbon nanotube dispersions, may prove difficult to integrate into existing electronic fabrication processes. Compared to all the other materials, thin metal films demonstrate excellent electro-mechanical performance, even when subjected to mechanical strains.[7,8,9]

Flexible interconnects are also crucial for biomedical application that require interconnectivity between implantable microchips, various sensors and electronic devices. At a minimum, the flexible electrodes are required to have low electrical resistance, provide mechanical flexibility with no appreciable change in conductivity and, in cases that involve implantation into biological materials, be nontoxic and biocompatible. Interconnects for these hybrid electro-biological systems should also be smooth, robust, lightweight and properly insulated.[10]

In this chapter, we describe a robust metallization scheme ideal for electrodes and interconnects suitable for flexible display applications[11,12] and possibly for bioelectronic applications as well. The metallization scheme consists of a multilayered structure of thin granular metal films deposited on a flexible substrate. The multilayered stack includes a granular discontinuous ductile indium layer that shows minimal resistance changes even when subjected to large mechanical strains and repeated low-strain fatigue loading. To fully demonstrate the potential of the interconnect scheme, we fabricated a simple flexible prototype display by incorporating patterned electrodes made by a standard photolithographic process.

2. Flexible interconnect materials and processing

Thin-metal film interconnects, containing chromium, indium and gold, were deposited onto dog bone-shaped polyethylene terephthalate (PET) and highly formable film (HFF) substrates, carefully prepared to minimize edge defects which could potentially lead to a localization of the applied strain and lead to a premature rupture in the films. Figure 2 shows a photograph of the testing system and a scanning electron micrograph (SEM) of a Cr/In film. The HFF substrates exhibit thermal properties similar to conventional PET, however, they can be stretched much more easily and their formability can be improved by heating. Such HFF substrates are particularly suitable for conformal display application, because they are engineered for bending and embossing (even 90° corners), which may not be achievable with standard PET. For both substrates, two types of films were fabricated: films deposited on the entire surface of a dog-bone shaped substrates and films patterned as thin lines using a shadow mask in the gauge length of the dog-bone shaped substrate, as shown in Fig. 2(a). The samples were loaded in a Minimat Miniature Tensile tester system, which can provide uniaxial tensile loading. The multilayered metal interconnects deposited on the polymer substrates consisted of a 5 nm chromium (Cr) adhesion layer, followed by a 50 nm granular indium (In) film and in some cases an additional 10 nm thick gold layer to increase conductivity and encapsulate the indium layer. All the metal films were deposited by electron-beam evaporation at room temperature at a deposition rate of ~0.1 nm/sec. The indium islands cover 75% of the total film area, with average diameter of 500 nm, as shown in Figure 2(b). The electro-mechanical performance of different multilayer structures based on thin films of chromium, indium and gold deposited on PET and HFF substrates was investigated.

3. Uniaxial tensile strain test and results

The electro-mechanical performance of thin-metal films patterned as interconnects on polymer substrates was evaluated by monitoring the electrical resistance as a uniaxial tensile force was applied to both ends of the film. Figure 3 shows an

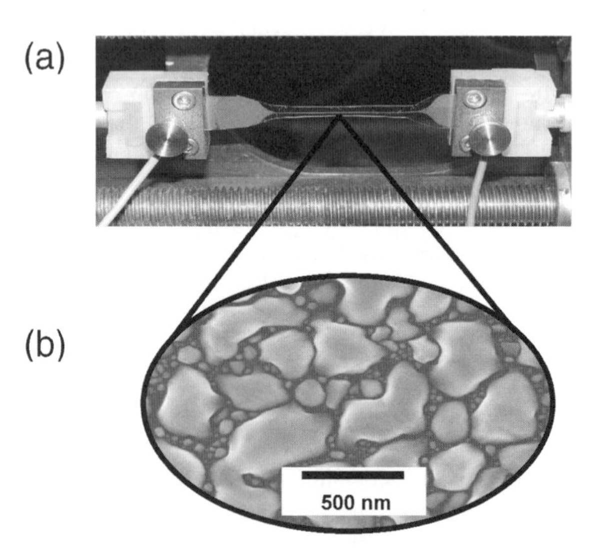

Figure 2. (a) Photograph of a Cr/In (5/50 nm) interconnect sample loaded in tensile strain. (b) SEM micrograph showing the granular indium layer deposited on a Cr-coated PET substrate.

illustration of the thin-metal-coated polymer film before and after a uniaxial tensile strain was applied. As depicted in the illustration, cracks were observed to form and propagate in the stiffer underlying chromium adhesion layer. The presence of In islands bridging the cracks, evident in the SEM micrographs of Fig. 4, maintains the electrical integrity of the metal line. The top Au layer can be used to further improve the electrical conductivity of the film.

Although the electrical resistance of the interconnects was observed to increase with increasing strain, there was no abrupt change or loss of electrical integrity over a wide range of strain values for all samples tested. Interconnects fabricated on PET substrates were observed to deform uniformly over the entire gauge-length of the dog bone-shaped sample, and remained conducting until the PET substrate itself ruptured at strain $\varepsilon = 38\%$. In the HFF samples, on the other hand, the deformation was concentrated in the mid-section of the gauge-length, thus causing a localization of the stress in the interconnect. Notwithstanding this peculiar deformation characteristic of the HFF substrate, the thin-metal film interconnects were observed to be electrically conductive even after applying a strain of >100% to the HFF sample. Figure 4 shows the stress-strain curves together with the normalized change in resistance $\Delta R/R_0 \equiv (R-R_0)/R_0$, for Cr/In/Au interconnects deposited on the two different substrates.[11,12]

The change in resistance of the interconnects deposited on a PET substrates, was observed to be <100% at an applied external strain of 38%. The resistance increase is a result of crack formation and geometry changes in the films. Cracks form perpendicular to the loading direction and increase in size and density with increasing applied strain for continuous films. The island structure of the indium layer however, deforms continuously through an elastic mechanism in which they

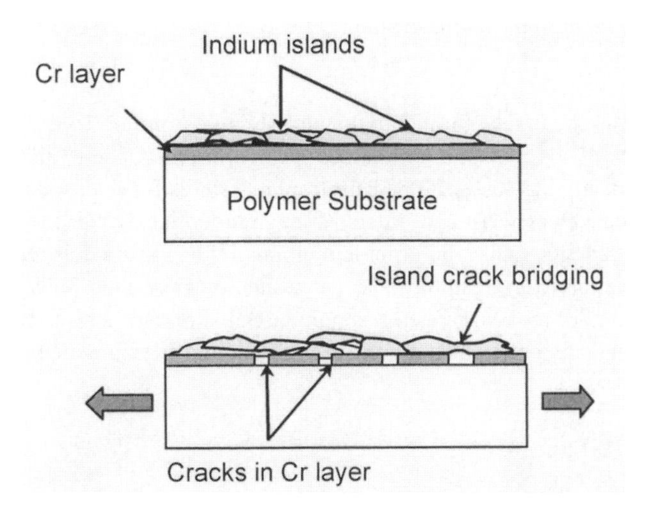

Figure 3. Schematic illustration of thin granular film structure (a) before and (b) after a tensile force is applied at both ends of the sample.

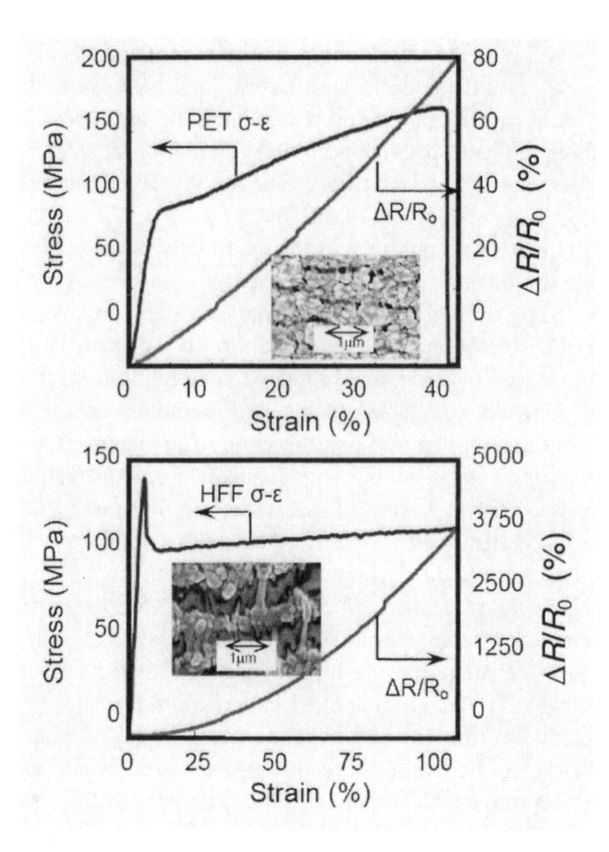

Figure 4. Stress-strain curve with change in resistance for interconnects on a PET (top) and HFF (bottom) substrates; insets show SEM micrographs of the bridging mechanism after applied tensile strain.

extend across the cracks formed in the underlying chromium layer, as shown in the SEM insets of Fig. 4. The normalized change in resistance of the Cr/In/Au film showed a steady increase with no catastrophic failures for both substrates, although the recorded values were two orders of magnitude higher with the HFF substrate. The observed geometric deformation in the HFF substrate is dominated by a necking phenomenon, resulting in local strains of more than 300% and relatively bigger change of the resistance. It should be noted that the strains applied in Fig. 4 vastly exceed any anticipated deformation required of flexible interconnects.

4. Cyclic fatigue loading

Flexible electronic applications designed to undergo continuous flexing, rolling and bending during fabrication or while in use, must also be robust enough under repeated low-strain fatigue conditions. Samples of the indium island-containing films were tested on a custom-built fatigue tensile-tester for 10,000 bend cycles. In this test, the thin films deposited on both PET and HFF substrates were repeatedly rolled and unraveled around a half-inch diameter cylindrical mandrel (resulting in a strain $\varepsilon \sim 2\%$). The fatigue test system as well as the results from this test is presented in Fig. 5.

The results clearly show that there is negligible change in the resistance of the interconnects due to fatigue loading. The very small apparent fluctuations in the measured resistance can be attributed to the resolution and inherent vibrations present in the test system.

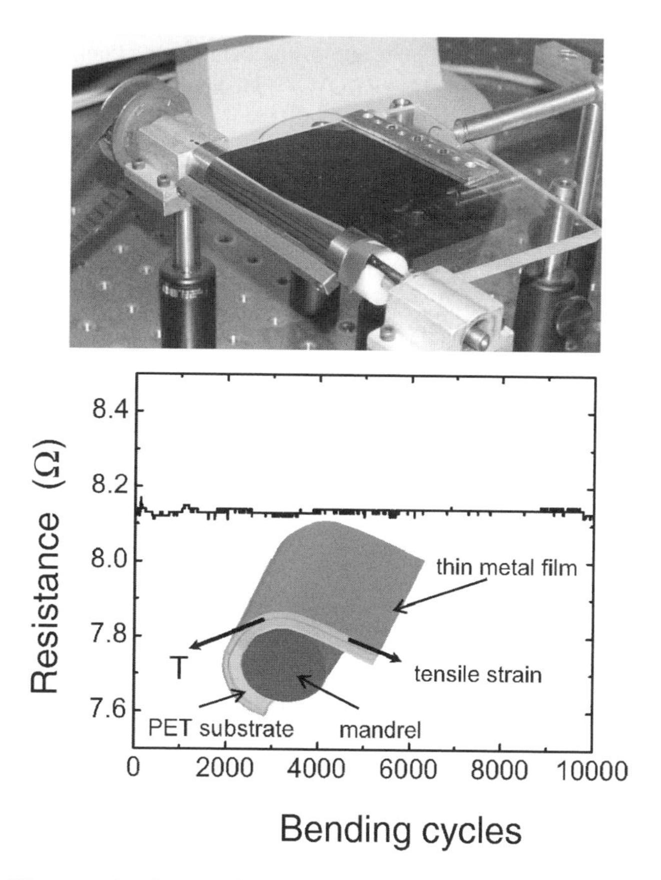

Figure 5. Photograph of cyclic fatigue testing apparatus (top) and the measured two-point resistance (bottom) of a Cr/In/Au film deposited on a PET substrate after 10,000 cycles showing less than a 0.01% change in electrical resistance.[12]

5. Prototype flexible display

We have fabricated a simple, proof-of-concept single-pixel flexible display on a 5×5 cm PET substrate. A 5/50/10 nm Cr/In/Au interdigitated electrode was fabricated by conventional lithography, e-beam metal evaporation, and lift-off, as shown in Fig. 6. A conventional polymer dispersed liquid crystal (PDLC) film deposited on top of the flexible electrode resulted in a fully functional and bendable display. The PDLC film was made on the patterned electrodes and using a polymer-induced phase separation to obtain micron-sized liquid crystals trapped within an extended polymer matrix. Figures 7(a) and (b) shows photographs of the flat device in both a power-on and power-off state, after being bent to ~1 cm radius of curvature for 100 times. In the "on" state (a), the film is transparent, showing the "BROWN" image printed on the back; while in the "off" state the film is opaque. Figure 7(c) shows the "on" state of the bent device bent to ~1 cm radius of curvature, again displaying the "BROWN" image. There is no fundamental obstacle to scaling this type of display to many pixels via lithographic interconnection of small PDLC pixels with our flexible metal lines.

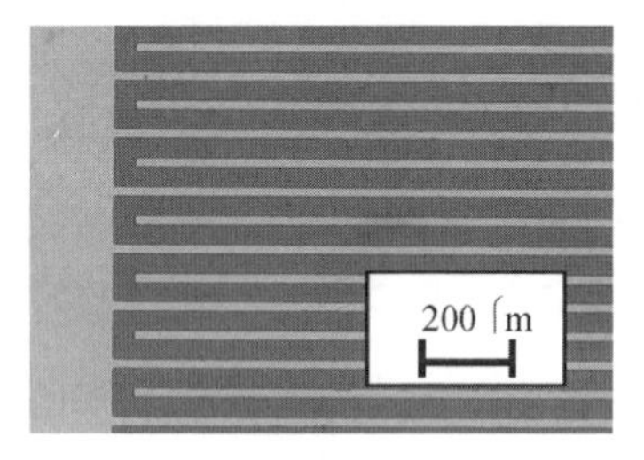

Figure 6. An optical micrograph of interdigitated electrodes fabricated on a PET substrate showing sharp features.

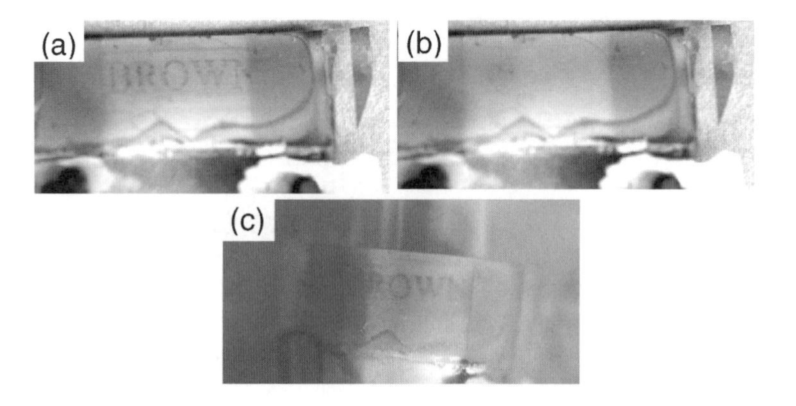

Figure 7. (a) Transparent "on" state of a flat device. (b) Opaque "off" state of a flat device. (c) Transparent "on" state of the device bent to ~1 cm radius of curvature.

7. Potential biomedical applications

Beside flexible display application, flexible interconnects hold promise for biomedical applications. Metal film interconnects have in recent years been incorporated into devices for interfacing with parts of the central or peripheral nervous system for neuronal disorders,[10] have enabled electronic stimulation of retinal neurons,[13] and have been integrated into various neural biohybrid interfaces.[14]

For instance, the flat interface nerve electrode (FINE) chip[15] can functionally stimulate a selected peripheral nerve, while reshaping the nerve into a configuration of axons closer to the stimulating contacts by increasing the nerve surface area. During this process, an elastomer clip will deform both by opening to go over the fiber and by bending in a slight arc as it is closed to put pressure on the bundle. One challenge is to interconnect the various functional blocks and the electrodes on the elastomer clip. At the same time, the interconnect must withstand the complex body environment.

As an initial test, our Cr/In films were dipped into a physiological saline solution for a week. After that, the resistance was measured and the samples were observed under SEM, with no change found. After a week in saline, Cr/In (5/500 nm) films deposited on PET were tested with the Minimat tester system. The resistance slightly changed from 2 Ω/square to 3.4 Ω/square at $\varepsilon = 35\%$ strain, which is sufficient for most biomedical applications. Although these preliminary results are promising, further research is needed. Thus far, we have mainly focused our research on PET and HFF substrates suitable for displays, while for biomedical application elastomer (silicone) substrates are widely used. Compared to PET, elastomers have rather different mechanical properties, with a much smaller Young's modulus and a much rougher surface, making film adhesion problematic. Further, the biocompatibility of In has not been fully investigated.

8. Conclusions

The inclusion of a granular layer of indium islands enhances the ultimate stretchability and reliability of thin-film interconnects and electrodes compatible with compliant display substrates. We have demonstrated the patternability of this robust and flexible electrode scheme on PET by using a standard photolithography process. Additionally, we have shown that there is no degradation in performance of the electrode or the display when repeatedly bent around a ~1 cm radius of curvature.

Our preliminary results have shown the compatibility of Cr/In films and physiological saline solution. The benefits of the stretchable interconnects present a great opportunity for bioelectronic applications that require patternable interconnect materials to connect electronic devices to biological materials.

Acknowledgements

The work at Brown University was supported by the MRSEC/NSF (DMR-0079964) and by NSF ECS-0223943 and NSF CCF-0403958. The authors acknowledge the use of the Microelectronics Central Facility at Brown, supported by the NSF MRSEC (DMR-0079964).

References

1. G. P. Crawford, ed., *Flexible Flat Panel Displays*, New York: Wiley, 2005.
2. B. S. Kim, M. Hong, Y. U. Lee, *et al.*, "Developments of transmissive a-Si TFT-LCD using low temperature processes on plastic substrate," *Soc. Inform. Display Digest* **35**, 19 (2004).
3. T. Whitesides, M. Walls, R. Paolini, *et al.*, "Towards video-rate micro-encapsulated dual-particle electrophoretic displays," *Soc. Inform. Display Digest* **35**, 133 (2004).
4. J. Innocenzo and E. I. Dupont, "Roll to roll PLED process development," *Soc. Inform. Display Digest* **33**, 884 (2002).
5. M. Nakata, M. Sato, Y. Matsuo, S. Maeda, and S. Hayashi, "Hollow fibers containing various display elements: A novel structure for electronic paper," *J. Soc. Inform. Display* **14**, 729 (2006).
6. S. J. Gorkhali, D. R. Cairns, and G. P. Crawford, "Reliability of transparent conducting substrates for rollable displays: A cyclic loading investigation," *J. Soc. Inform. Display* **12**, 45 (2004).
7. T. Li, Z. Huang, Z. Suo, S. P. Lacour, and S. Wagner, "Stretchability of thin metal films on elastomer substrates", *Appl. Phys. Lett.* **85**, 3435 (2004).
8. S. P. Lacour, J. Jones, S. Wagner, T. Li, and Z. Suo, "Stretchable interconnects for elastic electronic surfaces," *Proc. IEEE* **93**, 1459 (2005).
9. D. S. Gray, J. Tien, and C. S. Chen, "High-conductivity elastomeric electronics," *Adv. Mater.* **16**, 393 (2004).
10. T. Stieglitz, H. Beutel, M. Schuettler, and J.-U. Meyer, "Micromachined, polyimide-based devices for flexible neural interfaces", *Biomed. Microdev.* **2**, 283 (2000).
11. D. P. Wang, F. Y. Biga, A. Zaslavsky, and G. P. Crawford, "Electrical resistance of island-containing thin metal interconnects on polymer substrates under high strain," *J. Appl. Phys.* **98**, 086107 (2005).
12. D. P. Wang, F. Y. Biga, A. Zaslavsky, and G. P. Crawford, "Robust-stretchable interconnects for flexible display applications", to appear in *Soc. Inform. Display Digest* **37** (2006).
13. E. T. Kim, J. M. Seo, J. A. Zhou, H. Jung, and S. J. Kim, "A retinal implant technology based on flexible polymer electrode and optical/electrical stimulation," *IEEE Intern. Workshop BioCAS* (2004), S1.8-12.
14. T. Stieglitz, H. H. Ruf, M. Gross, M. Schuettler, and J.-U. Meyer, "A biohybrid system to interface peripheral nerves after traumatic lesions: Design

of a high channel sieve electrode," *Biosensors Bioelectronics* **17**, 685 (2002).
15. D. J. Tyler and D. M. Durand, "Functionally selective peripheral nerve stimulation with a flat interface nerve electrode," *IEEE Trans. Neural Syst. Rehab. Eng.* **10**, 294 (2002).

Part IV

Photonics: Light to the Rescue

4 Photonics: Light to the Rescue

We conclude the book with a part on photonics – a subject seemingly orthogonal to microelectronics. Electrons, being highly localized, are naturally suited for information processing, whereas photons are highly delocalized and thereby ideal for information transmission. Traditionally, electronics and photonics were pursued separately by different specialists. In the modern context, they can remain complementary but need not to stay orthogonal. As will be powerfully argued through a series of in-depth analyses by several leading researchers, photonics can both offer solutions to some of the most difficult and demanding problems in microelectronics, interconnects being one example, and open new spaces of applications, such as photovoltaic power sources, sensing, and real-time transmissive imaging and spectroscopy.

Contributors

4.1 D. A. B. Miller

4.2 M. N. Abedin, I. Bhat, S. D. Gunapala, S. V. Bandara, T. F. Refaat, S. P. Sandford, and U. N. Singh

4.3 Q. Hu, B. S. Williams, S. Kumar, A. W. M. Lee, Q. Qin J. L. Reno, H. C. Liu, and Z. R. Wasilewski

4.4 E. H. Linfield, J. E. Cunningham, and A. G. Davies

4.5 S. Suchalkin, M. Kisin, S. Luryi, G. Belenky, F. Towner, J. D. Bruno, C. Monroy, and R. L. Tober

4.6 D. Botez, M. D'Souza, G. Tsvid, A. Khandekhar, D. Xu, J. C. Shin, T. Kuech, A. Lyakh and P. Zory

4.7 M. A. Green

4.8 N. N. Ledentsov, V. A. Shchukin, and D. Bimberg

4.9 E. Paspalakis, M. Tsaousidou, and A. F. Terzis

Future Trends in Microelectronics. Edited by Serge Luryi, Jimmy Xu, and Alex Zaslavsky
ISBN 0-471-48 © 2007 John Wiley & Sons, Inc.

Silicon Photonics – Optics to the Chip at Last?

D. A. B. Miller
Ginzton Laboratory, Stanford University, Stanford, CA 94305, U.S.A.

1. Introduction

Optics dominates long distance communications, but will it ever be useful on silicon chips or their successors? In this chapter, we will discuss why we might be interested in the use of optics for such shorter interconnections, what technology we might need, and what new technological opportunities are emerging that could make such use practical or even ubiquitous.

Optics for use in handling information has been around for several decades now, with clear successes in optical fiber communications and in removable disk storage (compact discs and DVDs, for example). Early interest in optics for logic faded as integrated circuits advanced and limitations in energy for optical logic became clearer,[1] but optics for communication and interconnect has become increasingly interesting. There, optics is competing against copper, not silicon. Basic issues of physics favor optics for communication anywhere a high density of information has to be communicated over any substantial distance.[2,3]

2. Problems of wires

The physical problems and limitations of electrical wired interconnects are many and substantial. They perform increasingly poorly at higher frequencies, showing both signal attenuation and distortion. One surprising aspect of the performance is that it is essentially scale invariant: that is, once one has filled all the available space with wires, then, at least for simple on-off signaling, one cannot get any more information down the wiring system either by miniaturizing it or by making it bigger (see Fig. 1).[4] This argument is straightforward to derive for an "*RC*" line from the resistance and capacitance of wires, but, surprisingly, it also applies to "*LC*" wires or transmission lines. It leads to a capacity limit $B \propto A/\ell^2$ bits/second, where A is the total cross-sectional area of the wiring system and ℓ is the wire length. The proportionality constant is $\sim 10^{16}$ for *RC* lines (as typically found on chip) and $\sim 10^{15}$ for *LC* lines. Hence, wire capacity in large dense systems can suddenly become a problem at all size scales, including long on-chip lines and between chips. We routinely run into this limit in long on-chip wires and in coaxial cables, for example. Optics completely avoids this scaling limit, and is therefore particularly attractive wherever we require dense high-speed wiring of any substantial relative length.[4]

 Future Trends in Microelectronics. Edited by Serge Luryi, Jimmy Xu, and Alex Zaslavsky

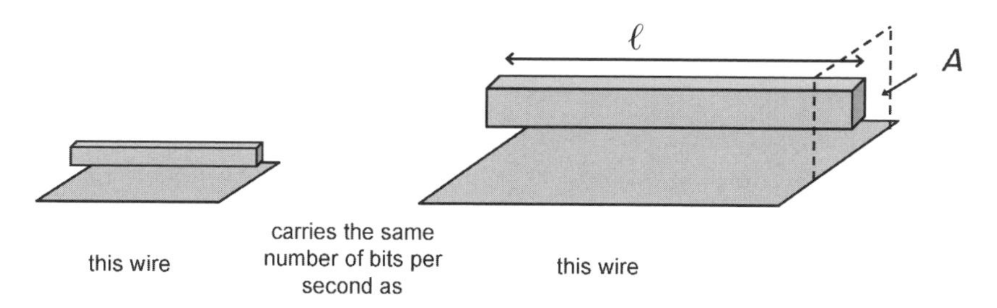

Figure 1. Illustration of the scaling of wire capacity for simple on-off signaling. Scaling a wire in all three dimensions leaves its information capacity the same, though scaling transistors in the same way makes them faster. Hence wiring progressively becomes the problem.

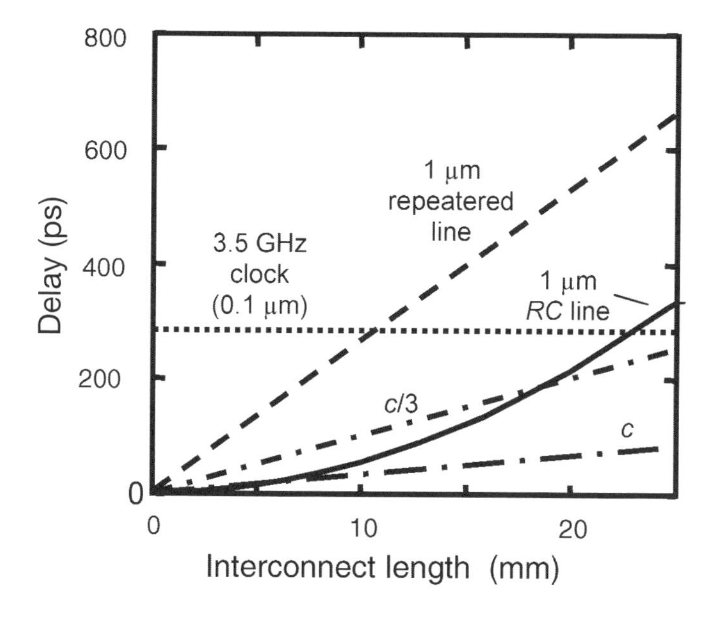

Figure 2. Delay on repeatered lines on chip compared with delay from propagation at the velocity of light c.[3]

Wires also have unavoidably low impedances (*e.g.*, $Z = 50\ \Omega$) and/or high capacitances C per unit length (*e.g.*, several pF/cm); Z and C both scale roughly logarithmically with the ratio between conductor size and conductor spacing, and so there is little that can be done to change such numbers substantially. Low impedance and high capacitance both contribute to substantial power dissipation. With sufficiently well-integrated optoelectronic devices, optics might be able to solve that power dissipation problem through a process that can be called quantum-impedance conversion,[5] which is actually inherent in any optical link.

Even on chips themselves, where repeaters can be incorporated to break wires into shorter lengths to mitigate the above scaling problem, there are problems with wires. The effective propagation velocity of signals on chips, for example, is generally much slower than the velocity of light, as shown in Fig. 2.[3]

Another major feature of optics is its ability to deliver very precise timing to electronic circuits, based on the relative ease with which short optical pulses can be generated and propagated over substantial distances. Clocking[6] of digital logic circuits with sub-picosecond precision has been demonstrated,[7] and optics has been used to trigger analog-to-digital converters with timing precision as good as 80 fs.[8]

3. Technologies for optical interconnects to chips

Despite physical advantages, and despite growing problems with interconnection at all levels, optics has never made any impact at the level of integrated circuits. Why? As discussed above, the physical arguments for using optics, especially off-chip, are particularly strong, with possible substantial reductions in power and increases in communication density, but the cost targets are daunting for introducing a new technology, such as optics. The practical targets for power dissipation are also very severe.

Given the increasing dominance of power dissipation as the limit on the performance of information processing machines, optics must certainly use no more power than the electrical systems it would replace, and likely must promise significantly lower power if it is to convince engineers to adopt it. A not unreasonable starting target for off-chip interconnects would be something ~1 mW/Gb/s (1 pJ/bit) for the total optical link, if it is to replace electrical chip-to-chip or chip-to-board technologies. Then 100 links each running at 10 Gb/s would consume 1 W. For long on-chip optical interconnects to be clearly superior to current electrical systems, dissipation of ~100 fJ/bit for the entire link is likely required. (Short optical interconnections on chip likely do not make much sense since wires are extremely good at such connections.)

Both the cost and the power requirements mean that, for such applications, the optoelectronic devices are going to have to be very well integrated with the electronics. For example, if a link is to dissipate no more than 100 fJ/bit, and the components such as detectors, modulators, and driver transistors are to swing through the supply voltage of 1 V, the total capacitance involved in the link cannot possibly exceed 100 fF based on the energy to charge such a capacitance. In fact, capacitance would likely have to be lower because there will be other sources of energy dissipation in the system. Capacitances of this order or smaller absolutely require very well integrated technologies. Prior optoelectronic technologies for communications have never been integrated well with silicon. Silicon itself is a frustrating optoelectronic material, because of its indirect gap, whereas III-V materials, which are good for optoelectronic devices, are not easy to integrate with silicon processing.

The idea of on-chip optics on silicon is gaining momentum in research, however. Significant advances have been made recently in silicon optical systems (waveguides, couplers), and in active optoelectronic devices (modulators).[9-12] Recently, quantum-confined Stark effect (QCSE) electroabsorption[13] has been observed in Ge quantum wells grown on silicon-germanium buffers on silicon substrates, in processes that are likely compatible with silicon CMOS manufacturing.[11,12] Figure 3 shows electroabsorption spectra for such wells, showing clear strong shifts of the absorption with field, as required for compact optical modulators. This mechanism is likely the strongest high-speed optical modulator mechanism known, and is routinely used with III-V materials to make optical modulators integrated with lasers for telecommunications. The importance of the observation, at telecommunications wavelengths near 1.5 microns, of the QCSE in Ge is that it may finally allow group IV optoelectronics with performance comparable with III-V's, avoiding the difficult materials integration issues of attempting III-V integration on Si. Hence, there is now serious hope for a very high-performance group IV CMOS-compatible optoelectronics technology capable of low-cost, low-power optoelectronics for interconnects and other applications.

The other required aspects for low-cost Si-compatible optical connections are also becoming more realistic. Serious attempts are being made at commercializing optoelectronic chips made entirely in a CMOS platform, such as recent work by

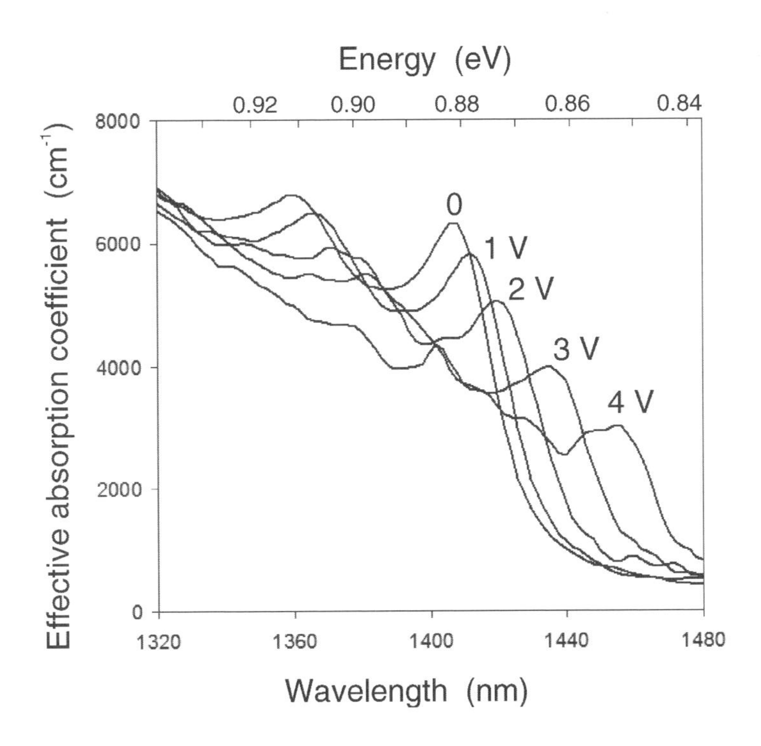

Figure 3. Effective absorption coefficient spectra of Ge/SiGe (10 nm Ge well and 16 nm Ge/$Si_{0.15}Ge_{0.85}$ barrier) quantum wells on a relaxed $Si_{0.1}Ge_{0.9}$ buffer.[11,12]

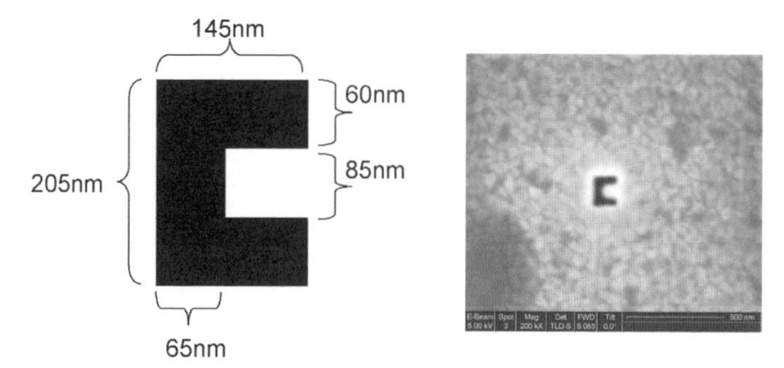

Figure 4. Schematic of a nanoaperture in a gold film on Ge (left) and picture of the fabricated structure (right). For light shining on the top of this structure, the metallic structure acts like an antenna to generate an intense local spot in the middle of the C-aperture. The photocurrent per unit active volume is enhanced by ~10x compared to the corresponding current in a simple piece of Ge illuminated by the same intensity of light.[16]

Luxtera Inc. A possible path for the introduction of optics to Si is the progressive evolution of integrated optoelectronics on Si for telecommunications and data communications transceivers, leading to technology we may be able to use for other applications, such as possibly optical interconnects on chip.

At the same time, radical ideas are emerging from nanophotonics, in dielectrics, semiconductors, and now also nanometallics, all in principle compatible with CMOS. With such nanotechnologies, it is possible to make optical devices much more compact than before, and to contemplate completely new kinds of structures, such as miniaturized wavelength[14] or mode[15] splitters, miniaturized metallic optical antennas[16] and 50 nm sized waveguides[17,18] that could concentrate and guide light into high-speed, low-capacitance photodetectors the same size as current transistors. Figure 4 shows a recent structure used to demonstrate photodetection enhanced by a metallic nanostructure,[16] in this case a sub-wavelength C-shaped aperture in a gold film on Ge.

4. Conclusions

Optics has many strong physical reasons for seriously considering it for interconnects, now possibly all the way to silicon chips themselves. The implementation of optics for such interconnects is very challenging, but recent breakthroughs are very promising for the ultimate construction of an integrated, low-cost, low-power technology. Advances also in nanophotonics structures, themselves enabled by the nanotechnologies of silicon electronic manufacturing, offer additional exciting opportunities for optics and optoelectronics well beyond even current devices.

References

1. R. W. Keyes, "Power dissipation in information processing," *Science* **168**, 796 (1970).
2. D. A. B. Miller, "Physical reasons for optical interconnection," *Intern. J. Optoelectronics* **11**, 155 (1997).
3. D. A. B. Miller, "Rationale and challenges for optical interconnects to electronic chips," *Proc. IEEE* **88**, 728 (2000).
4. D. A. B. Miller and H. M. Ozaktas, "Limit to the bit-rate capacity of electrical interconnects from the aspect ratio of the system architecture," *J. Parallel Distributed Comp.* **41**, 4252 (1997).
5. D. A. B. Miller, "Optics for low-energy communication inside digital processors: Quantum detectors, sources, and modulators as efficient impedance converters," *Optics Lett.* **14**, 146 (1989).
6. C. Debaes, A. Bhatnagar, D. Agarwal, *et al.*, "Receiver-less optical clock injection for clock distribution networks," *IEEE J. Selected Topics Quantum Electronics* **9**, 400 (2003).
7. D. A. B. Miller, A. Bhatnagar, S. Palermo, A. Emami-Neyestanak, and M. A. Horowitz, "Opportunities for optics in integrated circuits applications," *Dig. Tech. Papers ISSCC 2005*, paper 4.6, pp. 86–87.
8. L. Y. Nathawad, R. Urata, B. A. Wooley, and D. A. B. Miller, "A 40-GHz-bandwidth, 4-bit, time-interleaved A/D converter using photoconductive sampling," *IEEE J. Solid-State Circ.* **38**, 2021 (2003).
9. A. Liu, R. Jones, L. Liao, *et al.*, "A high-speed silicon optical modulator based on a metal-oxide-semiconductor capacitor," *Nature* **427**, 615 (2004).
10. Q. Xu, B. Schmidt, S. Pradhan, and M. Lipson, "Micrometer-scale silicon electro- optic modulator," *Nature* **435**, 325 (2005).
11. Y.-H. Kuo, Y.-K. Lee, Y. Ge, *et al.*, "Strong quantum-confined Stark effect in germanium quantum-well structures on silicon," *Nature* **437**, 1334 (2005).
12. Y.-H. Kuo, Y.-K. Lee, Y. Ge, *et al.*, "Quantum-confined Stark effect in Ge/SiGe quantum wells on Si for optical modulators," to appear in *IEEE J. Special Topics Quantum Electronics* (2007).
13. D. A. B. Miller, D. S. Chemla, T. C. Damen, A. C. Gossard, W. Wiegmann, T. H. Wood, and C. A. Burrus, "Electric field dependence of optical absorption near the bandgap of quantum well structures," *Phys. Rev. B* **32**, 1043 (1985).
14. M. Gerken and D. A. B. Miller, "Wavelength demultiplexer using the spatial dispersion of multilayer thin-film structures," *IEEE Photonics Technol. Lett.* **15**, 1097 (2003).
15. Y. Jiao, S. H. Fan, and D. A. B. Miller, "Demonstration of systematic photonic crystal device design and optimization by low rank adjustments: An extremely compact mode separator," *Optics Lett.* **30**, 141 (2005).
16. L. Tang, D. A. B. Miller, A. K. Okyay, *et al.*, "C-shaped nanoaperture-enhanced germanium photodetector," *Optics Lett.* **31**, 1519 (2006).

17. R. Zia, M. D. Selker, P. B. Catrysse, and M. L. Brongersma, "Geometries and materials for subwavelength surface plasmon modes," *J. Opt. Soc. Am. A* **21**, 2442 (2004) .
18. G. Veronis and S. Fan, "Guided subwavelength plasmonic mode supported by a slot in a thin metal film," *Optics Lett.* **30**, 3359 (2005).

The Future of Single- to Multi-Band Detector Technologies

M. N. Abedin
NASA Langley Research Center, Hampton, VA 23681, U.S.A.

I. Bhat
Rensselaer Polytechnic Institute, Troy, NY 12180, U.S.A.

S. D. Gunapala, S. V. Bandara
Jet Propulsion Laboratory, Pasadena, CA 91109, U.S.A.

T. F. Refaat
Old Dominion University, Norfolk, VA 23529, U.S.A.

S. P. Sandford and U. N. Singh
NASA Langley Research Center, Hampton, VA 23681, U.S.A.

1. Introduction

Using classical optical components such as filters, prisms, and gratings to separate the desired wavelengths before they reach the detectors results in complex optical systems composed of heavy components. A simpler approach might rely on a single optical system and a detector that responds separately to each wavelength band. Therefore, the development and fabrication of reliable detector arrays that respond to multiple wavelength regions has been a continuous endeavor. In this chapter, we will review the state-of-the-art single and multicolor detector technologies over a wide spectral range, for use in space-based and airborne remote sensing applications. Our discussion will be focused on current and most recently developed focal plane arrays (FPAs), in addition to emphasizing future development in UV-to-far infrared (IR) multicolor FPA detectors for next generation space-based instruments to measure water vapor and greenhouse gases.

Multi-band detector technology is progressing towards a number of exciting applications, such as remote sensing and imaging, military, and medical imaging. Development of this technology in the UV-to-far infrared region is critical and remains one of the important building blocks for the successful development of active and passive remote sensing. Such novel detectors will make instruments more efficient while reducing complexity and associated electronics and weight.

A multi-band detector consists of a stacked arrangement of different materials in which the shorter wavelength detector is placed top of the longer wavelength detector. The top detector material absorbs only the shorter wavelength radiation and transmits the longer wavelength radiation to the bottom detector. This

Future Trends in Microelectronics. Edited by Serge Luryi, Jimmy Xu, and Alex Zaslavsky
ISBN 0-471-48

arrangement is preferable to using lenses, prisms and gratings to separate the individual wavelength before it reaches the detectors.

Mercury cadmium telluride (HgCdTe) and quantum well infrared photodetector (QWIP) technology is expanding from single-color to multicolor detectors. Multicolor detector arrays based on QWIPs have recently been demonstrated.[1-3] Development of multi-band arrays will continue[4-5] and two-band detectors have been demonstrated based on HgCdTe FPAs.[6-10] A joint effort among NASA Langley, JPL, and RPI is developing a multicolor FPA with three-dimensional (3D) structure to provide high-resolution spectroscopy for imaging Fourier-transform interferometers. An increasing interest in multicolor imaging technology is materializing in the UV-to-far IR with the goal of producing large and 3D FPAs.

At present, the availability of multicolor detectors is limited to the 3–15 μm spectral range and no multicolor detector technology has been reported spanning the entire UV-to-far IR spectrum.[11,12] The progress in multicolor detector development is driven by the need to detect water vapor and greenhouse gases over a broad wavelength range. This technology will be very valuable to the NASA Earth and Space Science Enterprises and Planetary Exploration Program, allowing critical measurements at improved accuracy with greatly reduced system complexity, weight, and cost. A simpler system would result by utilizing a single optical system and a detector that responds separately to each wavelength band as shown in Fig. 1. The multicolor detector approach is likely to prove useful for several other terrestrial and planetary remote sensing applications.

To achieve the wavelengths either less than 1 μm or greater than 15 μm, one can extend the operating wavelengths by using Si microbolometers and shorten the operating wavelengths by using phosphor coating on Si material.[12] However, these approaches face significant technical difficulties. Quaternary layers of GaInAsSb with wavelength longer than 2.5 μm have been grown by metal-organic chemical vapor deposition (MOCVD) with good optical and electrical properties. These layers are lattice-matched to GaSb or InAs substrates with cutoff wavelengths for absorption tailored from 1.7 to 3.5 μm. Epitaxial growth on a lattice-matched substrate is crucial to reducing unwanted defects that could lead to higher noise in the detector. There are at present few materials with varying band gaps that cover the ultraviolet to far IR wavelength band with lattice-matched conditions.

Therefore, a multi-band detector array, sensitive in broadband range with different bands covering the 0.4 μm-to-far IR wavelength range is urgently needed.[11,12] To achieve this, we anticipate making use of Si, GaSb, InAs, III-V semiconductor ternary and quaternary alloy-based materials, and of vertically integrated *p-i-n* heterostructure diodes grown by metal-organic vapor phase epitaxy (MOVPE) with wafer bonding used when necessary. Adding up to 4–5 materials expands the flexibility of the semiconductor material system so that photo-excitation band gaps and lattice parameters can be adjusted independently, allowing us to find a lattice-matched substrate on which to grow these multicolor detector stacks monolithically, or lattice-mismatched detector stacks using wafer bonding technology.

2. Multicolor single element detector structures

The integration of silicon technology with the antimonide-based material system allows for the simultaneous detection of multiple wavelengths in the visible and the IR. The materials of interest – GaSb, AlSb and InAs – are closely lattice-matched, but have widely different bandgaps: GaSb has a bandgap of 0.72 eV, AlSb has an indirect band gap of 1.63 eV, and InAs has a bandgap of 0.356 eV. The lattice constants of these materials are 6.0959Å (GaSb), 6.136Å (AlSb), and 6.0583Å (InAs). There is no perfect lattice match, and hence we can expect a high density of dislocations when one material is grown on the other. However, one can grow ternary and quaternary layers lattice-matched to one of the binaries, and the band gap of the quaternary layers can be tuned to any desired wavelength, still maintaining the lattice-matched condition. This allows great flexibility in device design, but the material growth becomes more complicated. Still, recent advances in MOVPE growth technique can allow reproducible growth of such quaternary layers. Growth of high quality quaternary layers was demonstrated very recently for use in thermophotovoltaic systems.[13,14]

The multi-wavelength device structure in the UV-to-IR band is shown schematically in Fig. 1. An *n*-type GaSb substrate is chosen on which a *p*-on-*n* InAs device structure is grown, followed by an *n*-on-*p* second IR detector in a quaternary layer of InGaAsSb. The band gap of the quaternary layer can be changed from 0.29 eV to 0.35 eV or from 0.7 eV to 0.5 eV (wavelength cut-off from 4.3 to 3.5 μm or from 1.7 to 2.5 μm), while still maintaining the lattice match with the substrate.

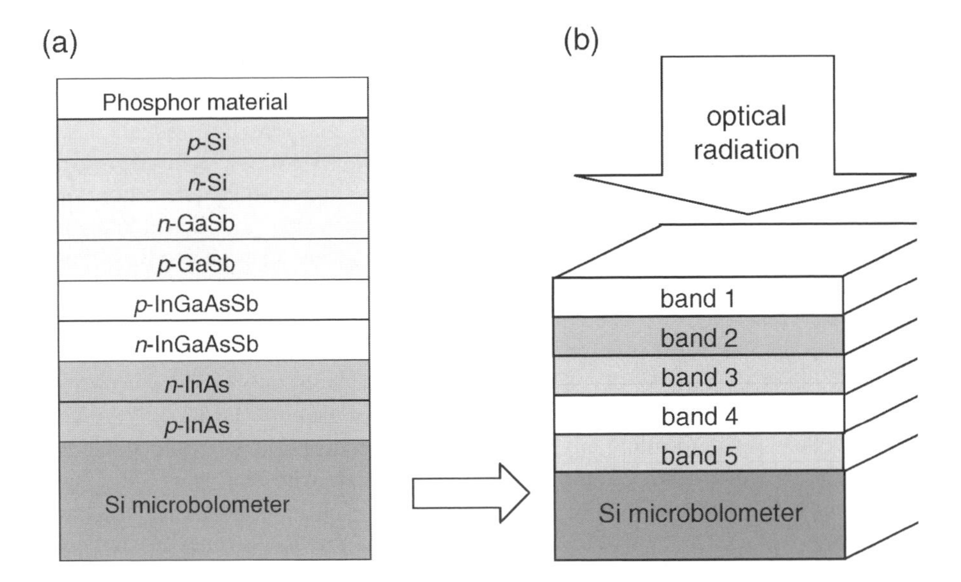

Figure 1. Schematic of multi-layer bias selectable multi-wavelength IR detector: (a) single element layered structures, and (b) 3D focal plane array detector structures.[12]

One important point to remember in the above structure is the band alignment. The InAs/GaInAsSb heterostructure has a type-II band alignment. In the above configuration, *i.e.*, *n*-*p*(InAs)–*p*-*n*(GaInAsSb) configuration, the minority carrier electrons in *p*-GaInAsSb will not be collected by the lower bandgap junction. However, the electrons from the lower bandgap *p*-type layer can be collected by the common junction, and hence there will be a small amount of crosstalk. Independently biasing the two junctions will alleviate this problem, and in addition, the *p*-type lower bandgap material thickness should be kept to a minimum to reduce this cross talk further.

We propose to use wafer-bonding approach to overcome the lattice mismatch problem present in ternary layers. This allows us to optimize the detectors for each region of operation and achieve high quantum efficiency. The proposed multi-color detector is based on the integration of Si microbolometer with the Si and the III-V compound-based IR detectors to cover the wavelengths of interest. The IR detector structures used are grown on a GaSb substrate using MOVPE. The composite single detector layered structures are shown in Fig. 1(a).

Figure 1(b) shows the 3D detector sketch with input radiation from the optical system impinging the detector stack. Light of the appropriate wavelengths of interest is absorbed in the first detector in this "sandwich", while longer wavelengths are passed into the second detector, then the third, *etc.* The desired absorption wavelengths in each detector can be tailored for each atmospheric or geophysical sounding application, making for a very efficient instrument, as necessary spectral lines or spectral bands are processed.

3. State-of-the-art multicolor focal plane array detector technologies

Many different device structures, such as HgCdTe, GaAs/AlGaAs quantum well photodetectors, strained layer InAs/GaInSb superlattices (SLSs), Schottky barriers on Si, SiGe heterojunctions, pyroelectric detectors, silicon bolometers, and high temperature superconductors are used for the detection of UV-to-far IR radiation. But the bulk single-crystal semiconductor remains the material of choice for efficient detection of electromagnetic radiation in the UV- long wavelength ranges.[11]

At present, HgCdTe photodiodes and quantum well infrared photoconductors (QWIPs) present multicolor capability in the MWIR–LWIR wavelength range. HgCdTe is based on II-VI and QWIP is based on the well-developed III-V material systems, which have some advantages and disadvantages. HgCdTe FPAs have higher operating temperature, higher quantum efficiency, but lower yield and higher cost. On the other hand, QWIPs are easier to fabricate with high operability, good uniformity, high yield, and lower cost, but have lower quantum efficiency and lower operating temperature. In this section, we discuss the state-of-the-art multicolor detector based on II-VI (HgCdTe), III-V (quantum well) materials, and also future prospects of single-crystal material systems.[11]

- *II-VI (HgCdTe) material systems*

There is considerable interest in multispectral HgCdTe detectors employing liquid phase epitaxy (LPE), molecular beam epitaxy (MBE) and metal-organic chemical vapor deposition (MOCVD) for the growth of a variety of devices by research groups at BAE Systems, Hughes Research Laboratory, Rockwell Scientific Company, LETI LIR, SOFRADIR, Raytheon Vision Systems, and AIG AEG INFRAROT.[15-26] Devices for the sequential and simultaneous detection of two closely spaced subbands in the MWIR and LWIR ranges have been demonstrated.

An early example of two-color detector is the one described in references,[20,21] where an *n-p-n* three-layer structure in HgCdTe was epitaxially grown on CdTe substrates as shown in Fig. 2. The detector array is illuminated from the backside through the substrate, since CdTe has a bandgap much higher than the wavelengths of interest. The wavelength of interest can be selected by selecting the appropriate bias. This bias-selectable two-color detector affords perfect spatial collocation of the two detectors, but it has inherent drawback of not allowing temporal simultaneity of detection. Either one or the other photodiode is functioning depending on the bias polarity applied across the back-to-back pair. Reine and co-workers[15] demonstrated a two-color detector with a four-layer *P-N-n-p* structure was grown epitaxially on CdTe substrate (the capitalized doping designations refer to the higher bandgap material, *i.e.* the *P-N* junction refers to the higher bandgap HgCdTe, whereas the second *n-p* junction refers to lower bandgap HgCdTe). The signal enters through the substrate and the signal processing electronics is In-bonded to the top surface. This allows independent simultaneous detection of two colors. Two color detection in the 3–5 μm and 8–12 μm ranges was demonstrated, with the layers were grown by

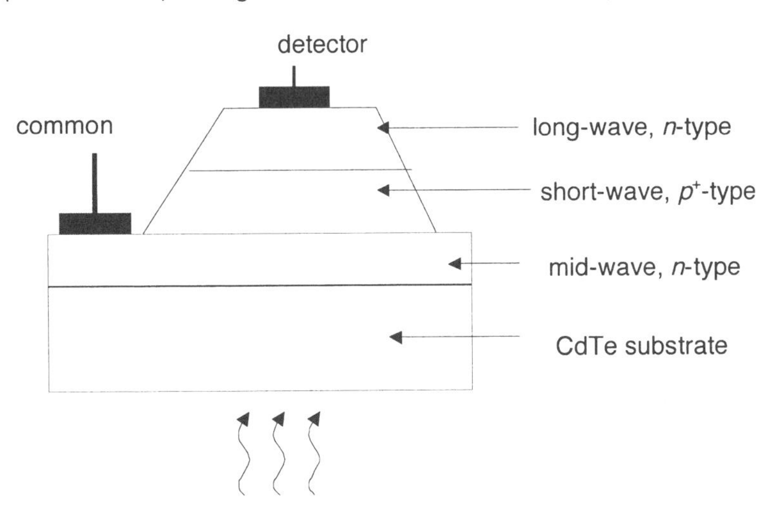

Figure 2. Schematic cross-section of triple-layer bias-selectable dual-wavelength HgCdTe IR detector.[20]

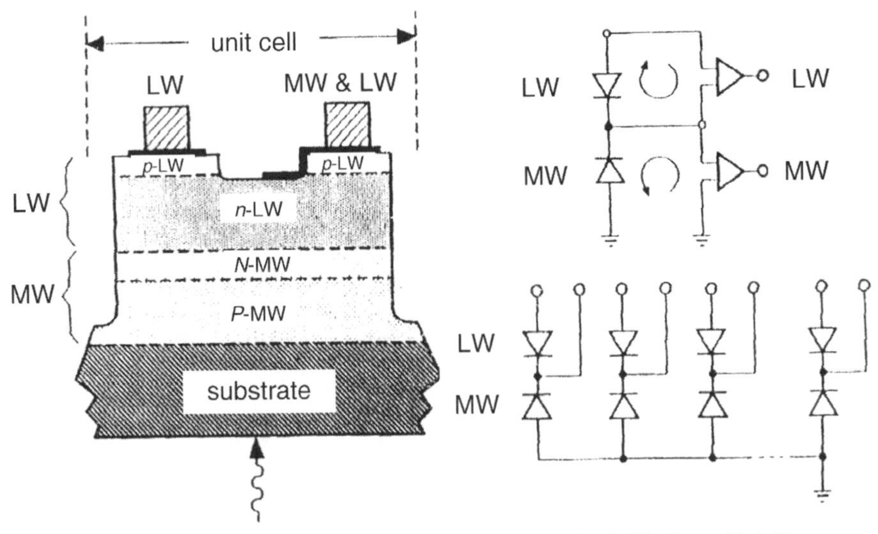

Figure 3. Schematic diagram of dual-band detectors in HgCdTe (from Ref. 6).

MOVPE. The cross-section of the independently accessed back-to-back photodiode dual-band detector is shown in Fig. 3. There are three ohmic contacts in this structure, one to the *p*-LW layer, one to the *n*-LW and *N*-MW layers, and one to the *P*-MW layer (not shown). The common terminal is the contact to the *n*-LW and *N*-MW layer. The *p*-type regions of all the MW photodiodes in the dual band array are electrically connected and are accessed at the edge of the array through a common array ground contact.

The above two-color detectors are possible in HgCdTe since the bandgap of HgCdTe can be continuously varied from 0 to 1.4 eV without any changes in the lattice constant. Hence, it is possible to grow high quality materials with minimum defects on CdTe substrates. Since CdTe has higher bandgap, backside illumination is possible without extra processing steps.

A 64x64 MW/LW dual-band MOCVD HgCdTe array has been demonstrated in Ref. 6 and also discussed in Ref. 22. A unit cell size of 75x75 μm^2 of these arrays was hybridized to a dual-band silicon multiplexer readout chip. This allowed the MW and LW photocurrents to be integrated simultaneously and independently. The average cut-off wavelengths of the MW and LW were in the 4.27–4.35 μm and 10.1–10.5 μm ranges at $T = 77$ K, respectively. High average quantum efficiencies of 79% and 67%, high median detectivities of 4.8×10^{11} cm·√Hz/W and 7.1×10^{10} cm·√Hz/W, and low median noise equivalent differential temperatures (NEDTs) of 20 mK and 7.5 mK were demonstrated for MW and LW infrared dual-band detector at $T_{scene} =$ 295 K and *f*/2.9. Tissot at CEA/LETI[27] reported a high performance MWIR two-color detector by superposition of HgCdTe layers with different composition. This two-color detector structure allows simultaneous detection of 3 μm cutoff wavelength for band 1 and 5 μm cutoff wavelength for band 2 radiation in the same pixel.

The average resistance R_0A values of 10^7 $\Omega \cdot cm^2$ for the band 1 and 4×10^5 $\Omega \cdot cm^2$ for the band 2 diodes at $T = 77$ K were reported, as well as quantum efficiencies of 75% for band 2 and 50% for band 1 without anti-reflection (AR) coating. The crosstalk between band 1 and 2 was measured to be 2%. Tennant and co-workers[19] obtained two-color FPAs with background-limited detectivity performance for band 1 (MWIR: 3–5 μm) devices at $T > 130$ K and band 2 (LWIR: 8–10 μm) devices at $T \sim 80$ K. An FPA of 128x128 pixels was fabricated from these devices with 40 μm pixel pitch, with low NEDTs of 9.3 mK for band 1 and 13.3 mK for band 2 – values similar to good quality single-color FPAs. Finally, this group demonstrated 79% of broadband quantum efficiency for band 1 with only a 4% variation over the pixels and 75% for band 2 with a 2% variation over the pixels without AR coating. The authors speculated that the larger variation in band 1 was due to the effect of crosstalk on pixels adjacent to a defective pixel.

Recently, Raytheon Vision Systems developed two-color detectors with MWIR and LWIR cut-off wavelengths of 5.5 μm and 10.5 μm with quantum efficiency >70% for both MW and LW performance at 78 K.[9,10] Large 1280x720 MWIR/LWIR FPAs with 20 μm unit cells and 99% pixel operability have been demonstrated, with high-quality simultaneous imaging of the spectral bands achieved by mating the FPA to a readout integrated circuit. Smith and co-workers[9,10] have demonstrated large-format megapixel 2048x2948 FPAs with 20 μm unit cells and 2560x512 FPAs with 25 μm unit-cell by using double layer heterojunction HgCdTe growth on Si substrates in the short wavelength infrared (SWIR) and MWIR spectral ranges.

- *III-V compound material systems*

Quantum well infrared photodetectors based on III-V compound materials have been extensively investigated for high-background applications by using large area highly uniform FPAs. This technology holds considerable promise for the fabrication of multicolor FPAs.[28-33] Among the different types of QWIPs, GaAs/AlGaAs multiple quantum well detectors are the most mature.[34-36] Sundaram and co-workers at BAE Systems demonstrated three separate two-color FPAs with MW/MW (4.0/4.7 μm), MW/LW (5.1/8.5 μm), and LW/LW (8.3/11.2 μm) simultaneous detection.[30,37] Typical operating temperature for these QWIP detectors is in the $T = 40$–100 K range. Peak responsivities of both blue and red QWIPs (LW/LW) are around 320 and 480 mA/W at $T = 40$ K at an operating bias of –2.0 V; 35 and ~95 mA/W for the blue and red QWIPS (LW/MW) at an operating bias –2.0 V; and ~12 and ~27 mA/W for the blue and red QWIPs (MW/MW) at an operating bias –2.0 V. This group has demonstrated high operability (>99%) and reasonably high peak responsivity in both colors in several different color combinations (MW/MW, MW/LW, LW/LW FPAs). Large format single-color 640x480 pixel QWIP FPAs with 22 μm squares on 24 μm centers, giving a fill factor of 84%, were fabricated by BAE Systems for U.S. Army Research Laboratory (ARL). Goldberg and co-workers[38,39] at ARL also demonstrated the potential of dual-band MWIR/LWIR imaging (developed by BAE Systems) for tactical environments by the application of a color image-fusion technique. Recently, Goldberg and co-workers[31] presented results of a large-format megapixel 1024x1024 QWIP FPA, which was produced by QWIP Technologies, Inc. This device was tested at the ARL and results obtained including conversion efficiency close to 10%, the spectral response peak at 8.55 μm, and a dark current level low enough to allow background-limited infrared performance (BLIP) at 76 K.

Gunapala and co-workers[28] have demonstrated LWIR and very LWIR (8–9 and 14–15 μm) two-color imaging camera based on a 640x486 dual-band QWIP FPAs. This dual-band QWIP device can be processed into simultaneously readable dual-band FPAs with triple contacts to access the CMOS readout multiplexer[40] or interlace readable dual-band FPA (*i.e.*, odd rows for one color and the even rows for the other color). The first approach requires a special dual-band readout multiplexer that contains two readout cells per detector unit cell, whereas the second approach needs only a single-color readout multiplexer. The advantages of this scheme are that it provides simultaneous data readout and allows the use of currently available single-color CMOS readout multiplexers. However, the disadvantage is that it does not provide a full fill factor for both wavelength bands. The device structure, shown in Fig. 4,[22,28] consists of a 30 period stack (500 Å AlGaAs barrier and a 60 Å GaAs well) of VLWIR structure and a second 18 period stack (500 Å AlGaAs barrier and a 40 Å GaAs well) of LWIR structure, separated by a heavily-doped 0.5 μm thick intermediate GaAs contact layer. The VLWIR QWIP structure has been designed to have a bound-to-quasibound intersubband absorption peak at 14.5 μm, whereas the LWIR QWIP structure has been designed to have a bound-to-continuum intersubband absorption peak at 8.5 μm, since photocurrent and dark current of the LWIR device structure is relatively small compared to the VLWIR portion of the device structure. The performance of dual-band FPAs has been tested at a background temperature of 300 K, with *f*/2 cold stop, and at 30 Hz frame rate and are discussed in Ref. 30. The mean values of quantum efficiency at operating $T = 40$ K and –2V bias are 12.9% and 8.9% in LW and VLW spectral range, respectively. The estimated NEDTs of LWIR and VLWIR detectors at 40 K are 36 and 44 mK,

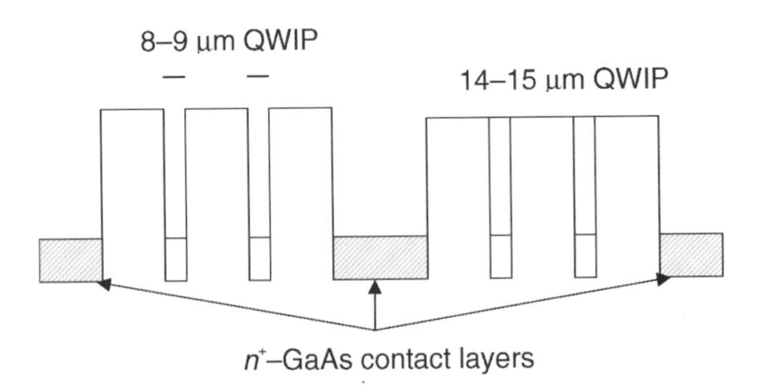

Figure 4. Conduction band energy diagram of the long-wavelength two-color infrared detector. The long-wavelength (8–9 μm) sensitive multiquantum well stack utilizes the bound-to-continuum intersubband absorption. The very long-wavelength (14–15 μm) stack utilizes the bound-to-quasibound intersubband absorption.[28]

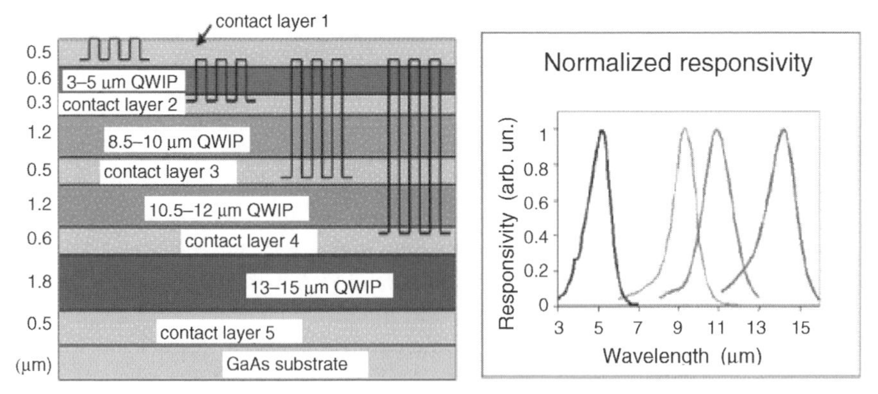

Figure 5. JPL completed a four-color 640x512 imaging focal plane array (3–15 μm) for the hyperspectral QWIP imaging system.[1]

respectively. Recently, Gunapala and co-workers[32,33] demonstrated MWIR and LWIR 1024x1024 pixel QWIP FPAs with excellent imaging performance. The MWIR and LWIR prototype cameras with similar optics have shown BLIP performance at operating $T = 90$ K and 70 K respectively, at 300 K background.

A joint effort between Goddard Space Flight Center, Jet Propulsion Laboratory, and ARL demonstrated a four-band, hyperspectral 640x512 QWIP array for the NASA Earth Science mission.[1,41] Recently, Bandara and co-workers[2,3] have reported remarkable progress in realizing large-format multi-band and broadband FPAs. These QWIP FPAs have been developed for an imaging interferometer based on the InGaAs/GaAs/AlGaAs material system. The spectral range is from 3 to 15.4 μm. This FPA consists of four independently readable IR bands: (1) 3–5 μm, (2) 8.5–10 μm, (3) 10–12 μm, and (4) 14–15.4 μm. Each band occupies a 640x128 pixel area within the single imaging array. Figure 5 illustrates the conceptual design and spectral response of the four-band multiquantum well QWIP.[1]

Most recently, NASA Langley Research Center with partnerships at RPI and JPL proposed to develop the technical capability to reliably fabricate detector arrays that respond to multiple wavelength regions.[11,12] We plan to use epitaxial growth followed by wafer bonding to overcome the lattice mismatch problem. This will allow us to optimize the detectors for each region of operation to achieve high quantum efficiency. The proposed multicolor detector will be based on the integration of silicon detectors with the antimonide-based IR detectors to cover the wavelengths of interest. The IR detector structures will be grown on GaSb or InAs substrates by MOVPE and these detector structures will be transferred to the silicon substrates using wafer bonding technology. The composite single-element detector structure is shown in Fig. 1.

4. Conclusions

Multicolor detector technology is rapidly advancing towards a number of exciting applications, such as remote sensing and imaging, military, and medical imaging. HgCdTe and QWIP technology is expanding from single-band to multi-band detectors and, recently, QWIPs have been integrated into multi-band detector arrays. Development of multi-band arrays will continue and three-band detectors will be soon demonstrated based on HgCdTe FPAs. A joint effort between NASA Langley, RPI, and JPL is currently working on a multi-band FPA with 3D structure to provide high-resolution spectroscopic imaging for military, medical, hyperspectral and Fourier-transform interferometers. An increasing interest in multi-band imaging technology is materializing in the UV-to-far IR with the feasibility of focal planes founded on large and three-dimensional arrays.

References

1. S. V. Bandara, S. D. Gunapala, J. K. Liu, *et al.*, "Four-band quantum well infrared photodetector array," *Infrared Phys. Technol.* **44**, 369 (2003).
2. S. V. Bandara, S. Gunapala, C. Hill, *et al.*, "Multi-band infrared detectors based on III-V materials," *Proc. SPIE* **5543**, 1 (2004).
3. S. V. Bandara, S. D. Gunapala, J. K. Liu, *et al.*, "Multi-band and broad-band infrared detectors based on III-V materials for spectral imaging instruments," *Infrared Phys. Technol.* **47**, 15 (2005).
4. L. Becker, "Multicolor LWIR focal plane array technology for space and ground based applications," *Proc. SPIE* **5564**, 1 (2004).
5. A. K. Sood, J. E. Egerton, Y. R. Puri, *et al.*, "Design and development of multicolor MWIR/LWIR and LWIR/VLWIR detector arrays," *J. Electron. Mater.* **34**, 909 (2006).
6. M. B. Reine, A. Hairston, P. O'Dette, *et al.*, "Simultaneous MW/LW dual-band MOCVD HgCdTe 64x64 FPAs," *Proc. SPIE* **3379**, 200 (1998).
7. E. Smith, L. Pham, G. Venzor, *et al.*, "Two-color HgCdTe infrared staring focal plane arrays," *Proc. SPIE* **5209**, 1 (2003).
8. J. Giess, M. A. Glover, N. T. Gordon, *et al.*, "Dual-waveband infrared focal plane arrays using MCT grown by MOVPE on Si substrate," *Proc. SPIE* **5783**, 316 (2005).
9. E. P. G. Smith, E. A. Patten, P. M. Goetz, *et al.*, "Fabrication and characterization of two-color mid-wavelength/long-wavelength HgCdTe infra-red detectors," *J. Electron. Mater.* **35**, 1145 (2006).
10. E. P. G. Smith, R. E. Bornfreund, I. Kasai, *et al.*, "Status of two-color and large format HgCdTe FPA technology at Raytheon Vision Systems," *Proc. SPIE* **6127**, 61271F-1 (2006).
11. M. N. Abedin, T. F. Refaat, J. M. Zawodny, *et al.*, "Multicolor focal plane array detector: A review," *Proc. SPIE* **5152**, 279 (2003).

12. M. N. Abedin, T. F. Refaat, Y. Xiao, and I. Bhat, "Characterization of dual-band infrared detectors for application to remote sensing," *Proc. SPIE* **5883**, 588307-1 (2005).
13. C. Hitchcock, R. Gutmann, J. Borrego, I. Bhat, and G. Charache, "Antimonide based devices for thermophotovoltaic applications," *IEEE Trans. Electron Dev.* **46**, 2154 (1999).
14. C. A. Wang, H. K. Choi, and S. L. Ransom, "High-quantum-efficiency 0.5 eV GaInAsSb/GaSb thermophotovoltaic devices," *Appl. Phys. Lett.* **75**, 1305 (1999).
15. M. B. Reine, P. W. Norton, R. Starr, *et al.*, "Independently accessed back-to-back HgCdTe photodiodes: A new dual-band infrared detector," *J. Electron. Mater.* **24**, 669 (1995).
16. R. D. Rajavel, P. D. Brewer, D. M. Jamba, *et al.*, "Status of HgCdTe-MBE technology for producing dual-band infrared detectors," *J. Crystal Growth* **214/215**, 1100 (2000).
17. S. M. Johnson, J. L. Johnson, W. J. Hamilton, *et al.*, "HgCdZnTe quaternary materials for lattice matched two-color detectors," *J. Electron. Mater.* **29**, 680 (2000).
18. W. E. Tennant, M. Thomas, L. J. Kozlowski, *et al.*, "A novel simultaneous unipolar multispectral integrated technology approach for HgCdTe IR detectors and focal plane arrays," *J. Electron. Mater.* **30**, 590 (2001).
19. P. Ferret, J. P. Zanatta, R. Hamelin, S. Cremer, A. Million, M. Wolny, and G. Destefanis, "Status of the MBE technology at LETI LIR for the manufacturing of HgCdTe focal plane arrays," *J. Electron. Mater.* **29**, 641 (2000).
20. K. Kosai, "Status and application of HgCdTe device modeling," *J. Electron. Mater.* **24**, 635 (1995).
21. E. R. Blazejewski, J. M. Arias, G. M. Williams, W. McLevige, M. Zandian, and J. Pasko, "Bias-switchable dual-band HgCdTe infrared photodetector," *J. Vac. Sci. Technol. B* **10**, 1626 (1992).
22. A. Rogalski, "Infrared detectors: An overview," *Infrared Phys. Technol.* **43**, 187 (2002).
23. J. Baylet, P. Ballet, P. Castelein, *et al.*, "TV/4 dual-band HgCdTe infrared focal plane arrays with a 25-micron pitch and spatial coherence," *J. Electron. Mater.* **35**, 1153 (2006).
24. P. Tribolet and G. Destefanis, "Third generation and multi-color IRFPA development: A unique approach based on DEFIR," *Proc. SPIE* **5783**, 350 (2005).
25. W. A. Radford, E. A. Patten, D. F. King, *et al.*, "Third generation FPA development status at Raytheon Vision Systems," *Proc. SPIE* **5783**, 331 (2005).
26. R. Breiter, W. Cabanski, K.-H. Mauk, W. Rode, J. Ziegler, H. Schneider, and M. Walther, "Multicolor and dual-band IR camera for missile warning and automatic target recognition," *Proc. SPIE* **4718**, 280 (2002).
27. J. L. Tissot, "Advanced IR detector technology development at CEA/LETI," *Infrared Phys. Technol.* **43**, 223 (2002).

28. S. D. Gunapala, S. V. Bandara, A. Singh, *et al.*, "8–9 and 14–15-micron two-color 640x486 quantum well infrared photodetector (QWIP) focal plane array camera," *Proc. SPIE* **3698**, 687 (1999).
29. P. Bois, E. Costard, X. Marcadet, and E. Herniou, "Development of quantum well infrared photodetectors in France," *Infrared Phys. Technol.* **42**, 291 (2001).
30. M. Sundaram, S. C. Wang, M. F. Taylor, *et al.*, "Two-color quantum well infrared photodetector focal plane arrays," *Infrared Phys. Technol.* **42**, 301 (2001).
31. A. Goldberg, K. K. Choi, E. Cho, and B. McQuiston, "Laboratory and field performance of megapixel QWIP focal plane arrays," *Infrared Phys. Technol.* **47**, 91 (2005).
32. S. D. Gunapala, S. V. Bandara, J. K. Liu, *et al.*, "Development of mid-wavelength and long-wavelength megapixel portable QWIP imaging cameras," *Infrared Phys. Technol.* **47**, 67 (2005).
33. S. D. Gunapala, S. V. Bandara, J. K. Liu, *et al.*, "1024x1024 pixel MWIR and LWIR QWIP focal plane arrays and 320x256 MWIR:LWIR pixel collocated simultaneous dualband QWIP focal plane arrays," *Proc. SPIE* **5783**, 789 (2005).
34. B. F. Levine, "Quantum well infrared photodetectors," *J. Appl. Phys.* **74**, R1 (1993).
35. Y. Zhang, D. S. Jiang, J. B. Xia, L. Q. Cui, C. Y. Song, Z. Q. Zhou, and W. K. Ge, "A voltage-controlled tunable two-color infrared photodetector using GaAs/AlAs/GaAlAs and GaAs/GaAlAs stacked multiquantum wells," *Appl. Phys. Lett.* **68**, 2114 (1996).
36. A. Rogalski, "Assessment of HgCdTe photodiodes and quantum well infrared photodetectors for long wavelength focal plane arrays," *Infrared Phys. Technol.* **40**, 279 (1999).
37. M. Sundaram and S. C. Wang, "2-color QWIP FPAs," *Proc. SPIE* **4028**, 311 (2000).
38. A. Goldberg, T. Fischer, S. Kennerly, W. Beck, V. Ramirez, and K. Garner, "Laboratory and field imaging test results on single-color and dual-band QWIP focal plane arrays," *Infrared Phys. Technol.* **42**, 309 (2001).
39. A. C. Goldberg, S. W. Kennerly, J. W. Little, *et al.*, "Comparison of HgCdTe and QWIP dual-band focal plane arrays," *Proc. SPIE* **4369**, 532 (2001).
40. S. D. Gunapala, S. V. Bandara, A. Singh, *et al.*, "Quantum well infrared photodetectors for low background applications," *Proc. SPIE* **3379**, 225 (1998).
41. M. Jhabvala, "Application of GaAs quantum well infrared photoconductors at the NASA/Goddard Space Flight Center," *Infrared Phys. Technol.* **42**, 363 (2001).

Terahertz Quantum Cascade Lasers and Real-Time T-Ray Imaging

Q. Hu, B. S. Williams, S. Kumar, A. W. M. Lee, Q. Qin
Department of Electrical Engineering and Computer Science
Massachusetts Institute of Technology, Cambridge, MA 02139, U.S.A.

J. L. Reno
Sandia National Labs, Albuquerque, NM 87185-1303, U.S.A.

H. C. Liu and Z. R. Wasilewski
Institute for Microstructural Sciences
National Research Council, Ottawa K1A 0R6, Canada

1. Introduction

Terahertz (1–10 THz, $\hbar\omega$ = 4–40 meV, and λ = 30–300 μm) frequencies are among the most underdeveloped in the electromagnetic spectrum, even though their potential applications are promising in detection of chemical and biological agents, imaging for medical and security applications, astrophysics, plasma diagnostics, end-point detection in dry etching processes, remote atmospheric sensing and monitoring, noninvasive inspection of semiconductor wafers, high-bandwidth free-space communications, and ultrahigh-speed signal processing.[1] This under-development is primarily due to the lack of coherent solid-state THz sources that can provide high radiation intensities (greater than a mW) and continuous-wave (cw) operation. This is because the THz frequency falls between two other frequency ranges in which conventional semiconductor devices have been well developed. One is the microwave and millimeter-wave frequency range, and the other is the near-infrared and optical frequency range. Semiconductor electronic devices that utilize freely moving electrons (such as transistors, Gunn oscillators, Schottky-diode frequency multipliers, and photomixers) are limited by the transit time and parasitic *RC* time constants. Consequently, the power level of these electronic devices decreases as $1/f^4$, or even faster, as the frequency f increases above 1 THz. Semiconductor photonic devices based on quantum-mechanical interband transitions (such as bipolar laser diodes), however, are limited to frequencies higher than those corresponding to the semiconductor energy gap, which is higher than 10 THz even for narrow-gap lead-salt materials. Thus, the 1–10 THz frequency range is inaccessible for conventional semiconductor devices.

Semiconductor quantum wells are human-made quantum-mechanical systems in which the energy levels can be designed and engineered to be of any value.

Future Trends in Microelectronics. Edited by Serge Luryi, Jimmy Xu, and Alex Zaslavsky
ISBN 0-471-48

Consequently, unipolar lasers based on intersubband transitions (electrons that make lasing transitions between subband levels) were proposed for long-wavelength sources as early as the 1970's.[2] This device concept has been realized in the successful development of quantum-cascade lasers (QCLs) at mid-infrared wavelengths.[3] Since then, impressive improvements in performance have been made in terms of power levels, operating temperatures, and frequency characteristics at mid-infrared frequencies.

In contrast to the remarkable development of mid-infrared QCLs, the development of THz QCLs below the *Reststrahlen* band (~8–9 THz in GaAs) turned out to be much more difficult than initially expected, because of two unique challenges at THz frequencies. First, the energy level separations that correspond to THz frequencies are quite narrow (~10 meV). Thus, the selective depopulation mechanism based on energy-sensitive LO-phonon scattering, which has been successfully implemented in mid-infrared QCLs, is not applicable. Second, mode confinement, which is essential for any laser oscillation, is difficult to achieve at THz frequencies. Conventional dielectric-waveguide confinement is not useful because the evanescent field penetration, which is proportional to the wavelength and is on the order of several tens of microns, is much greater than the active gain medium of several microns.

In October 2001, almost eight years after the initial development of QCLs, the first QCL operating at 4.4 THz, below the *Reststrahlen* band, was developed.[4] This laser was based on a chirped superlattice structure that had been successfully developed at mid-infrared frequencies. Mode confinement in this THz QCL was achieved using a double-surface plasmon waveguide grown on a semi-insulating (SI) GaAs substrate. Shortly after this breakthrough, a THz QCL based on bound-to-continuum intersubband transition was developed at ~3.4 THz.[5]

Our group has pursued a different approach to achieve lasing at THz frequencies. We have investigated possibilities of using fast LO-phonon scattering to depopulate the lower radiative level,[6-8] and using double-sided metal wave-guides for THz mode confinement.[9] After an extensive investigation, these efforts finally bore fruit. In November of 2002, a 3.4 THz QCL was developed in which the depopulation of the lower radiative level was achieved through resonant LO-phonon scattering.[10] The performance of this laser device is promising, especially its operating temperatures. One of the laser devices was operated in the pulse mode up to 87 K, above liquid-nitrogen temperature.[11] When fabricated with the double-sided metal-metal waveguides, THz QCLs based on similar quantum-well structures have demonstrated the highest pulsed operating temperature of ~170 K, the highest cw operating temperature of 117 K, and the longest wavelength of ~188 μm. Using a high-power THz QCL and a 240×320 focal-plane array camera, we are now able to perform real-time THz imaging at video rate, that is, taking movies in "T-rays". These rapid developments indicate great potentials for THz QCLs in various applications. In the following sections, we summarize key results of these investigations.[10-36]

2. THz gain medium based on resonant LO-phonon scattering for depopulation

The unipolar intersubband lasers are known to yield a high value of gain, because of a large joint density of states as a result of the two subbands tracking each other in the *k*-space; thus electrons emit photons at the same energy regardless of their initial momentum. Consequently, the peak gain is related to the inverted population density Δn_{3D} in a simple linear fashion:

$$g_{\text{peak}} = (e^2\omega/\pi\hbar\varepsilon_r^{1/2}\varepsilon_0 c_0)\Delta n_{3D}\, z_{ij}^2/\Delta f = (e^2/2\pi\hbar\varepsilon_r^{1/2}\varepsilon_0 c_0 m^*)\Delta n_{3D}\,[f_{ij}/\Delta f] \sim$$
$$\sim 67\,[\Delta n_{3D}/10^{15}\text{ cm}^{-3}]\,[f_{ij}/\Delta f\text{ THz}]\text{ cm}^{-1}. \quad (1)$$

In Eq. (1), Δn_{3D} is the three-dimensional inverted population density in the active region; $z_{ij} = \langle i|z|j\rangle$ is the dipole moment and $f_{ij} = (2m^*\omega z_{ij}^2)/\hbar$ is the dimensionless oscillator strength of the $i \rightarrow j$ transition; and Δf is the FWHM linewidth of spontaneous emission in units of Hz (but in units of THz in the last numerical part of Eq. (1)). Clearly from Eq. (1), there are only *three* parameters that determine the peak material gain: Δn_{3D}, Δf, and f_{ij}. All the other parameters are either fundamental constants or material parameters that are not subject to engineering and manipulation. In our structures, the measured spontaneous emission linewidth is typically ~1 THz (~4 meV). We need to mostly concentrate on the optimization of the remaining two parameters, Δn_{3D} and f_{ij}.

The first successful THz QCLs were designed around chirped superlattice structures,[4] which are characterized by large oscillator strengths. However, the depopulation of the lower lasing level relies on resonant tunneling and electron-electron scattering, which could suffer from thermal backfilling because of narrow subband separations within the miniband of a superlattice. The direct use of LO-phonon scattering for depopulation of the lower state offers several distinctive advantages. First, when a collector (ground) state is separated from the lower lasing level by at least the LO-phonon energy $\hbar\omega_{LO}$ =36 meV (~9 THz), depopulation can be extremely fast, and it does not depend much on temperature or the electron distribution. Second, the large energy separation provides intrinsic protection against thermal backfilling of the lower radiative state. Both properties are important in allowing higher temperature operation of lasers with high output power levels.

While fast scattering out of the lower lasing level is necessary to achieve population inversion, a long upper-state lifetime τ_u is also highly desirable to increase the level of population inversion. Our previous designs addressed this problem by making the optical transition diagonal (*i.e.* between states in adjacent wells), so as to reduce upper-state overlap with the collector state.[6] However, this resulted in a small oscillator strength, and in a broad emission linewidth due to interface roughness. A second design featured a vertical radiative transition,[7] which improved the radiative overlap and had a relatively narrow linewidth (~2 meV ≈ 0.5 THz), but depopulation was nonselective and slow, due to the thick barrier needed to reduce parasitic scattering from the upper state.

The key element in the present design is the use of *resonant LO-phonon scattering* to selectively depopulate the lower radiative level while maintaining a long upper level lifetime, and it resulted in a breakthrough success. Figure 1 shows the conduction band profile and subband wavefunctions under the designed bias of 64 mV/module. Each module contains four quantum wells, shown inside the dashed box, and 175 such modules are connected in series to form the quantum cascade laser. Under this bias, the electrons are injected from the injector level 1' to the upper lasing level 5 through resonant tunneling. The radiative transition between levels 5 and 4 is spatially vertical, yielding a relatively large oscillator strength ($f_{54} \approx 0.96$). The depopulation is highly selective and fast, as only the lower level 4 is at resonance with a level 3 in the adjacent well, where fast LO-phonon scattering ($\tau_4 \approx 0.5$ ps) takes place to transfer electrons in levels 3 and 4 to the injector doublet 1 and 2. The scattering time of the upper level 5 to the ground states 2 and 1, due to a relatively thick barrier, is quite long ($\tau_{5\rightarrow2,1} \approx 7$ ps), which is important to maintain a population inversion between levels 5 and 4. Electrons in level 1 are then injected to level 5 of the following module (not shown), completing the cascade pumping scheme.

Mode confinement in this laser device was achieved using a double-surface plasmon waveguide formed between the top metallic contact and the bottom heavily doped GaAs layer, as in the case of other THz QCLs.[4] Lasing was obtained in this device and the emission frequency of 3.4 THz corresponds to a photon energy of 14.2 meV, close to the calculated value of 13.9 meV. Pulsed lasing operation from devices fabricated from this first laser wafer is observed up to 87 K.[11]

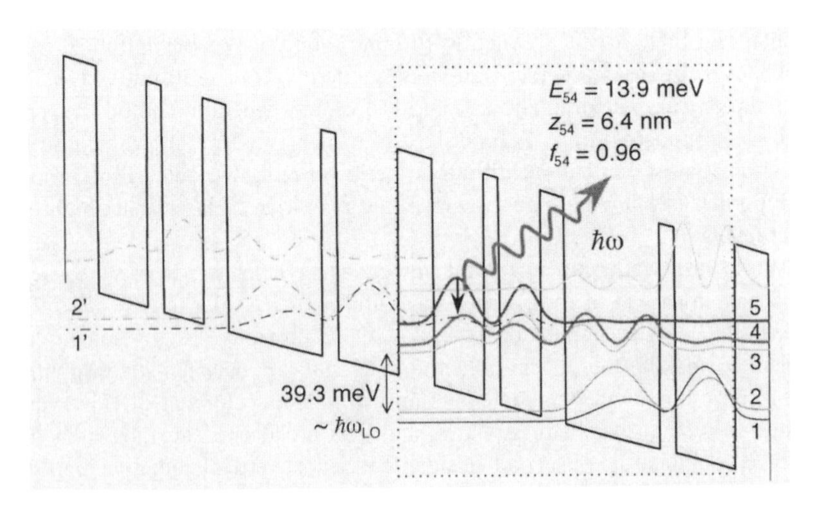

Figure 1. Conduction band profile of a cascade module calculated using a self-consistent Schrödinger and Poisson solver biased at 64 mV/module.

3. THz mode confinement using double-side metal waveguides

Because of the amphoteric nature of silicon dopants in GaAs materials, the maximum attainable electron density is approximately $5\times10^{18}/cm^3$. At this carrier concentration, the penetration depth is on the order of 1 μm at THz frequencies, causing a significant cavity loss if the lower side of the mode confinement is provided by heavily doped GaAs layers. The first successful development of THz QCLs utilized a double surface plasmon layer grown on semi-insulating GaAs substrate for mode confinement. This structure is easy to grow and fabricate, and it provides adequate mode confinement for most of the THz QCLs. However, the mode confinement factor Γ in this scheme is far below unity ($\Gamma \sim 0.2$–0.5 for reported lasers).

Following our initial success in developing the 3.4 THz laser, we demonstrated the first THz QCL that uses a double-sided metal-metal waveguide for mode confinement.[13] This metal-semiconductor-metal structure is essentially the same as the microstrip transmission lines that are widely used for waveguiding at microwave and millimeter-wave frequencies, and the geometry is compatible with the TM polarization of intersubband transitions. Due to the shallow skin depth in the metal (several tens of nm), the waveguide can be made with very low losses and a confinement factor close to unity. Our current THz lasers with metal-metal waveguides were fabricated using copper-to-copper wafer bonding followed by substrate removal. The schematic of the bonding and substrate removal process is illustrated in Fig. 2.

Based on this metal-metal waveguide structure and using improved gain media that reduced the lasing threshold current densities, we have achieved several records in the performance of THz QCLs. These include but not limited to: the highest pulsed operating temperature of ~169 K, the first CW THz QCL operating above the important liquid nitrogen temperature of 77 K (T_{max} = 117 K), and the longest wavelength QCL to date without the assistance of magnetic fields ($\lambda \approx 188$ μm, corresponding to 1.59 THz), as illustrated in Figs. 3 and 4.

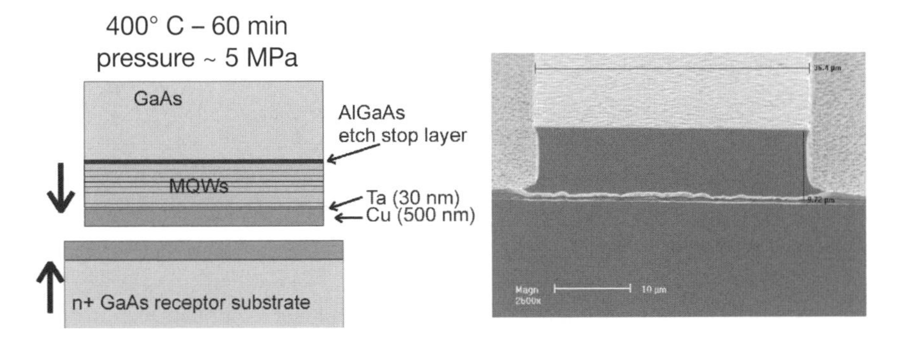

Figure 2. Schematic of the wafer bonding process for double-side metal-metal waveguide (left) and SEM picture of the fabricated device (right).

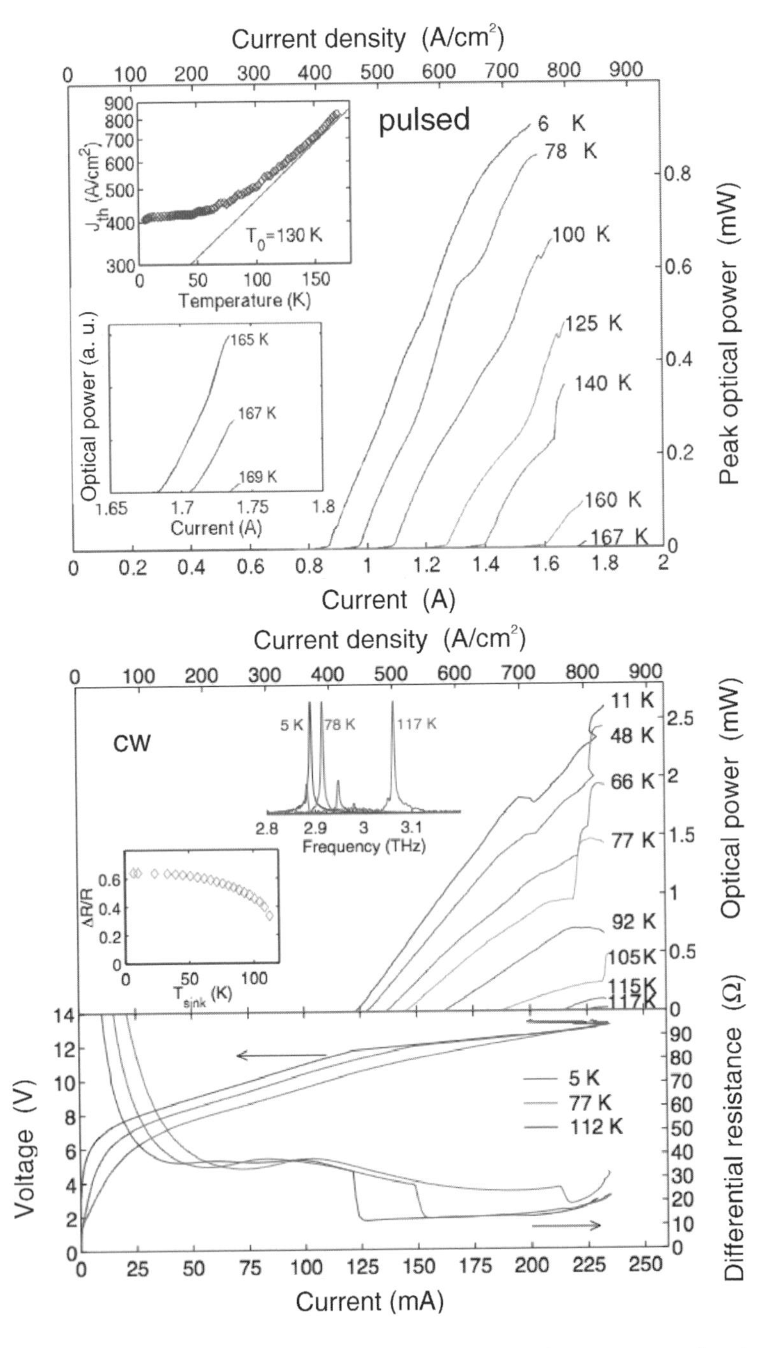

Figure 3. Pulsed power-current and voltage-current relations measured up to ~169 K heat-sink temperature (upper) and cw power-current and voltage-current relations measured up to ~117 K heat-sink temperature (lower), with lower inset showing the differential resistance-current relations of the device.

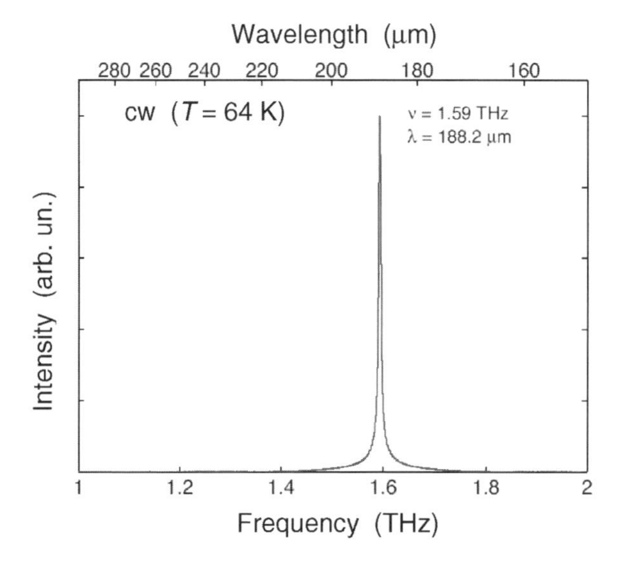

Figure 4. Continuous wave lasing emission spectra at 1.59 THz.

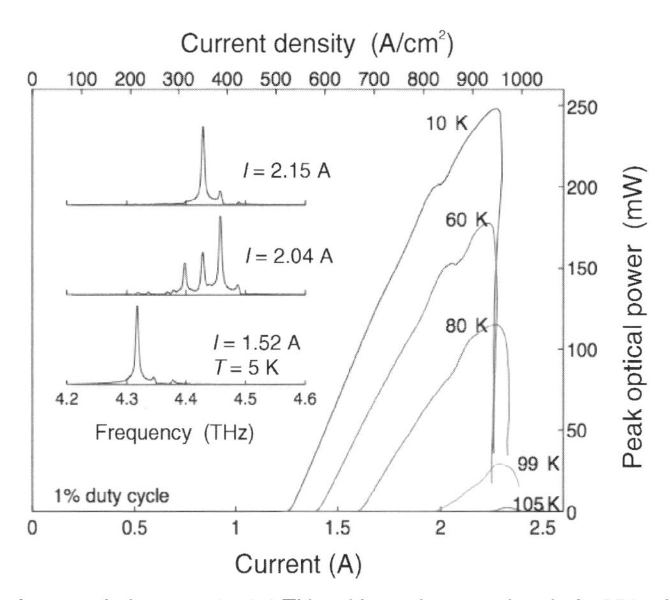

Figure 5. A cascade laser at $f \sim 4.4$ THz with peak power level of ~250 mW.

In addition to the record performance in operating temperatures and wavelength, we have recently developed high-power THz QCLs that produce ~250 mW of power, as shown in Fig. 5. Using these high-power lasers, we are now able to perform THz imaging in real time at a video rate of ~20 frames/second, that is, making movies in T-rays.

4. Real-time THz imaging using QCLs and focal-plane array cameras

Imaging using radiation in the terahertz frequency range, 0.3–10 THz, has demonstrated the ability to see the details within visibly opaque objects, such as integrated circuits packages, leaves, teeth, thin tissue samples, and illicit drugs in envelopes. The vast majority of THz imaging has been done by linearly scanning an object through a tightly focused THz beam – a practice that limits the acquisition time to the mechanical scan rate of the system. With upper limits of hundreds of pixels/second for mechanical scanning, a complete image takes minutes to acquire.

Real-time imaging (~30 frames per second or more) has previously been demonstrated by using an electro-optic crystal for frequency up-conversion so that THz images can be viewed with a CCD focal-plane camera. However, this set-up requires precise timing of the optical and THz pulses, necessitating a scanning delay mechanism, adding to its complexity. Furthermore, because of the short THz pulses (<1 ps), this scheme is inherently broadband (>1 THz). In applications such as drug detection, where the acquistion of a narrow-band fingerprint is required, a coherent narrow-band illumination source is crucial. Because of their compact sizes, many THz quantum-cascade lasers with different frequencies, corresponding to different chemical absorption bands, can be packaged tightly, forming a frequency agile coherent radiation source. In combination with a focal-plane imager, such a system can perform frequency-sensitive THz imaging far faster than previous methods, allowing real-time THz monitoring and screening.

We have recently demonstrated the first real-time cw terahertz imaging system using THz QCLs and a focal-plane array camera. The experimental arrangement is shown in Fig. 6. The THz QCL, cooled by a cryogen-free pulsed-tube thermomechanical cooler, produces ~50 mW of power at ~30 K. As shown in the figure, imaging experiments in both transmission and reflection mode can be performed. Since the microbolometer camera was initially designed for the 10-μm wavelength range for night-vision applications, we developed a differential scheme to subtract the strong ambient background at ~300 K and reduce $1/f$ noise.[34]

An example of the real-time imaging experiment is shown in Fig. 7, in which several hand-written characters inside a regular mail envelope are clearly visible in THz imaging, in both transmission and reflection mode. It should be pointed out that this particular imaging application cannot be done at other frequencies: x-rays lack contrast; millimeter waves do not provide sufficient spatial resolution; and infrared radiation is heavily scattered and/or absorbed by fibrous materials. While these still images are recognizable, the integration of the eye and pattern recognition of the brain aids recognition tremendously when real-time video is used. With additional QCLs the system will allow analytic, real-time multi-frequency imaging. Very recently, by carefully designing and fabricating QCLs with the desired frequency characteristics, we have demonstrated real-time, long-range (>25 meters), transmission mode imaging.[36] This demonstration is a necessary step towards the ultimate goal of a standoff imaging system.

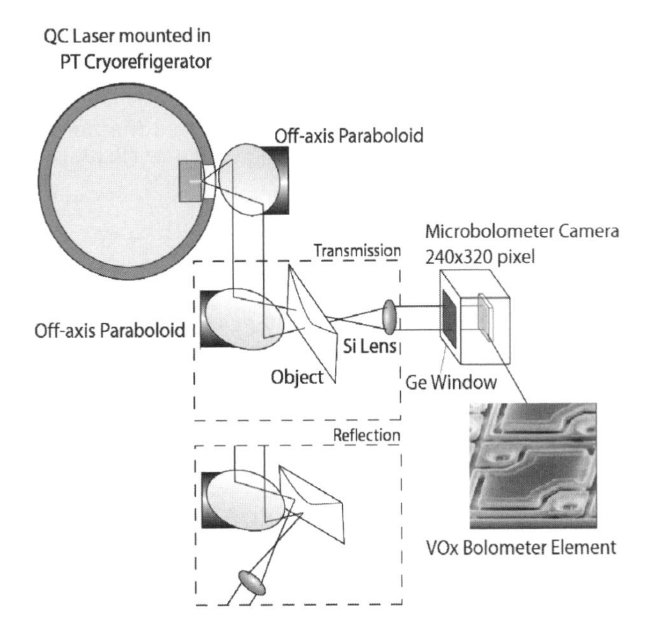

Figure 6. Experimental setup of the THz imaging system. The photo shows a vanadium oxide microbolometer (courtesy of BAE Systems, Lexington, MA). Cutaway depicts alternate reflection mode setup.

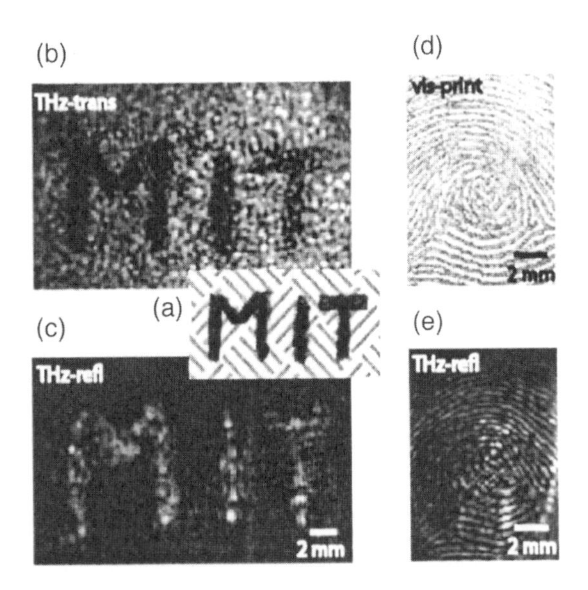

Figure 7. Pencil letters written on the inside of a paper security envelope at visible frequencies (a), in THz transmission mode (b, 1 frame, 1/20 second) and THz reflection mode (c, 20 frames, 1 second). Visible frequency thumb print (d) and THz reflection mode image the thumb of the leading author (e, 20 frames).

Acknowledgments

This work is supported by AFOSR, NASA, and NSF. Sandia is a multiprogram laboratory operated by Sandia Corporation, a Lockheed Martin company, for the United States Department of Energy under Contract DE-AC04-94AL85000.

References

1. Special issue on "The terahertz gap: The generation of far-infrared radiation and its applications," *Phil. Trans. Royal Soc. London A* **362** (2004), pp. 195–414.
2. R. F. Kazarinov and R. A. Suris, "Possibility of amplification of electromagnetic waves in a semiconductor with a superlattice," *Sov. Phys. Semicond.* **5**, 707 (1971).
3. J. Faist, F. Capasso, D. L. Sivco, C. Sirtori, A. L. Hutchinson, and A. Y. Cho, "Quantum cascade laser," *Science* **264**, 477 (1994).
4. R. Köhler, A. Tredicucci, F. Beltram, *et al.*, "Terahertz semiconductor-heterostructure laser," *Nature* **417**, 156 (2002).
5. G. Scalari, L. Ajili, J. Faist, H. Beere, G. Davies, E. Linfield, and D. Ritchie, "Far infrared ($\lambda \approx$ 86 μm) quantum-cascade lasers based on bound-to-continuum transition with operation temperature up to 90 K," *Appl. Phys. Lett.* **82**, 3165 (2003).
6. B. Xu, Q. Hu, and M. R. Melloch, "Electrically pumped tunable terahertz emitter based on intersubband transition," *Appl. Phys. Lett.* **71**, 440 (1997).
7. B. S. Williams, B. Xu, Q. Hu, and M. R. Melloch, "Narrow-linewidth terahertz intersubband emission from three-level systems," *Appl. Phys. Lett.* **75**, 2927 (1999).
8. B. S. Williams and Q. Hu, "Optimized energy separation for phonon scattering in three-level terahertz intersubband lasers," *J. Appl. Phys.* **90**, 5504 (2001).
9. B. Xu, *Development of Intersubband Terahertz Lasers Using Multiple Quantum Well Structures*, Ph.D. thesis, MIT, 1998.
10. B. S. Williams, H. Callebaut, S. Kumar, Q. Hu, and J. L. Reno, "3.4-THz quantum cascade laser based on LO-phonon scattering for depopulation," *Appl. Phys. Lett.* **82**, 1015 (2003).
11. B. S. Williams, S. Kumar, H. Callebaut, Q. Hu, and J. L. Reno, "3.4 THz quantum cascade laser operating above liquid nitrogen temperature," *Electronics Lett.* **39**, 915 (2003).
12. H. Callebaut, S. Kumar, B. S. Williams, and J. L. Reno, "Analysis of transport and thermal properties of THz quantum cascade lasers," *Appl. Phys. Lett.* **83**, 207 (2003).
13. B. S. Williams, S. Kumar, H. Callebaut, Q. Hu, and J. L. Reno, "Terahertz quantum cascade laser at $\lambda \approx$ 100 μm using metal waveguide for mode confinement," *Appl. Phys. Lett.* **83**, 2124 (2003).

14. Q. Hu, B. S. Williams, S. Kumar, H. Callebaut, and J. L. Reno, "Terahertz quantum cascade lasers based on resonant phonon scattering for depopulation," *Phil. Trans. Royal Soc. London A* **362**, 233 (2004).
15. B. S. Williams, S. Kumar, H. Callebaut, Q. Hu, and J. L. Reno, "Terahertz quantum cascade lasers operating up to 137 K," *Appl. Phys. Lett.* **83**, 5142 (2003).
16. H. Callebaut, S. Kumar, B. S. Williams, Q. Hu, and J. L. Reno, "Importance of electron-impurity scattering for electron transport in THz quantum-cascade lasers," *Appl. Phys. Lett.* **84**, 645 (2004).
17. S. Kumar, B. S. Williams, S. Kohen, Q. Hu, and J. L. Reno, "Continuous-wave operation of terahertz quantum-cascade lasers above liquid-nitrogen temperature," *Appl. Phys. Lett.* **84**, 2494 (2004).
18. B. S. Williams, S. Kumar, Q. Hu, and J. L. Reno, "Resonant-phonon terahertz quantum-cascade laser operating at 2.1 THz ($\lambda \approx 141$ μm)," *Electronics Lett.* **40**, 431 (2004).
19. Q. Hu, B. S. Williams, S. Kumar, H. Callebaut, S. Kohen, and J. L. Reno, "Resonant-phonon-assisted THz quantum cascade lasers with metal-metal waveguides," *Semicond. Sci. Technol.* **20**, S228 (2005).
20. S. Kohen, B. S. Williams, and Q. Hu, "Electromagnetic modeling of terahertz quantum cascade laser waveguides and resonators," *J. Appl. Phys.* **97**, 053106 (2005).
21. M. S. Vitiello, G. Scamarcio, B. S. Williams, S. Kumar, Q. Hu, and J. L. Reno, "Measurement of subband electronic temperatures and population inversion in THz quantum cascade lasers," *Appl. Phys. Lett.* **86**, 111115 (2005).
22. J. R. Gao, J. N. Hovenier, Z. Q. Yang, *et al.*, "A terahertz heterodyne receiver based on a quantum cascade laser and a superconducting bolometer," *Appl. Phys. Lett.* **86**, 244104 (2005).
23. H. C. Liu, M. Wächter, D. Ban, *et al.*, "Effect of doping concentration on the performance of terahertz quantum-cascade lasers," *Appl. Phys. Lett.* **87**, 141102 (2005).
24. A. L. Betz, R. T. Boreiko, B. S. Williams, S. Kumar, Q. Hu, and J. L. Reno, "Frequency and phaselock control of a 3-THz quantum cascade laser," *Optics Lett.* **30**, 1837 (2005).
25. A. W. M. Lee and Q. Hu, "Real-time, continuous-wave terahertz imaging using a microbolometer focal-plane array," *Optics Lett.* **30**, 2563 (2005).
26. B. S. Williams, S. Kumar, Q. Hu, and J. L. Reno, "Operation of terahertz quantum-cascade lasers at 164 K in pulsed mode and at 117 K in continuous-wave mode," *Optics Express* **13**, 3331 (2005).
27. B. S. Williams, S. Kumar, Q. Hu, and J. L. Reno, "Distributed-feedback terahertz quantum-cascade lasers using laterally corrugated metal waveguides," *Optics Lett.* **30**, 2909 (2005).
28. H. Callebaut and Q. Hu, "Importance of coherence for electron transport in terahertz quantum cascade lasers," *J. Appl. Phys.* **98**, 104505 (2005).

29. E. E. Orlova, J. N. Hovenier, T. O. Klaassen, *et al.*, "Antenna model for wire lasers," *Phys. Rev. Lett.* **96**, 173904 (2006).
30. B. S. Williams, S. Kumar, Q. Hu, and J. L. Reno, "High-power terahertz quantum-cascade lasers," *Electronics Lett.* **42**, 89 (2006).
31. A. J. L. Adam, I. Kašalynas, J. N. Hovenier, *et al.*, "Beam pattern of terahertz quantum cascade lasers with sub-wavelength cavity dimensions," *Appl. Phys. Lett.* **88**, 151105 (2006).
32. S. Kumar, B. S. Williams, Q. Hu, and J. L. Reno, "1.9-THz quantum-cascade lasers with one-well injector," *Appl. Phys. Lett.* **88**, 121123 (2006).
33. A. Baryshev, J. N. Hovenier, A. J. L. Adam, *et al.*, "Phase-lock and free-running linewidth of a two-mode terahertz quantum cascade laser," *Appl. Phys. Lett.* **89**, 031115 (2006).
34. A. W. M. Lee, B. S. Williams, S. Kumar, Q. Hu, and J. L. Reno, "Real-time imaging using a 4.3-THz quantum cascade laser and a 320×240 micro-bolometer focal-plane array," *IEEE Photonics Technol. Lett.* **18**, 1415 (2006).
35. B. S. Williams, S. Kumar, Q. Qin, Q. Hu, and J. L. Reno, "Terahertz quantum cascade lasers with double resonant-phonon depopulation," *Appl. Phys. Lett.* **88**, 261101 (2006).
36. A. W. M. Lee, Q. Qin, S. Kumar, B. S. Williams, Q. Hu, and J. L. Reno, "Real-time terahertz imaging over a standoff distance (>25 meters)," *Appl. Phys. Lett.* **89**, 141125 (2006).

Terahertz Spectroscopy and Imaging

E. H. Linfield, J. E. Cunningham, and A. G. Davies
School of Electronic and Electrical Engineering
University of Leeds, Leeds LS2 9JT, U.K.

1. Introduction

The terahertz (THz) region (0.3–10 THz) of the electromagnetic spectrum spans the frequency range between the infrared and millimeter/microwave, as illustrated in Fig. 1. Historically this region has not been fully exploited owing to the very limited number of suitable (in particular, coherent) radiation sources and detectors that have been available. In recent years,[1] outstanding progress has been made in developing THz components and systems. Highlights include the fabrication of the first THz quantum cascade laser,[2] together with the commercialization of THz spectroscopy and imaging systems based on femtosecond-laser technology,[3] notably for non-destructive testing in the pharmaceutical industry (for example, investigating polymorphic transformations, and drug distributions in tablets[4]). Further, three years ago an ultra-broadband THz spectroscopy system was proposed, making use of photoconductive emitters in a reflection geometry. This offered a bandwidth in excess of 20 THz for spectroscopy applications,[5] and its performance was demonstrated through measurements of the absorption of polycrystalline adenosine.[1] Good agreement was obtained with spectra acquired using the complementary technologies of Fourier-transform infrared, Raman, and neutron inelastic spectroscopy.

Despite these exciting developments, when the previous conference treatise[1] reviewed THz technology, the field of THz imaging and spectroscopy was still in its infancy: extensive research was needed on the basic interactions of THz radiation with materials. Even so, a diverse range of applications had been identified across the physical, biological and medical sciences, including medical

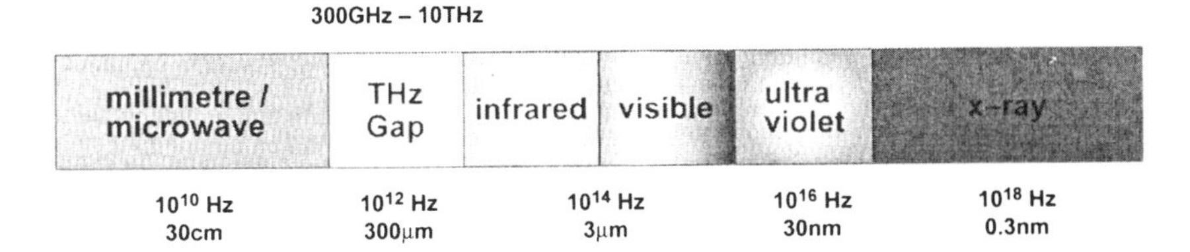

Figure 1. Schematic diagram showing the electromagnetic spectrum, and illustrating the "THz gap" for which there has been, until recently, a lack of convenient solid-state sources and detectors.

Future Trends in Microelectronics. Edited by Serge Luryi, Jimmy Xu, and Alex Zaslavsky
ISBN 0-471-48

and dental imaging, the detection of concealed explosives and drugs, biomedical spectroscopy, atmospheric sensing, process monitoring, non-destructive testing of composite materials, diagnostic testing, high-bandwidth communications, astronomy, and condensed matter physics, *inter alia*.

Over the past three years, the field of THz spectroscopy and imaging has undergone extremely rapid growth, with major developments being seen in THz components, THz systems and potential application areas. To give just three examples:

- THz spectroscopy is being considered by security agencies around the world as a potential technique for identifying drugs-of-abuse and explosives;
- on-chip THz systems have been extensively developed, and offer an exciting prospect both for biomedical sensing with femtomole sensitivity, and the detailed study of condensed matter systems at cryogenic temperatures;
- THz quantum cascade lasers have been demonstrated to operate at frequencies below 2 THz, and at temperatures above 160 K. They are also being incorporated into THz imaging systems.

Here, we discuss two of these developments in greater detail: the development of on-chip THz systems, and the applications of THz spectroscopy for studying drugs-of-abuse and explosives; the latest developments in the field of THz quantum cascade lasers are discussed elsewhere in this volume. We conclude this chapter by considering just a few of the possible developments that may be seen in the field of THz spectroscopy and imaging over the next three years and beyond.

2. Broadband THz spectroscopy of drugs-of-abuse and explosives

Figure 2 shows an experimental configuration allowing THz radiation with frequency components over 20 THz to be obtained using a biased and asymmetrically excited GaAs photoconductive emitter. Its maximum useful bandwidth exceeds that seen in most THz systems based on photoconductive emitters, which are generally limited to ~4 THz.

In brief, the emitter comprises two vacuum-evaporated NiCr/Au electrodes separated by a 0.4 mm gap, deposited on a low-temperature-grown (LT) GaAs layer. We chose LT-GaAs because it provides a short photocarrier recombination lifetime (0.4 ps in this case), together with both high resistivity and high carrier mobility. A bias voltage of ±120 V, modulated at 31 kHz, was applied across the emitter. A pulsed Ti:sapphire laser of 300 mW average power (12 fs pulse width, 800 nm centre wavelength, and 76 MHz repetition rate) was focused to a 40 μm spot diameter on the edge of one of the two NiCr/Au electrodes of the LT-GaAs emitter, to generate THz pulses. In contrast to previous experiments, where the THz radiation was collected after being transmitted through the GaAs substrate, THz radiation was collected, in this case, in the same direction as the reflected pump laser beam. As a result, the absorption and dispersion of the THz pulses in the GaAs substrate were minimized. The emitted THz pulses were collimated and focused onto the sample by a pair of parabolic mirrors. The transmitted THz pulse was then collected and focused using another pair of parabolic mirrors onto a 20 μm-thick (110) ZnTe crystal glued onto a 1 mm-thick wedged (100) ZnTe crystal for electro-optic detection. A variable delay stage provided the time delay between the THz and probe pulses, and the system had a spectral resolution of 75 GHz (2.5 cm^{-1}). Using a lock-in detection scheme referenced to the frequency of the bias across the GaAs emitter, a signal-to-noise ratio of over 1000 was achieved. The apparatus was enclosed in a vacuum-tight box, and purged with dry nitrogen gas to reduce the effects of water vapor absorption.

Figure 3 shows absorption spectra observed at room temperature with this system for pellets of the explosives 2,4,6-trinitrotoluene (TNT), hexahydro-1,3,5-trinitro-1,3,5-triazine (RDX), and pentaerythritol tetranitrate (PETN),[6] diluted in a 20% weight-to-weight ratio with the inert matrix PTFE. Absorption peaks in the 0.3–3 THz region can be easily seen and correspond well with previously reported data.[7] Significantly, absorption bands can also be seen in the 3–5.5 THz region, with the bandwidth of the system in this case being limited by the PTFE phonon resonance at 6 THz. The higher frequency absorption is particularly prominent in the spectra of TNT, where the bands above 3 THz are much more intense than those in the 0.1–3 THz region. This is important, because the lower frequency peaks are difficult to observe above the background noise at low concentrations of explosive.

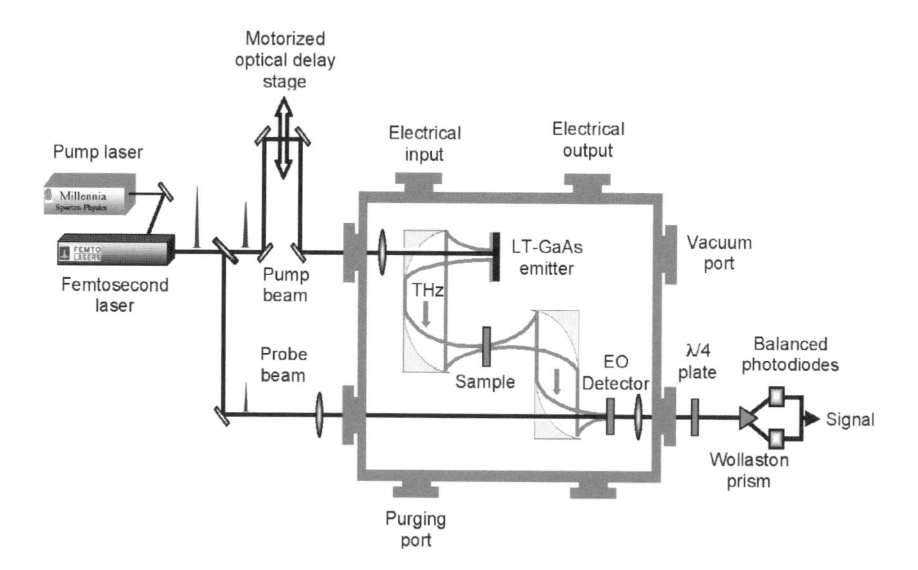

Figure 2. Schematic diagram illustrating a femtosecond laser-based ultrabroad-band THz spectroscopy system.

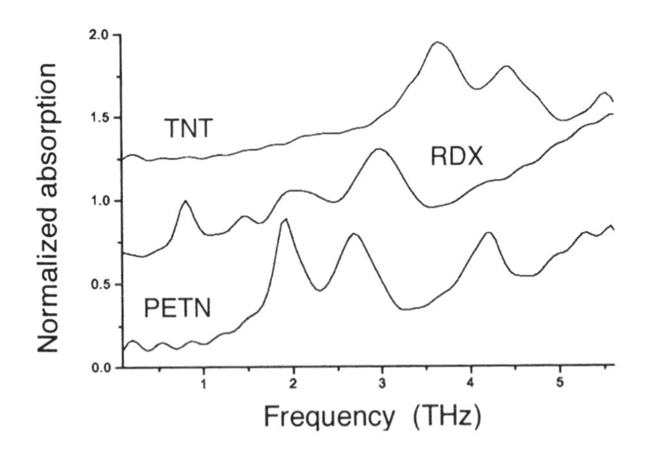

Figure 3. THz absorption spectra of TNT, RDX and PETN, diluted in a 20% weight-to-weight ratio with PTFE. Spectra for TNT and RDX are offset for clarity.[6]

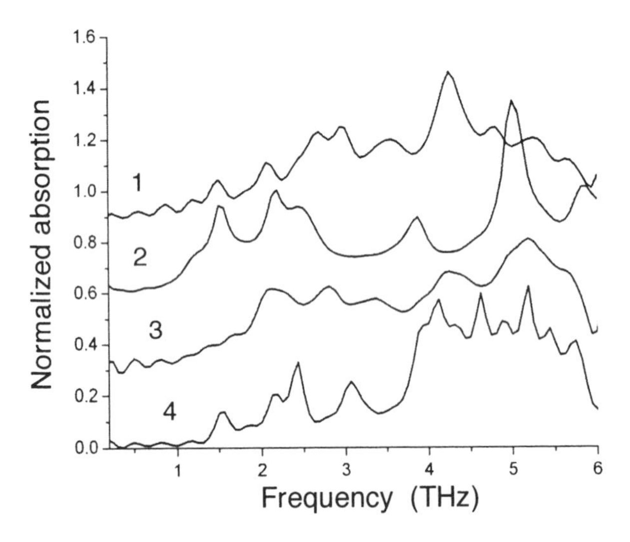

Figure 4. THz spectra of a number of drugs-of-abuse: (1) cocaine hydrochloride 25%; (2) cocaine free base 100%; (3) morphine sulfate pentahydrate 25%; and (4) ephedrine hydrochloride 12.5%. The spectra are offset for clarity; the percentages refer to weight-to-weight ratios of drug to PTFE matrix in the pellets, chosen to optimize the absorption spectra. There are marked differences in spectral features, even for similar molecular structures.[6]

Figure 4 shows spectra for three drugs-of-abuse: cocaine hydrochloride (Sigma-Aldrich lot no: 123k1339); cocaine free base (lot no: 044k0854); and morphine sulfate pentahydrate (lot no: 044k0736); together with ephedrine hydrochloride (lot no: 12019cu-034) that, although not classified as a drug-of-abuse, can act as a precursor to amphetamine and ecstasy. By mixing the drug samples with PTFE and optimizing the concentrations, good definition of spectral peaks was obtained over the full bandwidth of the spectrometer (no difference in the spectral positions of absorption features was observed for different concentrations). It is notable that large differences in spectra can be seen for small differences in molecular structure (for example, between cocaine free base and cocaine hydrochloride).

These data demonstrate that broadband THz spectroscopy can be used to analyze materials of security relevance, although considerable further development is needed before such technology can be used in real-life applications. In particular, there is a need to demonstrate a similar discrimination in "street" samples, which are likely to contain a range of contaminants.

3. The development of on-chip THz systems

The majority of THz spectroscopy, to date, has been undertaken using free-space reflection or transmission measurements, in which THz radiation is generated in free space by near-infrared excitation of photoconductive or electro-optic emitters, focused onto a sample using off-axis parabolic mirrors, and subsequently detected using either photoconductive or electro-optic detectors, in schemes similar to that presented in Section 2 above. Whilst this has been enormously successful in allowing a broad range of applications to be investigated, it has some clear limitations. For example, spectral resolution typically exceeds 20 GHz – too large to obtain the required peak resolution for many applications, and too much material is often required for promising areas such as the development of biosensors.

An extremely promising alternative is to generate and detect THz radiation on-chip.[8] A similar photoconductive excitation and detection mechanism is used as in the free-space systems, but the excited THz pulses are now directly coupled into

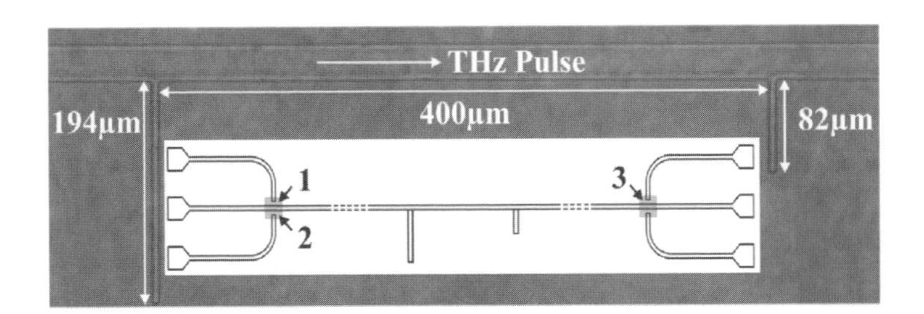

Figure 5. Schematic diagram (inset) of an on-chip guided wave THz system. A pulse of THz radiation is emitted by photoconductive antennae at position 1 or 2, and transmitted along the strip-line to a detector at position 3. Band-stop filters, in this case 82 μm and 194 μm long (optical photograph), remove selected frequencies from the transmitted THz pulse.

lithographically defined transmission lines, which are formed in either microstrip or strip-line/coplanar geometries.

A recent on-chip system is illustrated in Fig. 5.[9] Terahertz radiation is generated with a femtosecond laser pulse by exciting a photoconductive emitter (marked 1, Fig. 5) deposited on LT-GaAs. This excites a THz transient response, which propagates along the transmission line (a 4 mm-long strip-line in this case), and is detected in a photoconductive detector (marked 3, Fig. 5), which is gated by part of the initial femtosecond laser pulse. Figure 6 shows typical data when a ~ps duration pulse has been excited in an on-chip device, and then detected, both before and after propagation down 4 mm of microstrip. In this case, the transmitted pulse shows a ringing associated with the presence of a band-stop filter attached to the microstrip line (see Fig. 5). Fourier transformation of the acquired time-domain spectra then allows the broadband transmitted THz pulse to be determined in the frequency domain, giving similar data to that shown in Fig. 3, albeit over a narrower bandwidth.

Low-loss materials, for example benzocyclobutene (BCB) or PTFE, can be used as appropriate dielectrics in such microstrip geometries. Signal conductors are typically patterned as gold tracks that extend over several mm on the dielectric surface, before terminating at a photoconductive emitter/detector regions. In order to use on-chip systems for useful spectroscopy, the evanescent field lines around the transmission line can be exploited. To increase sensitivity, filter structures can be fabricated in the waveguide whose resonant frequency is sensitively dependent on the local dielectric environment around the filter. By monitoring detuning of the filter when overlaid materials are deposited, one can determine an unknown dielectric's permittivity at the filter frequency (provided the thickness can also be measured, which is a limitation of this technique). Several filter geometries have been studied for this application, including bandpass,[8] bandstop filters,[9] and ring resonators.[10]

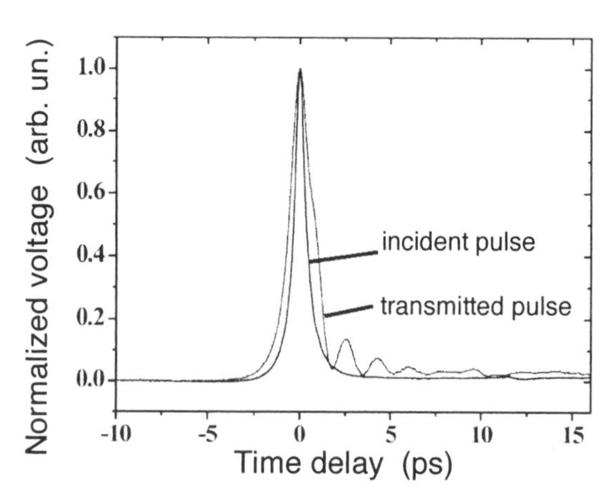

Figure 6. Typical pulse data obtained in the microstrip system shown in Fig. 5, with BCB used as the dielectric. Incident (before transmission down microstrip-line) and transmitted pulses (after transmission through the filter) are both shown.

Typical results for an overlaid dielectric material on dual-bandstop filter systems are shown in Fig. 7, where the frequency shift of each filter depends critically on the different thicknesses of dielectric material deposited. These measurements of frequency shift have been compared with full 3D electromagnetic simulations[11] and, using a value for relative permittivity ε_r that had been independently measured using a free-space THz time domain spectroscopy system, good agreement is obtained with the frequency shifts found experimentally.

A significant feature of on-chip systems is that the sample volume required is miniscule compared with the free-space techniques, typically a factor of several hundred lower. Furthermore, since reflected pulses can be designed out of the system (by choosing several mm-long probe arms and a microstrip signal conductor), the frequency resolution of an on-chip system has the potential to exceed free-space systems (where resolution is typically limited by back-reflections from sample, emitter, and detector surfaces). Indeed, a resolution of 2 GHz has recently been demonstrated in on-chip Fourier transformed data, limited by the resolution of linear translation in time-delay stages.

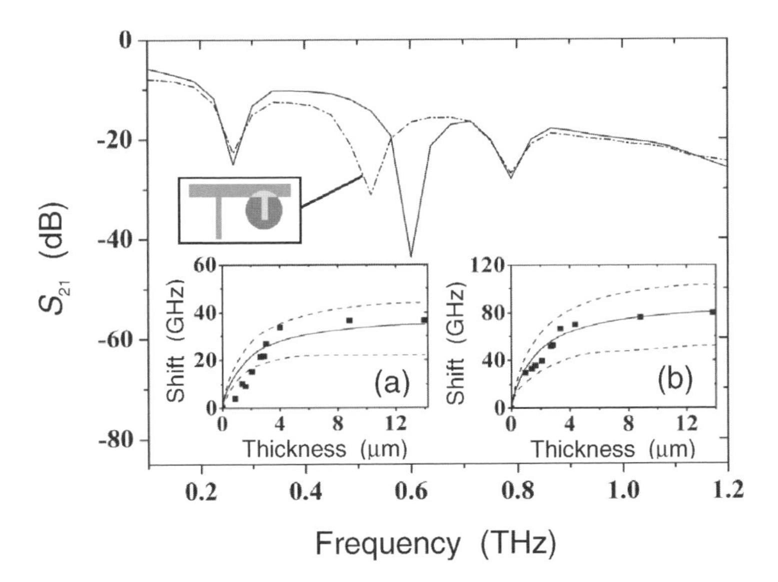

Figure 7. Fourier transforms of pulsed transmission data obtained from a two bandstop filter array, before (solid) and after (dashed) the deposition of a circular film of dielectric material (Shipley S1813 photoresist) around the higher frequency filter (600 GHz), as shown in the small inset. A resulting shift to lower frequencies in the response of the 600 GHz filter can be seen, leaving the centre frequency of the 260 GHz filter and its third harmonic unaffected. Insets show response of the 260 GHz filter (a) and 600 GHz filter (b) to different thickness of S1813 dielectric. In both insets, squares represent experimental data, lines are simulations using ε_r = 2.75 (solid) obtained from free-space measurement for S1813 at 600 GHz, ε_r = 2 (dashed, lower) and ε_r = 3.5 (dashed, upper).

In contrast to free-space propagation, the potential of THz guided wave devices for imaging has not yet been explored. Evanescent field imaging is a method common in the GHz range,[12] in which samples are scanned in the region of an evanescent electric field that extends above a planar microwave circuit. The transmission parameters of the circuit are affected by this perturbation, and these changes are recorded as the sample is scanned in a 2D plane in order to give a dielectric contrast map of the sample. The work on evanescent fields above planar THz filter systems described here offers a potential route to performing similar microscopic studies at THz frequencies.

A further advantage of on-chip systems is that it should be possible to undertake THz spectroscopy on condensed matter physics based systems, such as high mobility two-dimensional electron gases in the quantum Hall regime. In particular, it offers a far greater interaction length with the device under test compared with free-space based techniques. A key requirement for this type of investigation is that the micro-strip geometry should operate successfully not only at room temperature, but also at cryogenic temperatures. Figure 8 demonstrates that this can be successfully accomplished,[13] with the detected pulse at 4 K actually narrowing compared to the measured pulse at 290 K. This technology thus opens the pathway for high-frequency (THz) studies of the dynamic conductivity in solid-state systems at cryogenic temperatures, and in high magnetic fields.

4. Future directions

Outstanding progress has been made, over the last three years, in developing THz components and systems. In particular, intense international interest has arisen

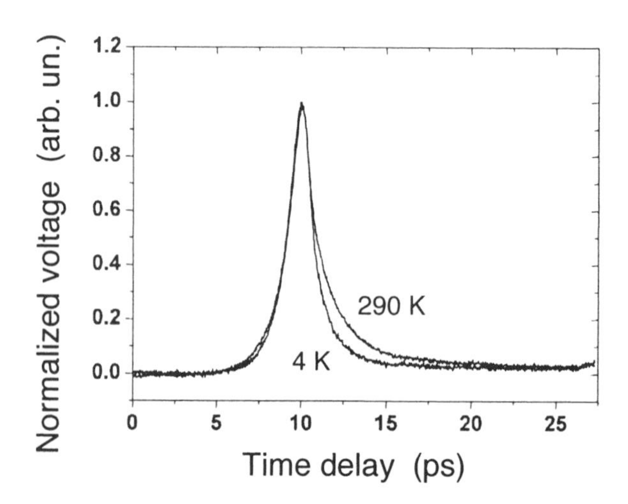

Figure 8. Detected THz pulse along a microstrip line based on a BCB dielectric, and LT-GaAs photoconductive emitters and detectors.[13] Narrowing of the pulse width is observed as T is lowered from 290 K to 4 K.

in the prospects of designing THz spectroscopy and imaging systems that can detect materials of security interest. In parallel, there has been striking progress in the design of on-chip THz time domain spectroscopy systems, which offer a 10x higher frequency resolution, and are sensitive to far smaller sample volumes. These may find potential applications as biochip DNA sensors or in the study of condensed matter physics systems.

Despite this remarkable progress, there remain many key issues still to be resolved before the full impact of THz technology becomes clear over the next three years and beyond. These include:

- despite the immense progress seen in quantum cascade laser design, will spectroscopy and imaging systems be constructed that compete long-term with the already established commercial pulsed (femtosecond) systems?
- can the cost of commercial pulsed (femtosecond) spectroscopy and imaging systems be reduced to allow a greater market uptake? For example, can incorporation of 1.55 μm InP telecom technology, or the use of diode laser mixing schemes, supplant existing femtosecond laser technology?
- can on-chip THz systems demonstrate their true potential over free-space alternatives, especially for biomedical spectroscopy and condensed matter physics?
- will we see THz systems being used for security applications at airports, and will we see stand-off detection being implemented in real-life situations, or will such systems remain simply a fundamental research tool in vibrational spectroscopy?
- what are the real prospects for using THz radiation in medical and dental imaging applications today, given the initial demonstrations that were made over five years ago?
- what new applications may emerge?
- can realistic simulations be obtained of the THz vibrational modes of complex molecular crystal structures?

Much research still has to be undertaken, but there is clearly an exciting future for THz technology and its applications, which is set to continue over the next decade and beyond. From a fundamental viewpoint, it is anticipated that extensive research will be undertaken on THz interactions with condensed matter physics systems and nanostructures, with a range of new phenomena being revealed in a high-frequency regime that has hitherto been inaccessible to solid state researchers. As examples, the energy spacings in quantum dots, transit times in submicron transistors, phonon modes in nanostructures, and torsional modes in DNA are all expected to lie within the THz spectral range.

Acknowledgments

This work was supported by the U.K. Engineering and Physical Sciences Research Council (EPSRC), and Research Councils UK.

References

1. S. Luryi, J. Xu, and A. Zaslavsky, eds., *Future Trends in Microelectronics: The Nano, the Giga, and the Ultra*, New York: Wiley, 2004.
2. R. Köhler, A. Tredicucci, F. Beltram, *et al.*, "Terahertz semiconductor-heterostructure laser," *Nature* **417**, 156 (2002).
3. D. H. Auston, K. P. Cheung, and P. R. Smith, "Picosecond photoconducting Hertzian dipoles," *Appl. Phys. Lett.* **45**, 284 (1984).
4. P. Taday, I. V. Bradley, D. D. Arnone, and M. Pepper, "Using terahertz pulse spectroscopy to study the crystalline structure of a drug: A case study of the polymorphs of ranitidine hydrochloride," *J. Pharm. Science* **92**, 831 (2003).
5. Y. C. Shen, P. C. Upadhya, E. H. Linfield, H. E. Beere, and A. G. Davies, "Ultrabroadband terahertz radiation from low-temperature-grown GaAs photoconductive emitters," *Appl. Phys. Lett.* **83**, 3117 (2003).
6. A. Burnett, W. Fan, P. Upadhya, *et al.*, "Analysis of drugs of abuse and explosives using terahertz time domain and Raman spectroscopy", *Proc. SPIE* **6120**, 155 (2006).
7. Y. C. Shen, T. Lo, P. F. Taday, B. E. Cole, W. R. Tribe, and M. C. Kemp, "Detection and identification of explosives using terahertz pulsed spectroscopic imaging," *Appl. Phys. Lett.* **86**, 24116 (2005).
8. M. Nagel, P. Haring-Bolivar, M. Brucherseifer, H. Kurz, A. Bosserhoff, and R. Büttner, "Integrated THz technology for label-free genetic diagnostics," *Appl. Phys. Lett.* **80**, 154 (2002).
9. J. Cunningham, C. D. Wood, A. G. Davies, K. C. Tiang, P. A. Tosch, E. H. Linfield, and I. C. Hunter, "Multiple-frequency terahertz pulsed sensing of dielectric films," *Appl. Phys. Lett.* **88**, 071112 (2006).
10. M. Nagel, F. Richter, P. Haring-Bolívar and H. Kurz, "A functionalized THz sensor for marker-free DNA analysis," *Phys. Med. Biol.* **48**, 3625 (2003).
11. C. K. Tiang, J. Cunningham, C. D. Wood, I. C. Hunter, and A. G. Davies, "Electromagnetic simulation of terahertz frequency range filters for genetic sensing," *J. Appl. Phys.* **100**, 066105 (2006).
12. M. Tabib-Azar, P. S. Pathak, and G. Ponchak, "Nondestructive super-resolution imaging of defects and nonuniformities in metals, semiconductors, dielectrics, composites, and plants using evanescent microwaves" *Rev. Sci. Instrum.* **70**, 2783 (1999) and references therein.
13. C. D. Wood, J. Cunningham, P. Upadhya, E. H. Linfield, I. C. Hunter, A. G. Davies, and M. Missous, "On-chip photoconductive excitation and detection of pulsed terahertz radiation at cryogenic temperatures," *Appl. Phys. Lett.* **88**, 142103 (2006).

Wavelength Tuning of Interband Cascade Lasers Based on the Stark Effect

S. Suchalkin, M. Kisin, S. Luryi, G. Belenky
State University of New York at Stony Brook, Stony Brook, NY 11794-2350, U.S.A.

F. Towner, J. D. Bruno
Maxion Technologies, Inc., Hyattsville, MD 20782-2003, U.S.A.

C. Monroy and R. L. Tober
Army Research Laboratory, Adelphi, MD 20873-1197, U.S.A.

1. Introduction

Tunable mid-IR lasers are in high demand for various military and civilian applications, such as free space communication, remote sensing, and environmental monitoring. The Stark effect, that is the energy level shift in an external electric field, provides an attractive mechanism for semiconductor laser tuning in that it allows an ultrafast and fully electrical control over a broad range of emission wavelengths. This mechanism is especially promising for cascade lasers, both type-I quantum cascade lasers (QCLs) and type-II interband cascade lasers (ICLs), where the spatially indirect lasing transition can be easily tuned with a strong first-order Stark shift. There is one major obstacle, however, on the way to a practical implementation of the Stark-effect tunable device. The laser generation condition (gain equals loss) usually pins the carrier concentration in the laser active region above threshold at a fixed level (the so-called concentration clamping effect). The concentration clamping does not allow any changes of the electric field in optically active layers by pinning the field to its threshold value and, therefore, prevents the Stark shift of the lasing levels.

Threshold clamping can be circumvented by a special laser design. Partitioning the laser structure in multiple sections[1] is the most straightforward way to achieve above-threshold laser tuning. In this approach, the additional gain/loss from the auxiliary section of the laser waveguide modifies the threshold condition in the main lasing section. The different threshold current implies a different threshold concentration in the active region and, correspondingly, different electric field that results in a Stark shift of the lasing transition. Shortcomings of this approach are also straightforward – multi-sectioned design implies a complicated multi-electrode control of the laser action. As an alternative, a complementary section can be incorporated directly into the laser active region with some control functions that should depend on the total bias current. This approach was first suggested for type-I quantum well (QW) laser diodes.[2] The controlling sections

Future Trends in Microelectronics. Edited by Serge Luryi, Jimmy Xu, and Alex Zaslavsky

ISBN 0-471-48

were represented by additional accumulation QWs located in the active region on each side of the optically active QW. Such outside carrier accumulation is not restricted in any way by the threshold condition and hence it is capable of unclamping the electric field in the active layers. The tuning of the main lasing transition was, therefore, achieved by the bias injection current alone. The tunability range, however, was quite small because of the weak second-order Stark effect in active QWs.

In this chapter, we apply the concept of outside accumulation to a type-II ICL design, taking full advantage of the strong first-order Stark effect in a cascade structure. We show that this approach allows efficient all-electrical control of the peak gain position and demonstrate high-speed tuning of the laser wavelength.

2. The tuning concept

Type-II ICL is especially promising for electrical tuning due to the inherently strong first-order Stark effect in interband cascade structures. Indeed, in a type II structure, the recombining electrons and holes are spatially separated, so that there is a nonzero dipole moment in the growth direction even in the absence of an applied field. The key element of our approach is a specially designed tunnel barrier that separates injected electrons into two groups. One group of electrons is directly involved in the optical transition between electron and hole states of the optically active region, which in our design includes an asymmetric type-II heterojunction or a type-II double-QW heterostructure – see Fig. 1, layers 1 and 2. The other group of the injected electrons is accumulated in a specially provided QW (layer 3) located outside the optically active layers. The tunnel barrier 4 separates the accumulation QW from optically active layers. The electric charge accumulated in the accumulation QW provides an additional electric field that modifies the electron energy spectrum in the optically active layers 1 and 2 through the linear Stark effect.

As the injection current increases above threshold, the density of electrons in the accumulation QW grows in proportion to the current. Simultaneously, holes accumulate in the hole QW (layer 1). This charge separation results in an electric field increasing with the injection current after the laser threshold. The energy level positions E_2 and E_1 (and hence the emission wavelength λ) become dependent on the bias current that controls the electric field via the charge accumulation in the electron accumulation QW (layer 3) and hole QW (layer 1). Separation of the electron charge accumulation layer 3 from the optically active layers 1 and 2 enables the wavelength tuning above the laser threshold. At current densities $J \geq J_{TH}$, the maximum net modal optical gain is zero:

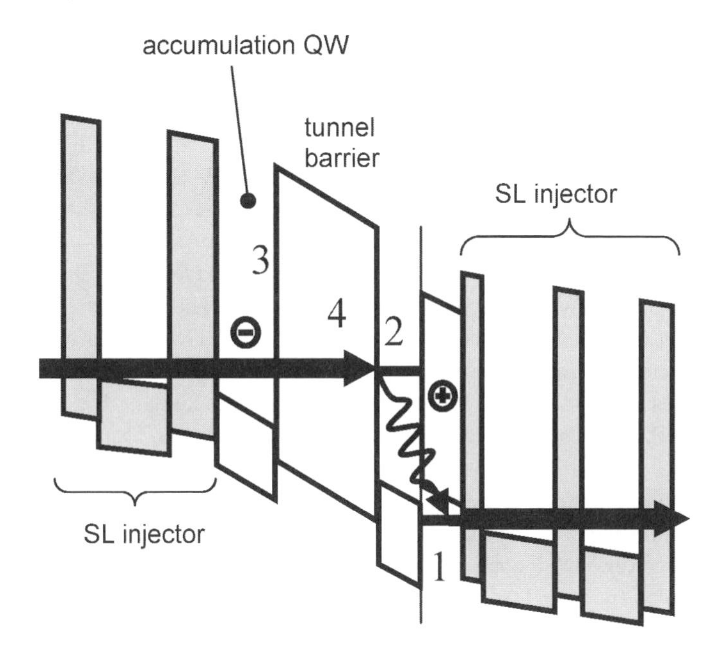

Figure 1. Band diagram of the structure under bias.

$$g_M^{max} \equiv G_0[1 - \exp(-n_2^{(c)}/N_{2D}) - \exp(-n_1^{(h)}/P_{2D})] - \alpha = 0\ . \qquad (1)$$

Here, $n_2^{(c)}$ is the electron concentration in the electron quantum well 2, $n_1^{(h)}$ is the hole concentration in the hole quantum well 1, $N_{2D} = 2kTm_e^*/\pi\hbar^2$ and $P_{2D} = kTm_h^*2/\pi\hbar^2$ are effective 2D densities of states in the conduction and valence subbands, G_0 is the saturation gain, and α is the optical loss.[3] The concentration in the electron accumulation layer 3 can be estimated in a steady state regime by $n_3^{(c)} = J\tau_{tun}/q$, where τ_{tun} is the tunneling time and q is the electron charge. Using the charge neutrality condition, $n_1^{(h)} = n_3^{(c)} + n_2^{(c)}$, the electron concentration in the active quantum well is related to the injection current density J by the equation:

$$\exp(-n_2^{(c)}/N_{2D}) + \exp(-(n_2^{(c)} + J\tau_{tun}/q)/P_{2D}) = 1 - \alpha/G_0\ . \qquad (2)$$

It is essential to note that while the generation condition (1) clamps the gain, it no longer pins either of the two concentrations $n_1^{(h)}$ and $n_2^{(c)}$, which remain current-dependent, as seen explicitly from Eq. (2). As the concentration in the electron accumulation layer $n_3^{(c)}$ increases with the current, so does the hole concentration $n_1^{(h)}$; both the overall charge neutrality and the modal gain pinning are satisfied without preventing the increase of the electric field in the active type-II heterojunction region. The latter, in turn, affects the laser wavelength tuning above threshold.

3. Tunable structure

The tunable ICL structure was grown by MBE on a *p*-doped GaSb substrate. The laser core contained a cascade of 14 periods. Each period included a digitally graded InAs/AlSb injector with accumulation QW and InAs/$Ga_{0.8}In_{0.2}Sb$ type II optically active double quantum well heterostructure, separated from each other by a 4 nm AlSb tunneling barrier. The widths of the InAs and $Ga_{0.8}In_{0.2}Sb$ layers were 2.1 nm and 3.1 nm, respectively. The $Ga_{0.8}In_{0.2}Sb$ layer was followed by a *p*-doped 5.8 nm GaSb QW that served as a hole reservoir. The active area was sandwiched between InAs/AlSb superlattice cladding layers. The devices were fabricated as deep-etched mesas and soldered, epilayer side up, to Au-coated copper mounts. The mesas were 35 μm wide with 0.5-mm-long cavity. Both facets were left uncoated. The mounts were attached to the cold finger of a liquid N_2 or Helitran cryostat. The emission was collected with reflection optics and analyzed with an FTIR spectrometer. We compared the tuning characteristics of this structure with those of a regular 18-cascade ICL.[4] The latter had an active region with type II W-like quantum wells and contained no special tunnel barriers for charge accumulation.

The current–voltage dependence of the structure is shown in Fig. 2. The experimental turn-on voltage is ~5.2 V and agrees well with the calculated value 4.9 V for an ideal 14-period cascade structure. Calculation shows that a voltage drop of 0.35 V per each injector region provides for injector level alignment. For 56 nm injectors this corresponds to the turn-on internal electric field of ~65 kV/cm.

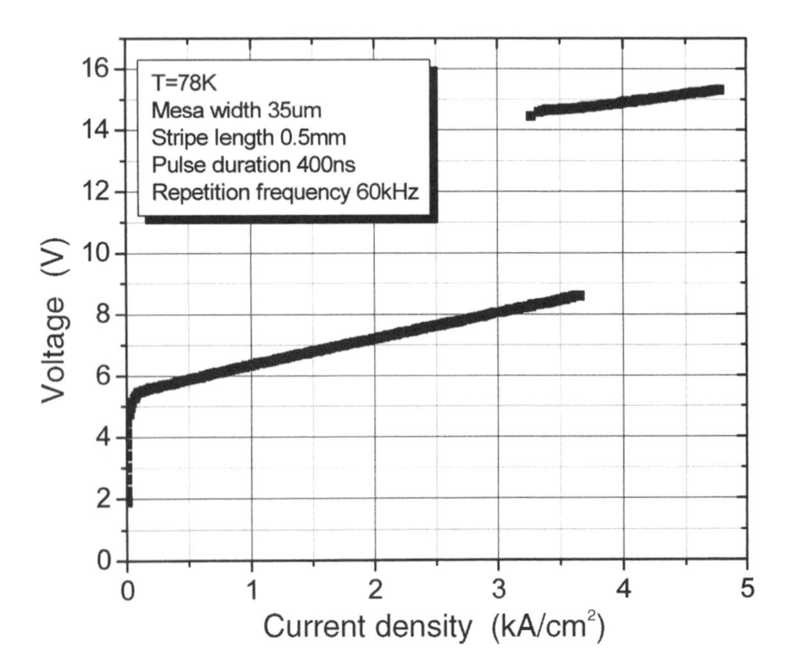

Figure 2. Current-voltage dependence of the tunable cascade laser.

As the bias current reaches ~650 mA, the voltage jumps from ~8.5 V to ~14.5 V. There is no such a jump in the regular IC laser. This effect indicates switching into another conducting state and is indicative of the negative differential resistance instability associated with the resonant tunneling between layers 3 and 2. The switching may be accompanied by charge build-up in the accumulation quantum well. The switching is reversible and the bias current can be increased further after the voltage jump until the electrical breakdown of the dielectric separating the upper metal contact from the substrate occurs. A possible explanation of this behavior is that at higher bias current the charge accumulation and subsequent voltage drop leads to the energy alignment of the accumulation level and the second subband of the InAs/GaInSb laser quantum well. According to calculations, this subband is ~0.5 eV higher than the upper lasing level. The width of the tunnel barrier 4 in our design was chosen as 4 nm to ensure the alignment of the accumulation level 3 and upper lasing level 2 at turn-on voltage (see Fig. 1). As the charge builds up in the accumulation well, this alignment breaks due to the voltage drop across the barrier. This does not necessarily mean immediate disruption of the carrier injection into the upper lasing level. Effective injection can be maintained within a finite range of the energy mismatch. This range is determined by the width of the injector miniband, formed at turn-on voltage, as well as inelastic tunneling processes.[5] As the energy mismatch increases further, the net tunneling rate decreases in spite of the increasing carrier density build-up in the accumulation quantum well. Enforcing the bias current at these conditions may provide a positive feedback for carrier accumulation and, as a result, uncontrollable increase of the voltage drop until higher energy states in the type II quantum well are aligned with the accumulation level. Assuming that these states belong to the second subband in the InAs/GaInSb double quantum well, we can estimate the total voltage jump as ~7 V for a 14 cascade structure. This agrees well with the observed voltage jump of ~6V.

4. Optical characteristics

The electroluminescence (EL) spectrum at low current is shown in the inset of Fig. 3. The emission energy quantum of 0.34 eV agrees with the calculated value of 0.32 eV. The EL spectral maximum energy increases linearly with the bias voltage. Since this dependence was measured in the subthreshold pumping region, the linear shift should be attributed to the Stark effect that results from charge accumulation in the type II quantum wells of the laser active area. A similar effect has been observed in a regular interband cascade laser (see Fig. 3), though there the effect was weaker due to the smaller electron-hole dipole in active W-quantum wells.

The optical gain spectra measured with the Hakki-Paoli technique are shown in Fig. 4. The modulation of the gain spectra is a manifestation of "leaky" modes.[6] This effect also manifests itself in the modulation of the amplified spontaneous emission (ASE) spectra and mode grouping in the lasing spectra. The observed threshold current density is 91 A/cm^2 at 80 K. The lasers demonstrate cw operation up to 120 K and pulsed operation up to 200 K (pulse duration = 400 ns, duty cycle = 2.4%). The external quantum efficiency is ~250% (at 80 K). The internal loss is ~10 cm^{-1} (Fig. 4).

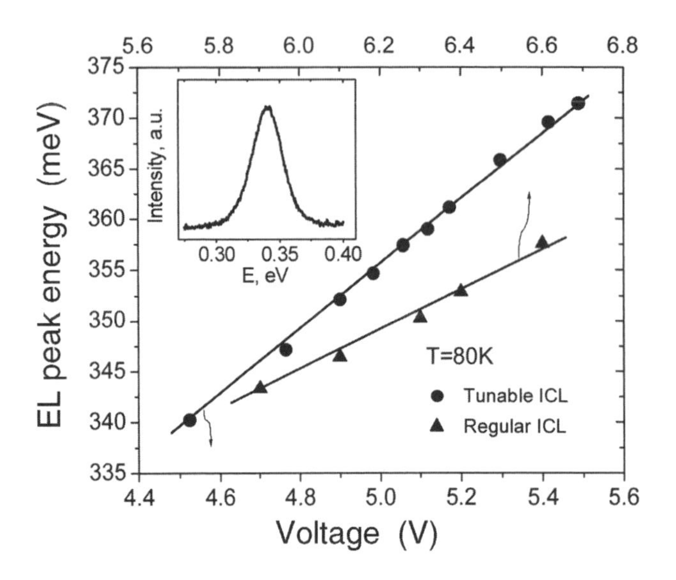

Figure 3. Dependence of the electroluminescence (EL) energy quantum on the bias voltage for a regular IC laser (triangles) and tunable IC laser (circles). Inset shows the EL spectrum at low bias current.

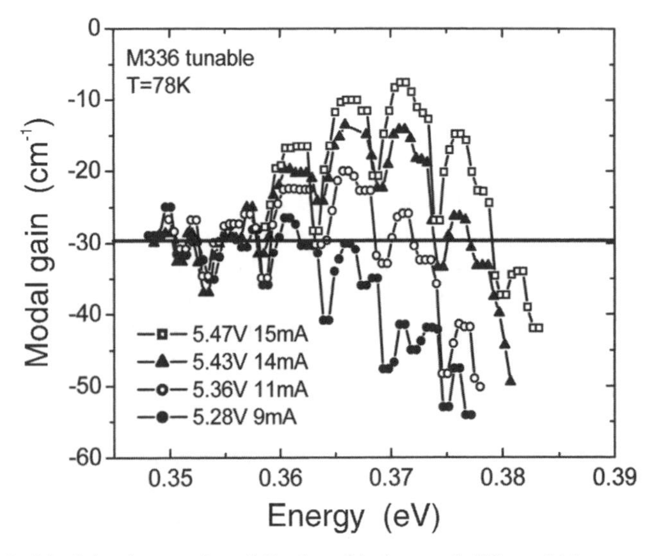

Figure 4. Modal gain spectra of the tunable laser at different bias currents.

5. Laser tunability

Amplified spontaneous emission and lasing spectra of both the tunable and the regular IC lasers are shown in Figure 5. In the regular laser, in spite of the ASE blue shift, the laser line spectral position is stable up to high bias currents (~220 times J_{TH}). This is as expected from the pinning of concentration and, hence, of electric field in the active area quantum wells.

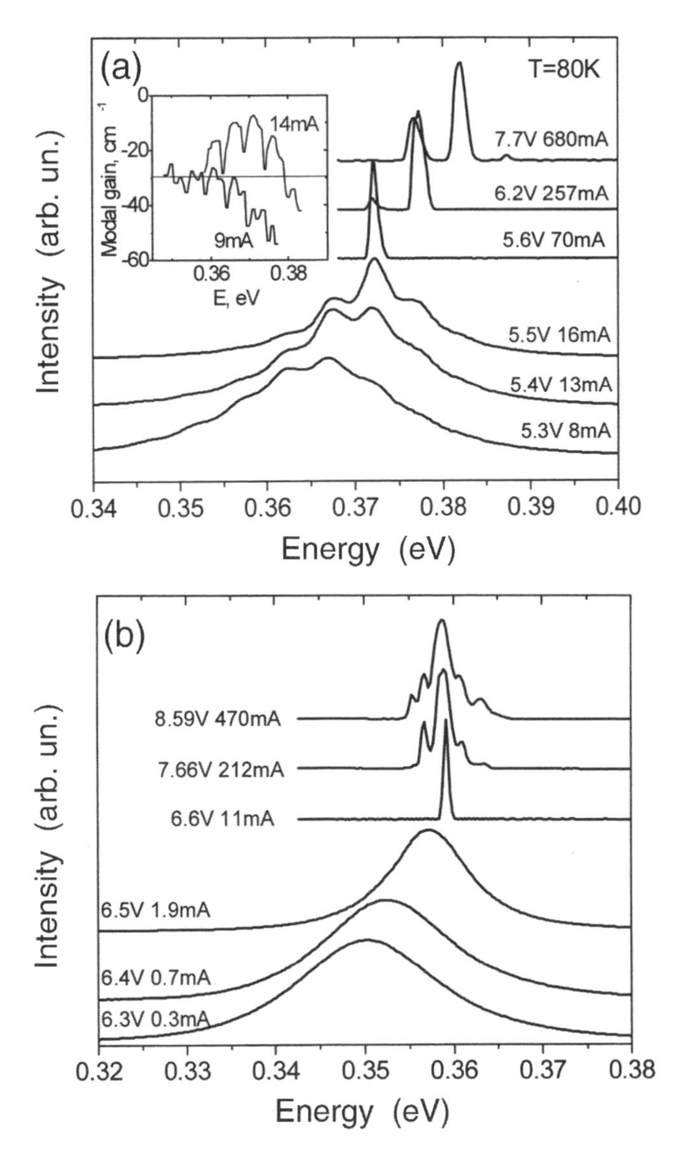

Figure 5. ASE and laser spectra of the tunable IC laser (a) and regular IC laser (b) at different bias currents. The inset in the panel (a) shows the modal gain spectrum of the tunable IC laser.

The lasing spectrum of the tunable laser (see Fig. 5(a), three upper curves) clearly demonstrates a blue shift. The periodic modulation of the ASE spectrum in the tunable laser is consistent with the observed strong modulation of the modal gain spectrum with the same period (see Fig. 4 and Fig. 5(a), inset). The spectral positions of the gain modulation maxima and minima are determined by the substrate thickness, as well as by the effective refractive indices of the active area, cladding layers, and substrate. The dependence of the modulation peak positions on the bias current is weaker than the Stark shift of the gain spectrum. As the bias current increases, the material gain curve shifts with respect to the modulation extremes and consequently, the modal gain maximum shows a discrete blue shift with the increments equal to the leaky mode modulation period. This behavior takes place at pumping levels far above the laser threshold. The lasing spectrum of the tunable laser demonstrates a blue shift with increasing bias current. The rate of this shift with respect to the bias current (and voltage) is slower than the rate of ASE tuning in the subthreshold region (approximately 5 meV/V *vs.* 30 meV/V). This indicates an abrupt change in the tuning mechanism as the bias current exceeds the laser threshold. In the subthreshold regime, the wavelength shift is determined primarily by the carrier accumulation in the optically active quantum wells and is related to the corresponding increase of the internal electric field in the type–II heterojunction.

After the laser threshold has been reached, the wavelength tuning becomes determined by the charge buildup in accumulation quantum well 3, which, in turn, depends on the electron tunneling rate through the barrier 4. Figure 6 demonstrates the laser spectrum shift recorded throughout the whole range of the injection current (1–42 threshold values) without noticeable saturation. The maximum value of the laser emission tuning range is 15 meV or 120 nm.

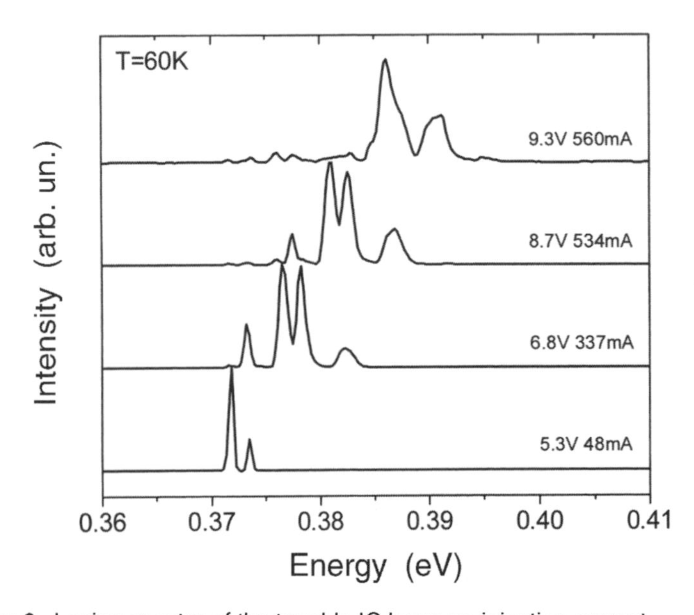

Figure 6. Lasing spectra of the tunable IC laser *vs.* injection current.

6. Modulation response characteristics

One of the principal advantages of the presented tuning scheme in comparison with temperature tuning is the possibility of fast wavelength modulation. To show the potential of the Stark tuning for high-speed applications, we measured the frequency response of the laser tuning. To do this, we combined the above-threshold dc bias component with ac bias component. The frequency of the ac component was varied in the range 0.01–2 GHz. While the bias modulation period is larger than the tuning response time of the laser, the laser spectrum shape is changing simultaneously with the bias. Since the spectrometer response time is larger than the bias modulation period over the whole modulation frequency range, the measured spectrum is the result of time averaging over all spectral shapes corresponding to the bias values within the modulation amplitude. As the bias modulation period becomes smaller than the laser tuning response time, the laser wavelength is not affected anymore by the bias modulation and the measured spectrum corresponds to dc component of the bias. By monitoring the change of the laser spectrum shape with the bias modulation frequency, we can estimate the tuning modulation bandwidth.

For the measurements of the frequency dependence of the wavelength tuning we used the 0.5 mm long laser with the mesa width of 8 μm and threshold current of 6 mA at 80 K. The differential resistance of the laser at the given dc bias is 6.9 Ω. A high frequency 47 Ω resistor was connected in series with the laser for the cable impedance matching. The dc bias current was 31 mA; the ac modulation amplitude was ~25 mA. For the 31 mA bias, the laser spectrum consists of a single line (further referred to as line 1). As the bias current increases to ~40 mA, another line (line 2) corresponding to the next "leaky mode maximum" appears due to the gain curve shift. The presence of li.1e 2 in the time-averaged spectrum means that the tuning response time is shorter that the laser modulation period. At higher frequencies line 2 disappears and the spectrum returns to that without modulation (Fig. 7). The relative amplitude of the second line as a function of frequency in Fig. 7 provides a rough estimation for the modulation bandwidth. The sharp decrease of the line 2 intensity takes place at ~1.6 GHz. The vanishing of the line 2 means that the peak bias current decreased from 56 mA to ~40 mA so the modulation amplitude decreased by a factor of ~2.5.

The bandwidth for the Stark wavelength modulation can be estimated as $\nu \sim (2\pi RC)^{-1}$, where C is the capacitance of the device and R is the series resistance of the laser. The capacitance is the sum of the parasitic capacitance of the laser package C_P and internal capacitance of the laser structure C_L. To estimate C_L we consider the active areas in each period as plate capacitors connected in series by the conducting injectors. The electron and hole accumulation quantum wells represent the capacitor plates. The 8 × 500 μm plates are separated by the distance of ~80 Å. Taking the relative dielectric constant $\varepsilon \sim 12$, we obtain $C_L \sim 4$ pF for the 14 cascade laser. At the series resistance of R ~ 6.9 Ω, the tuning bandwidth is ~6 GHz. In our case the cutoff is most probably determined by C_P, which is ~20 pF. This corresponds to the bandwidth of ~1.2 GHz, in a good agreement with the experimental results.

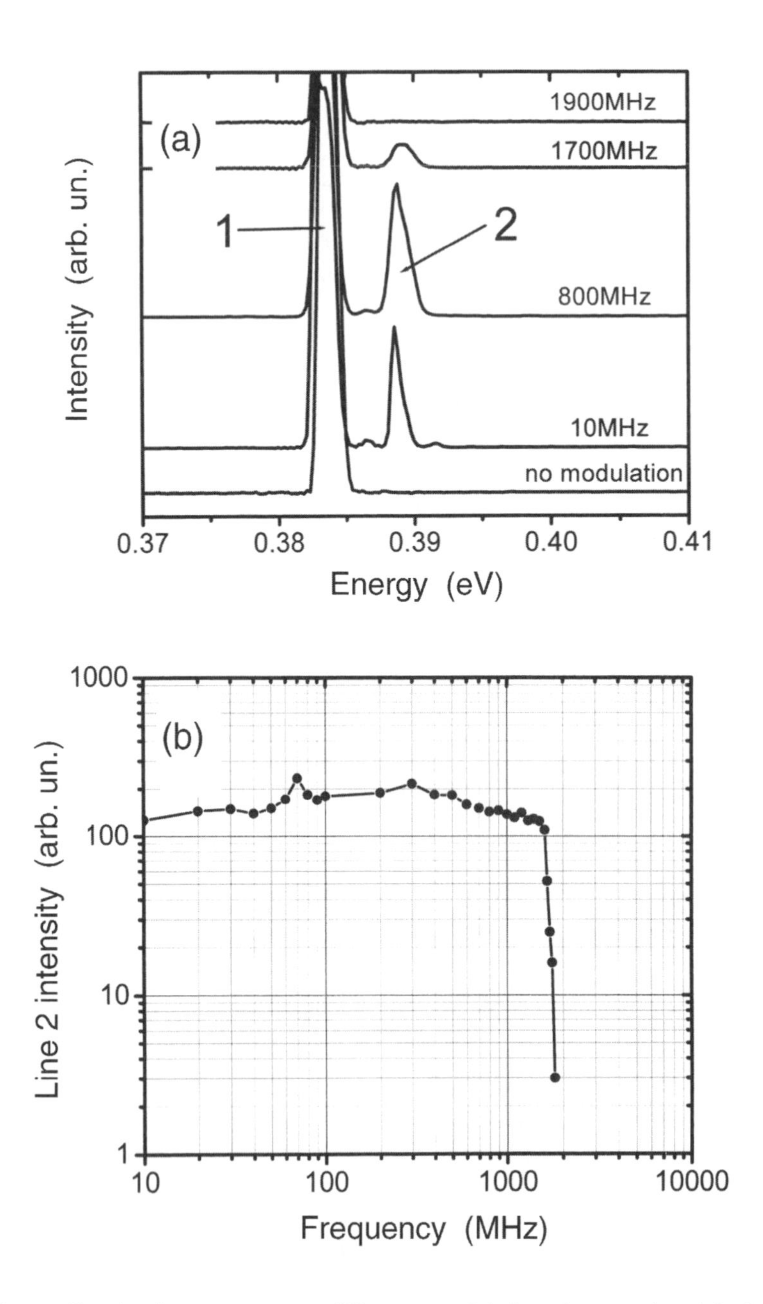

Figure 7. Lasing spectra at different modulation frequencies (a); frequency dependence of the intensity of line 2 (b).

7. Conclusions

The concept of outside carrier accumulation has been applied to a type-II ICL design to take full advantage of the strong first-order Stark effect in a cascade structure. Our design of the Stark-effect tunable ICL includes an additional electron accumulation layer located outside the optically active type-II quantum well heterostructure. This eliminates the pinning of the electron-hole concentration to the threshold value and enables ultra-wide Stark shift of the optical gain spectrum. We experimentally demonstrate an electrically tunable ICL operating in the mid-IR spectral range with tuning range of 120 nm (starting from the initial lasing wavelength $\lambda \sim 3.33$ μm), or 120 cm^{-1}. We also demonstrate wavelength modulation at frequencies up to 1.6 GHz. The ICL tuning performance can be further improved by suppressing the gain spectrum modulation due to the substrate leaky modes.

Acknowledgements

This work was supported by ARO grant DAAD 190310259.

References

1. J. Faist, F. Capasso, C. Sirtori, D. L. Sivco, A. L. Hutchinson, and A. Y. Cho, "Laser action by tuning the oscillator strength," *Nature* **387**, 782 (1997).
2. N. Le Thomas, N. T. Pelekanos, Z. Hatzopoulos, E. Aperathitis, and R. Hamelin, "Tunable diode lasers by Stark effect", *Appl. Phys. Lett.* **83**, 1304 (2003).
3. V. B. Gorfinkel and S. Luryi, "Fundamental limits of linearity for CATV lasers", *J. Lightwave Technol.* **13**, 252 (1995).
4. J. L. Bradshaw, N. P. Breznay, J. D. Bruno, *et al.*, "Recent progress in the development of type II interband cascade lasers", *Physica E* **20**, 479 (2004).
5. M. V. Kisin, S. Suchalkin, and G. Belenky, "Stark effect tunable QCL", *8th Intern. Conf. Intersubband Transitions Quantum Wells ITQW-2005*, North Falmouth, MA (September, 2006).
6. D. Westerfeld, S. Suchalkin, M. V. Kisin, G. Belenky, J. Bruno, and R. Tober, "Experimental study of optical gain and loss in 3.4–3.6 μm interband cascade lasers", *IEE Proc. Optoelectron.* **150**, 293 (2003).

Intersubband Quantum-Box Lasers: An Update

D. Botez, M. D'Souza, G. Tsvid, A. Khandekhar, D. Xu, J. C. Shin, T. Kuech
Reed Center for Photonics, University of Wisconsin, Madison, WI 53706, U.S.A.

A. Lyakh and P. Zory
University of Florida, Gainesville, FL 32611, U.S.A.

1. Introduction

Semiconductor lasers operating continuous wave (cw) at or near room temperature (RT) and emitting in the mid- and far-infrared wavelength ranges, 3–13 μm, are critically needed for a vast array of applications. Intersubband (IS) transition emitters are the most likely solution. The first implementation of the concept for using IS transitions for laser action,[1] was realized in early 1994[2] and named the quantum cascade (QC) laser. Current QC devices have demonstrated room-temperature cw operation at 4.3 μm and 4.8 μm,[3,4] but with very low wallplug-efficiency values (< 2.5%) due to inherently high operating voltages (10–11 V). Furthermore, these devices have extremely temperature-sensitive characteristics[3,4] at and near RT, due to thermal runaway triggered by the backfilling effect,[5,6] which raises serious issues of device reliability. In fact, no device reliability has been demonstrated to date for any type of QC laser.

Intersubband-QC lasers have fundamentally poor radiative efficiencies since the nonradiative, LO-phonon-assisted relaxation time for electrons in the upper laser states is about 1.8 ps,[5] whereas the radiative relaxation time is 4.2 ns. That is, nonradiative processes are about 2300 times faster than radiative processes. Since there are good reasons to believe that the LO-phonon-assisted relaxation time will substantially increase if the relaxing electrons are confined in quantum boxes,[7-16] the radiative efficiency problem can be overcome by replacing the quantum-well (QW) active regions of a QC laser with a quantum-box (QB) 2D array[6,17] or a 2D array of cascaded QBs.[18]

2. Intersubband quantum box lasers

In QW structures electron relaxation between subbands occurs[2,5] in about 1–2 ps, primarily via LO-phonon absorption or emission.[5,7] Making QBs causes discrete states in the subbands,[6,17] which in turn causes the LO-phonon-assisted electron relaxation time to increase[7,9] by a factor β.[6] Experimental results[10–14] from *unipolar* QBs (where electron-hole scattering[10] does not circumvent the phonon bottleneck) and photocurrent-response/dark-current measurements from QB IR detectors[11,15] indicate electron-relaxation times of the order of 100 ps, in good agreement with theory.[16] As the temperature increases to RT the relaxation times decrease[12,13] due to inherent carrier losses from self-assembled QBs to the wetting layers.[19] The proposed QBs are not self-assembled and thus can be made deep, resulting in negligible carrier leakage with increasing temperature. Therefore, for deep InGaAs/GaAs QBs β may well be as high as 50 at RT. Based on the experiments in Refs. 10–14, we feel confident to assume that for deep QBs $\beta > 30$.

The proposed device is schematically shown in Fig. 1: an intersubband-transition laser with an active region composed of a 2D array of QB ministacks, called "active boxes", separated by current-blocking material (CBM). Each mini-stack is composed of 2 to 3 QBs. The QBs will be fabricated by *in-situ* gas etching and regrowth, allowing for tight carrier confinement, unlike the inherently weak carrier confinement of self-assembled QBs. The active-boxes array, together with lightly doped *n*-GaAs layers constitute the core of an optical waveguide with heavily doped n^+-GaAs cladding layers. A standard four-level laser theory has been used by us[6] to estimate the threshold-current density J_{TH}, differential quantum efficiency η_D, and the wallplug efficiency η_P, of IQB and QC lasers ($\lambda = 4.5$ μm). We obtained η_P values of 24% at RT for single-stage IQB devices; that is at least an order of magnitude higher than for conventional QC devices[3,4] emitting in the 4.0–4.8 μm range.

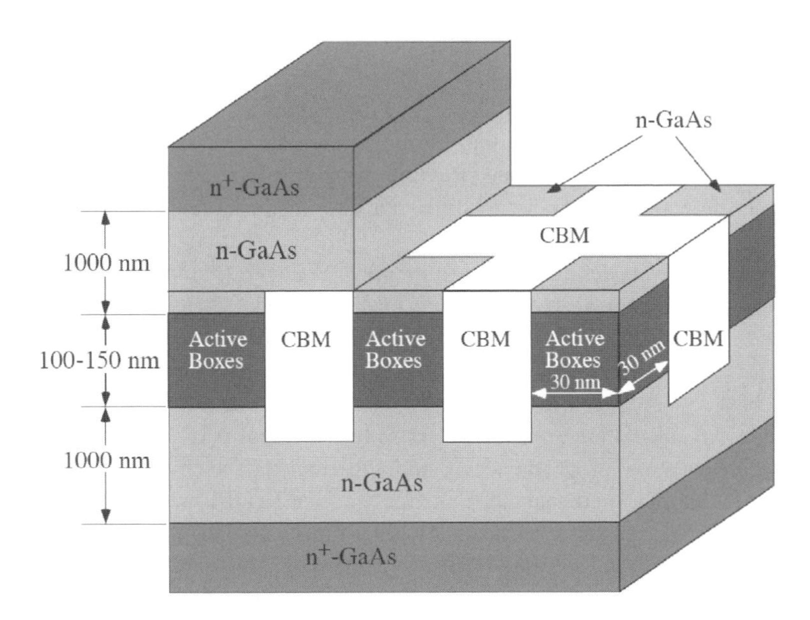

Figure 1. Representation of the intersubband quantum box structure, where CBM denotes current blocking material.

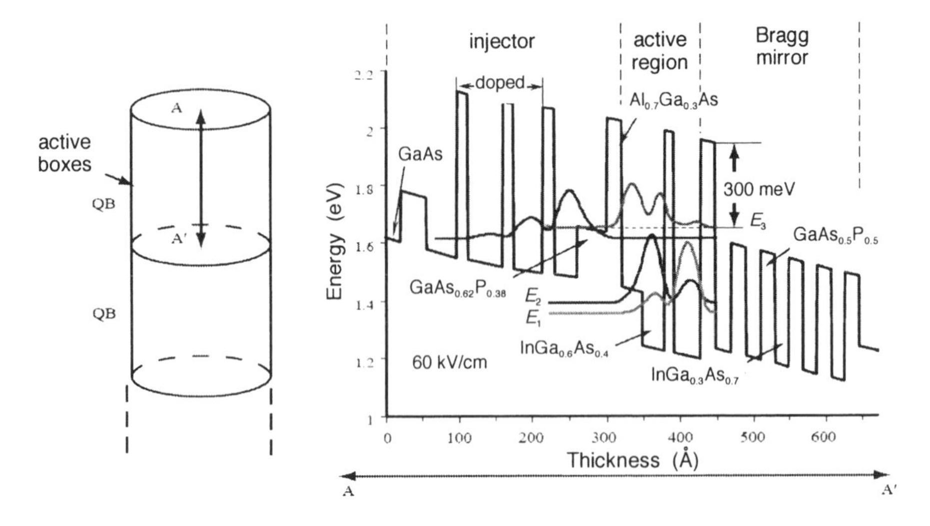

Figure 2. Schematic representation of a QB ministack (*i.e.* active boxes) and the conduction-band energy diagram inside one QB ($\lambda = 4.8\ \mu$m).

To achieve enough gain for higher wallplug efficiencies (larger than 25%) one requires ministacks of QBs: 2-3 in a vertical row, each of which having the conduction-band energy diagram shown in Fig. 2. The active region has two deep wells,[20] resulting in a high energy barrier for electrons in the upper energy state E_3 (~300 meV in Fig. 2) that suppresses thermionic carrier leakage. Furthermore, unlike QC lasers, the injector and Bragg mirror regions are separate, which ensures the suppression of carrier backfilling in multi-stage devices. The design of Fig. 2 incorporates a strain-compensated structure using tensile-strained $GaAs_{1-x}P_x$ layers to prevent defect formation due to strain relaxation[21] during the etch and regrowth process.

Let us now consider a 2D array of double QB stacks. A transverse waveguide of low loss coefficient $\alpha_W = 1.5\ cm^{-1}$ has been designed. Then, for 2 mm-long, 10 μm-aperture devices with 10% and 90% front- and back-facet reflectivities and taking $\beta = 30$, the calculated cw wallplug efficiency η_P at RT reaches a maximum of 50.3% at $J \sim 6J_{TH}$, see Fig. 3. If $\beta = 50$, η_P reaches a maximum of 54% at $\sim 8J_{TH}$. For the cw η_p calculations, we took a series resistance of 0.16 Ω consistent with a comprehensive study of QC devices,[22] and a thermal resistance $R_T = 1.5$ K/W, also consistent with measurements in QC devices[22] as well as the fact that bulk material has an R_T value at least a factor of five times smaller than that for thick superlattices (*i.e.* 30–40 stages).[23] The J_{TH} value is ~ 0.2 kA/cm^2, while the voltage is < 0.9 V, in sharp contrast to typical QC-device voltages of 10–11 V.

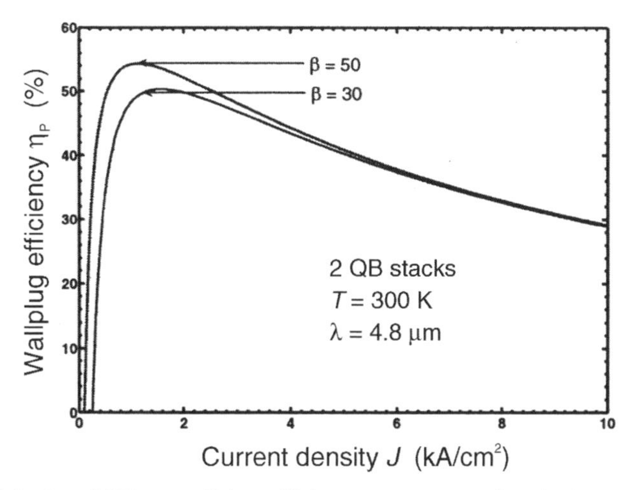

Figure 3. Calculated RT cw wallplug efficiency *vs.* current density for two QB-stack devices at different β values.

We note that the use of self-assembled QBs in QC-like structures has been proposed[18,24] and demonstrated.[25] Room-temperature J_{TH} values as low as 10 A/cm^2 have been predicted[18] for 10-stage devices. Intersubband luminescence was observed[25] at $\lambda = 22$ μm, but represented only 0.8% of the total luminescence, due most probably to the inherent problem of self-assembled QBs: carrier transitions involving the wetting layers. The proposed QBs do not involve self-assembly, but actual *in-situ* fabrication,[26] thus allowing for tighter confinement of carriers to the QBs. First of all, due to the deep-QW proposed design (Fig. 2) carrier leakage will be suppressed in the direction normal to the QBs, as already experimentally demonstrated for single-stage QW devices.[20] What is left is to confine the carriers radially. That can be achieved by regrowth of similar high-Al-content material as used for transverse confinement, see Fig. 4. Thus, truly deep QBs could be realized for the first time, which in turn will allow for temperature-insensitive characteristics and subsequent high wallplug efficiencies *and* device reliability.

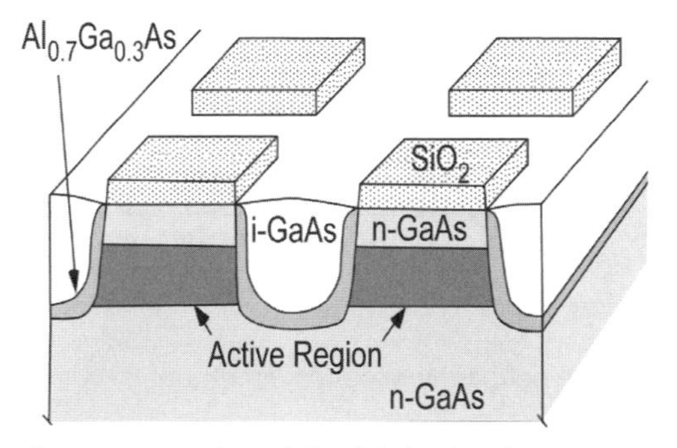

Figure 4. Schematic representation of the fabrication for a quantum-box array.

3. Preliminary experimental results

- *Single-stage intersubband emitters*

Optimization of the QW material for efficient emission involves optimizing the QW structure shown in Fig. 2. We have already developed[20] such a deep-well light-emitting structure, thus demonstrating the first room-temperature emission in the mid-IR ($\lambda = 4.7\mu m$) from a single-stage IS device – see Fig. 5. Electrons are injected, via resonant tunneling, from the ground level of the injector miniband into the upper level E_3 of the active region. We have significant experience with resonant-tunneling structures, having demonstrated[27] the first deep-well resonant tunneling diodes (RTD's); that is, double-barrier diodes for which the quantum well, being compressively strained, has a conduction-band edge lower in energy than that for the injector and emitter regions. We have extended the deep-well approach to light-emitting intersubband devices[20] in order to tightly confine the carriers to the active quantum well(s). In turn, carrier losses via thermionic emission and/or tunneling through the outer barrier are significantly suppressed. The FWHM spectral linewidths are 19 meV and 25 meV at 80 K and 300 K, respectively. These values are the same or better than the best results reported[28] from conventional 5 μm-emitting QC structures grown by MBE. Thus the results confirm that we can grow IS light-emitting structures by MOCVD equal in crystalline quality to those grown by MBE.

More recently, we have grown strain-compensated 25-stage QC structures using the single-stage design of Fig. 2 and obtained excellent surface morphology and sharp x-ray diffraction spectra, shown in Fig. 6, indicative of excellent material quality and nearly perfect lattice match to the GaAs substrate.

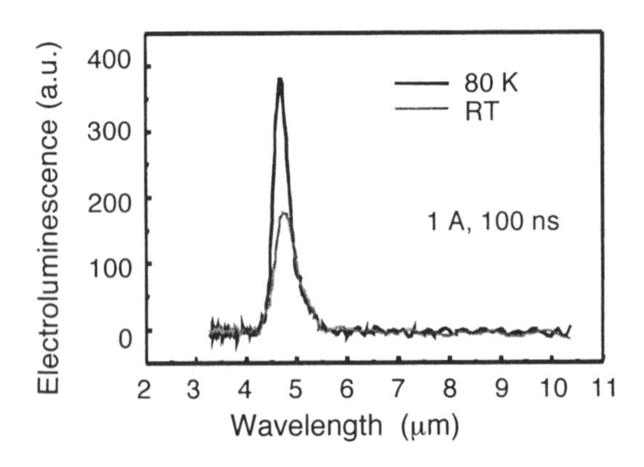

Figure 5. Intersubband electroluminescence spectra at 80 K and at RT.[20]

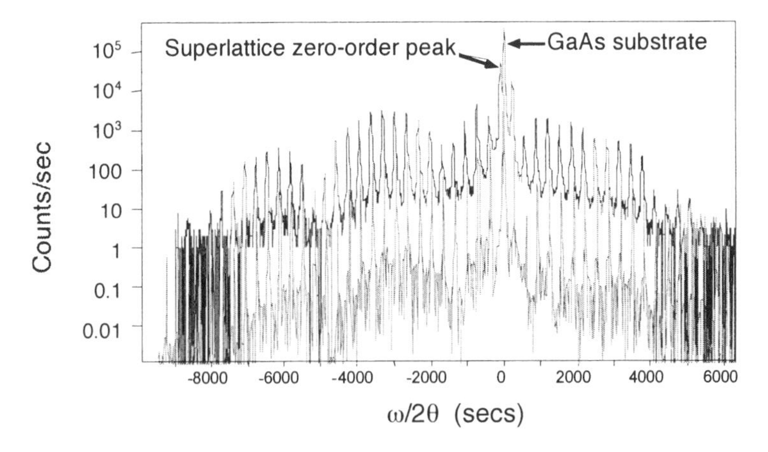

Figure 6. Experimental (top) and simulated (bottom) x-ray diffraction patterns for a 25-stage, strain-compensated, deep-well QC structure.

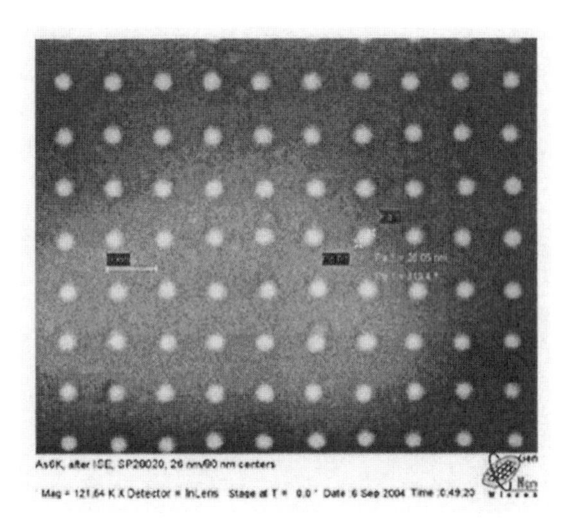

Figure 7. SEM of 26 nm-diameter SiO_2 posts spaced 80 nm center-to-center.

- *Quantum-box patterning*

The eventual design involves a QB array of 30 nm diameter boxes spaced 60 nm center-to-center (Fig. 1), a nontrivial task. Therefore, the patterning has to be done in progressive steps. We used e-beam direct writing employing a novel hydrogen silsesquioxane resist; which was found best for generating high-quality 2D dot patterns: 33 nm-diameter dots spaced 80 nm center-to-center.[26] This represents a 13% QB fill factor, well on the way to the 20% target value. Recently we achieved transfer of such patterns into SiO_2, a dielectric thought suitable as a mask for *in-situ* etching and regrowth. The resulting 26 nm diameter, ~50 nm tall posts were obtained with 80 nm center-to-center spacing, see Fig. 7.

- *Nanopost etching, in-situ gas etching and in-situ GaAs regrowth*

Initial work involved an SiO_2 mask on patterned GaAs substrates and dilute-HCl gas flow for *in-situ* etching.[29] Controlled etching (~3 nm/sec) through narrow, SiO_2-defined stripes reproducibly provided ~40 nm-deep trenches. However, when using structures with thin (2–3 nm), buried high-Al-content layers, the gas etching stopped at those layers due to oxidation[29] most likely because residual oxygen in the etching ambient caused almost instantaneous AlO_x formation on the surface, thus suppressing gas-phase etching. Therefore we proceeded to etch trenches in a dry-chlorine or BCl_3-based plasma environment followed by a mild *in-situ* etching, to get rid of dry-etch-induced damage prior to regrowth. In particular, BCl_3-based etching was found to provide the vertical walls necessary for the formation of the nanoposts required for (2–3)-QB stack devices. As shown in Fig. 8, 57 nm-diameter GaAs posts of 170 nm height were formed, spaced 200 nm center-to-center.

Scanning electron microscopy pictures of etched and regrown structures are shown in Fig. 9. The 33 nm-wide, 80 nm-tall ridges (Si_3N_4 + GaAs) were dry-etched, see inset of Fig. 9(a). Then, *in-situ* gas etching and regrowth were performed, see Fig. 9(a). Similarly, 35 nm-diameter, 80 nm-tall posts (Si_3N_4 + GaAs) were also subjected to *in-situ* etch and regrowth, see Fig. 9(b). AFM measurements confirmed that 40 nm-tall Si_3N_4 posts are left after regrowth. For the actual device fabrication the Si_3N_4 posts would be removed via etching and the growth continued in order to obtain the structure shown in Fig. 1.

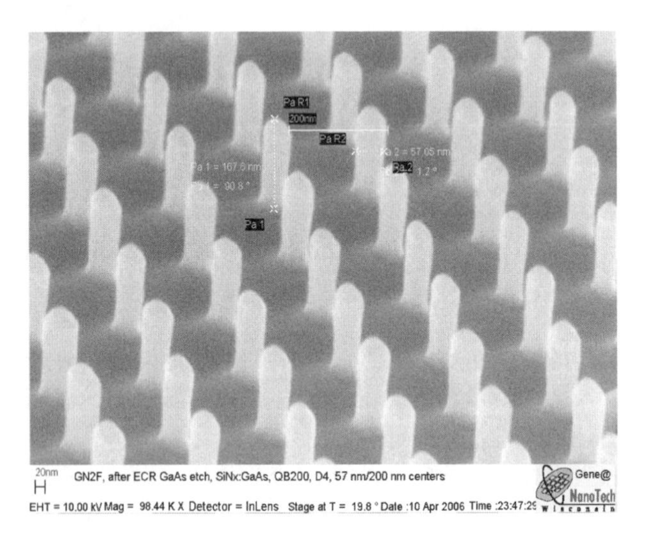

Figure 8. SEM of dry-etched GaAs 57 nm-diameter posts, spaced 200 nm center-to-center.

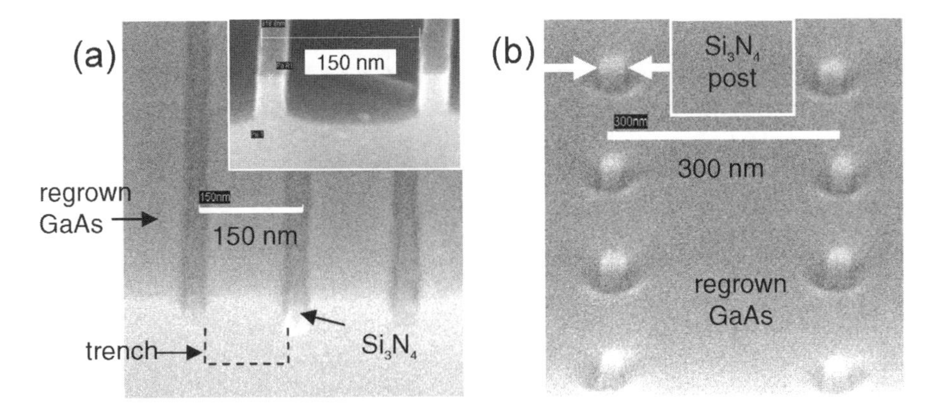

Figure 9. Structures subjected to *in-situ* etch and regrowth: (a) 33 nm-wide ridges, with dry-etched pattern before regrowth shown in the inset; (b) 35 nm-diameter Si_3N_4 posts left after regrowth.

- *Fermi-level pinning elimination at in-situ etched and regrown interfaces*

For unipolar intersubband devices such as QC lasers one need not be concerned about loss of carriers to defects at exposed surfaces, since the transition energies involved are much smaller than the energy between the midgap and the conduction-band edge at those exposed surfaces. Quantum cascade lasers with exposed 10 μm-wide ridges operate quite well. However, for nanostructures with in-plane dimensions of ≤ 50 nm the defect density needs to be drastically reduced at the device edges, since Fermi-level pinning would cause full depletion across the devices.[30] Therefore, our QB formation will be done such that no charge-trapping states are formed at the QB edges, thus eliminating Fermi-level pinning.

We have carried out experiments on (110)-oriented GaAs substrates; that is on crystalline planes equivalent to the side edge(s) of the QBs to be formed. *N*-type (2×10^{17} cm^{-3}) GaAs was grown and dry-plasma etched. The sample was then dipped in HCl to remove any residue from plasma, annealed in arsine in an MOCVD reactor, followed by an *in-situ* etch using HCl gas. A thin *n*-GaAs (1×10^{17} cm^{-3}) film was then grown, as shown schematically in Fig. 10(a), and capacitance-voltage characterization carried out. The capacitance was recorded as a function of the applied voltage as the depletion width W moved across the etched and regrown interface. Figure 10(b) shows the variation of the carrier concentration N *vs.* W. Presence of trap states at the interface would result in a decrease and subsequent increase in the apparent carrier concentration. Lack of any features in the N *vs.* W curve at the interface (that is, at $W = 0.18$ μm) indicates no trap states. This means that Fermi-level pinning has been eliminated for the proposed fabrication of intersubband QBs. It should be emphasized that samples without HCl dip and arsine annealing did exhibit trap states.

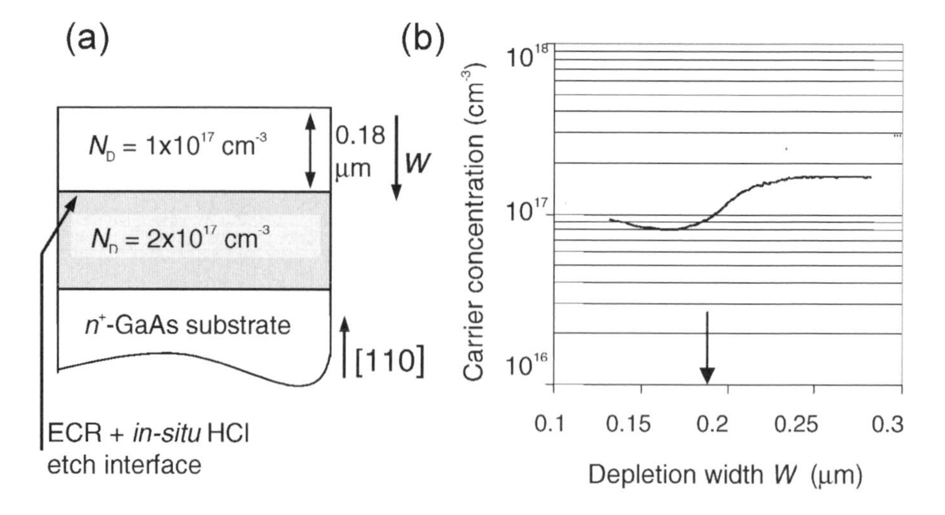

Figure 10. *In-situ* etching and regrowth on a (110) GaAs electron-cyclotron-resonance (ECR)-etched wafer: (a) schematic of the sample; (b) carrier concentration *vs.* depletion width W (arrow indicates the interface at $W = 0.18$ μm).

4. Conclusions

Significant suppression of phonon-assisted electron relaxation times in deep unipolar quantum boxes will allow the fabrication of intersubband lasers emitting cw at room temperature in the 3–13 μm wavelength range with high (> 40%) wallplug efficiency. Since the backfilling effect characteristic of cascaded QW lasers is a moot issue, IQB lasers are unlikely to suffer from the thermal runaway that currently mars the performance of cw QC lasers[3,4,23] and thus should prove to be reliable devices. Preliminary results include the demonstration at the nanoscale of the *in-situ* (gas) etch and regrowth processes necessary for the IQB formation, and the development of a technique for preventing Fermi-level pinning during the QB fabrication.

Acknowledgments

This work was supported in part by SPAWAR under a DARPA contract.

References

1. R. F. Kazarinov and R. A. Suris, "Possibility of the amplification of electromagnetic waves in a semiconductor with a superlattice," *Sov. Phys. Semicond.* **5**, 207 (1971).
2 J. Faist, F. Capasso, D. L. Sivco, C. Sirtori, A. L. Hutchinson, and A. Y. Cho, "Quantum cascade laser," *Science* **264**, 553 (1994).
3. A. Evans, J. S. Yu, S. Slivken, and M. Razeghi, "Continuous-wave operation at $\lambda \sim 4.8$ μm quantum-cascade lasers at room temperature," *Appl. Phys. Lett.* **85**, 2166 (2004).
4. J. S. Yu, A. Evans, S. Slivken, S. R. Darvish, and M. Razeghi, "Short wavelength ($\lambda \sim 4.3$ μm) high-performance continuous-wave quantum-cascade lasers," *IEEE Photonics Technol. Lett.* **17**, 1154 (2005).
5. J. Faist, A. Tredicucci, F. Capasso, *et al.*, "High-power continuous-wave quantum cascade lasers," *IEEE J. Quantum Electronics* **36**, 336 (1998).
6. C.-F. Hsu, J.-S. O, P. Zory, and D. Botez, "Intersubband quantum-box semiconductor lasers," *IEEE J. Selected Topics Quantum Electronics* **6**, 491 (2000).
7. U. Bockelmann and G. Bastard, "Phonon scattering and energy relaxation in two-, one-, and zero-dimensional electron gases," *Phys. Rev. B* **42**, 8947 (1990).
8. H. Benisty, C.M. Sotomayor-Torres, and C. Weisbuch, "Intrinsic mechanism for the poor luminescence properties of quantum-box systems," *Phys. Rev. B* **44**, 10945 (1991).
9. R. Heitz, M. Grundmann, N. N. Ledentsov, *et al.*, "Multiphonon-relaxation processes in self-organized InAs/GaAs quantum dots," *Appl. Phys. Lett.* **68**, 361 (1996).
10. J. Urayama, T. B. Norris, J. Singh, and P. Bhattacharya, "Observation of phonon bottleneck in quantum dot electronic relaxation," *Phys. Rev. Lett.* **86**, 4930 (2001).
11. L. Rebohle, F. F. Schrey, S. Hofer, G. Strasser, and K. Unterrainer, "Energy level engineering in InAs quantum dot nanostructures," *Appl. Phys. Lett.* **81**, 2079 (2002).
12. S. Sauvage, P. Boucaud, R. P. Lobo, *et al.*, "Long polaron lifetime in InAs-GaAs self-assembled quantum dots," *Phys. Rev. Lett.* **88**, 177402 (2002).
13. Z.–K. John Wu, H. Choi, T. B. Norris, A. Stiff-Roberts, and P. Bhattacharya, "Ultrafast electronic dynamics in unipolar *n*-doped InAs/GaAs quantum dot structures," paper QThC6 in: *Proc. 2005 CLEO/QELS*, Baltimore, MD (2005).
14. Z.–K. John Wu, H. Choi, T. B. Norris, S. Chakrabarti, X. Su, and P. Bhattacharya, "Electron dynamics in *n*-doped $In_{0.4}Ga_{0.6}As$/GaAs quantum dot infrared detector structures," paper QTuK3 in: *Proc. 2006 CLEO/QELS*, Long Beach, CA (2006).
15. B. Kochman, A. D. Stiff-Roberts, S. Chakrabarti, *et al.*, "Absorption, carrier lifetime, and gain in InAs-GaAs quantum-dot infrared photodetectors," *IEEE J. Quantum Electronics* **39**, 459 (2003).

16. L. Jacak, J. Krasnyi, D. Jacak, and P. Machnikoski, "Magnetopolaron in a weakly elliptical InAs/GaAs quantum dot," *Phys. Rev. B* **67**, 035303 (2003).
17. C.-F. Hsu, J.-S. O, P. S. Zory, and D. Botez, "Intersubband laser design using a quantum box array," *Proc. SPIE* **3001**, 271 (1997).
18. I. A. Dmitriev and R. A. Suris, "Quantum cascade lasers based on quantum dot superlattice," *Phys. Stat. Sol. A* **202**, 987 (2005).
19. I. C. Sandall, P. M. Smowton, C. L. Walker, H. Y. Liu, M. Hopkinson, and D. J. Mowbray, "Recombination mechanisms in 1.3 μm InAs quantum-dot lasers," *IEEE Photonics Technol. Lett.* **18**, 965 (2006).
20. D. P. Xu, A. Mirabedini, M. D'Souza, *et al.*, "Room-temperature, mid-infrared (λ = 4.7 μm) electroluminescence from single-stage, intersubband GaAs-based edge emitters," *Appl. Phys. Lett.* **85**, 4573 (2004).
21. H. Yagi, K. Muranushi, N. Nunoya, T. Sano, S. Tamura, and S. Arai, "GaInAsP/InP strain-compensated quantum-wire lasers fabricated by CH_4/H_2 dry etching and organometallic vapor-phase-epitaxial regrowth," *Japan. J. Appl. Phys.* **41**, Part 2, L186 (2002).
22. C. Gmachl, F. Capasso, A. Tredicucci, D.L. Sivco, R. Kohler and A.Y. Cho, "Dependence of the device performance on the number of stages in quantum-cascade lasers," *IEEE J. Selected Topics Quantum Electronics* **5**, 808 (1999).
23. M. Beck, D. Hofstetter, T. Aellen, *et al.*, "Continuous wave operation of a mid-infrared semiconductor laser at room temperature," *Science* **295**, 301 (2002).
24. N. S. Wingreen and C. A. Stafford, "Quantum-dot cascade laser: Proposal for an ultralow-threshold semiconductor laser," *IEEE. J. Quantum Electronics* **33**, 1170 (1997).
25. C. H. Fischer, P. Bhattacharya, and P.-C. Yu, "Intersublevel electro-luminescence from $In_{0.4}Ga_{0.6}As$/GaAs quantum dots in quantum cascade heterostructure with GaAsN/GaAs superlattice," *Electronics Lett.* **39**, 1537 (2003).
26. G. Tsvid, M. D'Souza, D. Botez, *et al.*, "Towards intersubband quantum box lasers: E-beam lithography update," *J. Vac. Sci. Technol. B* **22**, 3214 (2004).
27. A. Mirabedini, L. J. Mawst, D. Botez, and R. Marsland, "High peak-current-density strained-layer $In_{0.3}Ga_{0.7}As/Al_{0.8}Ga_{0.2}$ resonant tunneling diode grown by metal organic chemical vapor deposition," *Appl. Phys. Lett.* **70**, 2867 (1997).
28. J. Faist, F. Capasso, C. Sirtori, *et al.*, "High-power mid-infrared (λ = 5 μm) quantum cascade lasers operating above room temperature," *Appl. Phys. Lett.* **68**, 3680 (1996).
29. D. Botez, F. Cerrina, T. Kuech, and P. Zory, "Intersubband quantum-box (IQB) lasers," *Final Rep. DARPA/SPAWAR Grant No 660-7558-203-2003524*, 2004.
30. M. A. Reed, J. N. Randall, R. J. Aggarwal, R. J. Matyi, T. M. Moore and A. E. Wetsel, "Observation of discrete electronic states in a zero-dimensional semiconductor nanostructure," *Phys. Rev. Lett.* **60**, 535 (1988).

A New Class of Semiconductors Using Quantum Confinement of Silicon in a Dielectric Matrix

Martin A. Green
ARC Photovoltaics Centre of Excellence
University of New South Wales, Sydney 2052, Australia

1. Introduction

For photovoltaic use, control of the semiconductor bandgap is important for high energy conversion efficiency designs, such as those based on tandem stacks of cells responding to different spectral ranges. Traditionally, such bandgap control has been obtained using alloys of III–V semiconductors or by Si/Ge alloys in the case of amorphous material. This chapter explores the use of quantum confinement to produce high bandgap material within the crystalline silicon materials system. The optical and transport properties of a new form of synthesized material involving silicon quantum dots in a dielectric matrix of silicon oxide, nitride or carbide are described. These materials may have applications in areas other than photovoltaics where low mobility is tolerable.

The photovoltaics industry is currently booming, with annual sales growing at over 40%/year. This growth is expected to continue into the next decade.[1] Most solar sells are based on silicon wafers similar to those used in microelectronics. In 2006, a turning point was reached where the volume of silicon used in photovoltaics exceeded that used in microelectronics for the first time. Traditional polysilicon suppliers are now rapidly increasing capacity to meet the rapidly increasing demand.[1] With over 2 gigawatts of new capacity expected to be installed in 2006, new annual capacity additions of photovoltaics will exceed those of nuclear plants within the next year or two.

However, to wean the world from coal as a way of meeting spiralling power demand, photovoltaic costs have to decrease significantly. A transition from wafers to a stable, durable thin-film technology would improve prospects for such a large cost reduction. Incorporating high efficiency features into these thin films, *e.g.* by using tandem stacks of cells, has the potential to further reduce costs.[2]

In the tandem cell design, cells of different bandgaps are stacked on top of one another, with the highest bandgap cell uppermost (see Fig. 1). High-energy photons are absorbed in this cell, with the remainder passing through to the next highest bandgap cell, and so on. In this way, photons automatically are converted by the cell with the most appropriate bandgap. The cells are generally connected in series, requiring careful bandgap selection to ensure current matching.

Future Trends in Microelectronics. Edited by Serge Luryi, Jimmy Xu, and Alex Zaslavsky
ISBN 0-471-48 © 2007 John Wiley & Sons, Inc.

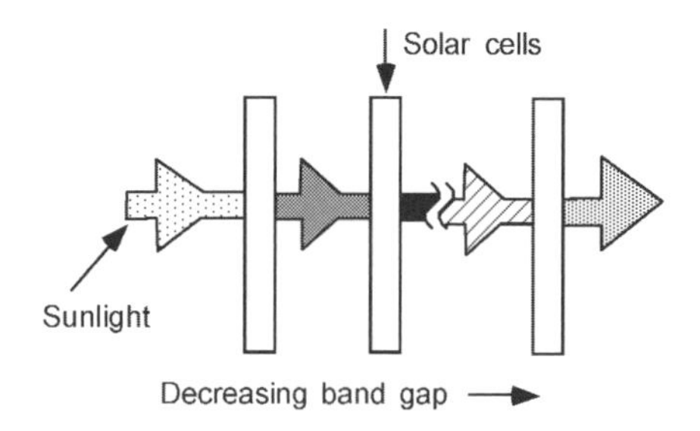

Figure 1. Tandem cell concept.

This concept has been well exploited in the group III-V materials system. Here, alloying is utilized for the required bandgap control. The highest performance devices to date are based on a GaInP/GaInAs/Ge 3-cell tandem stack. Alloying has also been used in the amorphous silicon system, with 3-cell stacks of a-Si/a-SiGe/a-SiGe cells commercially available. Another commercial combination is a 2-cell stack based on a-Si (1.7 eV bandgap) and microcrystalline Si (1.1eV).

The III-V tandem cells are used in spacecraft due to their favorable power to weight ratio. They are, however, too expensive for terrestrial use unless used in concentrating photovoltaic systems. The amorphous silicon variants suffer from low efficiency and poor stability, due to the unravelling of the hydrogen passivation in these materials.

A promising new thin-film solar call material is crystalline silicon.[3] The low optical absorption in silicon can be offset by optical designs that scatter incoming light into as many as possible of the available optical modes. Such "light-trapping" allows good performance from silicon films only 1–1.5 μm thick.[4]

To use such a cell in a tandem stack most effectively, some method is required for achieving high-bandgap material in a compatible materials system without compromising the inherent advantages of silicon (abundance, non-toxicity, stability, durability). The approach being explored in this work is to use quantum confinement in silicon to increase its effective bandgap to produce all-silicon superlattice cells as indicated in Fig. 2.

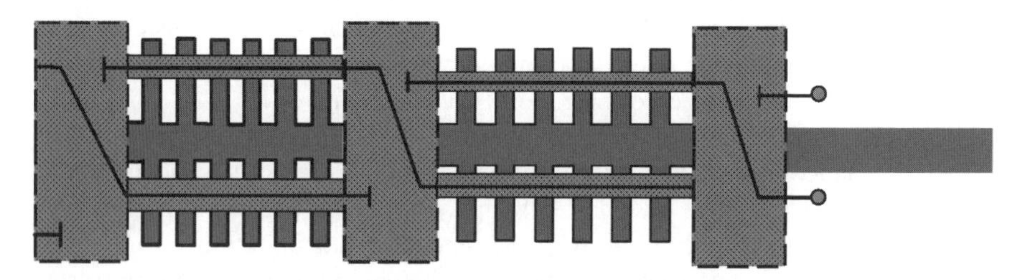

Figure 2. All silicon tandem cell structure using quantum superlattices as the uppermost cells to increase silicon's effective bandgap. The rectangular boxes between the cells represent the conduction to valence band connections between cells (normally implemented using tunnelling or highly defective junctions).

2. Proof of concept

Our initial work involved studies of confinement in high quality silicon quantum wells to ensure that quantum confinement in silicon posed no special problems. Starting with high quality ELTRAN silicon-on-insulator (SOI) wafers,[1] the original 52 nm thick silicon layer was carefully oxidized as in Fig. 3 (almost completely through) to produce thin silicon layers in the 1–2 nm thickness range enshrouded by high-quality silicon dioxide.[6] Oxidation conditions and subsequent treatments benefited from our group's experience in producing record-performance silicon solar cells[7] as well as exceptionally high photoluminescence efficiency (10% external quantum efficiency) from bulk silicon wafers.[8]

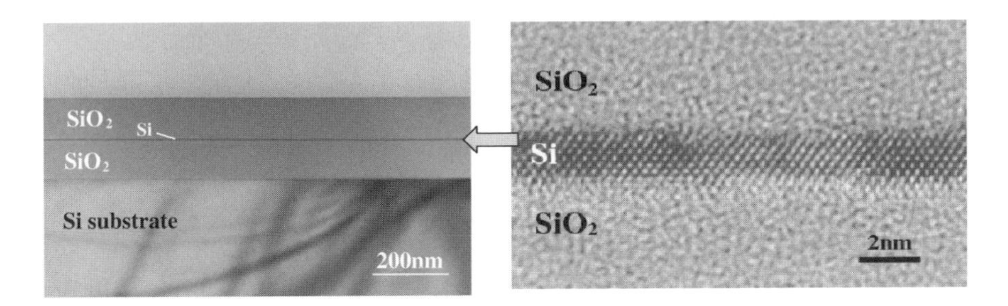

Figure 3. Silicon quantum well formed by oxidizing a 52 nm thick silicon layer on a SOI wafer at two difference magnifications.

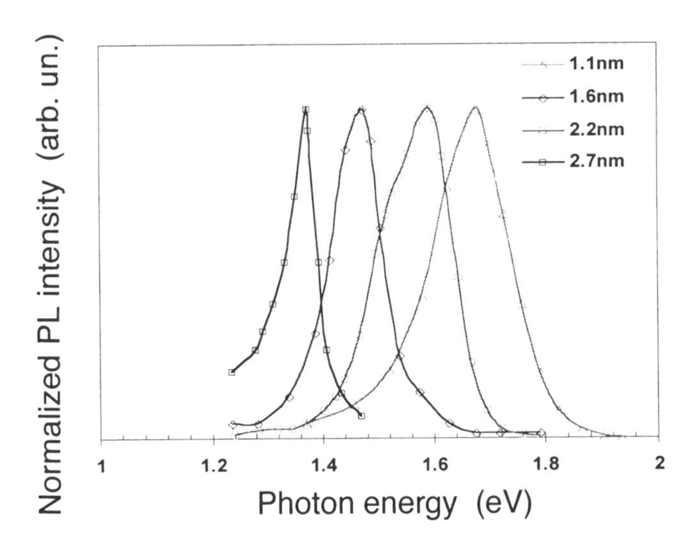

Figure 4. Photoluminescence (PL) from silicon quantum wells as a function of well thickness.

Low-temperature photoluminescence results from such silicon quantum wells are shown in Fig. 4. In accordance with simple confinement theory, the peak emission energy shifted upwards as the thickness of the final layer decreased, to values as high as 1.7 eV for layers of about 1 nm thickness.[6] There was no evidence of oxide-mediated transitions that featured prominently in earlier work. The observed broadening could be well described by assuming a contribution to the signal from regions one atomic layer thicker and thinner than that primarily responsible for the observed response. Not shown in Fig. 4 is the increase in intensity by a factor of 30 as the dot size decreased from 2.7 nm to 1.1 nm, due to the strengthening of optical processes by confinement, as also expected.

3. Experimental approach

Our initial superlattice work involved deposition of alternate layers of dielectric followed by layers of amorphous silicon. These were furnace crystallized with laser treatment explored as a way of improving crystallographic quality. However, a slight modification to this approach suggested by Zacharias *et al.*[9] which results in silicon quantum dots, has been the focus of subsequent work (Fig. 5).

As shown in Fig. 5, the deposition of each dielectric layer is followed by the deposition of a silicon-rich layer of the same dielectric, rather than pure silicon. On heating, silicon precipitates out of the silicon-rich layer in the form of silicon quantum dots, in what appears to be a 2D growth process, provided the layers are less than about 5 nm thick. This approach provides a way of controlling both the diameter of the dots and one of their spatial coordinates, as shown in Fig. 6.

Photoluminescence from such material shows the same general features as from quantum wells, with peak emission energy increasing from 1.3 eV to 1.7 eV as dot diameter decreased from 5 nm to 2 nm. Again, emission intensity strengthens as dot size decreases, offset by a counteracting tendency for the smallest dots.

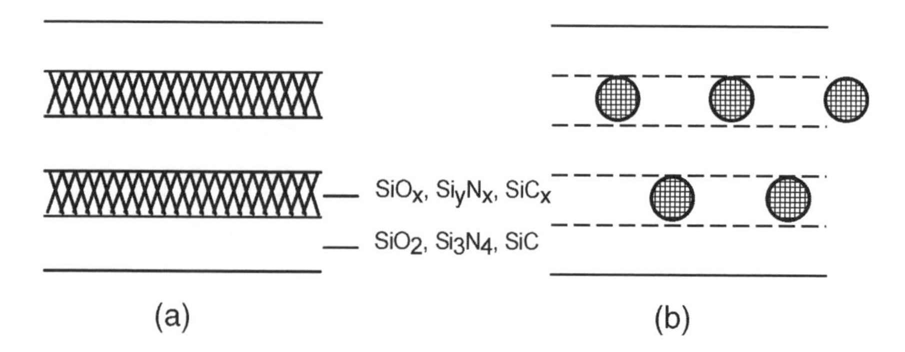

Figure 5. Method of forming silicon quantum dots in a dielectric matrix.[9]

Figure 6. (a) Si quantum dots in a SiO_2 matrix formed using the approach of Fig. 5; (b) high-resolution image of a dot within material deposited onto a Si substrate.

4. Requirements for tandem cells

The key requirement for use in tandem cells is bandgap control, for which evidence is provided by the photoluminescence results. A bonus is the increased optical strength due to quantum confinement, reducing the required thickness of the quantum dot cells. Additionally, reasonable carrier transport properties are required, as represented by mobility-lifetime products greater than 10^{-9} cm^2/V. Mechanisms to encourage directional transport within the cells are also required as provided by a *pn* junction in conventional cells. Control of the work function of the quantum dot material by doping or some other mechanism is therefore required. Finally, some mechanism of providing transport between the conduction and valence bands of adjacent cells is also required (Fig. 2). Tunnel junctions are generally used in good quality III-V cells to achieve this interconnection, while heavily doped junctions in amorphous material are sufficiently defective that they provide this connection via shunting. Progress towards demonstrating these requirements is described in the following sections.

5. Mobility and conductivity

Our group's initial investigation of issues relevant to mobility assumed the quantum dots were uniform in size and were positioned on a regular cubic lattice. With some analytical simplification including an effective mass approximation, analytical solutions were derived similar to those for atomic lattices.[10] By assuming a scattering time for electrons ten times smaller than the bulk silicon, idealized "Bloch" mobilities could be calculated as in Fig. 7.

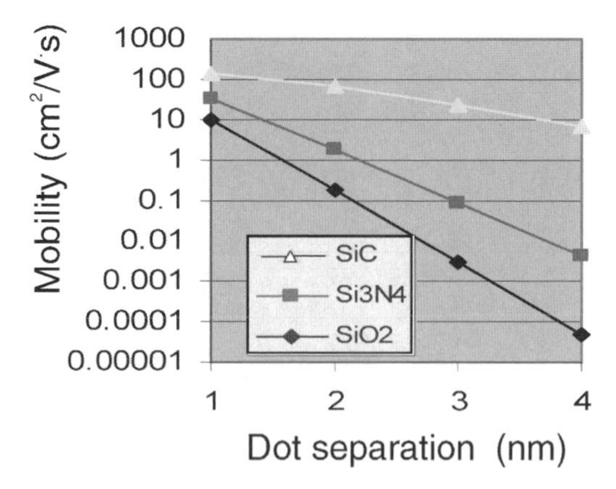

Figure 7. Calculated Bloch mobilities for electrons in a cubic lattice of cubic silicon quantum dots (2 nm edges) as a function of lattice spacing for different silicon-based matrices.

These results suggested it might be difficult to obtain targeted mobilities above $1cm^2/V{\cdot}s$ using an oxide matrix, but that this might be easier with nitride and carbide matrices. However, it was relatively straightforward to extend the previous analysis to include random variations in both dot position away from the regular lattice and in dot size. It was found that the calculated mobility was not particularly sensitive to departures from the idealised spacing, with random variations up to about 30% tolerable without significant impact. However, the calculated mobility was extremely sensitive to dot size. Variations of less than 1% in diameter caused a significant decrease in mobility.

Essentially, for such low-mobility superlattices, 1% diameter variation pushed the energy of the confined state within extreme-sized dots outside the range of energies corresponding to the miniband in the idealized material (Fig. 2).

Since such tight control of dot size is not feasible experimentally, this suggests that Bloch transport will not be an important mechanism in our experimental material. Phonon-assisted hopping transport may be a more likely transport process. Experimental conductivity versus inverse temperature for nominally undoped quantum dots in oxide and nitride matrices are shown in Fig. 8. These confirm the expectation that transport will be easier in the nitride matrix, although these nitride results are preliminary.

6. Doping

The situation regarding doping of silicon quantum dots is unclear. Experimental work of Fujii and co-workers[11] shows clear evidence for increasing free carrier concentration as phosphorus doping of the dots is increased, while theoretical work suggests that confinement increases dopant ionization energy to levels where ionization is unlikely at room temperature.[12] For example, Fig. 9 shows calculated dopant ionization energies for the materials system of current interest calculated using the expressions derived by Lannoo *et al.*[13]

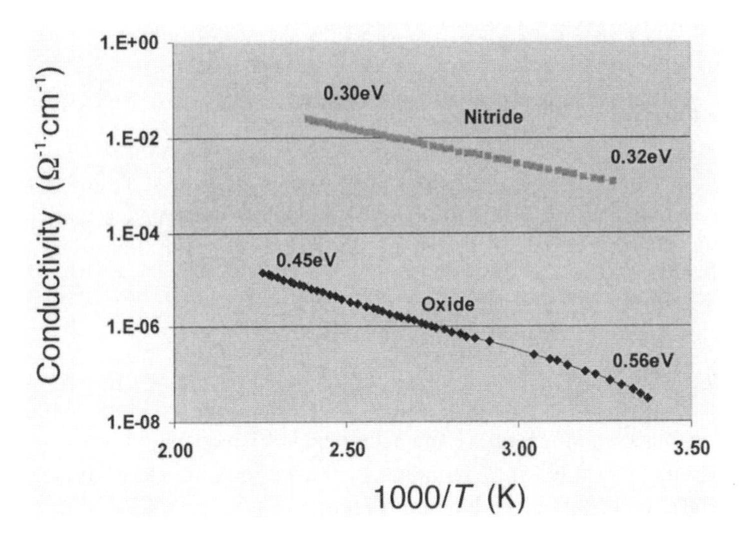

Figure 8. Lateral conductivity of silicon quantum dot material in oxide and nitride matrices *vs.* inverse temperature (nitride results are preliminary); corresponding activation energies are also shown.

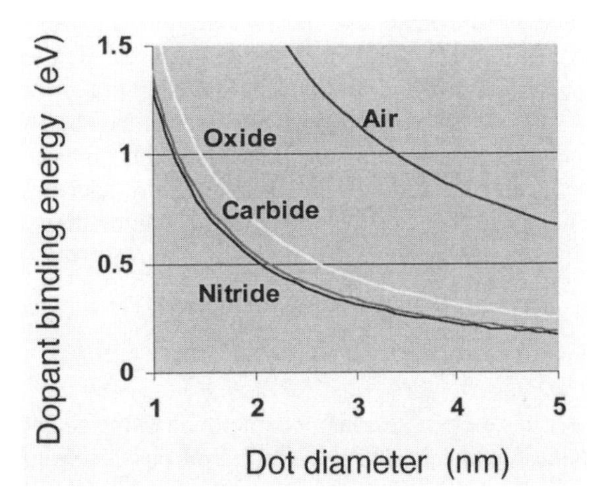

Figure 9. Calculated dopant ionization energy for a substitutional dopant centrally located in a silicon quantum dot embedded in the matrices indicated.

Embedding the dots in a high dielectric constant matrix tends to reduce the dopant ionization energy, although values are still too highly for significant ionization to be expected at room temperature. A high density of dots will increase the effective dielectric constant further, tending to further reduce this energy.

Our group is using *ab-initio* calculations to explore the properties of both conventional substitutional dopants as well as other strategies aimed at achieving the work function control provided by dopants in bulk material.

7. Excitons

The excitonic binding energy increases due to confinement in a similar way as the dopant binding energy.[9] In a bulk semiconductor, even under non-equilibrium conditions, there is a natural dynamic equilibrium concentration of excitons n_{xde} given by:[14,15]

$$n_{xde} = np/n^* \sim np \exp[E_{bx}/k_B T] , \qquad (1)$$

where np is the electron hole product and n^* is a material constant depending on the electron, hole and excitonic densities of states and the excitonic binding energy E_{bx} ($k_B T$ is the thermal energy). The actual excitonic concentration, n_x will deviate from this value to an extent determined by the association and disassociation rates that can be characterized by a binding lifetime τ_B:

$$B = [(np/n^*) - n_x]/\tau_B . \qquad (2)$$

For bulk silicon, τ_B is in the picosecond range,[12] and the excitonic concentration approaches its dynamic equilibrium value. Although excitons are present in reasonable densities in these materials even at room temperature, their equilibration with free carrier concentrations means they go largely unnoticed.

In these quantum dot materials, the relative concentration of excitons will increase as given by Eq. (1). Moreover, excitonic association and disassociation rates might be expected to slow, resulting in non-equilibration with the electron-hole product. As the size of quantum dots decreases, it may be possible to observe a transition from the situation in bulk semiconductors to one more similar to organic semiconductor solar cells. There, the rate of disassociation of excitons into free carriers is believed to be so slow that specific exciton disassociation features are incorporated into junction regions to allow full photocurrent capture.

8. Conclusions

This chapter describes the progress in our exploration of the potential of a new type of quasi-amorphous semiconductor material based on silicon quantum dots in a dielectric matrix. The immediate motivation for this work is the potential for bandgap control that would allow the implementation of low-cost tandem solar cell structures based entirely on silicon.

Control of dot diameter and spacing has been demonstrated, as has bandgap control, as demonstrated by photoluminescence spectra. Current work is directed at demonstrating the required levels of carrier transport, as well as at understanding doping and the consequence of increased excitonic binding energy upon device design.

Our material could be useful for other semiconductor device applications where high carrier mobility is not required.

Acknowledgments

This work is supported by the Australian Research Council (ARC) under its Research Centres of Excellence Program and by Stanford University under the Global Climate and Energy Project (GECEP). The author gratefully acknowledges the contribution of other members of the ARC Photovoltaics Centre of Excellence, particularly G. Conibeer, D. König, E. C. Cho, D. Song, Y. Cho, T. Fangsuwannarak, Y. Huang, G. Scardera, E. Pink, S. Huang, W. C. Jiang, T. Trupke, R. Corkish, and T. Puzzer.

References

1. M. Rogol, P. Choi, J. Conkling, A. Fotopolous, K. Peltzman, and S. Roberts, "Solar annual 2006: The gun has gone off," Photon Consulting, July 2006.
2. M. A. Green, *Third Generation Photovoltaics*, Berlin: Springer, 2003.
3. M. A. Green, P. A. Basore, N. Chang, *et al.*, "Crystalline silicon on glass (CSG) thin-film solar cell modules," *Solar Energy* 77, 857 (2004).
4. P. A. Basore, "CSG-2: Expanding the production of a new polycrystalline silicon PV technology," *21st Eur. Photovoltaic Solar Energy Conf.*, Dresden, (2006).
5. K. Yamagatu and T. Yonehara, "Bonding, splitting and thinning by porous Si in ELTRAN® SOI-epi wafer™," *MRS Spring Symp.*, San Francisco (2001).
6. E.-C. Cho, P. Reece, M. A. Green, J. Xia, R. Corkish, and M. Gal, "Clear quantum confined luminescence from crystalline silicon/SiO_2 single quantum wells," *Appl. Phys. Lett.* **84**, 2286 (2004).
7. M. A. Green, J. Zhao, A. Wang, P. J. Reece, and M. Gal, "Efficient silicon light emitting diodes," *Nature* **412**, 805 (2001).
8. T. Trupke, J. Zhao, A. Wang, R. Corkish, and M. A. Green, "Very efficient light emission from bulk crystalline silicon," *Appl. Phys. Lett.* **82**, 2996 (2003).
9. M. Zacharias, J. Heitmann, R. Scholz, U. Kahler, M. Schmidt, and J. Blaesing, "Size-controlled highly luminescent silicon nanocrystals: A Si/SiO_2 superlattice approach," *Appl. Phys. Lett.* **80**, 661 (2002).
10. C.-W. Jiang and M. A. Green, "Silicon quantum dot superlattices: Modelling of energy bands, densities of states and mobilities for silicon tandem solar cell applications," *J. Appl. Phys.* **99**, 114902 (2006).

11. M. Fujii, Y. Yamaguchi, Y. Takase, K. Ninomiya, and S. Hayashi, "Photoluminescence from impurity co-doped and compensated Si nanocrystals," *Appl. Phys. Lett.* **87**, 211919 (2005).
12. Z. Zhou, M. L. Steigerwald, R. A. Friesner, L. Brus, and M. S. Hybertsen, "Structural and chemical trends in doped silicon nanocrystals: First-principles calculations," *Phys. Rev. B* **71**, 245308 (2005).
13. M. Lannoo, C. Delerue and G. Allan, "Screening in semiconductor nanocrystallites and its consequences for porous silicon," *Phys. Rev. Lett.* **74**, 3415 (1995).
14. M. A. Green, *Silicon Solar Cells: Advanced Principles and Practice*, Sydney, Australia: University of New South Wales, 1995.
15. R. Corkish, D. S.-P. Chan, and M. A. Green, "Excitons in silicon diodes and solar cells – A three particle theory," *J. Appl. Phys.* **79**, 195 (1996).

Merging Nanoepitaxy and Nanophotonics

N. N. Ledentsov, V. A. Shchukin
Institut für Festkörperphysik
Tech. Univ. Berlin, D-10623 Berlin, Germany and
Abraham Ioffe Physical Technical Institute, St. Petersburg, Russia

D. Bimberg
Institut für Festkörperphysik
Tech. Univ. Berlin, D-10623 Berlin, Germany

1. Introduction

Nanostructures are rapidly penetrating optoelectronic devices, enabling novel active media concepts and operation principles. Applications of nanostructures occurs in the three main directions: *nanoepitaxy*, *nanophotonics* and *novel functionality*. Nanoepitaxy utilizes growth-related phenomena at the nanoscale level, including self-organized formation and control of epitaxial nanostructures, and techniques of defect reduction and defect engineering. Nanophotonics addresses modification of the key optical properties by advanced epitaxial design and/or lateral patterning of the surfaces and interfaces with nanoscale resolution. Novel functionality enables control over performance of nanophotonic devices by utilizing optical interference phenomena controlled by electrooptic effects. In this chapter we give an overview of recent results, concentrating on all-epitaxial approaches aimed at realizing a broad range of novel devices, as opposed to widely discussed nanoscale patterning concepts addressed in other chapters of this volume.

2. Nanoepitaxy: Self-organized growth of low-dimensional structures

Nanoepitaxy utilizes effects of spontaneous formation of *ordered nanostructures* at crystal surfaces[1] and can result in quantum wires, quantum dots (QDs) or vertical quantum well structures.[2] The key processes are: spontaneous formation of three-dimensional (3D) islands in lattice-mismatched growth, spontaneous surface periodic nanofaceting, and spontaneous lateral alloy phase separation into ordered nanodomains. These approaches are schematically presented in Fig. 1 and all are applied to the field of active media for advanced devices. The first lasing action in self-organized QDs was demonstrated in 1993.[3]

Future Trends in Microelectronics. Edited by Serge Luryi, Jimmy Xu, and Alex Zaslavsky
ISBN 0-471-48 © 2007 John Wiley & Sons, Inc.

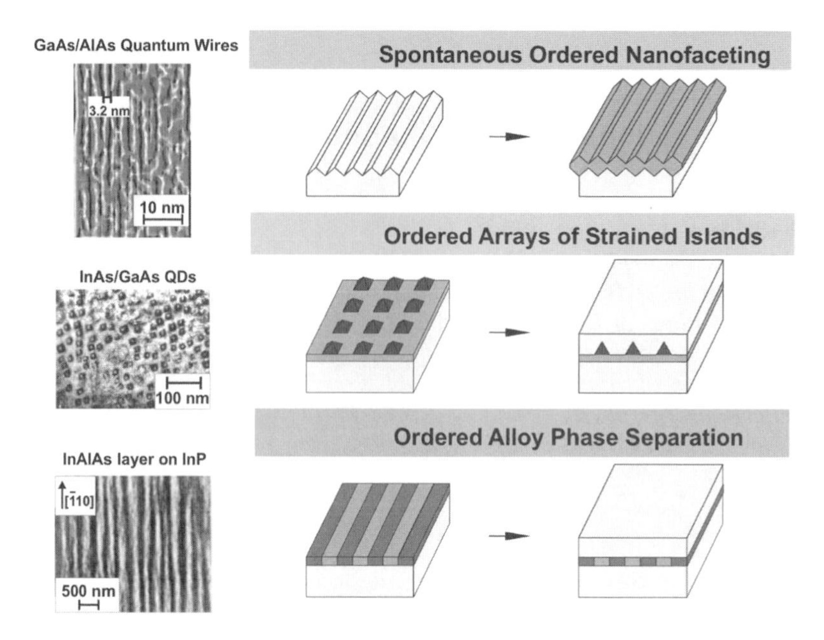

Figure 1. Schematic representation of the three key approaches to spontaneous formation of nanostructures at crystal surfaces (from Ref. 1).

- *Quantum wire devices*

Quantum well wire (QWW) structures are formed when growth takes place on nanofaceted or step-bunched surfaces.[1,2] It is well established that some high-index crystal planes with a high surface energy can be unstable with respect to spontaneous 1D or 2D nanofaceting, so ordered nanofaceting provides an energetically favorable state. Vicinal surfaces are also well known to undergo a transition to step bunching, with a characteristic example of a GaAs (100) surface misoriented by 10° towards the main diagonal direction, being effectively a high-index GaAs $(1\bar{1}8)$ surface. These substrates are broadly used for growth of (In,Ga,Al)P lasers for optical storage and display applications (620–660 nm). Elastic strain relaxation at the edges of the faceted surface enables alloy phase separation and spontaneous formation of natural superlattices during growth. A schematic representation of the growth mechanism is presented in Fig. 2.

Phase separation is a well-studied phenomenon in ternary and quaternary (In,Ga,Al)P layers. It originates from the intrinsic strain-stabilization mechanism in III-V alloys. Once the strain can relax at surface steps or step bunches, the alloys tend to separate into phases. To prevent growth-related instabilities resulting in bandgap and effective mass modulation across the layer, a significant surface misorientation of the GaAs substrate with respect to the exact (100) direction is used to control the phase separation effect and prevent formation of macroscopic domains.

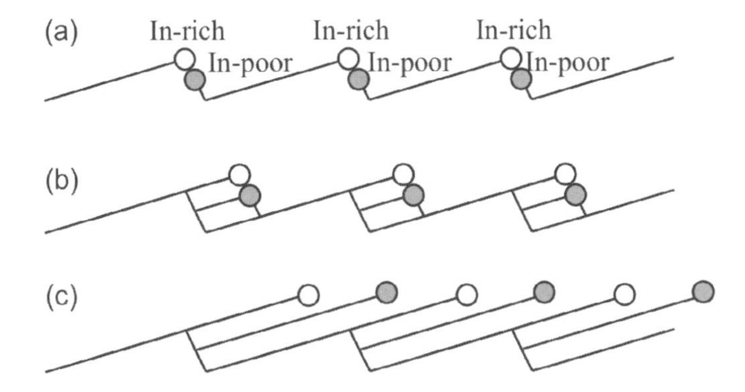

Figure 2. Schematic representation of epitaxial growth of an alloy layer on top of a step-bunched GaAs surface. Surface corrugation leads to the formation of a natural tilted superlattice.

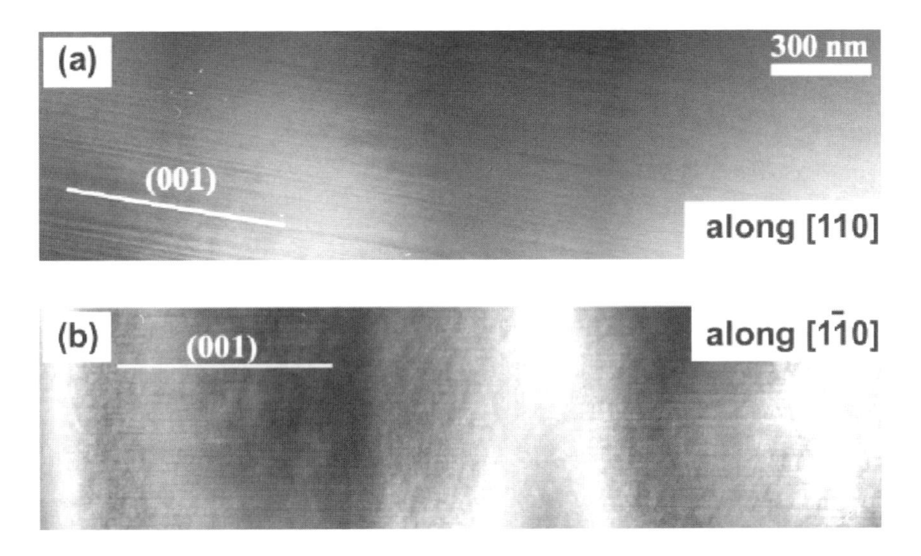

Figure 3. Natural superlattice formed by (In,Ga,Al)P layer growth on a step-bunched GaAs (100) surface misoriented by 10° towards <111>.

In Fig. 3 we show transmission electron microscopy (TEM) images of an (In,Ga,Al)P cladding layer of a DVD laser lattice-matched to the GaAs substrate. The images are taken in the direction parallel (a) and perpendicular (b) to the step bunches. From the vertical periodicity of the natural superlattice (~5–6 nm) one can extract the expected lateral periodicity of the step-bunched surface of ~30–35 nm. The interface between the (In,Ga,Al)P layer and an (In,Ga)P layer grown on top of it is shown in Fig. 4. One can see that, indeed, the lateral periodicity falls in the range expected from the stripe pattern revealed in the TEM images of Fig. 3.

Introduction of strain affects the lateral periodicity and the material distribution within the active region of the DVD laser diode.

To observe the structure of the active region of the DVD laser in detail, cross–section high-resolution TEM (HRTEM) studies have been performed. Figure 5(a) demonstrates the HRTEM image in the $(4\bar{4}1)$ cross-section.

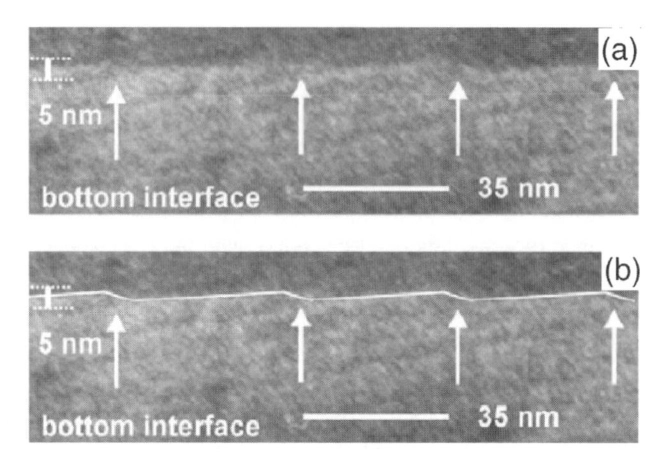

Figure 4. Natural superlattice formed by (In,Ga,Al)P and a corrugated interface (In,Ga,Al)P/InGaP formed by growth on a step-bunched GaAs surface.

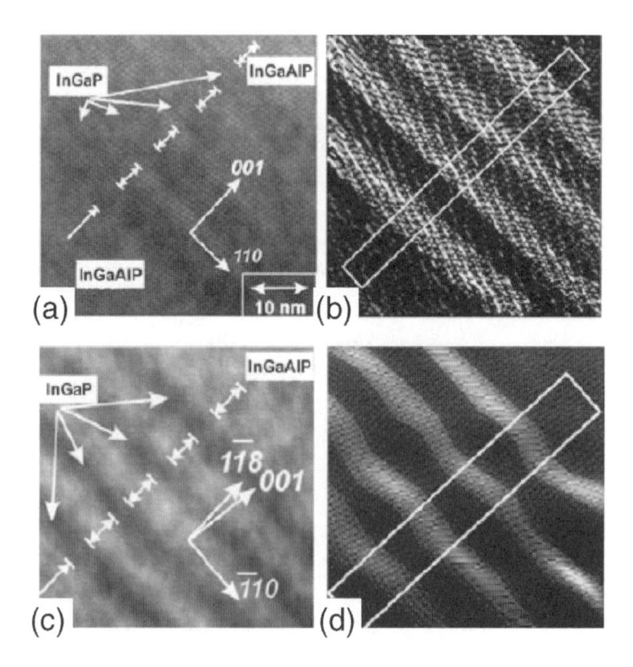

Figure 5. HRTEM and processed HRTEM images revealing interface corrugation of the active layers in the direction perpendicular to the step bunches.

The processed HRTEM image in Fig. 5(b) yields the lattice-mismatch map, where the gray scale is used to encode the lattice mismatch. Darker regions correspond to a smaller mismatch with respect to the substrate, *e.g.* regions with a lower In content, whereas brighter regions correspond to a larger mismatch with respect to the substrate, *e.g.* regions with a higher In composition. Figure 5(b) reveals planar interface structures, as it is taken perpendicular to the step bunches. In this direction the corrugation causes only broadening of the contrast and, thus, overestimated thickness is observed. In contrast, the image taken in the direction perpendicular to step bunches reveals the lateral periodicity of the corrugation of ~25 nm and a height of ~3 nm. The modification of the surface nanofaceting can be attributed to the renormalization of the surface energy by external strain resulting in a different equilibrium geometry of the faceted surface.[1,4]

The amplitude of the lattice mismatch modulation is about 3%, which corresponds to the variations of In content from 49% (barriers lattice-matched to the substrate) to 94% (local InGaP active layer regions that are close to pure InP). Thus, on top of interface corrugation, a dramatic alloy phase separation takes place, resulting in a quantum wire-like structures with a strong lateral confinement.

In Fig. 6 we show polarized luminescence spectra recorded from the heterostructure surface for directions parallel and perpendicular to the corrugation direction. One can see that the luminescence is predominantly polarized in the direction parallel to the corrugation direction, as expected for heavy-hole-like transitions. It follows from Fig. 6 that the degree of polarization approaches ~15% for the high-energy side of the luminescence spectrum, where the disorder-related effects are less important.

Interface and compositional modulations are particularly important for bright-red lasers for display and DVD applications, as the devices operate at wavelengths close to the Γ to X conduction band transition and a strong mixing between the states is needed to achieve efficient luminescence at the transition point.[5,6]

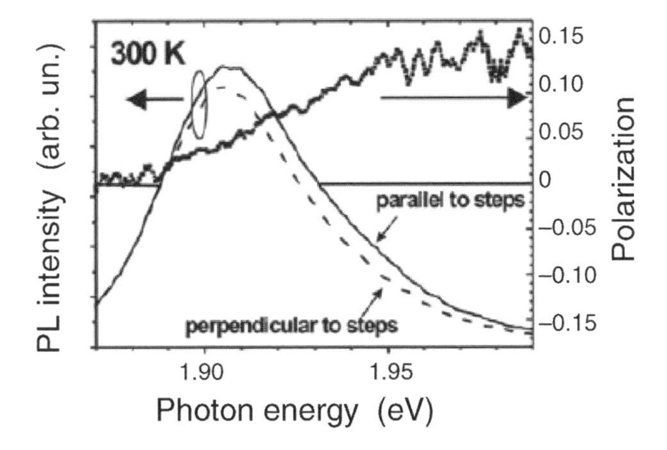

Figure 6. Room-temperature polarized surface photoluminescence spectrum of the active region of a DVD laser. Note the significant degree of polarization.

- *Quantum dot devices*

The next generation of lasers for optical storage is based on InGaN lasers in the violet (~410 nm). The InGaN light-emitting devices are also advantageous as sources for blue and green emission for projection displays. A unique property of these devices is spontaneous formation of InN-rich nanodomains (see Fig. 7), whose density and size increase dramatically as the substrate temperature is decreased. Evidence of QD-like InGaN structural and electronic properties was reported in a number of publications.[1,7-10] It was, for example, observed that when the luminescence spectra are recorded from small openings in metal masks, a broad spectrum of InGaN emission splits into discrete ultranarrow lines with a density corresponding to the density of quantum dots revealed in the HRTEM studies.[9]

Experiments also confirmed lack of lateral transport even at nanoscale dimensions of the openings. Resonant excitation along the contour of the InGaN luminescence band revealed the resonant nature of the resulting emission, with no significant spectral diffusion even after a long delay (>10 ns) after the excitation pulse.[10] Localization of carriers in InGaN QDs is a key requirement to achieve high degradation robustness of light-emitting diodes and lasers.

Quantum dot lasers based on InGaAs material system are also quickly progressing. High-power operation and long operating lifetimes, as well as ultrahigh-speed temperature-insensitive operation have been reported.[11] Lasers at 1300 nm with expected lifetime exceeding 10^6 hours at 40° heat sink temperature and 50 mW cw single-mode operation have been demonstrated.[11,12] Aging tests of the 1300 nm-range QD lasers are presented in Fig. 8. No sudden failure was observed and no laser reached the failure criterion (–20% of initial power) during a >2000 hour test. Stable spectral width and absence of any blue shift of the lasing wavelength are evidence of robustness of the QD region against In-Ga intermixing.

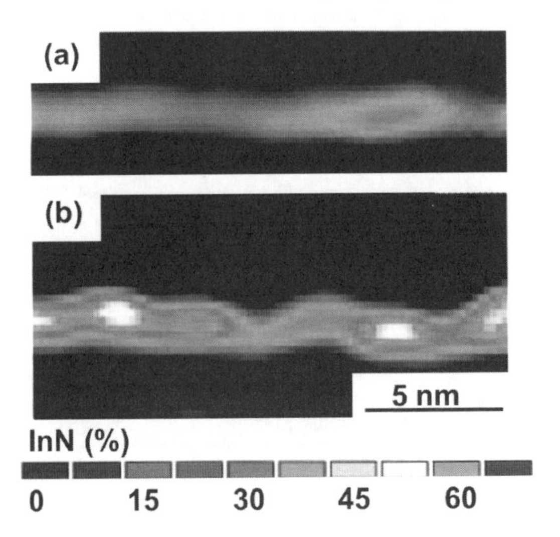

Figure 7. A grayscale map of InN distribution within an InGaN insertion in a GaN matrix, grown at 785 °C (a) and 755 °C (b), respectively. The dot size and density increase with a decrease in substrate temperature.

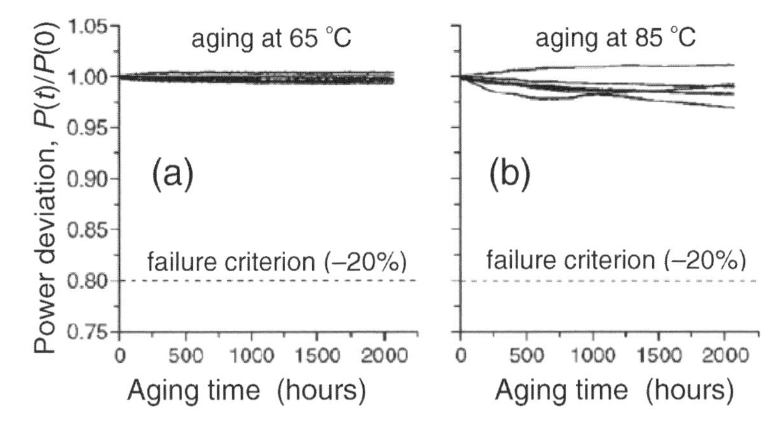

Figure 8. Output power *vs.* aging time at 65 °C (a) and 85 °C (b).

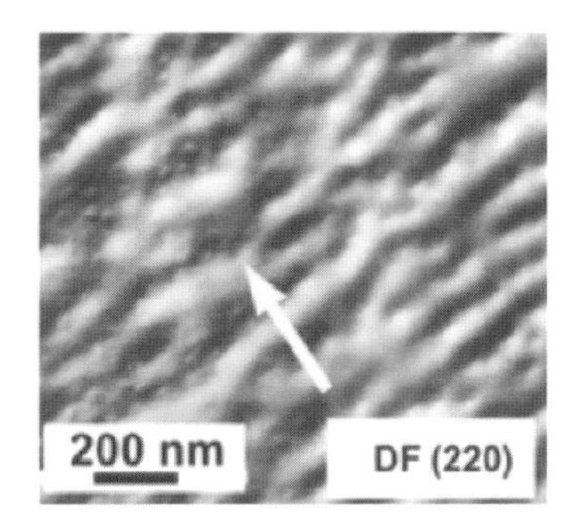

Figure 9. TEM image of low InAs composition InGaAs QDs formed in a submonolayer deposition mode.

Quantum dot 1300 nm lasers proved to be suitable for 10 GHz fully temperature-insensitive operation (20–70 °C) and enabled 5–100 GHz ultrashort-pulse temperature-robust mode locking and, also, distortion-free amplification of optical signals at 100 GHz and above.[12]

An important technique for QD formation is submonolayer growth, where QDs are "stacked" with a seed layer formed by submonolayer (SML) InAs deposition. This technique enables formation of QDs even in the case of relatively low average InAs composition in the insertion. A plan-view TEM image of such a structure is shown in Fig. 9.[13]

Submonolayer QDs can be successfully applied to conventional 800–1100 nm edge- and surface-emitting lasers, improving temperature performance, speed and degradation robustness. Ultrahigh temperature stability and ultrahigh-speed SML QD VCSELs are presented in Fig. 10.[13]

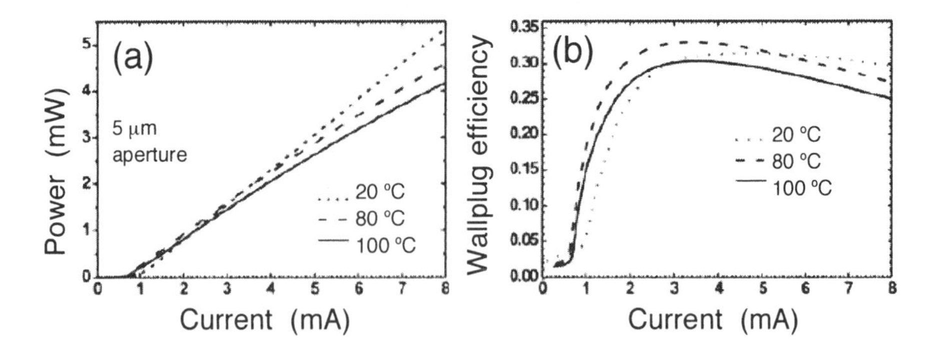

Figure 10. High-temperature performance of continuous wave 980 nm VCSELs based on submonolayer quantum dots.

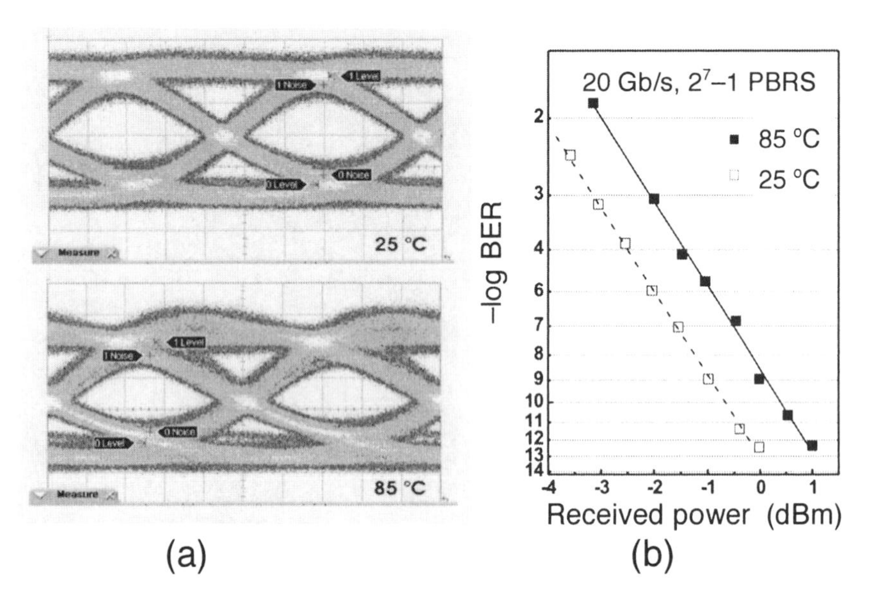

Figure 11. (a) Eye diagram for 20 GB/s NRZ 2^7-1 PRBS of a 6 μm SML QD VCSEL at 25 and 85 °C without change of the bias current and modulation voltage. (b) Bit-error-rate (BER) at 20 Gb/s with 2^7-1 PRBS at 25 and 85 °C for the same bias current.

Directly-modulated continuous wave SML QD VCSELs grown in the anti-waveguiding design[13,14] have demonstrated >20 Gb/s modulation in the 20–85°C temperature range. No adjustment of the modulation current and voltage was applied, as shown in Fig. 11. Furthermore, relaxation oscillation frequency of 28 GHz indicated potential to reach 40 Gb/s direct modulation with further design optimization.

3. Nanoepitaxy: Defect reduction

In this section we discuss the defect reduction (DR) advantages of nanoepitaxy.

- *Defect reduction in quantum dots*

Self-organized formation of QDs results in characteristic QD size and shape dispersion of typically 10–20%. Steps, defects, and impurities on the surface may stimulate formation of large, plastically relaxed islands, which degrade the device performance and reduce operating lifetime. In the 1970's and early 1980's, it was generally assumed that 3D InAs islands on GaAs[15,16] are severely dislocated.[16,17] It was therefore assumed that one had to avoid islands to obtain structurally perfect devices.[17]

Formation of dislocated InAs islands can be avoided, however, by proper adjustment of growth conditions and the InAs layer thickness.[18,19] For longer wavelength InAs QDs of larger sizes, this adjustment becomes increasingly difficult. For example, it was found that InAs/GaAs QDs formed at low substrate temperature may represent chains of laterally merged InAs islands emitting up to 1700 nm.[1] However, most of these objects are dislocated, resulting in extremely low integrated luminescence efficiency at room temperature. A possible way to remove the large dislocated islands is to use relatively thin GaAs caps to ensure that highly dislocated QDs remain uncapped. Then, growth interruption at high substrate temperature causes selective evaporation of uncapped InAs islands due to much higher temperature stability of GaAs.[20] At the same time, small dislocated islands may also exist. These islands, however, impose a different strain field in the GaAs cap layer after overgrowth, as compared to coherent QDs. Depositing an AlAs layer and annealing at an even higher temperature, high enough to enable GaAs evaporation (Fig. 12), leads to a drastic reduction in defect density and dramatic improvement in photoluminescence efficiency.[21]

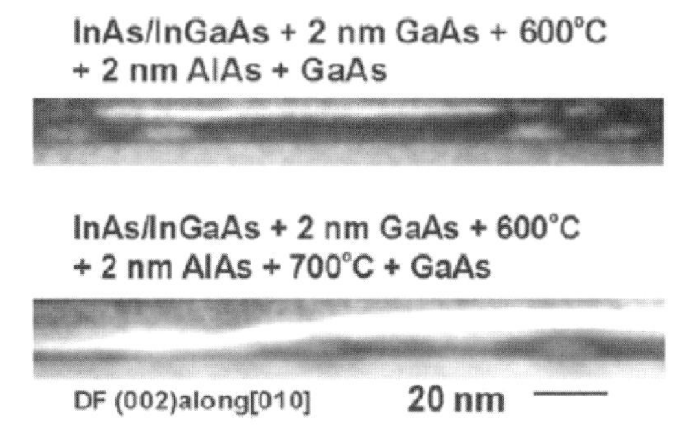

Figure 12. Cross-sectional TEM of InAs/InGaAs QDs emitting at 1.3 μm. The QDs are capped with 2 nm of GaAs, a 10 minute anneal at 600 °C is performed, then a 2 nm AlAs cap is deposited, followed by an additional 1 min anneal at 700 °C (bottom) before the GaAs overgrowth. AlAs deposition leaves defective QDs uncapped due to higher mismatch; they can be selectively removed by a 700 °C anneal.

- *Defect reduction in epilayers*

The concept of defect reduction can be further extended to layered structures containing both misfit and threading dislocations and to fabrication of thick defect-free layers.[22,23]

Misfit dislocations cause strain due to a missing (or additional) crystal plane originating at the epilayer interface in order to accommodate the lattice mismatch. As a result, the InGaAs dislocated layer becomes nonuniformly strained in the regions between the misfit dislocations. This causes a repulsive interaction between the GaAs/AlAs cap layer and the dislocated region (the effect opposite to the alignment of InAs QDs in stacked rows). Uncovered dislocated regions may be selectively removed by annealing, which does not affect the AlAs-capped regions[24] resulting in coherent nanopatterned epitaxial structures.

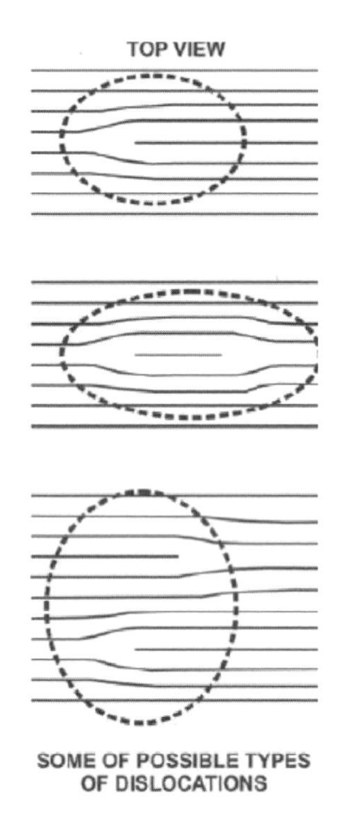

Figure 13. Schematic presentation of threading dislocations at the surface and their agglomerates.

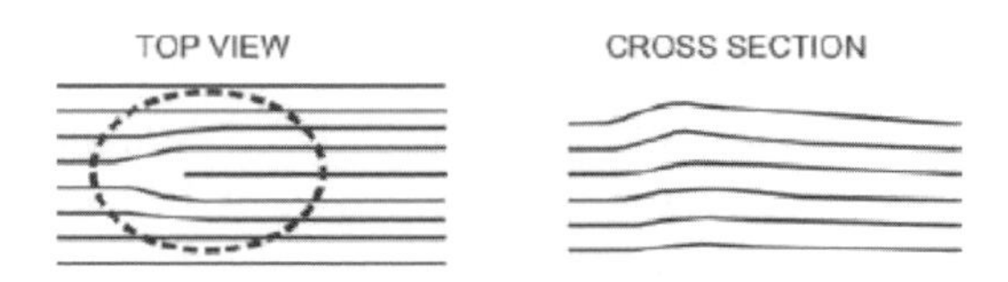

Figure 14. Surface morphology modulation in the vicinity of a threading dislocation.

Improved understanding of QD formation and defect removal stimulated work devoted to blocking the threading dislocations originating in thick epitaxial layers grown on top of lattice-mismatched substrate.

In the case of threading dislocations, additional lattice planes penetrate into the crystal, as shown in Fig. 13, and cause local strain fields in the vicinity of the exit point of the dislocation at the surface, as shown in Fig. 14. These defects can be fully removed by annealing only if they are correlated, as shown schematically in Fig. 13. Also, nanopipe defects in GaN, dislocation loops, local domains and nanoinclusions can be completely removed in certain cases.

For threading dislocations the most effective way of blocking them is their tilting towards in-plane penetration by selective annealing and overgrowth. Vertically propagating threading dislocations are driven by an underlying strain field, analogously to the vertically-correlated growth of strained quantum dots. In the case when this strain correlation can be removed and the growth front asymmetry is applied, bending of the dislocation becomes the most likely scenario.

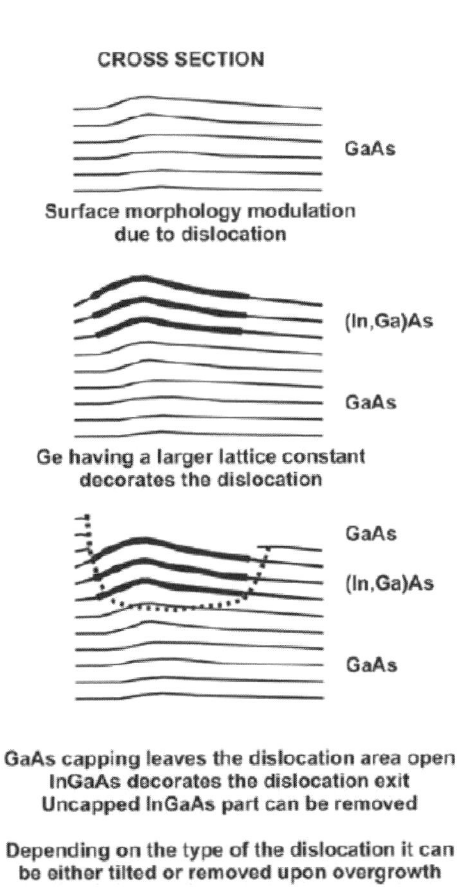

Figure 15. Surface morphology modulation at the vicinity of threading dislocation.

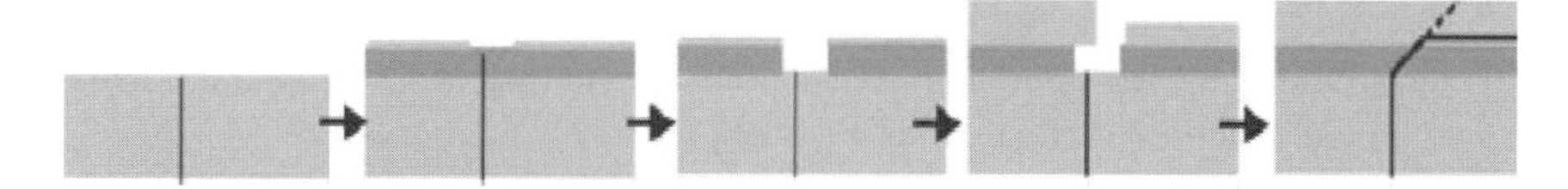

Figure 16. Schematic representation of possible routes of the defect reduction technique in case of a single threading dislocation.

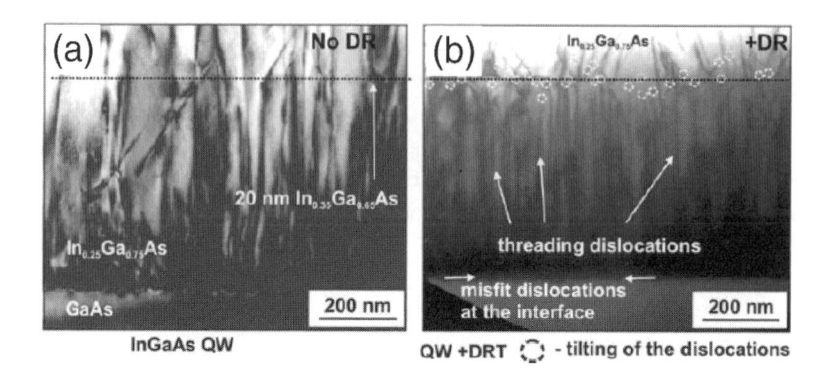

Figure 17. Cross-sectional TEM images illustrating defect reduction in an InGaAs layer grown on top of GaAs substrate at 500 °C. Annealing after 20 nm $In_{0.35}Ga_{0.65}As$ QW layer growth was not applied (a) or applied (b). One can see that annealing dramatically enhances the probability of dislocation tilting. The QW layer position is indicated by a dashed line.

In Figs. 15 and 16 we show a schematic representation of the effect of annealing on the defective region. Formation of nanovoids may occur,[26] further assisting the DR procedure. An example of applying the DR technique to $In_{0.3}Ga_{0.7}As$ epitaxial layers deposited at 500 °C on GaAs is shown in Fig. 17.

One can see that the efficiency of the procedure is very high and practically all the dislocations are tilted after its application. The tilted dislocations may annihilate much more effectively in the upper layers. Furthermore, the tilt angle may be enhanced by multiple application of the DR procedure, leading to a complete in-plane blocking of the dislocation. Thus, InGaAs buffers grown at lower temperatures demonstrate lower density of threading dislocations, as shown in Fig. 18, which also illustrates the blocking of dislocations.[25]

A similar approach can be applied to epitaxial growth of GaN on sapphire and other substrates.[25] In Fig. 19 we show defect blocking in GaN epitaxial layers, enabled by the defect reduction technique. We believe that further development of DR may also result in dramatic improvements in crystalline perfection of III-V epilayers grown on Si and other foreign substrates.

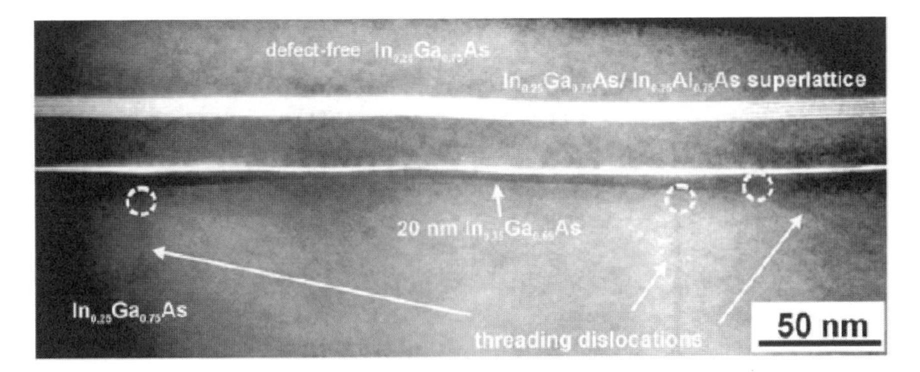

Figure 18. Cross-sectional TEM image illustrating application of the defect reduction technique to metamorphic heavily dislocated InGaAs layers grown on GaAs at 450 °C. In-plane blocking of dislocations is shown by circles.

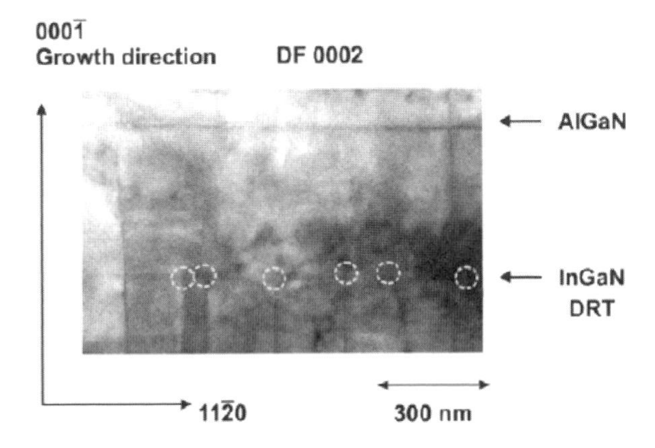

Figure 19. Cross-section TEM image illustrating application of DR to GaN epilayer on sapphire. Note in-plane blocking of threading dislocations at the DR regions.

- *Degradation robustness of metamorphic lasers*

Metamorphic (MM) QD lasers on GaAs are attractive for long-wavelength applications (1.4–3 μm), including VCSELs. Recently, edge-emitting QD lasers emitting near 1.5 μm demonstrated output power up to and above 7 W with high differential efficiency (>50%).[27] At high temperatures, the lasing wavelength was approaching ~1.52 μm. In Fig. 20 we show plan-view and cross-sectional TEM images of metamorphic 1500 nm QDs used in MM lasers. Narrow stripe lasers demonstrating cw 200 mW single-mode operation have also been realized.

The MM laser heterostructure studied for degradation robustness was grown at 450°C using elemental source molecular beam epitaxy (MBE) on Si-doped GaAs (100) substrate. The active region, consisting of 10 layers of InAs self-organized

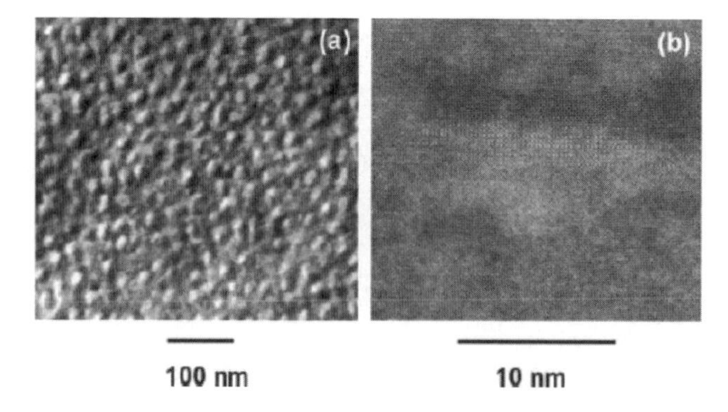

Figure 20. Plan-view (a) and cross-section (b) TEM images of InAs QDs inserted into a metamorphic $In_{0.2}Ga_{0.8}As$ layer on a GaAs substrate.

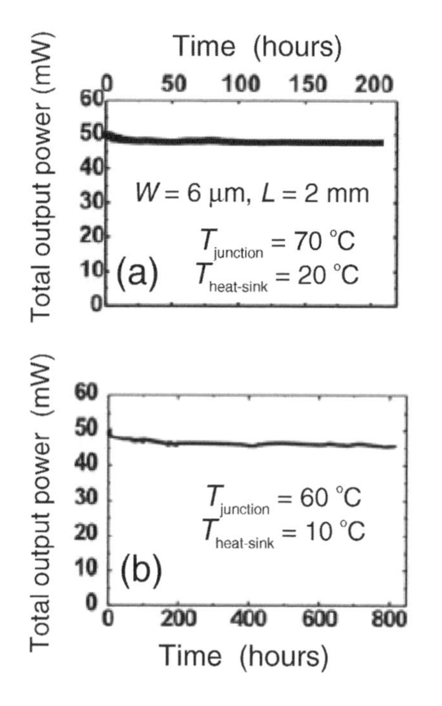

Figure 21. Degradation stability of long-wavelength MM lasers with as-cleaved facets *vs.* operation time, at 70 °C (a) and 60 °C (b) junction temperature. No significant degradation is been observed.

QDs, was placed in the middle of an $In_xGa_{1-x}As$ ($x \sim 0.21$) waveguide with a thickness of 0.8 μm. Each layer of QDs was overgrown by a thin $In_yGa_{1-y}As$ ($y \sim 0.41$) layer. The MM buffer consisted of a 1 μm thick InGaAs layer, followed by *n*-type cladding InAlGaAs layer (Al ~ 0.3) with a thickness of 1.5 μm.

Aging tests at 60 °C junction temperature (10 °C heat sink temperature) of several as-cleaved devices with a cavity length $L = 2$ mm were performed in the current-stabilized mode, see Fig. 21. The initial total cw output power was about 50 mW. During the first 20 hours lasers showed a burn-in effect, and the output power decreased by about 5%. After aging tests lasting >800 hours of cw operation at this power, the lasers showed nearly no degradation in the performance. Aging tests at 70 °C junction temperature (20 °C heat sink) of similar devices showed no measurable degradation after >200 hours.[28,29]

4. Nanophotonics

Novel applications based on nanophotonics using the all-epitaxial growth approach have been proposed recently. The combination of light refraction, reflection, and interference enables efficient light-wave engineering of tilted optical modes. Ultimate control over the beam quality and emission wavelength in planar waveguides became possible. In this section, we will discuss two device classes: photonic band crystal (PBC) lasers and tilted cavity lasers (TCLs).

- *Photonic band crystal laser*

The PBC laser, illustrated in Fig. 22, utilizes the filtering properties of multilayered media at a tilted angle of penetration with respect to the layer planes (longitudinal PBC effect).[30,31] The small but finite vertical component of the *k*-vector of the *in-plane* optical mode can be properly engineered to enable dramatic discrimination of the excited optical modes through leakage loss and the boundary conditions at the multilayer filter-substrate interface. As a result, ultrahigh performance lasers emitting in the 640–650 nm range having a vertical beam divergence (full width at half maximum, FWHM) of only ~8 degrees can be fabricated. We have demonstrated 20 W pulsed power at 85% differential efficiency in 100 μm wide devices.[32] Continuous wave single-mode operation in 4 μm wide stripes resulted in ~150 mW output power in devices without facet passivation, limited by catastrophic optical mirror damage (COMD).[33]

More recently, high power PBC lasers emitting at 850 nm have been fabricated.[34] Devices based on 2 μm wide stripe geometry with facet passivation demonstrated above 160 mW cw operation at 20 °C heat sink temperature, still limited by COMD. It is important to note that the broadening of the waveguide suppresses both lateral modes and beam filamentation, enabling ultrahigh-power single mode lasers.

Further reduction of the vertical far field is possible. We have realized devices emitting in the 920–940 nm range with a FWHM of the vertical far field as narrow as ~4 degrees, as shown in Fig. 22(c). Single-mode 5 μm wide and 1.5 mm long stripe geometry devices demonstrated >500 mW cw power with a round-shape beam (~4°x4°).

The longitudinal PBC concept can be applied in 2D laser arrays and VCSELs and, also, utilized in light emitting diodes.

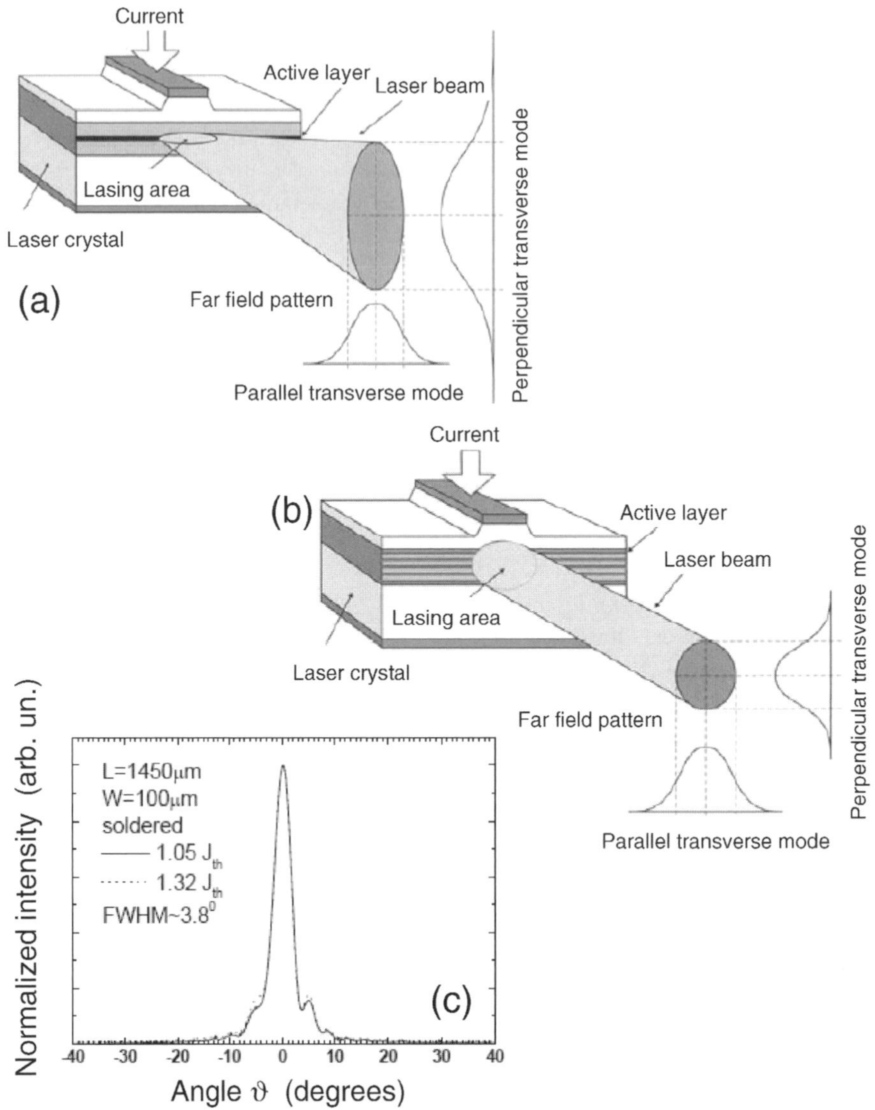

Figure 22. Conventional (a) and PBC (b) lasers. Note the multilayer waveguide region in (b), enabling stable spot size and a narrow far-field pattern. Vertical far field of the device emitting at 940 nm *vs.* angle ϑ is shown in (c).

- *Tilted cavity laser*

To address the problem of high-power wavelength-stabilized laser, the tilted cavity concept has been proposed.[30,35] A tilted cavity laser (TCL) is another realization of an all-epitaxial nanophotonic structure, where lasing occurs in a high-order vertical optical mode. The value of the lasing mode angle is intermediate between the mode angle of a typical edge-emitting laser and the mode angle of a typical VCSEL. Such laser can operate in both edge-emitting and surface-emitting geometries. A TCL operating in edge-emitting geometry combines advantages of an edge emitter (high power operation) and a surface emitter (wavelength stabilization at any requested power).

A typical edge-emitting TCL contains a photonic band crystal that plays the role of a multilayer interference reflector (MIR) and an optical defect localizing the optical mode. The defect plays a role of a high-finesse cavity. To illustrate the principles of operation of a TCL it is convenient first to separately examine the optical properties of two key elements:[30,35,36] the cavity, shown in Fig. 23(a), and the MIR, shown in Fig. 23(b).

Let us consider the reflectivity spectra of these two elements at oblique or tilted incidence of light, where we assume that light impinges on the element from a reference medium, and a semi-infinite GaAs substrate lies below. The cavity is basically a layer with a high refractive index sandwiched between two low-index layers. Figure 23(c) shows the reflectivity spectra of the cavity at three tilt angles ϑ. Each spectrum contains a dip that shifts quickly as a function of ϑ. Figure 23(d) depicts the reflectivity spectra of the MIR at the same three ϑ. Each spectrum contains a stopband, which shifts slowly as a function of ϑ.

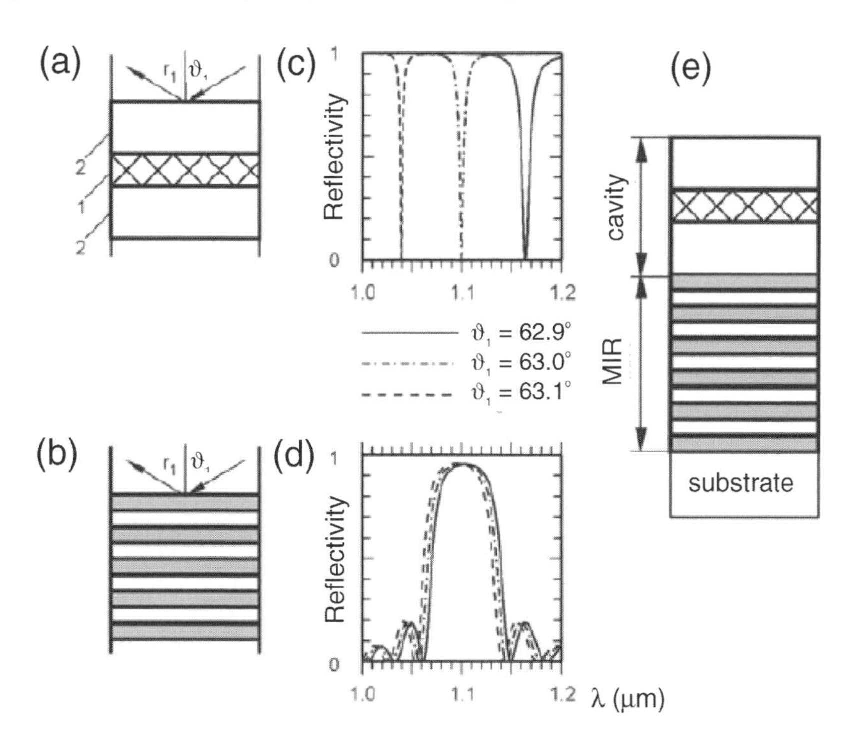

Figure 23. Principles of the wavelength selectivity in a tilted cavity laser (TCL): (a) cavity sandwiching a high refractive index layer between two low-index layers; (b) multilayer interference reflector (MIR); (c) reflectivity spectra of the cavity at three tilt angles ϑ; (d) reflectivity spectra of the MIR at the same three tilt angles ϑ; (e) schematic structure of a TCL.

Figure 24(a) shows the angular dependence of the two features of the tilted cavity structure, namely the reflectivity dip of the cavity and the stopband maximum of the MIR. Most important is that these two features shift at different rates as a function of tilt angle ϑ. Therefore the cavity and the MIR can be intentionally designed such that the spectral position of the cavity dip and the MIR stopband reflectivity maximum coincide only at a certain angle ϑ_0, and draw apart as the angle deviates from ϑ_0. Then the optical mode propagating at this angle ϑ_0 is strongly confined within the cavity and only weakly transmitted through the MIR. As the angle deviates from the optimum value, the wavelength of the optical mode shifts away from the MIR stopband reflectivity maximum, the transmission of the MIR increases and the optical mode starts leaking out of the structure, either into the substrate or into the top contact.

In other words, the optical mode of a tilted cavity structure at the optimum wavelength λ_0 should have a very low leakage loss, whereas the leakage loss will dramatically increase as the wavelength deviates from the optimum value. This promotes the wavelength selective operation of the laser. Moreover, the thermal shift of the optimum wavelength is governed by the temperature dependence of the refractive indices, like in VCSELs, rather than by the thermal shift of the electronic bandgap energy, like in conventional edge emitters. This feature of the device combines the advantages of the VCSEL and the edge emitter, resulting in a high-power wavelength-stabilized laser. The first experimental studies of the TCLs[36,37] have confirmed the wavelength stabilization and a reduced thermal shift of the lasing wavelength with respect to conventional edge emitters. In particular, a GaAs/GaAlAs TCL with strained GaInAs quantum wells operating around 980 nm exhibited a 0.2 nm/K thermal shift of the lasing wavelength, as opposed to the 0.4 nm/K value for conventional edge-emitting lasers.

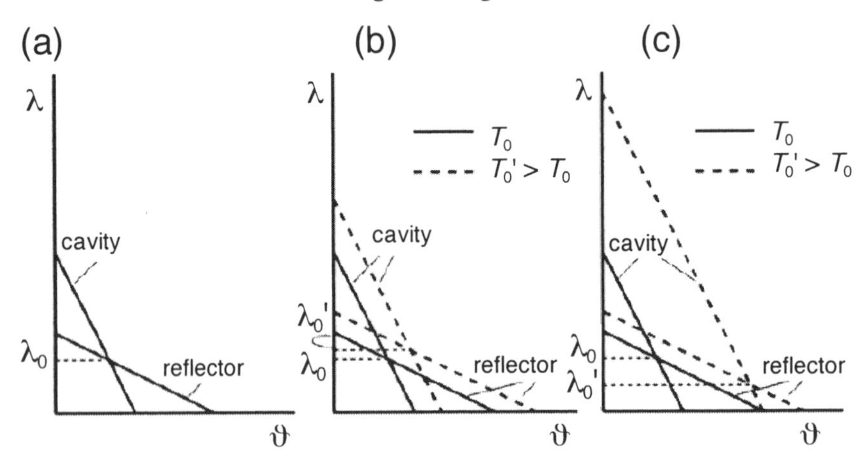

Figure 24. Optimum wavelength of a tilted cavity laser (TCL) and its thermal dependence. (a) Intersection of the angular dependences of the optical features of the cavity and MIR determines the lasing wavelength λ_0 of a TCL. A weak thermally-induced red shift of the cavity produces a red shift of λ_0 (b), whereas a strong red shift of the cavity produces a blue shift of λ_0 (c).

Figure 24(b) illustrates the origin of the thermal shift in the TCL optimum wavelength. As the refractive indices of semiconductor materials increase with temperature, all curves shift to longer wavelengths. The intersection curve in Fig. 24(b) shifts to larger angles and to longer wavelengths as well, as expected.

However, a possibility of controlling the thermal shift[34] is based on the strongly nonlinear behavior of the refractive index temperature coefficient dn/dT in semiconductor alloys. The temperature coefficient increases dramatically as the photon energy approaches resonance with the bandgap energy of the semiconductor. In GaAlAs alloys, the thermal coefficient dn/dT is especially high for GaAs. Then it is possible to replace, say, in one cavity layer, a homogeneous GaAlAs layer by a superlattice consisting of alternating layers of GaAlAs with a higher Al content and pure GaAs, such that the averaged refractive index at room temperature remains the same, and the temperature coefficient dn_{av}/dT is much larger. As a result, the curve (dip position *vs.* angle) for the cavity will shift at an elevated temperature to much longer wavelengths, *i.e.* will exhibit a much stronger red shift, as shown in Fig. 24(c). As the dependence of the stopband maximum reflectivity of the MIR *vs.* tilt angle remains the same, the intersection point will be shifted to larger angles. Such shift can eventually become so large, that the intersection at an elevated temperature will occur at a shorter wavelength $\lambda_0' < \lambda_0$. The corresponding superlattice in the cavity plays a role of the control element, allowing tuning of the thermal shift of λ_0.

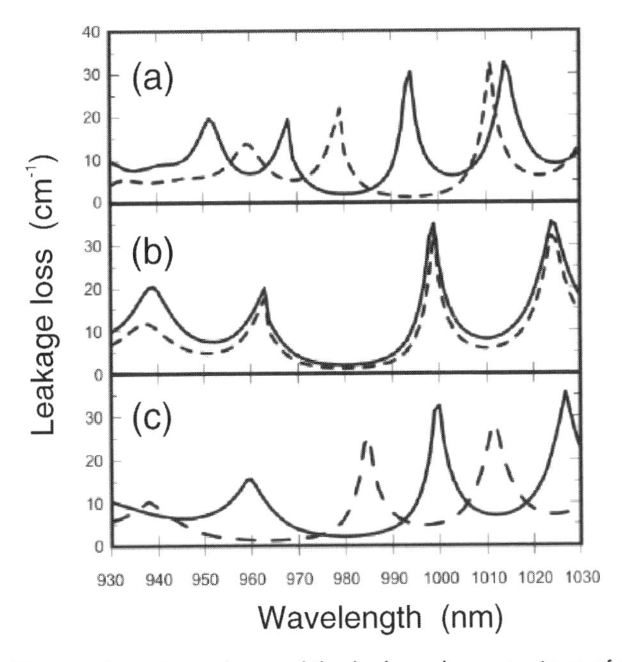

Figure 25. Temperature dependence of the leakage loss spectrum of a tilted cavity laser with three different realizations of the control element of the cavity at $T = 300$ (solid lines) and 400 K (dashed lines): (a) red shift of the loss minimum; (b) temperature independence of the loss minimum; (c) blue shift of the loss minimum.

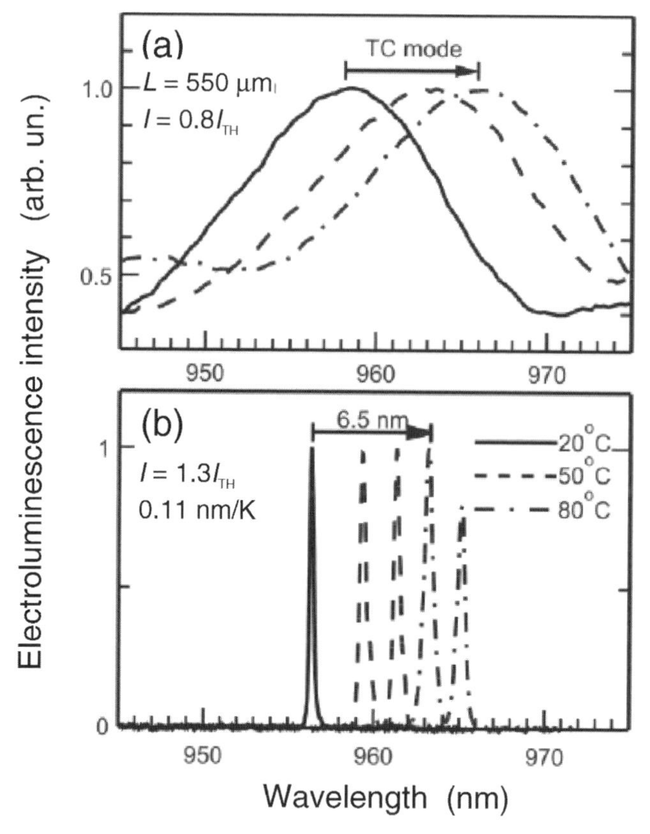

Figure 26. Thermal shift of the electroluminescence spectra (a) and lasing spectra (b) in a tilted cavity laser (TCL) using a GaAs/GaAlAs superlattice in the cavity.

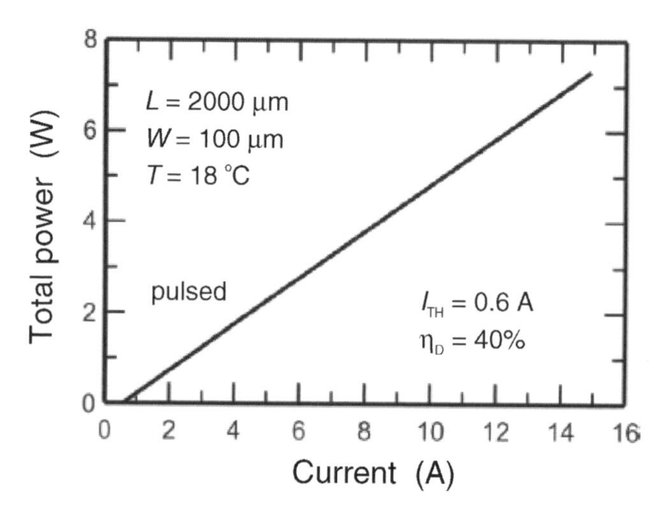

Figure 27. High-power operation of a tilted cavity laser (TCL) showing a reduced thermal shift (0.11–0.13 nm/K) of the lasing wavelength.

The cavity layers that can be replaced by various superlattices serve as control elements for the thermal shift of the lasing wavelength. Figure 25 shows how changing a control element in the same tilted cavity structure may yield a red shift of the optimum wavelength upon temperature increase, an absolute temperature stabilization, or a blue shift – see Figs. 25(a)–(c), respectively. It should be emphasized that the possibility of obtaining either a red or a blue shift of the lasing wavelength exists even if the refractive indices of all constituent layers increase with temperature. The possibility of absolute temperature stabilization of the lasing wavelength is unique compared to all known types of semiconductor lasers, including distributed feedback lasers and VCSELs.

A 980 nm tilted cavity laser with a GaAs/GaAlAs waveguide and strained GaInAs quantum well active region has been fabricated containing a superlattice as a part of the cavity.[36] Figure 26(a) shows a slow shift of the maximum of the subthreshold electroluminescence spectrum at a rate of 0.11 nm/K, whereas Fig. 26(b) illustrates the same slow shift of the lasing wavelength. Figure 27 illustrates the high-power operation of the TCL, limited by COMD, with the differential efficiency $\eta_D = 40\%$.

Theoretical results[34] predict that it should be possible to control the thermal shift of the lasing wavelength (for 980 nm TCL) over a broad interval, say, from – 0.15 nm/K to 0.2 nm/K. This means that absolute thermal stabilization of the lasing wavelength may be possible, opening new horizons for uncooled lasers in wavelength division multiplexing systems and wavelength-stabilized pumps.

Another approach of wavelength stabilization is based on the idea of coupling the waveguiding cavity to the transparent substrate.[38] Back-reflected emission undergoes constructive or destructive interference, as shown in Fig. 28, and for a

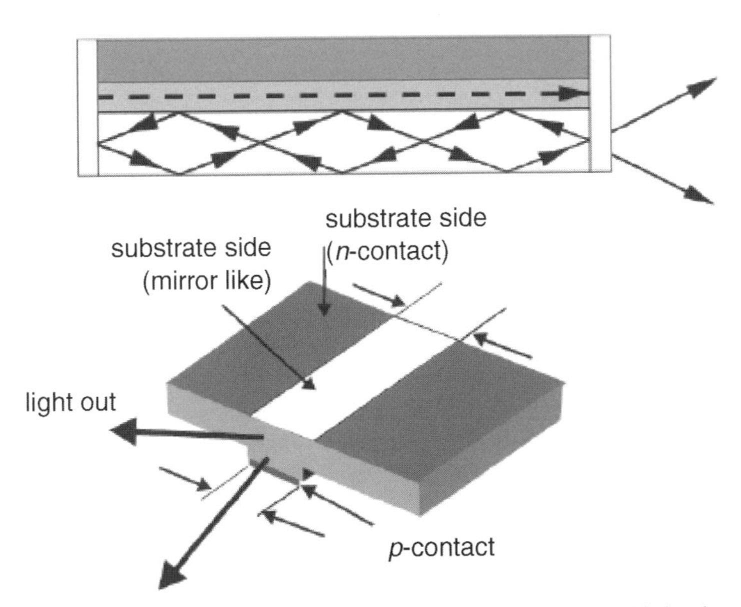

Figure 28. Schematic representation of the tilted cavity laser with back reflection from the substrate.

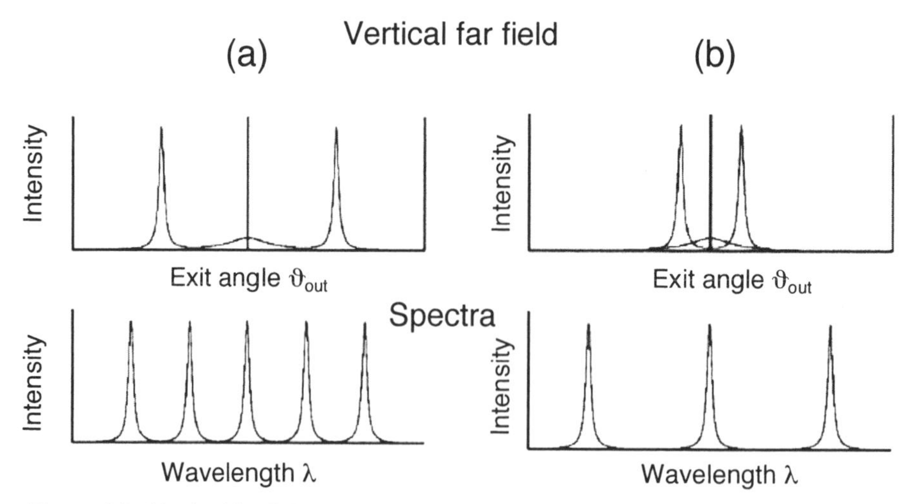

Figure 29. Vertical far field patterns and mode structure in the TCL with substrate reflection. Larger (a) and smaller (b) tilt angles are illustrated.

given angle of penetration in the substrate only certain resonant cavity wavelengths are possible. The smaller the angle of the light in the substrate, the larger the angular spacing between the modes, as shown in Fig. 29. Dielectric and metal multilayers, as well as lateral patterning, may be used to tune the wavelength, improve wavelength stabilization, and realize multi-wavelength on-chip arrays.

5. Novel functionality

Epitaxial nanophotonics offers a unique advantage of adding novel functionality to optoelectronic devices. Quantum size effects dramatically modify the optical properties of various optoelectronic media, *e.g.* enhancing the electrooptic effect by many orders of magnitude. Optical confinement and interference effects in multilayer structures also result in a strong modification of optical properties. When combined, these approaches result in qualitatively new opportunities. For example, by using an electrooptically-modulated (EOM) filter section it becomes possible to achieve ultrahigh-speed intensity modulation in VCSELs.[39] The EOM device utilizes resonant interaction between two cavities and can be realized in a surface-emitting geometry or as an edge emitter. The device contains a refractive-index-tunable element controlled by an applied voltage. Electrooptic modulation is achieved by using the quantum-confined Stark effect in narrow quantum wells and the related modulation of the refractive index. In this scenario, the electrooptic effect is quadratic, because the refractive index change is proportional to the square

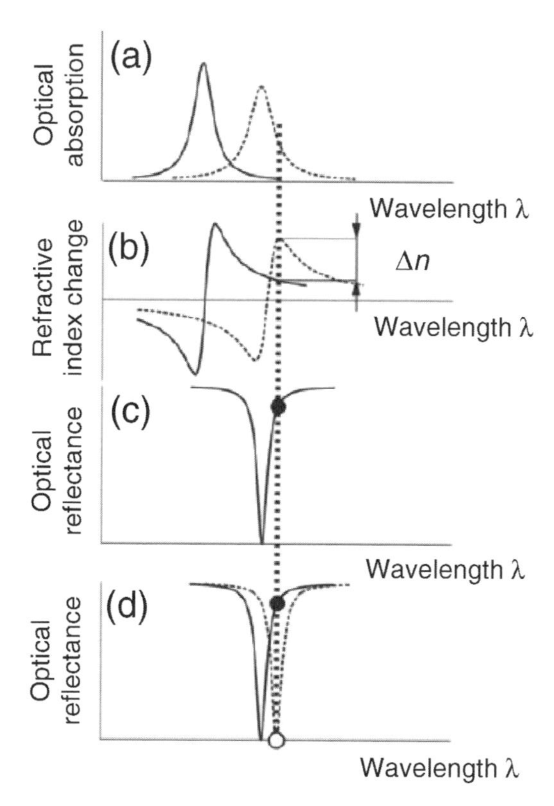

Fig. 30. Electrooptic effect in a vertical cavity. Application of reverse bias (a, b) causes the cavity mode to match the lasing wavelength (c, d) increasing the light output.

of the applied electric field, while being inversely proportional to the energy separation between the photon energy and the resonant absorber energy. A schematic illustration of the operating principle, wherein the modulation is realized by the tuning of the resonant wavelength of the "filter" cavity, is shown in Fig. 30.

Once the refractive index is tuned, the cavity mode of the EOM filter cavity can be shifted towards the resonance with the VCSEL cavity – see Fig. 30(d). The efficiency of light extraction is increased in this case. If the quality factor of the filtering cavity is low enough, no splitting of the cavity modes occurs and only the field intensity at the exit is modulated.

The device contains three electrical contacts to apply laser drive current and reverse modulator bias, as shown schematically in Fig. 31(a). When the bias is applied to the modulator section, the wavelength of the filter section mode is tuned to match the VCSEL cavity mode.

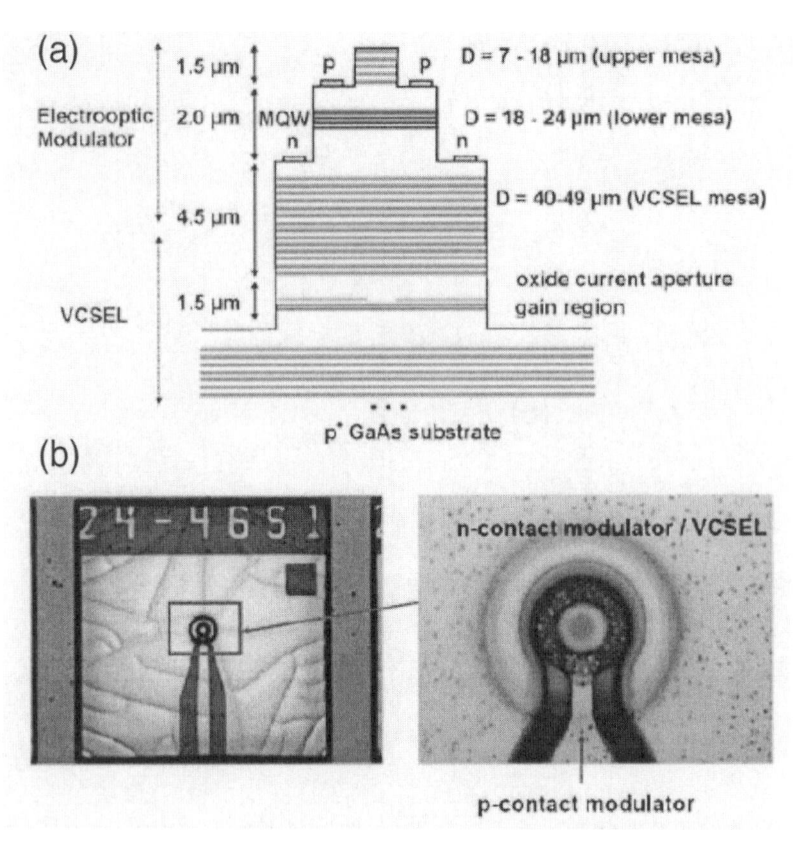

Figure 31. Schematic design (a) and optical microscope image (b) of the EOM VCSEL device.

The EOM VCSEL heterostructure was grown by MBE on a GaAs(100) *p*-doped substrate.[39] The *p*- and *n*-doped distributed Bragg reflectors were composed of $Al_{0.9}Ga_{0.1}As$ and GaAs layers. The VCSEL and the modulator cavity were designed to operate at ~960 nm with a blue shift of the modulator cavity by ~3–4 nm. The detuning of the cavities was measured by optical reflectance and photo-reflectance spectroscopy and varied across the wafer. Strain-compensated multiple quantum wells (MQWs) were used to provide VCSEL gain at ~960 nm. Strain-compensated MQWs with a strong absorption peak at ~930 nm were introduced in the modulator section. Due to growth nonuniformity, the detuning between the cavities varied from 2 to ~10 nm, but this detuning could be compensated by applying a reverse bias in the 2–20 V range.

The sizes of the modulator and VCSEL mesas were chosen to be compatible with >40 Gb/s modulation frequency. BCB planarization was employed for low capacitance contact pads. Wet thermal oxidation of the VCSEL aperture was performed at 400 °C. Optical microscope images of the device are shown in Fig. 31(b).

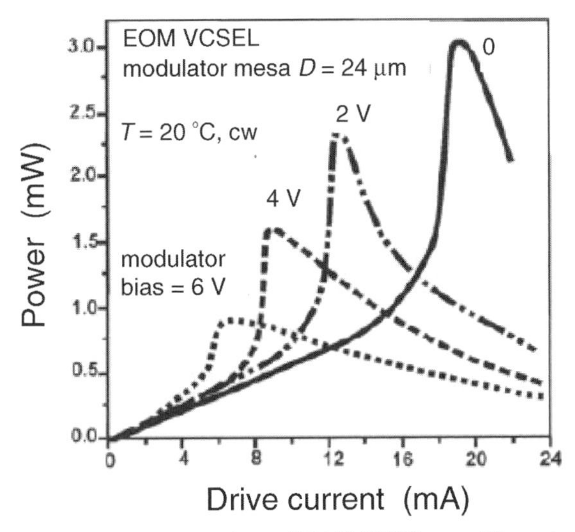

Figure 32. Static performance of an EOM/VCSEL at different reverse biases applied to the modulator section.

The static performance of on-wafer devices was evaluated by measuring light-current *L–I* curves at different modulation bias voltages, as well the emission spectra. Device characteristics are strongly affected by the detuning of the two resonant cavities. Figure 32 shows a typical static modulation performance of an EOM VCSEL device in the case of small detuning between the VCSEL and the modulator cavity. At zero bias, the *L–I* dependence shows a strong resonant feature. This effect originates from thermal shifting of the QW exciton absorption peak in the modulator section with cw current, and the related tuning of the modulator cavity mode. When the reverse bias is applied to the modulator section, there is an additional contribution of the refractive index modulation due to the EOM effect, which shifts the resonance feature towards smaller currents. Thus, we were able to achieve a 2.6 dB/V intensity modulation at very small bias voltages. The devices having a larger detuning between the cavities did not demonstrate such a feature and the application of reverse bias resulted in a modulation of the effective differential efficiency.[39]

To assess the high speed performance of the devices, optical modulation studies and S_{21} impedance measurements were performed under different bias voltages. Devices of similar geometry exhibited remarkably consistent characteristics. The estimated electrical modulation bandwidth improved as the modulator mesa siz decreased, from ~33 GHz for 24 μm mesas to ~59 GHz for 18 μm mesas, scaling with the modulator surface area – see Fig. 33.

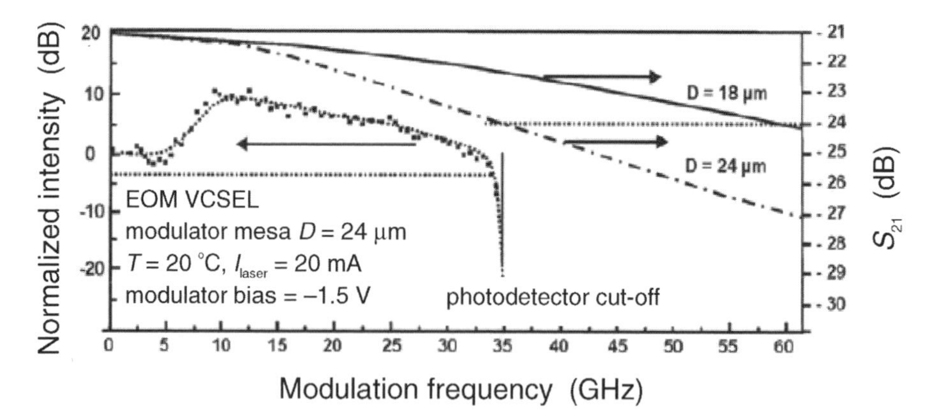

Figure 33. Optical and electrical modulation performance of the EOM/VCSEL, showing the dependence of electrical bandwidth on the modulator mesa diameter *D*. The speed scales with the surface of the modulator mesa, which defines the capacitance. The maximum optical bandwidth is system limited at 35 GHz.

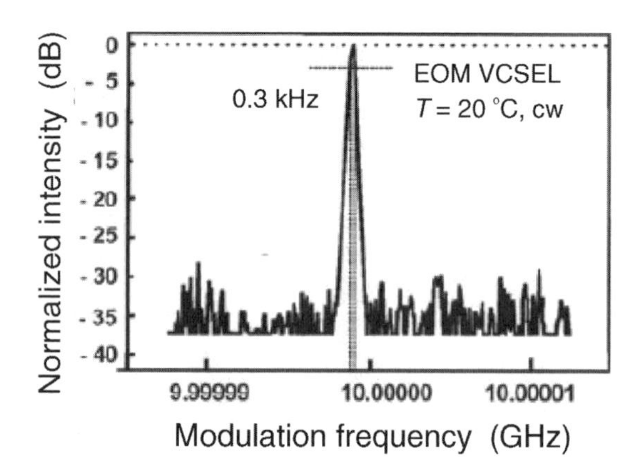

Figure 34. Optical response of the EOM VCSEL under 10 GHz electrical modulation.

Our results thus demonstrate the high-speed potential of these devices and confirms the compatibility of the fabrication technology with > 40 Gb/s modulation regime. We note that the performance is still capacitance-limited even at ~60 GHz. As the active part of the modulator interacting with the oxide-confined VCSEL mode(s) can be restricted down to a few μm, moderate modification of the processing, such as ion implantation into the passive modulator regions, can easily enhance the bandwidth to ~200 GHz or even more. The optical modulation performance of these devices was studied in the 0–40 GHz frequency range. For this on-wafer test, the electrical modulation signal was coupled to the device through a high-

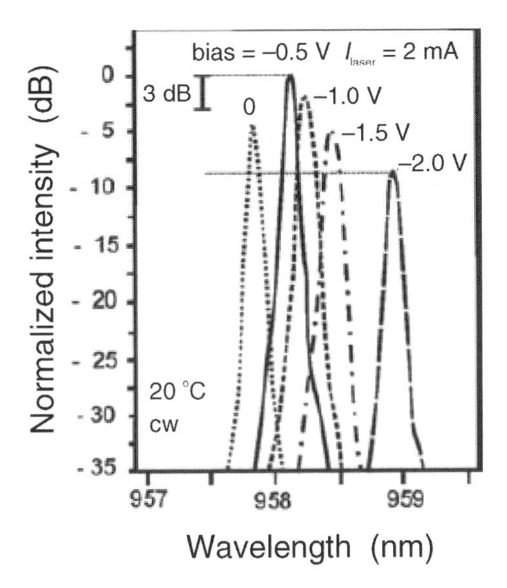

Figure 35. Intensity and spectral changes in the single mode EOM VCSEL device.

frequency contact head. Single mode *pin* diodes having a low sensitivity in the 960–980 nm range were used due to the lack of high-speed multimode 980 nm photodetectors. As shown in Fig. 33, the modulation bandwidth exceeded 35 GHz and was limited by the cut-off frequency of the photodetector.

In Fig. 34 we show an optical response of the detector system when the EOM VCSEL device is operated at 10 GHz. A ~35 dB suppression of parasitic frequencies and a quasi-ideal (<1 kHz) width of the optical response signal is observed. Finally, in Fig. 35 we show spectral evolution in a single mode EOM VCSEL with an oxide aperture size of ~1 μm, where a reverse voltage of 20 V was applied to the modulator section. The intensity is modulated by applying additional bias voltages, as indicated in Fig. 35. A side-mode suppression ratio of >35 dB is obtained. The modulation voltage efficiency factor is up to ~7–10 dB/V. In the vicinity of the resonant modulator voltage, the wavelength chirp is ~0.1 nm/V. Depending on the system application requirements, the design may be adjusted to meet the related specifications imposed by the application.

6. Conclusions

Epitaxial nanophotonics is presently targeting the mainstream applications in optoelectronics.

- *Self-organized growth: Quantum dots and quantum wires*

To date, the key practical applications of self-organized nanostructures are in InGaN light emitting diodes and lasers, where carrier localization in quantum dots enables degradation robustness and spectral control. Further migration of conventional III-V lasers towards higher powers, higher direct modulation speeds, lower optical nonlinearities and better degradation robustness will necessarily require using of QDs for all spectral ranges. Significant market penetration of epitaxial nanostructures is taking place in (In,Ga,Al)P red lasers used for DVD recorders, grown on corrugated surfaces formed on misoriented substrates. Formation of both natural superlattices and the related corrugated GaInP active regions is observed. Effective lateral confinement can be further enhanced by the alloy phase separation at the facets.

- *Self-organized growth: Defect blocking*

Threading dislocations produce significant strain, which can be used to mark the related regions in the vicinity of the dislocation termination at the epitaxial surface. This strain may give rise to numerous effects, such as decoration of the dislocation (in case of the alloy layer growth) or its interaction with self-organized nanoislands. Strain-related effects enable selective capping of the dislocation-related areas, with subsequent etching of trenches and strain-driven overgrowth. Depending on the type of the dislocation, either complete removal or tilting of the dislocation may be engineered in a controllable way. This enables heterogeneous material integration. The first ever robust high-power lasers on metamorphic substrates[28,29] were recently demonstrated using defect reduction. Given sufficient industrial R&D, it may become possible to grown defect-free III-V or SiGe layers on Si or Ge substrates. The III-V integration with Si is of interest in silicon technology to reduce the heat dissipation in high-speed processors. The need in material integration to realize high-efficiency multi-junction solar cells may be another driving force to further develop the DR technology.

- *Epitaxial nanophotonics*

Novel all-epitaxial approaches promise wavelength-stabilized high-power lasers and the devices with ultranarrow beam divergence. Furthermore, ultanarrow beam operation may be combined with complete wavelength stabilization by using the simplest epitaxial design of back-reflection from transparent substrates. At an advanced stage, light mode velocity engineering, optical nonlinearity, optical bistability and other effects may be engineered.

- *Novel functionality*

Epitaxial nanophotonics extends the opportunities towards novel functionality. Coupling of the laser cavity to electrooptically-tunable elements may provide ultrahigh-speed wavelength or intensity modulation. An even stronger effect may be realized when electrooptic tuning of the stopband edge of the multilayer interference reflector is applied. The immediate goal of these concepts is to increase dramatically the operating bandwidth and realize cost-effective on-chip arrays of wavelength-tunable devices. Dramatic increase in data traffic (presently at ~250-fold per decade) and the need to comply with ever accelerating productivity of modern computers demands rapid development of ultrahigh-speed low-cost data links. It is widely believed that at speeds beyond 20 Gb/s VCSEL-based optical interconnects will replace conventional copper links.

Acknowledgments

We appreciate collaboration with G. Fiol, F. Hopfer, T. Kettler, M. Kuntz, A. Mutig, and K. Posilovic (Technische Universität Berlin, Institut für Festkörperphysik, Berlin, Germany), Zh. I. Alferov, N. Yu. Gordeev, L. Ya. Karachinsky, M. B. Lifshits, M. V. Maximov, I. I. Novikov, M. Yu. Shernyakov, A. E. Zhukov, and V. M. Ustinov (A. F. Ioffe Physical-Technical Institute, St. Petersburg, Russia), P. Werner and N. Zakharov (Max-Planck Institute for Microstructure Physics, Halle, Germany), H. Krimse and W. Neumann (Humboldt-Universität Berlin, Institut für Physik, Berlin, Germany), A. R. Kovsh (NL-Nanosemiconductor, Germany), N. Grote, H. J. Hensel, H. Klein, W. D. Molzow, and A. Paraskevopoulos (Fraunhofer-Institute for Telecommunications Heinrich-Hertz-Institut, Berlin, Germany). Continuous support from BMBF, DFG, DLR (Germany) and the Russian Foundation for Basic Research is gratefully acknowledged. N.N.L. is a Mercator Professor (DFG).

References

1. V. A. Shchukin, N. N. Ledentsov, and D. Bimberg, *Epitaxy of Nanostructures*, Berlin: Springer, 2003.
2. D. Bimberg, M. Grundmann, and N. N. Ledentsov, *Quantum Dot Heterostructures*, Chichester, UK: Wiley, 1999.
3. N. N. Ledentsov, V. M. Ustinov, A. Yu. Egorov, A. E. Zhukov, M. V. Maximov, I. G. Tabatadze, and P. S. Kop'ev, "Optical properties of heterostructures with InGaAs-GaAs quantum clusters," *Semicond.* **28**, 832 (1994).
4. V. A. Shchukin, A. I. Borovkov, N. N. Ledentsov, and D. Bimberg, "Tuning and breakdown of faceting under externally applied stress," *Phys. Rev. B* **51**, 10104 (1995).
5. R. Nötzel, N. N. Ledentsov, L. Däweritz, K. Ploog, and M. Hohenstein, "Semiconductor quantum wire structures directly grown on high-index surfaces," *Phys. Rev. B* **45**, 3507 (1992).
6. R. Noetzel, N. N. Ledentsov, L. Daeweritz, and K. Ploog, "Method of fabricating a compositional semiconductor device," U.S. Patent 5,714,765 (1998).
7. S. Nakamura, M. Senoh, S. Nagahama, *et al.*, "Subband emissions of InGaN multi-quantum-well laser diodes under room-temperature continuous wave operation," *Appl. Phys. Lett.* **70**, 2753 (1997).
8. N. N. Ledentsov "Vertical-cavity surface-emitting lasers using InGaN quantum dots," *Compound Semicond.* **5**, 61 (1999).

9. R. Seguin, S. Rodt, A. Strittmatter, *et al.*, "Multi-excitonic complexes in single InGaN quantum dots," *Appl. Phys. Lett.* **84**, 2753 (2004).
10. I. L. Krestnikov, N. N. Ledentsov, A. Hoffmann, *et al.*, "Quantum dot origin of luminescence in InGaN-GaN structures," *Phys. Rev. B* **66**, 155310 (2002).
11. I. Krestnikov, D. Livshits, S. Mikhrin, A. Kozhukhov, A. Kovsh, N. N. Ledentsov, and A. Zhukov, "Reliability study of InAs-InGaAs quantum dot diode lasers," *Electronics Lett.* **41**,1330 (2005).
12. N. N. Ledentsov, A. R. Kovsh, V. A. Shchukin, *et al.*, "1.3–1.5 μm quantum dot lasers on foreign substrates: Growth using defect reduction technique, high-power cw operation, and degradation resistance," *Proc. SPIE* **6133**, 61330S (2006).
13. F. Hopfer, A. Mutig, G. Fiol, *et al.*, "20-Gb/s direct modulation of 980 nm VCSELs based on submonolayer deposition of quantum dots" *Proc. SPIE* **6350**, 6 (2006).
14. N. N. Ledentsov and V. Shchukin, "Optoelectronic device based on an antiwaveguiding cavity," U.S. Patent application 2005/0226294, published October 13, 2005.
15. L. Goldstein, F. Glas, J. Y. Marzin, M. N. Charasse, and G. Leroux, "Growth by molecular-beam epitaxy and characterization of InAs/GaAs strained-layer superlattices," *Appl. Phys. Lett.* **47**, 1099 (1985).
16. F. Glas, C. Guille, P. Hénoc, and F. Houzay "TEM study of the molecular beam epitaxy island growth of InAs on GaAs" in: A. G. Cullis and P. D. Augustus, eds., *Proc. Microscopy Semiconducting Mater.* (1987), Oxford: Institute of Physics, 1987.
17. O. Brandt, M. Ilg, and K. Ploog, "Coherently strained InAs insertions in GaAs: Do they form quantum wires and dots?" *Microelectronics J.* **26**, 861 (1995).
18. N. N. Ledentsov, M. Grundmann, N. Kirstaedter, *et al.*, "Ordered arrays of quantum dots: Formation, electronic spectra, relaxation phenomena, lasing," *Solid State Electronics* **40**, 785 (1996).
19. N. N. Ledentsov, V. A. Shchukin, D. Bimberg, *et al.*, "Reversibility of the island shape, volume, and density in Stranski-Krastanow growth," *Semicond. Sci. Technol.* **16**, 502 (2001).
20. N. N. Ledentsov, M. V. Maximov, D. Bimberg, *et al.*, "1.3 μm luminescence and gain from defect-free InGaAs-GaAs quantum dots grown by metal-organic chemical vapour deposition," *Semicond. Sci. Technol.* **15**, 604 (2000).
21. D. S. Sizov, M. V. Maksimov, A. F. Tsatsul'nikov, *et al.*, "The influence of heat treatment conditions on the evaporation of defect regions in structures with InGaAs quantum dots in the GaAs matrix," *Semicond.* **36**, 1020 (2002).
22. N. N. Ledentsov, "Semiconductor device and method of making same," U.S. Patent 6,653,166 (2003).
23. V. A. Shchukin and N. N. Ledentsov, "Defect-free semiconductor templates for epitaxial growth and method of making same," U.S. Patent 6,784,074 (2004).

24. N. N. Ledentsov and D. Bimberg, "Growth of self-organized quantum dots for optoelectronic applications: Nanostructures, nanoepitaxy, defect engineering," *J.* Cryst. Growth **255**, 68 (2003).
25. N. N. Ledentsov, A. R. Kovsh, V. A. Shchukin, *et al.*, "QD lasers: Physics and applications," *Proc. SPIE* **5624**, 335 (2005).
26. A. Lenz, H. Eisele, R. Timm, *et al.*, "Nanovoids in InGaAs/GaAs quantum dots observed by cross-sectional scanning tunneling microscopy," *Appl. Phys. Lett.* **85**, 3848 (2004).
27. N. N. Ledentsov, A. R. Kovsh, A. E. Zhukov, *et al.*, "High performance quantum dot lasers on GaAs substrates operating in 1.5 μm range," *Electronics Lett.* **39**, 1126 (2003).
28. T. Kettler, L. Ya. Karachinsky, N. N. Ledentsov, *et al.*, "Degradation-robust single mode continuous wave operation of 1.46 μm metamorphic quantum dot lasers on GaAs substrate," *Appl. Phys. Lett.* **89**, 041113 (2006).
29. L. Ya. Karachinsky, T. Kettler, N. Yu. Gordeev, *et al.*, "High-power single mode cw operation of 1.5 μm-range quantum dot GaAs-based laser," *Electronics Lett.* **41**, 478 (2005).
30. N. N.Ledentsov and V. A. Shchukin, "Novel concepts for injection lasers," *SPIE Optical Eng.* **41**, 3193 (2002).
31. V. A. Shchukin and N. N. Ledentsov, "Semiconductor laser based on the effect of photonic band gap crystal-mediated filtration of higher modes of laser radiation and method of making same," U.S. Patent 6,804,280 B2 (2004).
32. M. V.Maximov, Yu. M. Shernyakov, I. I. Novikov, *et al.*, "High power GaInP/AlGaInP visible lasers ($\lambda = 646$ nm) with narrow circular shaped far field pattern," *Electronics Lett.* **41**, 741 (2005).
33. I. I. Novikov, L. Ya. Karachinsky, M. V. Maximov, *et al.*, "Single mode cw operation of 658 nm AlGaInP lasers based on longitudinal photonic band gap crystal," *Appl. Phys. Lett.* **88**, 231108 (2006).
34. V. A. Shchukin, N. N. Ledentsov, N. Yu. Gordeev, *et al.*, "High brilliance photonic band crystal lasers," *Proc. SPIE* **6350**, 635005 (2006).
35. N. N. Ledentsov and V. A. Shchukin, "Tilted cavity semiconductor laser (TCSL) and method of making same," U.S. Patent 7,031,360 (2006).
36. N. N. Ledentsov, V. A. Shchukin, S. S. Mikhrin, *et al.*, "Wavelength-stabilized tilted cavity quantum dot laser," *Semicond. Sci. Technol.* **19**, 1183 (2004).
37. N. N. Ledentsov, V. A. Shchukin, A. R. Kovsh, *et al.*, "Edge and surface-emitting tilted cavity lasers," *Proc. SPIE* **5722**, 130 (2005).
38. N. N. Ledentsov and V. A. Shchukin, to be published.
39. A. Paraskevopoulos, H.-J. Hensel, W.-D. Molzow, *et al.*, "Ultra-high-bandwidth (>35 GHz) electrooptically-modulated VCSEL," paper PDP22 presented at *OFC/NFOEC 2006*, Anaheim, CA (March, 2006).

Quantum Control of the Dynamics of a Semiconductor Quantum Well

E. Paspalakis, M. Tsaousidou
Materials Science Department, University of Patras, Patras 26504, Greece

A. F. Terzis
Department of Physics, University of Patras, Patras 26504, Greece

1. Introduction

In recent years, the potential for coherent control of the electron dynamics in semiconductor quantum wells (QWs) that interact with strong oscillating electric fields has been a topic of increasing interest. In many studies the description of the intersubband transition dynamics has been based on atomic-like multi-level theoretical approaches. However, in a quantum well the many-body effects due to the macroscopic carrier density play a significant role and make the system behave quite differently from a simple noninteracting atomic-like system. These many-body effects have been studied in a significant number of theoretical papers,[1–14] where it has been shown that the optical response and the electron dynamics of QWs can be dramatically influenced by varying the carrier density.

In the present chapter, we examine the intersubband transition dynamics of a symmetric double quantum well (DQW), where the ground and the first excited states are coupled by a strong pulsed or a continuous wave (cw) electric field. The system dynamics can be described by the effective nonlinear Bloch equations (NBEs) derived recently by Olaya-Castro *et al.*[11] The nonlinearities in these equations arise from the electron-electron interactions. The NBEs are simplified by using the rotating wave approximation (RWA) and analytical solutions are obtained under specific conditions. The validity of the RWA is tested by comparing the analytical expressions with numerical results for a realistic DQW based on GaAs.

2. Effective nonlinear Bloch equations and the rotating wave approximation

We consider a symmetric AlGaAs/GaAs/AlGaAs DQW and assume that only the lowest two electron subbands contribute to the system dynamics. The QW interacts with a time-dependent electric field $E(t)=E_0 f(t)\sin[\omega t + \varphi(t)]$, where E_0 is the electric field amplitude, $f(t)$ is the dimensionless pulse envelope (for cw fields $f(t) = 1$), ω is the angular frequency, and $\varphi(t)$ is the time-dependent phase of the

 Future Trends in Microelectronics. Edited by Serge Luryi, Jimmy Xu, and Alex Zaslavsky

field. As shown by Olaya-Castro *et al.*,[11] the system dynamics can be described by the following effective NBE:

$$dS_1(t)/dt = [\omega_{10} - \gamma S_3(t)]S_2(t) - S_1(t)/T_2 \,, \tag{1a}$$

$$dS_2(t)/dt = -[\omega_{10} - \gamma S_3(t)]S_1(t) + 2[(\mu E(t)/\hbar) - \beta S_1(t)]S_3(t) - S_2(t)/T_2 \,, \tag{1b}$$

$$dS_3(t)/dt = -2[(\mu E(t)/\hbar) - \beta S_1(t)]S_2(t) - [S_3(t)+1]/T_1 \,, \tag{1c}$$

where $S_1(t)$ and $S_2(t)$ are, respectively, the mean real and imaginary parts of the polarization, while $S_3(t)$ is the mean population inversion per electron (difference of the occupation probabilities in the upper and lower subbands); μ is the electric dipole matrix element between the two subbands; and $\omega_{10} = \alpha + (\varepsilon_1 - \varepsilon_0)/\hbar$, where ε_0 and ε_1 are the energy eigenvalues of the ground and excited states in the QW, respectively. The parameters α, β and γ depend on the electron sheet density N, the dielectric constant of the host material, and the form factors L_{ijkl} with $i, j, k, l = 0$ or 1. Full expressions for these parameters are given in Ref. 11. Finally, T_1 and T_2 are, respectively, the population decay time and the dephasing time.

We can write Eqs. (1a)–(1c) in a more convenient form by introducing the variables $U(t)$, $V(t)$ and $Z(t)$ that are related to $S_1(t)$, $S_2(t)$ and $S_3(t)$ by the following linear equations:

$$S_1(t) = U(t)\sin[\omega t + \varphi(t)] - V(t)\cos[\omega t + \varphi(t)] \,, \tag{2a}$$

$$S_2(t) = U(t)\cos[\omega t + \varphi(t)] + V(t)\sin[\omega t + \varphi(t)] \,, \tag{2b}$$

$$S_3(t) = Z(t) \,. \tag{2c}$$

We now substitute the above set of equations into Eq. (1) and we employ the RWA, according to which the terms containing $\sin[2\omega t + 2\varphi(t)]$ and $\cos[2\omega t + 2\varphi(t)]$ are neglected. Then, we obtain

$$dU(t)/dt = [\delta - (d\varphi/dt) - (\gamma-\beta)Z(t)]V(t) - U(t)/T_2 \,, \tag{3a}$$

$$dV(t)/dt = -[\delta - (d\varphi/dt) - (\gamma-\beta)Z(t)]U(t) - \Omega_0 f(t)Z(t) - V(t)/T_2 \,, \tag{3b}$$

$$dZ(t)/dt = \Omega_0 f(t)V(t) - [Z(t)+1]/T_1 \,, \tag{3c}$$

where $\Omega_0 = -\mu E_0/\hbar$ is the Rabi frequency and $\delta = \omega_{10} - \omega$ is the detuning from resonance.

Analytical solutions for the nonlinear Eqs. (3a)–(3c) can be obtained for specific cases of pulsed and cw electric fields in the absence of relaxation processes ($T_1 \to \infty$ and $T_2 \to \infty$). In what follows, we assume that initially the ground subband is occupied, while the first excited subband is empty. Thus, the initial conditions for the system are $S_1(0) = S_2(0) = 0$, $S_3(0) = -1$ or, equivalently, $U(0) = V(0) = 0$, $Z(0) = -1$.

3. Analytical solutions of the nonlinear Bloch equations under the rotating wave approximation

- *Pulsed fields*

We neglect the last terms on the right hand side of Eqs. (3a)–(3c) associated with the relaxation processes. The system is excited at exact resonance, $\omega_{10} = \omega$, by a pulsed electric field of hyperbolic secant form, so $f(t) = \text{sech}[(t-t_0)/t_p]$. The characteristic times t_0 and t_p denote the center and the width of the pulse, respectively. The values for t_0 and t_p have been chosen such that the electric pulse is practically zero at $t = 0$ and $t = 2t_0$.

It can be shown[15,16] that if the phase $\varphi(t)$ is time-independent, and the Rabi frequency is taken to be $\Omega_0 = (d_p^2+1)^{1/2}/t_p$, where $d_p = (\gamma-\beta)t_p$, then the solution of the Eqs. (3a)–(3c) is given by

$$U(t) = [d_p/(d_p^2+1)^{1/2}]\,\text{sech}[(t-t_0)/t_p]\,, \tag{4a}$$

$$V(t) = [1/(d_p^2+1)^{1/2}]\,\text{sech}[(t-t_0)/t_p]\,, \tag{4b}$$

$$Z(t) = \tanh[(t-t_0)/t_p]\,. \tag{4c}$$

Inspection of Eq. (4c) shows that at $t = 2t_0$ all the electrons are transferred to the upper subband and complete inversion of the system is achieved ($Z(t) \to 1$). The analytical solution is shown in Fig. 1 (solid curve).

Complete inversion of the system at $t = 2t_0$ occurs also for the case where $\mathrm{d}\varphi/\mathrm{d}t = (\beta-\gamma)\tanh[(t-t_0)/t_p]$ and $\Omega_0 = 1/t_p$. In this case, the solution of the nonlinear equations (3a)–(3c) is[15]

$$U(t) = 0\,, \tag{5a}$$

$$V(t) = \text{sech}[(t-t_0)/t_p]\,, \tag{5b}$$

$$Z(t) = \tanh[(t-t_0)/t_p]\,. \tag{5c}$$

- *cw fields*

In the absence of relaxation processes, the system of the nonlinear equations (3a)–(3c) can be solved analytically for two different values of detuning: i) $\delta = 0$ and ii) $\delta = \beta-\gamma$. The phase of the field is assumed to be time-independent. Our results are based on methods used previously in problems of nonlinear optical waveguide directional couplers[17,18] and molecular dimers.[19,20]

For the first case, where the two-subband system is excited at exact resonance, the population inversion $Z(t)$ satisfies the following second-order differential equation:[21]

$$2\,\mathrm{d}^2Z(t)/\mathrm{d}t^2 = [(\gamma-\beta)^2 - 2\Omega_0^2]\,Z(t) - (\gamma-\beta)^2\,Z^3(t)\,. \tag{6}$$

By using the conditions that the population inversion and its derivative are, respectively, −1 and 0 at t = 0, the solution of Eq. (6) is an elliptic Jacobi function[17,19,20]

$$Z(t) = -\mathrm{cn}(\Omega_0 t|k) \tag{7}$$

where $k = (\gamma-\beta)^2/(4\Omega_0^2)$ is the parameter of the Jacobi function.

According to the above analytical result, if $\Omega_0 > (\gamma-\beta)/2$ the electron population oscillates between the two subbands and complete transfer of electrons to the upper subband can occur. For very large values of the Rabi frequency, $\Omega_0 >> (\gamma-\beta)/2$, the inversion can be approximated by $Z(t) = -\cos(\Omega_0 t)$ and the period of the Rabi oscillations can be approximated by the atomic Rabi oscillation period, $T \approx 2\pi/\Omega_0$. Moreover, if $\Omega_0 < (\gamma-\beta)/2$, the electrons still oscillate between the two subbands but without complete inversion and with more population on average in the lower subband. Finally, if $\Omega_0 = (\gamma-\beta)/2$, then $Z(t) = -\mathrm{sech}(\Omega_0 t)$, so no Rabi oscillations are expected and after an initial transient period the electron population is equally distributed in the two subbands. We note that the analytical expression in the region $\Omega_0 \approx (\gamma-\beta)/2$ is in strong disagreement with the numerical results.[21] Explicit expressions for the period of the Rabi oscillations in the regimes $\Omega_0 > (\gamma-\beta)/2$ and $\Omega_0 < (\gamma-\beta)/2$ are given in Ref. 21.

We now turn to the case where the detuning is $\delta = \beta-\gamma$. The manipulation of Eqs. (3a)–(3c) leads to the following differential equation for the inversion[21]

$$2\, d^2Z(t)/dt^2 = -[3(\gamma-\beta)^2 + 2\Omega_0^2]\, Z(t) - 3(\gamma-\beta)^2\, Z^2(t) - (\gamma-\beta)^2\, Z^3(t) - (\gamma-\beta)^2. \tag{8}$$

By using the same initial conditions as above, we obtain the solution[18]

$$Z(t) = -[(A+By_1)\mathrm{cn}(ht|k^2) + A-By_1]/[(A-B)\mathrm{cn}(ht|k^2) + A+B], \tag{9}$$

where $k = [(y_1+1)^2 - (A-B)^2]/4AB$, $h = [(\gamma-\beta)/2](AB)^{1/2}$, $A = [(y_1-\mathrm{Re}\{z\})^2 + \mathrm{Im}^2\{z\}]^{1/2}$ and $B = [(1+\mathrm{Re}\{z\})^2 + \mathrm{Im}^2\{z\}]^{1/2}$, with z, z^*, and y_1 being, respectively, the two complex and the real roots of the cubic equation $(x + 1)^3 + [4\Omega_0^2/(\gamma-\beta)^2](x - 1) = 0$.

When the Rabi frequency Ω_0 is larger than $(\gamma-\beta)$, we find Rabi oscillations with more population on average in the upper subband, whereas for $\Omega_0 < (\gamma-\beta)$ the electrons oscillate between the two subbands with more population on average in the lower subband. The period of oscillations decreases with the increase of Ω_0. If $\Omega_0 >> (\gamma-\beta)$, then $Z(t) \approx -\cos(\Omega_0 t)$.

4. Comparison between the analytical solutions and the numerical results obtained for a DQW based on GaAs

We will assess the accuracy of the analytical Eqs. (4c), (7) and (9) by comparing them with numerical solutions of the NBE (1a)–(1c) when $T_1 \to \infty$ and $T_2 \to \infty$. The DQW we consider consists of two GaAs square wells of width 5.5 nm and height 219 meV. The wells are separated by an AlGaAs barrier of width 1.1 nm. The electron sheet density is $N = 2\times10^{11}$ cm^{-2}. The energy separation between the ground and the first excited state is $\varepsilon_1 - \varepsilon_0$ = 44.955 meV. Moreover, for the parameters α, β and γ we obtain $\hbar\alpha = 0.412$ meV, $\hbar\beta = -1.56$ meV and $\hbar\gamma = 0.095$ meV. The parameters used in our calculations have been suitably chosen in order to obtain complete inversion of the two-subband system.

In Fig. 1 we present the numerical values (dashed curve) for the population inversion as a function of time, when the system is excited by a hyperbolic secant pulse at exact resonance. The center of the pulse is $t_0 = 1.5$ ps and its duration $t_p = 0.2$ ps. Complete inversion occurs in approximately 1.5 ps and the agreement with the analytical result (solid curve) is excellent. For $T_1 = 50$ ps and $T_2 = 5$ ps the numerical solution of the NBE (1a)–(1c) is shown as a dash-dotted curve in Fig. 1. We see that the effect of the relaxation mechanisms is quite small when the time required to achieve population inversion is much shorter than T_1 and T_2. Similar results are obtained for the case of Eq. (5c)[15] (not shown here).

We have also solved the NBE (1a)–(1c) numerically for cw excitation. For $\delta = 0$ we find Rabi oscillations in the population inversion that are in good agreement with the analytical result given by Eq. (7) when $\Omega_0 > (\gamma-\beta)/2$ and $\Omega_0 < (\gamma-\beta)/2$. However, around $\Omega_0 \approx (\gamma-\beta)/2$ the analytical solution fails. Details are given in Ref. 21. For the case $\delta = \beta-\gamma$, we find that the Rabi oscillations are accurately described by the analytical solution given by Eq. (9) as long as Ω_0 is not too large compared to $(\gamma-\beta)$. The analytical solution has been compared to the numerical result for several values of the Rabi frequency in the region $0.01(\gamma-\beta) < \Omega_0 < 10(\gamma-\beta)$ and good agreement has been found in all cases. But for $\Omega_0 >> (\gamma-\beta)$ the non-RWA terms that have been omitted from Eqs. (3a)–(3c) become significant and the analytical result does not describe the Rabi oscillations accurately.

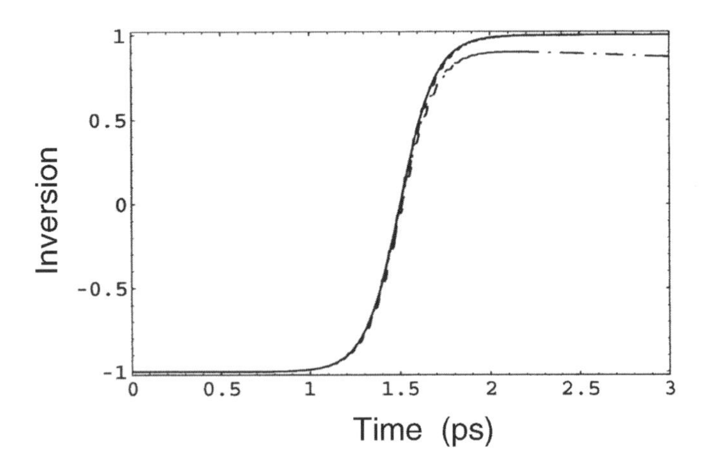

Figure 1. Time evolution of the inversion $Z(t)$ for a hyperbolic secant pulse with center $t_0 = 1.5$ ps and width $t_p = 0.2$ ps. The system is excited at resonance ($\omega_{10} = \omega$). The solid curve is the analytical solution given by Eq. (4c) and the dashed curve is the numerical solution of the NBE (1a)–(1c) when relaxation processes are neglected. The two curves are hardly distinguishable. The effect of the relaxation mechanisms for $T_1 = 50$ ps and $T_2 = 5$ ps is shown by the dashed-dotted curve.

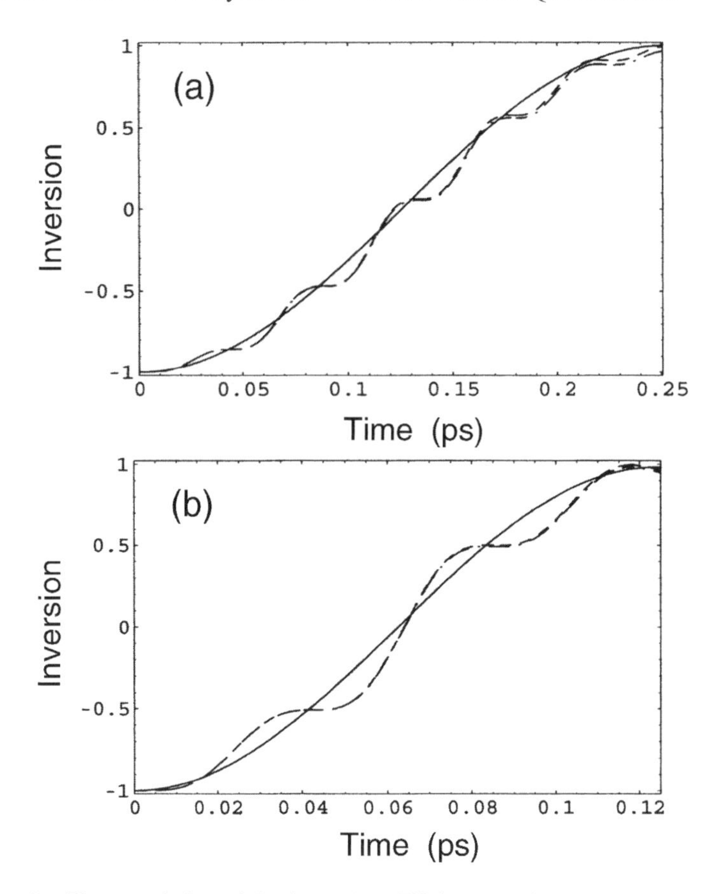

Figure 2. Time evolution of the inversion $Z(t)$ for a cw field. The parameters used are $\delta = 0$, $\Omega_0 = 5(\gamma-\beta)$ (a) and $\delta = \beta-\gamma$, $\Omega_0 = 10(\gamma-\beta)$ (b). The solid curve represents the analytical solutions given by Eqs. (7) and (9), respectively, as explained in the text. The dashed curve shows the numerical values obtained from the numerical solution of Eqs. (1a)–(1c) in the absence of relaxation processes. The effect of the relaxation mechanisms for $T_1 = 50$ ps and $T_2 = 5$ ps is so weak (dash-dotted curve) as to be nearly indistinguishable from the numerical solution obtained in the absence of these mechanisms.

In Fig. 2 we present the calculated values for the time evolution of the inversion for cw fields at a half period of the Rabi oscillations. The parameters used for the calculations are $\delta = 0$, $\Omega_0 = 5(\gamma-\beta)$ and $\delta = \beta-\gamma$, $\Omega_0 = 10(\gamma-\beta)$. The solid curves depict the values for $Z(t)$ obtained from the analytical expressions (7) and (9). As we can see complete inversion occurs in both cases. The numerical results are shown as dashed curves. We observe internal oscillations of small amplitude that are due to the non-RWA terms. The dash-dotted curves in Fig. 2 represent the numerical solutions of Eqs. (1a)–(1c) for $T_1 = 50$ ps and $T_2 = 5$ ps. Since T_1 and T_2

are much longer than the period of the Rabi oscillations, the effect of the relaxation processes is hardly noticeable.

5. Conclusions

We have studied the intersubband transition dynamics in a symmetric DQW that is coupled to strong pulsed and cw fields. Only the first two electron subbands have been taken into account. In the absence of relaxation processes we adopt the rotating wave approximation and provide analytical solutions of the NBE. Conditions for complete population inversion are presented. We show that significant inversion can be obtained even in the presence of relaxation processes. The effect of the relaxation mechanisms is very weak when the time required to reach inversion is much shorter than the population decay time and the dephasing time.

Acknowledgments

We thank the European Social Fund (ESF), Operational Program for Educational and Vocational Training II (EPEAEK II), and particularly the Program PYTHAGORAS II, for funding this work.

References

1. S. L. Chuang, M. S. C. Luo, S. Schmitt-Rink, and A. Pinczuk, "Many-body effects on intersubband transitions in semiconductor quantum-well structures," *Phys. Rev. B* **46**, 1897 (1992).
2. M. Zaluzny, "Influence of the depolarization effect on the nonlinear intersubband absorption spectra of quantum wells," *Phys. Rev. B* **47**, 3995 (1993).
3. M. Zaluzny, "Saturation of intersubband absorption and optical rectification in asymmetric quantum-wells," *J. Appl. Phys.* **74**, 4716 (1993).
4. B. Galdrikian and B. Birnir, "Period doubling and strange attractors in quantum wells," *Phys. Rev. Lett.* **76**, 3308 (1996).
5. D. E. Nikonov, A. Imamoglu, L. V. Butov, and H. Schmidt, "Collective intersubband excitations in quantum wells: Coulomb interaction versus subband dispersion," *Phys. Rev. Lett.* **79**, 4633 (1997).
6. C. A. Ullrich and G. Vignale, "Collective intersubband transitions in quantum wells: A comparative density-functional study," *Phys. Rev. B* **58**, 15756 (1998).
7. D. E. Nikonov, A. Imamoglu, and M. O. Scully, "Fano interference of collective excitations in semiconductor quantum wells and lasing without inversion," *Phys. Rev. B* **59**, 12212 (1999).

8. A. A. Batista, B. Birnir, and M. S. Sherwin, "Subharmonic generation in a driven asymmetric quantum well," *Phys. Rev. B* **61**, 15108 (2000).
9. A. A. Batista, P. I. Tamborenea, B. Birnir, M. S. Sherwin, and D. S. Citrin, "Nonlinear dynamics in far-infrared driven quantum-well intersubband transitions," *Phys. Rev. B* **66**, 195325 (2002).
10. J. Li and C. Z. Ning, "Interplay of collective excitations in quantum-well intersubband resonances," *Phys. Rev. Lett.* **91,** 097401 (2003).
11. A. Olaya-Castro, M. Korkusinski, P. Hawrylak, and M. Yu. Ivanov, "Effective Bloch equations for strongly driven modulation-doped quantum wells," *Phys. Rev. B* **68**, 155305 (2003).
12. P. Haljan, T. Fortier, P. Hawrylak, P. B. Corkum, and M. Yu. Ivanov, "High harmonic generation and level bifurcation in strongly driven quantum wells," *Laser Phys.* **13**, 452 (2003).
13. H. O. Wijewardane and C. A. Ullrich, "Coherent control of intersubband optical bistability in quantum wells," *Appl. Phys. Lett.* **84**, 3984 (2004).
14. A. A. Batista and D. S. Citrin, "Rabi flopping in a two-level system with a time-dependent energy renormalization: Intersubband transitions in quantum wells," *Phys. Rev. Lett.* **92**, 127404 (2004).
15. E. Paspalakis, M. Tsaousidou, and A. F. Terzis, "Coherent manipulation of a strongly driven semiconductor quantum well," *Phys. Rev. B* **73**, 125344 (2006).
16. M. E. Crenshaw and C. M. Bowden, "Quasiadiabatic following approximation for a dense medium of two-level atoms," *Phys. Rev. Lett.* **69**, 3475 (1992).
17. S. M. Jensen, "The non-linear coherent coupler," *IEEE J. Quantum Electronics* **18**, 1580 (1982).
18. S. Trillo and S. Wabnitz, "Nonlinear nonreciprocity in a coherent mismatched directional coupler," *Appl. Phys. Lett.* **49**, 752 (1986).
19. V. M. Kenkre and D. K. Campbell, "Self-trapping on a dimer: Time-dependent solutions of a discrete nonlinear Schrödinger equation," *Phys. Rev. B* **34**, 4959 (1986).
20. V. M. Kenkre and G. P. Tsironis, "Nonlinear effects in quasielastic neutron scattering: Exact line-shape calculation for a dimer," *Phys. Rev. B* **35**, 1473 (1987).
21. E. Paspalakis, M. Tsaousidou, and A. F. Terzis, "Rabi oscillations in a strongly driven semiconductor quantum well," *J. Appl. Phys.* **100**, 044312 (2006).

Contributors

Abedin, M. N.
NASA Langley Research Center, Hampton, VA, U.S.A.

Akarvardar, K.
Institut de Microélectronique, Electromagnétisme et Photonique
Minatec, Grenoble, France

Arnaud d'Avitaya, F.
CRMCN–CNRS, Campus de Luminy, Marseille, France

Austin, R. H.
Dept. of Physics, Princeton University, Princeton, NJ, U.S.A.

Bandara, S. V.
Jet Propulsion Laboratory, Pasadena, CA, U.S.A.

Banu, M.
MHI Consulting LLC, Murray Hill, NJ, U.S.A.

Bauer, G.
Inst. für Halbleiter und Festkörperphysik
Johannes Kepler University Linz, Linz, Austria

Belenky, G.
Dept. of Electrical Engineering, SUNY–Stony Brook, Stony Brook, NY, U.S.A.

Benisty, H.
Laboratoire Charles Fabry de l'Institut d'Optique, CNRS, Palaiseau, France

Benschop, J. P. H.
ASML Inc., Veldhoven, The Netherlands

Bhat, I.
Rensselaer Polytechnic Institute, Troy, NY, U.S.A.

Biga, F. Y.
Div. of Engineering, Brown University, Providence, RI, U.S.A.

Bimberg, D.
Institut für Festkörperphysik, Tech. Univ. Berlin, Berlin, Germany

Future Trends in Microelectronics. Edited by Serge Luryi, Jimmy Xu, and Alex Zaslavsky
ISBN 0-471-48 © 2007 John Wiley & Sons, Inc.

Botez, D.
Reed Center for Photonics, University of Wisconsin, Madison, WI, U.S.A.

Brillouët, M.
CEA-LETI, Grenoble, France

Bruno, J. D.
Maxion Technologies, Inc., Hyattsville, MD, U.S.A.

Bull, C. W.
Division of Engineering, Brown University, Providence, RI, U.S.A.

Cello, J.
School of Medicine, SUNY–Stony Brook, Stony Brook, NY, U.S.A.

Cerovic, G.
Genewave, Ecole Polytechnique, Palaiseau, France

Chardon, A.
Genewave, Ecole Polytechnique, Palaiseau, France

Choumane, H.
Genewave, Ecole Polytechnique, Palaiseau, France

Coleman, J. R.
School of Medicine, SUNY–Stony Brook, Stony Brook, NY, U.S.A.

Crawford, G. P.
Div. of Engineering, Brown University, Providence, RI, U.S.A.

Cristoloveanu, S.
Institut de Microélectronique, Electromagnétisme et Photonique
Minatec, Grenoble, France

Cunningham, J. E.
School of Electronic and Electrical Engineering, Univ. of Leeds, Leeds, U.K.

Dais, C.
Laboratory for Micro- and Nanotechnology
Paul Scherrer Institute, Villigen, Switzerland

David, A.
Materials Department, UC–Santa Barbara, Santa Barbara, CA, U.S.A.

Davies, A. G.
School of Electronic and Electrical Engineering, Univ. of Leeds, Leeds, U.K.

Deckhardt, E.
Laboratory for Micro- and Nanotechnology
Paul Scherrer Institute, Villigen, Switzerland

Donoghue, J. P.
Department of Neuroscience, Brown University, Providence, RI, U.S.A.

D'Souza, M.
Reed Center for Photonics, University of Wisconsin, Madison, WI, U.S.A.

Dwir, B.
Ecole Polytechnique Federale de Lausanne, Lausanne, Switzerland

Dyakonov, M. I.
Laboratoire de Physique Théorique et Astroparticules
Université Montpellier II, France

Efros, A. L.
Dept. of Physics, University of Utah, Salt Lake City, UT, U.S.A.

Ekinci, Y.
Laboratory for Micro- and Nanotechnology
Paul Scherrer Institute, Villigen, Switzerland

Eminente, S.
ARCES-DEIS, University of Bologna, Bologna, Italy

Engström, O.
Dept. of Microtechnology and Nanoscience
Chalmers University of Technology, Göteborg, Sweden

Esseni, D.
DIEGM, University of Udine, Udine, Italy

Fabian, J.
Institute for Theoretical Physics
University of Regensburg, Regensburg, Germany

Fiegna, C.
ARCES-DEIS, University of Bologna, Bologna, Italy

Filipe, A.
Spintron, Technopôle de Château-Gombert, Marseille, France

Fromherz, T.
Inst. für Halbleiter und Festkörperphysik
Johannes Kepler University Linz, Linz, Austria

Galajda, P.
Dept. of Physics, Princeton University, Princeton, NJ, U.S.A.

Gentil, P.
Institut de Microélectronique, Electromagnétisme et Photonique
Minatec, Grenoble, France

Goutel, G.
Genewave, Ecole Polytechnique, Palaiseau, France

Green, M. A.
ARC Photovoltaics Centre of Excellence
University of New South Wales, Sydney, Australia

Grützmacher, D.
Laboratory for Micro- and Nanotechnology
Paul Scherrer Institute, Villigen, Switzerland

Gunapala, S. D.
Jet Propulsion Laboratory, Pasadena, CA, U.S.A.

Ha, N.
Genewave, Ecole Polytechnique, Palaiseau, France

Hawrylak, P.
Institute for Microstructural Sciences
National Research Council of Canada, Ottawa, Ontario, Canada

Hofstraat, H.
Philips Research, High Tech Campus 34, Eindhoven, The Netherlands

Hu, Q.
Dept. of Electrical Engineering and Computer Science
Massachusetts Institute of Technology, Cambridge, MA, U.S.A.

Kapon, E.
Ecole Polytechnique Federale de Lausanne, Lausanne, Switzerland

Karpovski, M.
School of Physics and Astronomy, Tel Aviv University, Tel Aviv, Israel

Käser, P.
Laboratory for Micro- and Nanotechnology
Paul Scherrer Institute, Villigen, Switzerland

Kelly, M. J.
Centre for Advanced Photonics and Electronics,
Dept. of Engineering, University of Cambridge, Cambridge, U.K.

Keymer, J.
Dept. of Physics, Princeton University, Princeton, NJ, U.S.A.

Khandekhar, A.
Reed Center for Photonics, University of Wisconsin, Madison, WI, U.S.A.

Kisin, M.
Dept. of Electrical Engineering, SUNY–Stony Brook, Stony Brook, NY, U.S.A.

Kuech, T.
Reed Center for Photonics, University of Wisconsin, Madison, WI, U.S.A.

Kumar, S.
Dept. of Electrical Engineering and Computer Science
Massachusetts Institute of Technology, Cambridge, MA, U.S.A.

Ledentsov, N. N.
Institut für Festkörperphysik, Tech. Univ. Berlin, Berlin, Germany

Lee, A. W. M.
Dept. of Electrical Engineering and Computer Science
Massachusetts Institute of Technology, Cambridge, MA, U.S.A.

Levy, E.
School of Physics and Astronomy, Tel Aviv University, Tel Aviv, Israel

Linfield, E. H.
School of Electronic and Electrical Engineering, Univ. of Leeds, Leeds, U.K.

Liu, H. C.
Institute for Microstructural Sciences
National Research Council, Ottawa, Canada

Luryi, S.
Dept. of Electrical Engineering, SUNY–Stony Brook, Stony Brook, NY, U.S.A.

Lyakh, A.
University of Florida, Gainesville, FL, U.S.A.

Martinelli, L.
Genewave, Ecole Polytechnique, Palaiseau, France

Miller, D. A. B.
Ginzton Laboratory, Stanford University, Stanford, CA, U.S.A.

Monroy, C.
Army Research Laboratory, Adelphi, MD, U.S.A.

Mueller, S.
School of Medicine, SUNY–Stony Brook, Stony Brook, NY, U.S.A.

Müller, E.
Laboratory for Micro- and Nanotechnology
Paul Scherrer Institute, Villigen, Switzerland

Murota, J.
Laboratory for Nanoelectronics and Spintronics
RIEC, Tohoku University, Sendai, Japan

Nelep, C.
Genewave, Ecole Polytechnique, Palaiseau, France

Nishi, Y.
Dept. of Electrical Engineering, Stanford University, Stanford, CA, U.S.A.

Nurmikko, A. V.
Division of Engineering, Brown University, Providence, RI, U.S.A.

Oreg, Y.
Dept. of Condensed Matter Physics
The Weizmann Institute of Science, Rehovot, Israel

Palestri, P.
DIEGM, University of Udine, Udine, Italy

Palevski, A.
School of Physics and Astronomy, Tel Aviv University, Tel Aviv, Israel

Papamichail, D.
Dept. of Computer Science, SUNY–Stony Brook, Stony Brook, NY 11974, U.S.A.

Paspalakis, E.
Materials Science Department, University of Patras, Patras 26 504, Greece

Patterson, W. R.
Division of Engineering, Brown University, Providence, RI, U.S.A.

Paul, A.
School of Medicine, SUNY–Stony Brook, Stony Brook, NY, U.S.A.

Pelucchi, E.
Ecole Polytechnique Federale de Lausanne, Lausanne, Switzerland

Pinto, M. R.
Applied Materials Inc., Santa Clara, CA, U.S.A.

Prodanov, V.
MHI Consulting LLC, Murray Hill, NJ, U.S.A.

Qin, Q.
Dept. of Electrical Engineering and Computer Science
Massachusetts Institute of Technology, Cambridge, MA, U.S.A.

Rashba, E. I.
Dept. of Physics, Harvard University, Cambridge, MA, U.S.A.

Rattier, M.
Genewave, Ecole Polytechnique, Palaiseau, France

Refaat, T. F.
Old Dominion University, Norfolk, VA, U.S.A.

Reno, J. L.
Sandia National Labs, Albuquerque, NM, U.S.A.

Reymond, G.-O.
Genewave, Ecole Polytechnique, Palaiseau, France

Rudra, A.
Ecole Polytechnique Federale de Lausanne, Lausanne, Switzerland

Safarov, V.
CRMCN–CNRS, Campus de Luminy, Marseille, France

Sakuraba, M.
Laboratory for Nanoelectronics and Spintronics
RIEC, Tohoku University, Sendai, Japan

Sandford, S. P.
NASA Langley Research Center, Hampton, VA, U.S.A.

Sangiorgi, E.
ARCES-DEIS, University of Bologna, Bologna, Italy

Selmi, L.
DIEGM, University of Udine, Udine, Italy

Shchukin, V. A.
Institut für Festkörperphysik, Tech. Univ. Berlin, Berlin, Germany

Shin, J. C.
Reed Center for Photonics, University of Wisconsin, Madison, WI, U.S.A.

Shur, M. S.
Dept. of Electrical, Computer and Systems Engineering
Rensselaer Polytechnic Institute, Troy, NY, U.S.A.

Singh, U. N.
NASA Langley Research Center, Hampton, VA, U.S.A.

Skiena, S.
Dept. of Computer Science, SUNY–Stony Brook, Stony Brook, NY 11974, U.S.A.

Solak, H.
Laboratory for Micro- and Nanotechnology
Paul Scherrer Institute, Villigen, Switzerland

Solomon, P. M.
IBM T. J. Watson Research Center, Yorktown Heights, NY, U.S.A.

Song, Y.-K.
Division of Engineering, Brown University, Providence, RI, U.S.A.

Spivak, B.
Dept. of Physics, University of Washington, Seattle, WA, U.S.A.

Stangl, J.
Inst. für Halbleiter und Festkörperphysik
Johannes Kepler University Linz, Linz, Austria

Suchalkin, S.
Dept. of Electrical Engineering, SUNY–Stony Brook, Stony Brook, NY, U.S.A.

Suzuki, T.
Inst. für Halbleiter und Festkörperphysik
Johannes Kepler University Linz, Linz, Austria

Terzis, A. F.
Department of Physics, University of Patras, Patras, Greece

Tillack, B.
IHP, Frankfurt (Oder), Germany

Tober, R. L.
Army Research Laboratory, Adelphi, MD, U.S.A.

Towner, F.
Maxion Technologies, Inc., Hyattsville, MD, U.S.A.

Tsaousidou, M.
Materials Science Department, University of Patras, Patras, Greece

Tsukernik, A.
School of Physics and Astronomy, Tel Aviv University, Tel Aviv, Israel

Tsvid, G.
Reed Center for Photonics, University of Wisconsin, Madison, WI, U.S.A.

van Houten, H.
Philips Research, High Tech Campus 34, Eindhoven, The Netherlands

Wang, D. P.
Div. of Engineering, Brown University, Providence, RI, U.S.A.

Wasilewski, Z. R.
Institute for Microstructural Sciences
National Research Council, Ottawa, Canada

Weisbuch, C.
Ecole Polytechnique, Palaiseau, France

Williams, B. S.
Dept. of Electrical Engineering and Computer Science
Massachusetts Institute of Technology, Cambridge, MA, U.S.A

Wimmer, E.
School of Medicine, SUNY–Stony Brook, Stony Brook, NY, U.S.A.

Xu, D.
Reed Center for Photonics, University of Wisconsin, Madison, WI, U.S.A.

Xu, J. M.
Div. of Engineering, Brown University, Providence, RI, U.S.A.

Zaslavsky, A.
Div. of Engineering, Brown University, Providence, RI, U.S.A.

Zhitenev, N. B.
Bell Laboratories, Lucent Technologies, Murray Hill, NJ, U.S.A.

Zory, P.
University of Florida, Gainesville, FL, U.S.A.

Zutic, I.
Dept. of Physics, SUNY–Buffalo, Buffalo, NY, U.S.A.

Index

Future Trends in Microelectronics. Edited by Serge Luryi, Jimmy Xu, and Alex Zaslavsky